COMPUTATIONAL MECHANICS

Abstract (Volume II)

Abstracts of the Papers Presented
at the Regular Sessions of
the Sixth World Congress on Computational Mechanics
in conjunction with
the Second Asian-Pacific Congress
on Computational Mechanics

September 5-10, 2004, Beijing, China

Editors

Z. H. Yao
M. W. Yuan
W. X. Zhong

Tsinghua University Press
Beijing

Springer

图书在版编目（CIP）数据

计算力学摘要集. 2=Computational Mechanics Abstract. Volume II: Abstracts of the Papers Presented at the Regular Sessions of the WCCM VI in Conjunction with APCOM'04/姚振汉，袁明武，钟万勰主编.—北京：清华大学出版社，2004

ISBN 7-302-09342-3

I.计…　II.①姚…　②袁…　③钟…　III.计算力学–论文–文摘–英文　IV. Z89: 0302

中国版本图书馆 CIP 数据核字（2004）第 088038 号

出 版 者：清华大学出版社　　地　　址：北京清华大学学研大厦

http://www.tup.com.cn　　邮　　编：100084

社 总 机：010-62770175　　客户服务：010-62776969

责任编辑：陈朝晖

印 装 者：三河市春园印刷有限公司

发 行 者：新华书店总店北京发行所

开　　本：185×260　印张：36.75　字数：892 千字

版　　次：2004 年 8 月第 1 版　2004 年 8 月第 1 次印刷

书　　号：ISBN 7-302-09342-3/O・399

印　　数：1～1800

定　　价：150.00 元

本书如存在文字不清、漏印以及缺页、倒页、脱页等印装质量问题，请与清华大学出版社出版部联系调换。联系电话：(010)62770175-3103 或(010)62795704

PREFACE

The impact of Computational Mechanics on engineering and technological progress started in early 60's of the last century and continued to flourish during the entire 20th Century. From the latter part of the 20th Century and the beginning of the 21st Century till now, it also had an impressive influence on the mathematical, physical and biological sciences. The extensive research being carried out around the world in applying the concepts of computational mechanics to multi-physical, multi-scale problems, nano and micro mechanical simulations and biomechanics, is a testament to this.

The International Association for Computational Mechanics founded to promote the activities related to computational mechanics in the early 80's of the last century has also made an impressive progress. The most important scientific event of the Association is the World Congress. The first Congress was held in Austin (USA) in 1986 and then Stuttgart (Germany) in 1990, Chiba (Japan) in 1994, Buenos Aires (Argentina) in 1998 and Vienna (Austria) in 2002. The 6th World Congress on Computational Mechanics, WCCMVI is held in Beijing (China) in conjunction with 2nd Asian Pacific Congress on Computational Mechanics, APCOM'04. The first Asian Pacific Congress was held in Sydney (Australia) in 2001.The Beijing venue is the first time that both congresses are held together and it is a unique occasion for the world community of researchers in computational mechanics to get together.

The Chinese scientists in computational mechanics consider it a great honor to host the combined WCCM/APCOM in Beijing. The selection of Beijing (China) indicates the recognition of the world community, about the progress made by Chinese scientists in theory and practice of computational mechanics. China has made a fast economic growth during the past 25 years and it is the belief of the Chinese computational mechanics community that a significant growth of computational mechanics will be made with the international collaboration through this combined congresses.

The Congress Publications consist of the printed Proceedings, two volumes of abstracts and the Proceedings CDROM. The printed Proceedings contain the papers of 3 Plenary Lectures, 11 of 21 Semi-plenary Lectures and 108 of 155 Keynote lectures. The abstracts of other papers are included in two volumes of abstracts: 650 abstracts of the papers to be presented at 83 Minisymposia in volume 1, and 545 abstracts of the papers to be presented at Regular Sessions in volume 2. The Proceedings CDROM contains all the full length papers received in time, more than 900 papers, and the abstracts for other presentations. The editors acknowledge the help of the Paper Review Committee members in maintaining the high standard of the assessment of papers and the co-operation of the authors in complying with the requirement of the editors and the reviewers.

We would like to take this opportunity to thank the members of the Local Organizing Committee and the International Scientific Committee.

We acknowledge the financial support received from China National Science Foundation, China Ministry of Science and Technology, Peking University, Tsinghua University, Dalian University of Technology, Institute of Mechanics of China Academy of Sciences, South China University of Technology, Shanghai Jiaotong University, Hehai University, Wuhan University of Technology and personally from Prof. Zheng Yao, Zhejian University.

All the preparation work of the Congress is fully supported by the International Association for Computational Mechanics and its President, Prof. Eugenio Oñate. Prof. S. Valliappan, Vice President of IACM (Asia-Australia), and Secretary General of APACM gave us warm concerns, helpful comments and advice in details from the beginning to the end. Prof. W. K. Liu greatly helped us to organize minisymposia, papers and participants from the United States and other countries. We heartily thank them and all the friends who have provided help, advice and suggestions for the successful organization of WCCMVI/APCOM'04.

Finally the editors hope that the proceedings will reflect the success of WCCM VI in conjunction with APCOM'04.

Zhenhan Yao (Secretary General of WCCM VI)

Mingwu Yuan and Wanxie Zhong (Chairmen of WCCM VI)

June 12, 2004

CONTENTS

* denotes the presenter

Page number is the same as the serial number of the abstract

Computational Mathematics, Mathematical Modeling and Numerical Methods (0100)

High performance finite elements and their applications (0104)

Boundary element methods and fast multipole boundary element method (0105, 0106)

Meshless methods (0107)

Neural networks and soft computing in computational mechanics (0109)

Differential quadrature, generalized methods and related discrete element analysis methods (0110)

Computational Fluid Mechanics (0200)

Advances and challenges in flow simulation and modeling (0201)

Computational biofluiddynamics (0202)

Direct and large-eddy simulation of turbulent flows (0203)

Computational Solid and Structural Mechanics (0300)

Computational modeling of shells (0301)

Direct (non time-stepping) methods for limit state structural analysis (0304)

Structural dynamics (0307)

Contact problems (0308)

Computational fracture mechanics (0401)

Computational dynamic failure and fracture (0403)

Material modeling at various length scale (0404)

Computational modeling of particulates and multi-fracturing solids (0405)

Computational adaptive materials systems, structures and smart materials (0406)

Computational materials mechanics from atomistic simulation to continuum mechanics (0409)

Numerical simulation of composite materials (0410)

Computational geomechanics and geotechniques (0802)

Multi-disciplinary optimization (1001)

Evolutionary algorithms in structural optimization (1002)

Algorithm, Software Development and High Performance Computing (1100)

CAD, CAM and CAE (1102)

Computational methods in stochastic mechanics and reliability analysis (1201)

Structural health-monitoring (1203)

Industrial simulation of metal forming processes (1210)

High performance fracture approach to fatigue crack analysis and life prediction (1212)

Building and Construction (1213)

COMPUTATIONAL MECHANICS (Abstracts)
WCCM VI in conjunction with APCOM'04, *Sept. 5-10, 2004, Beijing, China*

Accuracy Gains from Exponential Discretization of the Diffusion-Advection Equation

El-Zein Abbas*

Centre for Geotechnical Research, Department of Civil Engineering, University of Sydney, NSW 2006, Australia
e-mail: az03@aub.edu.lb

Abstract

Polynomial interpolation functions introduce numerical dispersion erro rs into numerical solutions of the diffusion-advection equations. New finite-element and boundary-element methods for the solution of the diffusion-advection equation are proposed. The methods use non-isoparametric exponentially-varying interpolation funct ions, based on exact solutions of the homogeneous forms of the differential equation in the Laplace domain. Two eight-noded finite elements generated from one-dimensional (E8) and two-dimensional (E8C) solutions and one three-noded boundary element (E3) are developed and tested for convergence, stability, Peclet number, anisotropy, material heterogeneity, Dirichlet and Neumann boundary conditions and tolerance for mesh distortions. The performance of exponential finite elements is compared to that of conven tional, eight-noded (P8) and twelve-noded (P12) polynomial elements. The exponential boundary element is assessed against the conventional quadratic boundary element (P3). The solution is found in the Laplace domain and inverted back to the time domain num erically.

The exponential finite element based on two-dimensional analytical solutions (E8C) fails basic tests of convergence. The one based on one-dimensional solutions (E8) performs particularly well. It pushes the Peclet number limit for one-dimensional problems from 20 to 500. It reduces by about 75% the number of elements and degrees of freedom required for convergence, for a wide variety of two-dimensional problems, yielding an error that is one order of magnitude smaller than that of the eight-noded and twelve-noded polynomial elements. The exponential finite element E8 is stable and robust under relatively high degrees of heterogeneity, anisotropy and mesh distortions. The exponential boundary element (E3) speeds up convergence, albeit less spectacul arly than exponential finite elements, because improvements in boundary approximations have less impact on accuracy than those affecting domain approximations.

In conclusion, exponential discretization of the diffusion -advection equation are available and more computationally efficient alternatives to conventional polynomial elements.

COMPUTATIONAL MECHANICS (Abstracts)
WCCM VI in conjunction with APCOM'04, *Sept. 5-10, 2004, Beijing, China*

Derivation of Modified Stress-Strain Tensor for Wrinkled Membrane by Using Projection Matrix

Takeshi Akita [1*], Kyoichi Nakashino [1], M. C. Natori [1], K. C. Park [2]

[1] *Japan Aerospace Exploration Agency (JAXA), The Institute of Space and Astronautical Science, 3-1-1 Yoshinodai, Sagamihara, Kanagawa 229-8510, Japan*
[2] *Center for Aerospace Structures, University of Colorado, Boulder, Colorado, 80309-0429, USA*
e-mail: akita@taurus.eng.isas.jaxa.jp

Abstract

Recently, membranes attract great interest in the field of space structure s engineering due to such abilities as high packaging ratio and light weight proper ty. Since membranes have little resistance against compressive stresses, wrinkles may be formed over partial or entire regions of the membranes. The presence of such wrinkles could cause significant deterioration in the performance of membrane structures. Therefore, it is of great importance to develop efficient mathematical model accounting for wrinkling phenomenon.

Many studies have been conducted on wrinkling models that are applicable to finite element analysis. The majority of them take the approach to modify stress-strain tensor of membrane in order to simulate wrinkling phenomenon [1-2]. Numerical algorithms required in their approach are quite simple, and thus, can be easily integrated into finite element codes. However, the modified stress-strain tensors are derived from physical intuition, which somewhat obscures mathematical background of their wrinkling algorithms. In addition, nonlinearities originated from wrinkling phenomenon are not appropriately handled during solution processes. This may cause difficult convergence problems in some cases.

In this paper, we propose a mathematical method to construct modified stress-strain tensor that evaluates stress in wrinkled membrane. The method is based on the idea to "project out" wrinkle strain that develops no stress. A projection matrix is utilized in order to decompose the total strain into elastic and wrinkle components. The modified stress-strain tensor can be expressed as product of usual elasticity matrix and projection one. The incremental version of the modified stress-strain tensor is also derived for the construction of the precise tangent stiffness matrix. Numerical examples using the modified stress-strain tensor are presented to demonstrate the accuracy and effectiveness of the proposed method.

REFERENCES

[1] R. K. Miller, J. M. Hedgepeth, V. I. Weingarten, P. Das, S. Kahyai, *Finite element analysis of partly wrinkled membranes*, Computers & Structures, 20(1-3), (1985), 631-639.

[2] X. Liu, C. H. Jenkins, W. W. Shur, *Large deflection analysis of pneumatic envelopes using a penalty parameter modified material model*, Finite Elements in Analysis and Design, 37, (2001), 233-251.

COMPUTATIONAL MECHANICS (Abstracts)
WCCM VI in conjunction with APCOM'04, *Sept. 5-10, 2004, Beijing, China*

The Arlequin Method as a Multiscale Numerical Tool

Hachmi Ben Dhia *, Guillaume Rateau

LMSS-Mat, Ecole Centrale de Paris, Grande Voie des Vignes, Chatenay Malabry, 92295, France
e-mail: bendhia@mss.ecp.fr, rateau@mss.ecp.fr

Abstract

Changing locally the nature of a global mechanical model while saving resources is essential in the designing of engineering structures. From a computational stand point, one of the main difficulties to achieve this task lies in the lack of flexibility of classical numerical tools such as the Finite Element Method. Improved numerical methods have been developed during the last decade to tackle this difficulty. Let us mention in particular the Diffuse and Meshless methods, the Volumic Correction Multiscale approaches and the Partition of Unity and Generalized and Extended FE methods. However, to our best knowledge, none of these important numerical methods has reached the degree of generality and flexibility of the Arlequin method [1-2]. This approach consists in: 1) a superposition of mechanical states in subzones of a given material domain, 2) an energy distribution between the mechanical states in the sub-zones, 3) a weak, compatible and generally partial gluing of these states.

The possible discrepancy between the scales and the necessity to glue the models to each others have however to be treated with sufficient care. Various coupling operators [1]-[3] are discussed.

A second aspect of the approach is related to the fact that by definition, the Arlequin framework allows the co-existence of incompatible models, sharing the energies of the system in the superposition regions and linked to each others in the gluing subregions. These unusual energy distributions, both with the coupling of generally incompatible models, require special numerical and technical developments. We focus on these last aspects and give numerical simulations showing the efficiency of the method for mechanical examples as the one given by Figure 1.

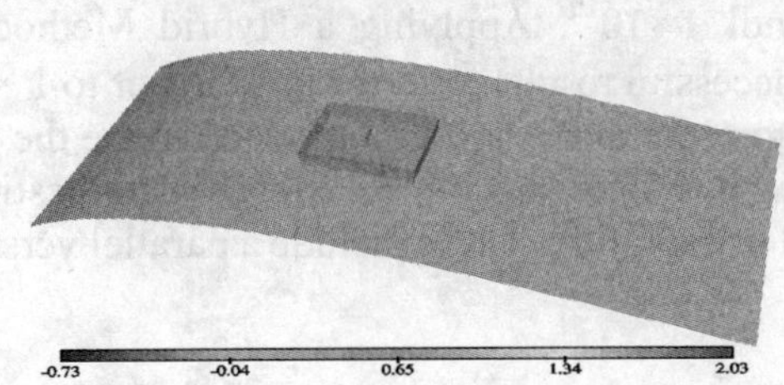

Fig. 1 A partition of a cracked 3D and a global plate models

REFERENCES

[1] H. Ben Dhia, *Multiscale mechnical problems: the Arlequin method*, C. R. Acad. Sci., Paris, Ser.II-b, 326, (1998), 899-904.

[2] H. Ben Dhia, *Numerical modelling of multiscale problems: the Arlequin method*, CD ECCM'99, Munchen, (1999).

[3] H. Ben Dhia, G. Rateau, *A mathematical analysis of the mixed Arlequin method*, C. R. Acad. Sci., Paris, Ser.I, 332, (2001), 649-654.

COMPUTATIONAL MECHANICS (Abstracts)
WCCM VI in conjunction with APCOM'04, *Sept. 5-10, 2004, Beijing, China*

An Algebraic Method for Numerical Stabilization of Transport Equation

Ruperto P. Bonet *

Department of Applied Physic, Technical University of Catalonia, Jordi Girona 1-3, Módulo B4, Barcelona, 08034, Spain

Abstract

A new methodology for evaluating unknown parameters in a numerical method for solving a partial differential equation is developed. The main result is the construction of several adap tive algorithms over the global matrix spectrum to yield "optimal" solutions over a parameter spaces. This method allows that the global matrix becomes in quasi self-adjoin matrix. Many algorithms based on the algebraic methods, such that Bisection Methods, Newton methods, or Truncated Newton methods have been explored with successful results. These procedures do not require exact or approximate, analytical solutions of the continuous problem, and are derived f rom the numerical solution of the algebraic equation that characterizes the zeros of the imaginary part of the global matrix spectrum. We call the attention that all operations are done over a set of finite -dimensional spaces. Also, a local procedure has been implemented to multiple spatial dimensions cases. As a simple illustration, it is applied to the linear advection-diffusion problem to yield an adaptive algebraic method. The numerical performance of this method in one and two dimensions is analyzed, a nd it is found that it yields results that are commensurate to the SUPG method, inclusive in cases of high anisotropy.
In the numerical experiments, we have found that the resulting diffusivity, which is a non -linear function of the numerical solution, con verges to the value obtained using the SUPG analysis. We show the convergence of Quasi-Newton method as a function of iteration number, for different values of Peclet numbers and error levels. For an error level equal to 1×10^{-2} in L^2-norm, this procedure converges to the SUPG method within about 10 iterations. For an error level equal to 1×10^{-8} in L^2-norm, we found that the iteration numbers decrease to 30 iterations while the values of Peclet numbers increase until 1×10^{10}. Applying a Hybrid Method the numerical results were improved. We have found successful results for error level lower to 1×10^{-4}, combining the Bisection Method and the Quasi-Newton Method. Also, a method to reduce the phase of largest eigenvalue of the resulting global matrix has been examined. Several interesting aspects of the proposed methodology will be studied in the future. These include a parallel version of this method and the use of gradient methods.

REFERENCES

[1] T. J. R. Hughes, A. Brooks, *A multidimensional upwind scheme with no crosswind diffusion*, volume 34 of AMD, *Finite element methods for convection dominated flows*, ASME, New York, (1979).

[2] Eugenio Oñate, *Derivation of stabilized equations for advective-diffusive transport and fluid flow problems*, Comput. Methods Appl. Mech. Engrg., 151(1/2), (1998), 233-267.

COMPUTATIONAL MECHANICS (Abstracts)
WCCM VI in conjunction with APCOM'04, *Sept. 5-10, 2004, Beijing, China*

Symplectic Computation of the Nonlinear Schrödinger Equation

Guidong Dai [*], Yifa Tang

LSEC, ICMSEC, Academy of Mathematics & Systems Science, Chinese Academy of Sciences, P.O. Box 2719, Beijing 100080, China
e-mail: daigd@lsec.cc.ac.cn, tyf@lsec.cc.ac.cn

Abstract

We consider the nonlinear cubic Schr ödinger equation (NLSE) with initial condition:

$$\begin{cases} iW_t + W_{xx} + a|W|^2 W = 0 \\ W(x,0) = W_0(x) \end{cases}$$

where $x \in \mathbb{R}$, a is a constant, $W(x,t))$ is a complex function and $W_0(\pm\infty) = 0$. We choose a popular spatial discretization:

$$iW_t^{(l)} + \frac{W^{(l+1)} - 2W^{(l)} + W^{(l-1)}}{h^2} + a\left|W^{(l)}\right|^2 W^{(l)} = 0$$

where h is the spatial step size and $W^{(l)}(t) = W(lh,t)$, $l = -n, \ldots, -1, 0, 1, \ldots, n$. The discrete model produces a Hamiltonian system:

$$\frac{dz}{dt} = J^{-1}\nabla H(z)$$

where $H(z) = H(p,q) = \frac{1}{2h^2}\left[p^T Bp + q^T Bq\right] + \frac{a}{4}\sum_{k=-n}^{n}\left[\left(p^{(k)}\right)^2 + \left(q^{(k)}\right)^2\right]^2$.

The Hamiltonian can be splitted into three parts, each of which can be explicitly integrated. We use some explicit symplectic scheme to s imulate the soliton motions. The conservativity of the scheme for the invariants of the invariants of NLSE and the formal energy of the scheme itself is also tested.

REFERENCES

[1] V. E. Zakharov, A. B. Shabat, *Interaction between solitons in a stable medium*, Sov. Phys. -JETP, 37(5), (1973), 823-828.

[2] Y. F. Tang, *Formal energy of a symplectic scheme for Hamiltonian systems and its applications* (*I*), Computers Math. Applic. 27(7), (1994), 31-39.

[3] Y. F. Tang, L. Vázquez, F. Zhang, V. M. Pérez-García, Symplectic Methods for the Nonlinear Schrödinger Equation, Computers Math. Applic., 32(5), (1996), 73-83.

COMPUTATIONAL MECHANICS (Abstracts)
WCCM VI in conjunction with APCOM'04, *Sept. 5-10, 2004, Beijing, China*

Computational Modeling of Penetration in Porous Soils Using Arbitrary Lagrangian-Eulerian Finite Element Method

Yuan Di [1*], Tadanobu Sato [2], Mingwu Yuan [3]

[1] *Department of Mechanics and Engineering Science, Peking University, Beijing, 100871, China*
[2] *Disaster Prevention Research Institute, Kyoto University, Uji, Kyoto, 6110011, Japan*
[3] *Department of Mechanics and Engineering Science, Peking University, Beijing, 100871, China*
e-mail: diyuan@catfish.dpri.kyoto-u.ac.jp, sato@catfish.dpri.kyoto-u.ac.jp, yuanmw@pku.edu.cn

Abstract

Combining the advantages of both Lagrangian and Eulerian approaches in a single analysis, the arbitrary Lagrangian-Eulerian (ALE) finite element formulation is nowadays standard and being increasingly used for large strain problems in nonlinear solid mechanics. Different from solid material, porous soils is here considered a two-phase material with a soil skeleton and a pore fluid phase. The soil skeleton yields according to an elasto-plastic model. The pore fluid phase flows through the pore space according to Darcy's law. An ALE method has been developed for large deformation analysis of saturated soils in [1-2].

The cone penetration test and standard penetration test are by far the oldest and most commonly used in site test in geotechnical engineering. They are essentially a large deformation and finite strain process. For computational modeling these penetration problems, the ALE finite element analysis needs to be performed. Due to the merits of convenient and efficient computation, an operator-split scheme proposed in [3] is adopted to implement the ALE analysis. Within every step of operator-split ALE analysis, the finite element mesh follows the material deformation in a Lagrangian step, and a pure Lagrangian procedure step is done. Then mesh smoothing is performed and the reference system is changed as desired. Finally, the state variables are remapped from the Lagrangian mesh to the new reference mesh to complete a Eulerian step. A frictional contact interface is used to represent interactions between the sampler and soils. An ALE formulation for computational modeling of the standard penetration test is proposed. A numerical example of modeling the cone penetration test is presented to illustrate the performance of the proposed method.

REFERENCES

[1] Y. Di, T. Sato and Y. Moon, *A new simulation method for liquefaction and induced ground flow*, Proceedings of 2nd Canadian Specialty Conference on Computer Applications in Geotechnique, Winnipeg, (2002), pp. 197-204.

[2] Y. Di and T. Sato, *Arbitrary Lagrangian-Eulerian Formulation for Large Deformation Analysis of Saturated Soils*, Proceedings of The 11th International Conference on Soil Dynamics & Earthquake Engineering & The 3rd International Conference on Earthquake Geotechnical Engineering, Berkeley, (2004), pp. 110-117.

[3] D. J. Benson, *An efficient, accurate, simple ALE method for nonlinear finite element programs*, Computer Methods in Applied Mechanics and Engineering, 72, (1989), 305-350.

COMPUTATIONAL MECHANICS (Abstracts)
WCCM VI in conjunction with APCOM'04, *Sept. 5-10, 2004, Beijing, China*

Artificial Dissipation for Strictly Stable Finite Volume Methods on Unstructured Meshes

Jing Gong [1*], Magnus Svärd [1], Jan Nordström [1, 2]

[1] *Department of Information Technology, University of Uppsala, Uppsala, SE-751 05, Sweden*
[2] *Department of Computational Physics, The Swedish Defense Research Agency, Stockholm, SE-172 90, Sweden*
e-mail: jing.gong@it.uu.se, Magnus.Svard@it.uu.se, Jan.Nordstrom@foi.se

Abstract

In computational fluid dynamics, unstructured finite volume methods are widely used to handle complex geometries and nonlinear phenomena. Strict stability which prevents error growth on realistic mesh sizes, is very important for calculation over long times. Following the development in [1-3], we have derived strictly stable unstructured finite volume methods for hyperbolic, parabolic and incompletely parabolic problems. These methods employ so ca lled summation by parts (SBP) operators and impose the boundary conditions weakly, see [4 -5].
For nonlinear hyperbolic problems such as the Euler equations it is often necessary to add artificial dissipation in order to remove the high frequency oscillatio ns of the unresolved modes. In [6], it was shown that the added artificial dissipation may destroy the stability of the basic scheme unless an energy estimate can be obtained.
In this paper we will derive an artificial dissipation operator which leads to a n energy estimate in the same norm as the main scheme, see also [6]. The artificial dissipation operator is based on the discrete Laplacian developed in [5]. Numerical calculations that support the theoretical conclusions will be provided.

REFERENCES

[1] M. H. Carpenter, J. Nordström, D. Gottlieb, *A stable and conservative interface treatment of arbitrary spatial accuracy*, Journal of Computational Physics, 148, (1999), 341 -365.
[2] J. Nordström, M. H. Carpenter, *Boundary and interface conditions for high order finite difference methods applied to the Euler and Navier-Stokes equations*. Journal of Computational Physics, 148, (1999), 621-645.
[3] J. Nordström, M. H. Carpenter, *High-order finite difference methods, multidimensional linear problems and curvilinear coordinates*, Journal of Computational Physics, 173(1), (2001), 149-174.
[4] J. Nordström, K. Forsberg, C. Adamsson, P. Eliasson, *Finite volume methods, unstructured meshes and strict stability*, Applied Numerical Mathematics, 48, (2003), 453 -473.
[5] M. Svärd, J. Nordström, *Stability of finite volume approximations for the Laplacian operator on quadrilateral and triangular grids*, Applied Numerical Mathematics, (accepted).
[6] K. Mattsson, M. Svärd, J. Nordström, *Stable and accurate artificial dissipation*, Journal of Scientific Computing, (accepted).

COMPUTATIONAL MECHANICS (Abstracts)
WCCM VI in conjunction with APCOM'04, *Sept. 5-10, 2004, Beijing, China*

Goal Oriented Recovery with Fundamental Solutions

Thomas Grätsch [1], Friedel Hartmann [2*], Thorsten Kunow [2]

[1] *Department of Mechanical Engineering, MIT, 77 Massachusetts Avenue, Building 3-359, Cambridge, MA 02139-4307, USA*
[2] *Department of Civil Engineering, Univ. of Kassel, Kurt-Wolters-Str. 3, Kassel, D-34109, Germany*
e-mail: graetsch@mit.edu, friedelhartmann@uni-kassel.de, t.kunow@uni-kassel.de

Abstract

Goal-oriented error estimation procedures have gained much attention in finite element analysis since they aim at an economical computation of arbitrary quantities of physical interest. They also serve to provide error bounds for the quantity of interest, and they are used to drive automatic adaptive refinement schemes [1-3].

We present a goal-oriented a posteriori error estimation technique for the pointwise error of finite element approximations using fundamental solutions. The approach is based on an integral representation of the pointwise quantity of interest using the corresponding Green's function, which is decomposed into an unknown regular part and a fundamental solution. This approach offers the following advantages: 1) The approach can be applied to both pure Neumann- and pure Dirichlet problems or problems with mixed boundary conditions and it can easily be extended to problems with piecewise constant coefficients. 2) The numerical examples show distinctive improvements compared to standard methods since the analytical singularity is built into the approximate Green's function. 3) No regularization of the Dirac Deltas and no extreme mesh refinement near the point of interest are necessary within the adaptive refinement process. 4) The procedure is efficient since the whole infrastructure of a finite element code can be used, and the approximation of the regular part needs only one additional calculation on the same finite element mesh with a different right hand side. 5) Lower and upper error bounds for the local quantity can be calculated with standard error estimation procedures. The technique is applied to second and fourth order problems of solid mechanics.

REFERENCES

[1] I. Babuska, T. Stroboulis, *The finite element method and its reliability,* Oxford Science Publications, Oxford, (2001).

[2] W. Bangerth, R. Rannacher, *Adaptive finite element methods for differential equations*, Birkhäuser, Basel, (2003).

[3] J. T. Oden, S. Prudhomme, *Goal oriented error estimation and adaptivity for the finite element method*, Comp. Math Appl, 41, (2001), 735-756.

COMPUTATIONAL MECHANICS (Abstracts)
WCCM VI in conjunction with APCOM'04, *Sept. 5-10, 2004, Beijing, China*

Some Applications of Stress Functions and Their Dual Theory in Finite Elements

Ruoyu Huang [1*], Changchun Wu [1], Wanxie Zhong [2]

[1] *School of Civil Engineering & Mechanics, Shanghai Jiao Tong University, Shanghai 200240 , China*
[2] *State Key Laboratory of Structural Analysis of Industrial Equipment, Dalian University of Technology, Dalian 116023, China*
e-mail: ruoyuhuang@sjtu.edu.cn, ccwu@sjtu.edu.cn, zwoffice@dlut.edu.cn

Abstract

In some continuum models such as thin plate bending or elasticity with couple -stress, there exists the C^1 continuity difficulty in finite element formulations when the elements are constructed in displacement spaces. Using Cosserat continuum, the reason for the occurrence of the C^1 continuity difficulty is discussed. Cosserat continuum is a generalized model of the classical elasticity. Many important continuum models can be taken as its special cases subjected to some geometric/ mechanical constrains. It is noted that when geometric or/and mechanical constraint(s) is (are) enforced upon Cosserat model there must exist C^1 continuity requirement for either displacement field or stress function field. And the key point is that when one of these two fields has the C^1 continuity requirement, the other is free from this difficulty. So it is a natural approach to avoid C^1 continuity difficulty in displacement space by using formulation in st ress function space. Nevertheless, the finite element constructed in stress function space is not convenient to apply because stress functions have no explicit physical meaning and it is difficult to appoint boundary conditions for them. For this practical reason, the dual theory of stress functions is discussed to provide an approach to transform the element with stress functions as degree of freedom (DOF) to the element with ordinary displacement as DOF. Based on this dual theory, a systematic way to construct the finite element using stress functions is discussed. As the applications of the new methodology, some new plate bending elements and elements of couple -stress are constructed. It is noticeable that any formulation in stress function space which ca n pass the patch test in the stress function space can also pass the corresponding patch test in displacement space after the DOF transformation. This fact ensures the convergence of the elements and makes the formulations in stress function space flexible. Plenty of numerical experiments show the new elements have satisfactory convergence and precision in both plate banding and plane couple -stress problems [1-2].

REFERENCES

[1] Ruoyu Huang, Changliang Zheng, Wanxie Zhong , Weian Yao, *A new quadrilateral thin plate element based on the membrane-plate similarity theory*, Applied Mathematics and Mechanics , 23, (2002), 259-270.

[2] Ruoyu Huang, Changchun Wu, Wanxie Zhong, *Finite element of elasticity with couple-stress using the analogy between plane couple-stress and Reissner/Mindlin plate bending*, Acta Mechanica Sinica, (in printing).

COMPUTATIONAL MECHANICS (Abstracts)
WCCM VI in conjunction with APCOM'04, *Sept. 5-10, 2004, Beijing, China*

Symplectic Solutions and Computational Method of Transversely Isotropic Elastic Cylinders

Hongzhi Jia [1*], Xinsheng Xu [1], Andrew Y. T. Leung [2]

[1] *State Key Laboratory of Structure Analysis of Industrial Equipment and Department of Engineering Mechanics, Dalian University of Technology, Dalian,116024, China*
[2] *Department of Building and Construction, City University of Hong Kong, Hong Kong, China*
e-mail: jia_h2@sina.com, xsxu@dlut.edu.cn, andrew.lenud@cityu.edu.hk

Abstract

In this paper, the three dimensional problem of transversely isotropic elastic cylinders is solved analytically in the symplectic space. The traditional method solves the problem using the semi-inverse method with local effects neglected according to the Saint-Venant principle in the Euclidian space by the Lagrange formulation. The semi-inverse method is no longer possible if transverse isotropy or local effects are of interest. Hamiltonian dual equation is an alternative. The analytical solutions of a transversely isotropic elastic cylinder are obtained by the method of separation of variables with an eigenvalue problem of a Hamiltonian matrix operator in the axial coordinate and partial differential equations in the cross-sectional coordinates. The zero eigenvalue is highly degenerate and its Jordan normal forms will be considered separately from the non-zero eigenvalue solutions. The structure of the zero eigenvalue solutions is of the Saint-Venant type (simple extension, free torsional, pure bending and shearing-bending) and the non-zero eigenvalue solutions correspond to the local boundary layer effects usually neglected according to the Saint-Venant principle. The solutions of the cylindrical problem can be the linear combinations of the eigenfunctions for both zero eigenvalues and non-zero eigenvalues respectively and they satisfy the adjoint symplectic orthogonal relationships. The problem boils down to solving eigenvalues and eigensolutions, which can be handled with the non-classic boundary conditions at the two ends. For simple problem (for example, circular cylinder), the eigenvalues and eigensolutions can be obtained with analytical method directly. And for complex problem, numerical method has to be used. Since the symplectic system is provided with the conservation, the numerical method ensures the exactness. It should point that the method can be applied to nonlinear problem and the finite element can be constructed. Finally, numerical examples are given. It is shown that the analytical results agree well with the available results for an isotropic problem.

REFERENCES

[1] B. Saint-Venant, *Memoire sur la flexion des prismes*, J. Math. Pures. Appl. 1, (1856), 89-189.
[2] W. X. Zhong, *A new systematic methodology for theory of elasticity*, Dalian University of Technology Press, Dalian, (1995), (in Chinese).
[3] X. S. Xu, W. X. Zhong, H. W. Zhang, *The Saint-Venant problem and principle in elasticity*, Int. J. Solids Structures, 34, (1997), 2815-2827.

COMPUTATIONAL MECHANICS (Abstracts)
WCCM VI in conjunction with APCOM'04, *Sept. 5-10, 2004, Beijing, China*

Reinterpretation and Simplified Implementation of a Discontinuous Galerkin Method for Hamilton-Jacobi Equations

Fengyan Li, Chi-Wang Shu *

Divison of Aplied Mathematics, Brown University, Providence, RI 02912, USA
e-mail: shu@smtp.dam.brown.edu

Abstract

In this talk, we reinterpret a discontinuous Galerkin method originally developed by Hu and Shu for solving Hamilton-Jacobi equations. By this reinterpretation, numerical solutions will automatically satisfy the curl-free property of the exact solutions inside each element. This new reinterpretation allows a method of lines formulation, which renders a more natural framework for stability analysis. Moreover, this reinterpretation renders a significantly simplified implementation with reduced cost, as only a smaller subspace of the original solution space in the original discontinuous Galerkin method is used and the least square procedure used before is completely avoided. We will start the talk with a general description of the discontinuous Galerkin method and its application to Hamilton-Jacobi equations, and end the talk with several numerical results .

COMPUTATIONAL MECHANICS (Abstracts)
WCCM VI in conjunction with APCOM'04, *Sept. 5-10, 2004, Beijing, China*

Discontinuous Galerkin Methods with Mesh Adaptivity

Ruo Li [1], Tao Tang [2*]

[1] *School of Mathematics, Peking University, Beijing, China*
[2] *Department of Mathematics, Hong Kong Baptist University, Hong Kong SAR, China & Institute of Computational Mathematics, Chinese Academy of Sciences, Beijing, China*
e-mail: rli@math.pku.edu.cn, ttang@hkbu.edu.hk

Abstract

The discontinuous Galerkin (DG) methods have been proved very powerful for solving many practical problems such as the nonlinear conservation laws and Maxwell equations. Since the solutions of these problems may be discontinuous, it is natural for them to be approximated in a discontinuous finite dimensional space. The moving mesh methods involve the solution of the underlying PDE for the physical problem in conjunction with a so-called moving mesh PDE for the mesh itself. The methods keep the total number of grid points unchanged, and can cluster more grid points to areas with singularities or large solution gradients. However, the mesh obtained by using the moving mesh methods may be very irregular. Since there are no continuity requirement among the elements for the DG methods, the geometry of each element can be very flexible, and as a result the DG methods can handle very irregular meshes. It is therefore natural to use the DG methods as the PDE evolution algorithms. In this talk, we will describe how to combine the DG methods with the moving mesh techniques. Several practical issues such as the choice of the monitor functions and the solution interpolation will be carefully investigated. The resulting scheme will be applied to solving nonlinear system of hyperbolic equations.

REFERENCES

[1] B. Cockburn, C. -W. Shu, *Runge-Kutta discontinuous Galerkin methods for convection-dominated problems*, Journal of Scientific Computing, 16, (2001), 173-261.
[2] R. Li, T. Tang, P.-W. Zhang, *A moving mesh finite element algorithm for singular problems in two and three space dimensions*, Journal of Computational Physics, 177, (2002), 365-393.
[3] Y. Di, R. Li, T. Tang, P. Zhang, *Moving mesh finite element methods for the incompressible Navier-Stokes equations*, SIAM J. Sci. Comput., (accepted).
http://www.math.hkbu.edu.hk/~ttang/MMmovie/NS/papers/mmm1.pdf

COMPUTATIONAL MECHANICS (Abstracts)
WCCM VI in conjunction with APCOM'04, *Sept. 5-10, 2004, Beijing, China*

Performance of a Geometric and an Algebraic Multilevel Preconditioner for Transport/Reaction Systems

P. T. Lin*, M. Sala, J. N. Shadid, R. S. Tuminaro

Sandia National Laboratories, P.O. Box 5800 MS 0316, Albuquerque, NM, 87185-0316, USA
e-mail: ptlin@sandia.gov

Abstract

This study investigates the performance of a geometric two-level domain decomposition preconditioner and an n-level algebraic multilevel preconditioner for accelerating the convergence of a parallel Newton-Krylov solution method for transport/reaction problems [1]. The system of equations includes the Navier-Stokes equations along with the thermal energy and mass transfer equations with non-equilibrium chemical reactions. The spatial discretization of the governing equations is based on a stabilized finite element formulation [2] for unstructured meshes. The geometric preconditioner [3] uses a two-level Schwarz domain decomposition approach. The fine mesh is produced from uniform refinement of the coarse mesh. The algebraic preconditioner [4] uses an approach that starts from a graph generated from the nonzeros of the matrix for the fine mesh and uses graph partitioning to produce the aggregates. A comparison of the scalability and efficiency of the two approaches is presented, along with a comparison with a 1-level domain decomposition method, on large-scale distributed-memory parallel machines.
The results demonstrate the influence of the grid transfer operators, the smoothers and the coarse grid solvers for a set of challenging 2D and 3D example problems. For more than two levels it is demonstrated that careful attention needs to be given to the balance of robustness and convergence rate for the smothers and the cost of applying these methods.
Sandia is a multiprogram laboratory operated by Sandia Corporation, a Lockheed Martin Company, for the United States Department of Energy's National Nuclear Security Administration under contract DE-AC04-94AL85000.

REFERENCES

[1] J. N. Shadid, S. A. Hutchinson, G. L. Hennigan, H. K. Moffet, K. D. Devine, A. G. Salinger, *Efficient parallel computation of unstructured finite element reacting flow solutions*, Parallel Computing, 23, (1997), 1307-1325.

[2] T. J. R. Hughes, L. P. Franca, M. Balestra, *A new finite element formulation for computational fluid dynamics: V. circumventing the Babuska-Brezzi condition: a stable Petrov-Galerkin formulation of the Stokes problem accommodating equal-order interpolations*, Comp. Meth. App. Mech. And Eng., 59, (1986), 85-99.

[3] R. S. Tuminaro, C. H. Tong, J. N. Shadid, K. D. Devine, D. M. Day, *On a multilevel preconditioning module for unstructured mesh Krylov solvers: two-level Schwarz*, Comm. Num. Method. Eng., 18, (2002), 383-389.

[4] M. Sala, *Domain Decomposition Preconditioners: Theoretical Properties, Application to the Compressible Euler Equations, Parallel Aspects*, Ph. D. Thesis, EPFL, Lausanne, (2003).

COMPUTATIONAL MECHANICS (Abstracts)
WCCM VI in conjunction with APCOM'04, *Sept. 5-10, 2004, Beijing, China*

Computational Model Validation under Uncertainty

Sankaran Mahadevan [*], Ramesh Rebba, Shuping Huang

Department of Civil and Environmental Engineering, Vanderbilt University, Box 1831 -B, Nashville, TN 37235, USA
e-mail: {sankaran.mahadevan, rebba.ramesh, shuping.huang}@vanderbilt.edu

Abstract

This paper presents methodologies for validating computational models under physical, informational and model uncertainties. A Bayesian methodology is proposed to develop validation metrics that integrate uncertainties in both computational simulation and empirical measurement. The simulation models that we address in this paper are primarily finite element-based structural analysis and limit state-based reliability prediction models.

Verification and validation under uncertainty involves quantifying the error in the model prediction and effectively comparing the prediction with the experimental result when both prediction and test data are stochastic. Several deterministic *a posteriori* error estimates are available in the literature for adaptive mesh refinement and model verification in finite element analysis. This paper presents a method to estimate the statistical distribution of discretization error in the prediction of finite element-based computational models. A collocation-based stochastic response surface method is used for computational efficiency in predicting the stochastic distribution of error. Next, a Bayesian methodology is developed for model validation [1]. The prior distribution of error in predicting the response is first computed, which is then updated based on experimental observation using Bayesian analysis. The prior and posterior error distributions are used to compute a validation metric that quantifies the confidence in the model prediction. Aggregate validation of a model, considering multiple response quantities and corresponding experimental observations, is achieved through a probabilistic principal component analysis. The noise in both data sets is first filtered, and in the transformed space of principal components, the two data vectors are compared in terms of distance and orientation to compute the validation metric.

Several components of computational prediction error, such as discretization error, element error, and stochastic analysis error are included. Two types of measurement error are included, in the context of model validation: error in the measurement of input variables that affects the model prediction, and error in the measurement of output variables. The concept of Bayesian hypothesis testing is extended to system-level problems where full-scale testing is impossible. Component-level validation results are used to derive a system-level validation measure. This derivation depends on the knowledge of inter-relationships between component modules. Bayes networks are used for the propagation of validation information from the component-level to system-level. The computational methods are illustrated with several numerical examples.

REFERENCES

[1] R. Zhang, S. Mahadevan, *Bayesian methodology for reliability model acceptance*, Reliability Engineering and System Safety, 80, (2003), 85-103.

COMPUTATIONAL MECHANICS (Abstracts)
WCCM VI in conjunction with APCOM'04, *Sept. 5-10, 2004, Beijing, China*

Reliable and Efficient Numerical Modeling of Elastic Wave Propagation in Engineering Structures by Discontinuous Galerkin FEM

B. Tie*, D. Aubry

LMSSMat, Laboratoire de Mécanique Sols-Structures-Matériaux, Ecole Centrale Paris, Grande Voie des Vignes, 92295, Châtenay Malabry, France
e-mail: tie@mssmat.ecp.fr, aubry@mssmat.ecp.fr

Abstract

In the last few years, the space-time Discontinuous Galerkin (DG) finite element method has become increasingly popular in the field of numerical modeling of wave propagation in elastic structures. This method has been proved to be more accurate than the conventional Newmark family schemes and its numerical damping allows filtering high frequency numerical noises and giving proper representation of wave fronts. Furthermore, it has been shown that the DG method is particularly suitable for managing mesh changes in time and implementing adaptive dynamic solvers. However, the drawback of the DG method, as for all the implicit methods, is that it is very CPU and memory consuming, which constitute a serious obstacle for its use in industrial applications.

This paper discusses implementation of the DG method in order to perform reliable and efficient numerical modeling of elastic wave propagation in engineering structures. Several important computational issues related to the solver are considered, such as the automatic mesh adaption, the error estimates and the parallel computing. Both implicit and explicit DG solvers are presented and their stability and convergence properties are discussed. Numerical examples are given to illustrate the interest of our solvers, which have been successfully applied to several industrial problems: shock wave modeling in space launcher structures under pyrotechnic cut loads and elastic wave propagation in honeycomb sandwich panels.

REFERENCES

[1] D. Aubry, D. Lucas, B. Tie, *Adaptive strategy for transient/coupled problems, applications to thermo-elasticity and elasto-dynamics,* Comp. Meth. Appl. Mech. Engrg., 176, (1999), 41-50.

[2] C. Johnson, P. Hansbo, *Adaptive finite elements methods in computational mechanics,* Comp. Meth. Appl. Mech. Engrg., 101, (1992), 143-181.

[3] B. Tie, D. Aubry, A. Boullard, *Adaptive computation for elastic wave propagation in plate/shell structures under moving loads,* European J. for F.E. Methods, 12(6), (2003), 717-736.

[4] N. E. Wiberg, X. Li, *Implementation and adaptivity of a space-time finite element method for structural dynamics,* Comp. Meth. Appl. Mech. Engrg, 156, (1998), 211-229.

COMPUTATIONAL MECHANICS (Abstracts)
WCCM VI in conjunction with APCOM'04, *Sept. 5-10, 2004, Beijing, China*

Numerical Model of Brick Structures Using the Rigid Bodies and Spring Model

Kazuto Yamamura [1*], Atsushi Kikuchi [2], Hideyuki Morihara [3]

[1] *Environment & Process Technology Center, Nippon Steel Corporation, 20-1 Shintomi, Futtsu City, Chiba Pref., 293-8511, Japan*
[2] *Computational Engineering Center, Nippon Steel Technoresearch, 20-1 Shintomi, Futtsu City, Chiba Pref., 293-8511, Japan*
[3] *Sanki Engineering Corporation, 46-59 Nakabaru, Tobata Ku, Kitakyushu City, Fukuoka Pref., 804-0002, Japan*
e-mail: yamamura@re.nsc.co.jp, a-kikuchi@re.nsc.co.jp, morihara@re.nsc.co.jp

Abstract

The finite element method (FEM), used generally for structure analysis, has limits in evaluation of brick structures such as behavior of joints and crack s in bricks. The rigid bodies and spring model (RBSM) suppose an element to be a rigid body and elements are connecting with springs, so it is technique suited for analysis of brick structures in order to be able to handle exfoliation and sliding between elements easily.

In this study, as new functions, a heat conduction analysis method is built in the RBSM and behavior model of joints such as exfoliation or sliding to be necessary for brick structure are developed. And we developed some new improvement of method under combined forces of thermal and mechanical load to the bricks.

By using this method, evaluation of brick structures becam e possible. Quantitative elucidations of behavior of brick structures and damage mechanism (prediction of joint opening or crack propagation), design optimization of brick structures (prevention structure of collapse or crack, design of dowel) are expected.

In this report, outline of the techniques and applications to basic structures are introduced.

REFERENCES

[1] T. Kawai, *A new discrete model for analysis of solid mechanics problems*, J. Seisan Kenkyu, Inst. Ind. Sci., University Tokyo, 29, (1977), 208 -210.

[2] A. Kikuchi, T. Kawai, N. Suzuki, *The rigid bodies spring model and their applications to three-dimensional crack problems*, Computer & Structures, 44(1 -2), (1992), 469-480.

[3] K. Takihara, H. Uchida, A. Kikuchi, Y. Toi, *Numerical simulation of nonlinear behaviors of three-dimensional block structures*, Journal of JSME, 1, (1995), 53 -54, (in Japanese).

[4] K. Yamamura, A. Kikuchi, K. Goto, *Analysis of brick structures using the rigid body and spring model*, UNITECR 2003 Congress Proceedings, (2003), 384-387.

COMPUTATIONAL MECHANICS (Abstracts)
WCCM VI in conjunction with APCOM'04, *Sept. 5-10, 2004, Beijing, China*

The Method of Precise Integration of Elliptic Functions

Zheng Yao *, Wanxie Zhong

Department of Engineering Mechanics, Dalian University of Technology, State Key Laboratory of structural analysis for industrial equipment, Dalian, 116023, China
e-mail: wind0411@student.dlut.edu.cn, wxzhong@dlut.edu.cn

Abstract

The precise integration method solves the time step integration for time invariant system first. For such problem, the precise integration gives highly precise numerical result, which approaches the full computer precision. It constitutes a 2^N algorithm. After the integration of time invariant system is solved, various approximate methods can be applied to other problems such as time variant or non-linear time integration.
Elliptic functions are a special kind of double period complex functions which require in engineering widely, especially in nonlinear problems. Numerical solution for elliptic functions meets great difficulties since the precision and efficiency in numerical method. In stead of original lookup table, many new numerical methods appear recent years. Such as Matlab and IMSL have functions which can compute the Jacobi elliptic functions.
Many elliptic functions in engineering problems are second order elliptic functions, and many complicated elliptic functions are found to be obtained from second order elliptic functions. The most familiar second order elliptic functions include Jacobi elliptic functions and Weierstrass elliptic functions. They can be expressed as power series expansion, directly calculating is discommodious. One of the most important properties of elliptic functions is the additional theorem, so that the method of precise integration can be invoked. In the present paper, the second order elliptic functions, both Jacobi and Weierstrass elliptic functions, are successfully computed via precise integration approach, which is shown an efficiency method for elliptic functions. Numerical experience points out that the method of precise integration has higher efficiency and stability than the counterpart of external functions in IMSL obviously.

REFERENCES

[1] Benqing Gao, *Elliptic functions and applications*, National Defence Industry Press, 1991, (in Chinese).
[2] Wanxie Zhong, *On precise integration method*, Journal of Computational and Applied Mathematics, 163, (2004).
[3] Wanxie Zhong, *Precise time integration for the equation of computational structural dynamics*, Journal of Dalian University of Technology, 43(2), (1994), 1865-1872, (in Chinese).
[4] Wanxie Zhong, *Duality system in applied mechanics and optimal control*, Kluwer, 2004.

COMPUTATIONAL MECHANICS (Abstracts)
WCCM VI in conjunction with APCOM'04, *Sept. 5-10, 2004, Beijing, China*

A Mixed Finite Element Method for a Thermally Nonlinear Coupled Problem

Jiang Zhu *, Abimael F. D. Loula

Laboratório Nacional de Computação Científica, MCT, Avenida Getúlio Vargas 333, Petrópolis, 25651-075 RJ, Brazil
e-mail: jiang@lncc.br, aloc@lncc.br

Abstract

In [1], we considered the following thermally coupled nonlinear elliptic problem:

$$-\nabla\cdot(\mu(\theta)\nabla u)=f \quad \text{in } \Omega$$

$$-\Delta\theta=\mu(\theta)(\nabla u)^2 \quad \text{in } \Omega$$

$$u=0 \quad \text{on } \Gamma$$

$$\theta=0 \quad \text{on } \Gamma$$

where $u:\Omega\to\mathbb{R}$ is the potential, $\theta:\Omega\to\mathbb{R}$ is the temperature, Ω - is a bounded open subset of $\mathbb{R}^d$, $d=2$ or 3, Γ its boundary.

As in [1], we admit for simplicity the homogeneous boundary condition and assume that the coupling function $\mu\in C(\mathbb{R})$ is bounded, i.e., there exist constants $K_2\geq K_1>0$ such that, for all $s\in\mathbb{R}$, $K_1\leq\mu(s)\leq K_2$.

In this work, first we improve a mathematical result presented in [1]. Next we propose a mixed method for the potential equation and establish existence, uniqueness and regularity of the weak solution of the problem. Then we apply a fixed point algorithm and propose a finite element approximation to the problem. Finally we present a convergence analysis and numerical implementations.

REFERENCES

[1] A. F. D. Loula, J. Zhu, *Finite element analysis of a coupled nonlinear system*, Comput. Appl. Math., 20, (2001), 321-339.

COMPUTATIONAL MECHANICS (Abstracts)
WCCM VI in conjunction with APCOM'04, *Sept. 5-10, 2004, Beijing, China*

Influence of Damping on Dynamic Forces Behavior in Large-Belt Conveyors

Gabriel Barrientos [*], Ramón Maldonado, Pedro Saavedra

Department of Mechanical Engineering, University of Concepción, Chile
e-mail: gbarrien@udec.cl

Abstract

In this work the dynamic forces produced in a conveyor belt of great length are evaluated, when it suffers a start and/or a sudden stop. In belts of this type, with steel cords inside, commonly used in the handling of minerals in the mining industry, the traditional design manuals [1] do not deliver information as to belts of great length, and because of this the designs can be under or over sized. Under-sizing can be the cause of catastrophic kinds of failures, causing great losses to the industry. For this reason the correct evaluation of dynamic forces is of great importance.
The Physical model used in this work is based on that proposed by Harrison [2] and corresponds to a bar in which, from the effect of a greater mass of material transported, from these start or sudden stops there begins a wave of stress in both senses (of traction and compression) which in some cases have managed to cut the belt. This phenomenon is represented by an equation of one dimensional wave propagation which is resolved by various methods. Among those which stand out: finite differences, finite elements, the wave propagation theory and impedance method.
Results are obtained for relative displacements, by which dynamic forces along the length of the belt are calculated. All the equations and methods used are programmed in MATLAB platform. Results are compared from all methods for a real problem in the Chilean mining industry, using data from a belt approximately five kilometers in length.
In the solution implemented, using the method of finite elements, there is included the effect of damping [3] and this is integrated numerically using Newark's provisional algorithm. In general this is used, depending on four cases: normal stop, abortive start, smooth stop, normal start.

REFERENCES

[1] Manual of design, 1997. *Belt conveyors for Bulk Materials*, Fifth Ed., *Published by CEMA.*
[2] Alex Harrison, *Criteria for minimizing transient stress in conveyor belts*, Mech. Eng. Trans. I. E. Aust., MES(3), (1983), 129-134.
[3] Ramón Maldonado, 2003, *Effect of damping in the dynamics forces produced in the large belt conveyors*, Inform of under graduation, University of Concepción, Chile, (in Spanish).

COMPUTATIONAL MECHANICS (Abstracts)
WCCM VI in conjunction with APCOM'04, *Sept. 5-10, 2004, Beijing, China*

Finite Element Based on Reduced Trefftz Type Concepts

Sulaiman Y. Abo Diab *

Department of Structural Mechanics, Tishreen University, Faculty of Civil Engineering, Lattakia, Syria
e-mail: sabodiab@scs-net.org

Abstract

Variational displacement and stress concepts associated with reduced Trefftz type approaches are presented in this paper. The displacement concept operates on the natural boundary conditions enforced in an integral sense. It recovers the conventional displacement and hybrid stress finite element concept. The element approximation basis can be constructed using a modified geometrical interpolation technique or using a weak hybrid technique as well as a boundary technique by applying the Gauss divergence theorem or the equivalency of virtual work between the entire boundary and the static and kinematic boundaries, respectively. The stress concept operates on the essential boundary conditions enforced in an integral sense. It recovers the stress and hybrid displacement finite element concept. The element approximation basis can be constructed in an analogous way to that of the displacement formulation in form of stress components.

A combination of the modified interpolation technique and the use of the Gauss divergence theorem or the equivalency of virtual work are possible.

In order to apply the reduced Trefftz-type approach in the framework of the finite element method it is suitable to select the approximation functions in an element local Cartesian coordinate system defined from the directions of the covariant base vectors and the perpendicular contravariant base vectors. This enables the formatting of complex element forms such quadrilateral or curved elements and facilitates considerably the satisfaction of the Lagrangian equation. We can take advantages from the numerical integration by transforming the approximation functions and their derivatives to a natural coordinate system using the isoparametric transformation and after that performing the element integrals in this system. It is not necessary to transform the energy expressions to the natural coordinate system in order to evaluate the element matrices and the load vectors. The energy expressions can be evaluated as they are related to the Cartesian coordinate system after replacing the Cartesian variables with their natural equivalents. Some numerical results produced by quadrilateral plate bending elements with exactly integrated stiffness matrix and load vectors are also presented.

REFERENCES

[1] S. Y. Abo Diab, *Formatting of Quadrilateral Finite Elements*, Fifth World Congress on Computational Mechanics (WCCM V), 7-12 July, Vienna (2002), Austria.

[2] S. Abo Diab, *Quadrilateral folded plate structure elements of reduced Trefftz type*, CAMES, 10(4), (2003), 391-406.

[3] J. Jirousek, A. Wrobelwski, *Alternative displacement frame formulations in hybrid-Trefftz Kirchhoff plate p-elements*, CAMES, 4, (1997), 417-451.

COMPUTATIONAL MECHANICS (Abstracts)
WCCM VI in conjunction with APCOM'04, *Sept. 5-10, 2004, Beijing, China*

A Mixed Eight-Node Hexahedral Element Applicable to Nonlinear Structural and Mechanics Problems

Yung-I Chen [1*], Nuan-Hsuan Ho [1], Guan-Yuan Wu [2]

[1] *Bureau of High Speed Rail, MOTC, 9F, No. 7, Sec. 2, Sianmin Blvd., Banchiao City, Taipei County 220, Taiwan, China*

[2] *Department of Fire Science, Central Police University, No. 56, Shu Jen Road, Kuei San, Taoyuan 333, Taiwan, China*

e-mail: yichen@hsr.gov.tw

Abstract

The mixed eight-node hexahedral element formulated utilizing the Hu -Washizu principle and the idea of the field extrapolation is presented in the context o f corotational formulation for large displacement but small strain problems. In this approach, each element of the deforming body is observed from the point of view of its own frame, which corotates with it. Large overall displacements of the deformed elem ent are separated into two parts. One is relative to large rigid body motions that cause no strain in the deformation process; the other relative to the corotating frame, termed corotational displacements, is the only one affecting strains. Even though, ge nerally, corotational displacements can have arbitrary magnitude, they are assumed to be small that linear strain-displacement relationships of the mixed hexahedral element can be used. In linear analysis, the mixed hexahedral element has been found very a ccurate and effective for both structural and mechanics problems even when severely distorted meshes and nearly incompressible materials are used. The element satisfies the patch test when each element of the patch is a parallelepiped, but fails if of arbitrary hexahedral shape. From fulfilling all convergence criteria point of view, if numerical simulations, particularly in nonlinear applications, generated by the mixed element converge to the correct results might be a questionable issue. As an intermedia te stage of investigating such issue, some benchmark nonlinear problems, which have proven difficult to solve using continuum elements, are analyzed to assess the element properties. It is shown that nonlinear structural behavior, including buckling and po st-buckling responses, is well cover by the proposed approach and competitive to some well -known elements.

REFERENCES

[1] Y. I Chen, G. Y. Wu, *A mixed 8-node hexahedral element based on the Hu-Washizu principle and the field extrapolation technique*, Struct. Engrg. Mech., 17, (2004), 113-140.

[2] H. K. Stolarski, T. Belytschko, S. H. Lee, *A review of shell finite elements and corotational theories*, Comp. Mech. Adv., 2, (1995), 125-212.

[3] M. A. Crisfield, *A fast incremental/iteration procedure that handle snap-through*, Comp. Struct., 13, (1981), 55-62.

COMPUTATIONAL MECHANICS (Abstracts)
WCCM VI in conjunction with APCOM'04, *Sept. 5-10, 2004, Beijing, China*

Hybrid Trefftz Plane Elements with Drilling Degrees of Freedom

Noori Choi *, Yeonseok Choo, Byungchai Lee

Department of mechanical engineering, Korea advanced institute of science and technology, 373-1, Guseong-dong, Yuseong- gu, Daejon, 305-701, Republic of Korea
e-mail: noori@kaist.ac.kr, ChooYeonSeok@kaist.ac.kr, bclee@sys713.kaist.ac.kr

Abstract

We introduce efficient and accurate plane elements with drilling degrees of freedom based on the hybrid Trefftz method. Hybrid Trefftz method connects two independent displacement fields : internal displacement fields satisfying governing equations of plane elasticity problems and boundary displacement fields consistent at the inter-element boundaries. In this study, we derive element stiffness of triangular and quadrilateral elements by using hybrid Trefftz variational formulation which is based on the total potential energy principle and boundary integrations for stiffness matrices. At the same time, we approximate the boundary displacement fields by using Allman type displacement fields for drilling degrees of freedom. However, we can easily apply it to arbitrary polygon elements because all the domain integrals are changed into boundary integrals. With these plane elements, we performed a series of tests and got good results in view of accuracy and robustness.

REFERENCES

[1] J. Jirousek, *Basis for development of large finite elements locally satisfying all field equations*, Comput. Meth. in Appl. Mech. and Engng, 14, (1978), 65-92.

[2] J. Jirousek and P. Teodorescu, *Large finite elements methods for the solution of problems in the theory of elasticity*, Comput. & Struct., 15, (1982), 575-587.

[3] V. Kompis, *Finite elements satisfying all governing equations inside the element*, Comput. & Struct., 50, (1984), 273-278.

[4] J. Jirouske and A. Venkatesh, *Hybrid Trefftz plane elasticity elements with p-method capabilities*, Int. J. for Num. Meth. in Engng, 35, (1992), 1443-1472.

[5] J. Jirousek, *Variational formulation of two complementary hybrid-Trefftz FE models*, Comm. in Num. Meth. in Engng, 9, (1993), 837-845.

[6] J. Jirouske and A. P. Zielinski, *Survey of Trefftz-type element formulations*, Comput. & Struct., 63(2), (1997), 225-242.

[7] D. J. Allman, *A compatible triangular element including vertex rotations for plane elasticity analysis*, Comput. & Struct., 19, (1984), 1-8.

[8] R. D. Cook, *On the Allman triangle and a related quadrilateral element*, Comput. & Struct., 22, (1986), 1065-1067.

[9] D. J. Allman, *A quadrilateral finite element including vertex rotations for plane elasticity analysis*, Int. J. for Num. Meth. in Engng, 26, (1988), 717-730.

COMPUTATIONAL MECHANICS (Abstracts)
WCCM VI in conjunction with APCOM'04, *Sept. 5-10, 2004, Beijing, China*

A Study of a Macro Membrane Triangle with Drilling Freedoms

Jaesung Eom*, Byungchai Lee

Department of mechanical engineering , Korea advanced institute of science and technology, 373-1, Guseong-dong, Yuseong- gu, Daejon, 305-701, Korea
e-mail: kuchen@kaist.ac.kr, bclee@sys713.kaist.ac.kr

Abstract

A simple self-refined triangle element with drilling d.o.f.'s is proposed for plane stress problems. Three-node triangular element has geometrical advantages in preprocessing but suffers from the bad performance like membrane locking and mesh directional favor (Cutting a quadrilateral to two triangles in the shortest-diagonal-cut) comparing to quadrilateral elements.
In the present study, the characteristic matrix of macro element is extracted by using algebraic form of matrices. Construction of macro element is dividing a three-node triangular mesh into three IET passing sub-triangular element and then merging these by condensing out the virtual node at center. Each sub-triangle stiffness comes from a three-node triangle using the optimal ANDES (Assumed Natural Deviatoric Strain) formulation [6] and the general framework of F.E. templates. The resulting macro element stiffness is decomposed into IET passing basic stiffness and self-refined property matrix. This macro property matrix takes care of the element's stability and enhances accuracy like higher-order stiffness part in the F.E. template and mesh directional insensitivity.
The macro element has invariant stiffness with no commutable zero energy mode and fits on the uses of conventional three-node element. Benchmark examples show proposed element in closed form stiffness from CAS (Computer algebra system) gives the improved results without more computational efforts than others.

REFERENCES

[1] R. D. Cook, *A plane hybrid element with rotational DOF and adjustable stiffness*, Int. J. for Num. Meth. Eng., 24, (1987), 1499-1508.
[2] K. Alvin, H. M. de la Fuente, B. Haugen, C. A. Felippa, *Membrane triangles with corner drilling freedoms, I. the EFF element*, Finite Elements Anal. Des., 12, (1992), 163-187.
[3] C. A. Felippa, C. Militello, *Membrane triangles with corner drilling freedoms, II. the ANDES element*, Finite Elements Anal. Des., 12, (1992), 189-201.
[4] C. A. Felippa, S. Alexander, *Membrane triangles with corner drilling freedoms, III. implementation and performance evaluation*, Finite Elements Anal. Des., 12, (1992), 203–239.
[5] C. A. Felippa,, B. Haugen et al, *From the individual element test to finite-element templates, evolution of the patch test*, Int. J. for Num. Meth. Eng., 38, (1995), 199-229.
[6] C. A. Felippa, *A study of optimal membrane triangles with drilling freedoms*, Computer Methods in Applied Mechanics and Engineering, 192, (2003), 2125-2168.
[7] C. A. Felippa, C. Militello, *Construction of optimal 3-node plate bending triangles by templates*, Computational Mechanics, 24(1), (1999), 1-13.

COMPUTATIONAL MECHANICS (Abstracts)
WCCM VI in conjunction with APCOM'04, *Sept. 5-10, 2004, Beijing, China*

Generalized Conforming Quadrilateral Membrane Elements Based on Analytical Trial Functions

Xiangrong Fu [1*], Yuqiu Long [2], Mingwu Yuan [1], Song Cen [3], Xiaoming Chen [2]

[1] *Department of Mechanics and Engineering Science, Peking University, Beijing, 100871, China*
[2] *Department of Civil Engineering, Tsinghua University, Beijing, 100084, China*
[3] *Department of Engineering Mechanics, Tsinghua University, Beijing, 100084, China*
e-mail: fuxr@pku.edu.cn, longyq@mail.tsinghua.edu.cn, censong@mail.tsinghua.edu.cn

Abstract

Based on the analytical trial functions, two generalized conforming 4-node quadrilateral membrane elements denoted as ATF-Q4(a) and ATF-Q4(b) have been derived and analyzed in this paper. In contradiction with traditional isoparametric elements, these elements select analytical solutions of the problem as trial functions. And several flexible generalized conforming conditions are introduced to determine all unknown parameters. Some benchmark problems are studied to examine their stability and convergence properties and robustness to mesh distortion. Numerical results show that the presented elements give reasonable way to overcome locking to trapezoidal shape and sensitivity to distorted mesh. Their high precision and well convergence emphasize the benefit of these new ATF type elements.

REFERENCES

[1] N. S. Lee, K. J. Bathe, *Effects of element distortion on the performance of isoparametric elements,* Int. J. Num. Meth. Eng., 36, (1993), 3553-3576.

[2] E. L. Wilson, R. L. Taylor, W. P. Doherty, T. Ghabussi, *Incompatible displacement models*, in S. T. Fenven et al ed. *Numerical and Computer Methods in Structural Mechanics*, New York, Academic Press, (1973), pp. 43-57.

[3] R. L. Taylor, P. J. Beresford, E. L. Wilson, *A non-conforming element for stress analysis,* Int. J. Num. Meth. Eng., 10, (1976), 1211-1219.

[4] E. L. Wachspress, *Incompatible quadrilateral basis functions,* Int. J. Num. Meth. Eng., 12, (1978), 589-595.

[5] Y. Q. Long, M. F. Huang, *Generalized conforming isoparametric element,* Applied Mathematics and Mechanics, 9, (1988), 871-877, (in Chinese).

[6] T. H. H. Pian, K. Sumihara, *Rational approach for assumed stress fin ite elements,* Int. J. Num. Meth. Eng., 20, (1984), 1685-1695.

COMPUTATIONAL MECHANICS (Abstracts)
WCCM VI in conjunction with APCOM'04, *Sept. 5-10, 2004, Beijing, China*

Finite Element Method for Smeared and Embedded Reinforcements in Object Oriented Programming

André Luís Gamino *, Túlio Nogueira Bittencourt

[1] *Department of Foundation and Structural Engineering, University of São Paulo, Av. Prof. Almeida Prado n.271, São Paulo, 05508-900, Brazil*
e-mail: andre.gamino@poli.usp.br, tulio.bittencourt@poli.usp.br

Abstract

The use of discrete reinforcements in concrete structures modelling by finite elements method provides some disadvantages as the dependence of the mesh for the positioning of the bars in agreement comments of Elwi, Hrudey [1]. This problem can be solved by means of the use of embedded bars to the representative finite element of the concrete with the advantage not to be introduced new incognito to the problem. The formularizations for introduction of embedded bars are presented considering perfect bond or bond-slip adequate model. Solutions for implementation of distributed armors (on the basis of the proposals of Chang, Taniguchi and Chen [2] and Hartl [3]) are presented and argued. Finally the results gotten for a platform of finite elements developed in the paradigm of the object oriented programming are analyzed.

REFERENCES

[1] A. E. Elwi, T. M. Hrudey, *Finite element model for curved embedded reinforcement*, Journal of Engineering Mechanics, 115, (1989), 740-754.

[2] T. Y. Chang, H. Taniguchi, W. F. Chen, *Nonlinear finite element analysis of reinforced concrete panels*, Journal of Structural Engineering, 113, (1987), 122-140.

[3] H. Hartl, *Development of a continuum-mechanics-based for 3D finite element analysis of reinforced concrete structures and application to problems of soil-structure interaction*, PhD Thesis, Technische Universität Graz, Austria, (2002).

COMPUTATIONAL MECHANICS (Abstracts)
WCCM VI in conjunction with APCOM'04, *Sept. 5-10, 2004, Beijing, China*

Mixed Finite Element for Nonlinear Material Problems

R. Grimaldi, D. Addessi, V. Ciampi *

Department of Structural Engineering, University of Rome 'La Sapienza', Via Eudossiana 18, Rome, 00184, Italy
e-mail: vincenzo.ciampi@uniroma1.it

Abstract

Mixed and enhanced finite elements have been widely used in the context of linear structural analyses, mainly to avoid locking that arises from classical displacement based methods. More recently, mixed-hybrid and enhanced strain formulations have been tested to model, with improved accuracy, the stress and strain distributions due to various nonlinear material behaviours, such as plasticity or damage. Indeed, when standard displacement approaches are adopted in the presence of material nonlinearity, a relevant increase in computational effort is required.
Here, the capability of such alternative methods to describe strain localization, due to adoption of constitutive relationships with softening, is investigated.
Two different mixed finite elements and an enhanced strain method are formulated, as an extension to two dimensional problems of hybrid models used for the beam problem. The first mixed model is deduced from the Hu-Washizu variational principle, adopting as element variables stress and strain, while displacement are assumed as global variable. The second mixed formulation is deduced from the Hellinger-Reissner variational principle, assuming as independent fields stress and global displacement. Finally, the Simo-Rifai enhanced strain element is extended by adopting higher order self-strain shape functions, which do not affect the element performance in the linear case, while are able to better model nonlinear behaviours.
For the three models 4 node as well as 8 node quadrilateral element have been developed. All the models take advantage of internal d.o.f. condensation techniques, in order to limit the size of the global problem and to make possible their use in the framework of general displacement based finite element codes.
A number of numerical tests and comparisons among the different models, with particular reference to the ones based on classical displacement based formulations, are presented.

REFERENCES

[1] A. Capsoni, L. Corradi, *A mixed model for plane strain elastic-plastic analysis, part I. formulation and assesment of the overall behaviour*, Comp. Meth. in Appl. Mech. And Engrg, 141, (1996), 67-79.

[2] C. Comi, U. Perego, *A unified approach for variationally consistent finite elements in elasto-plasticity,* Comp. Meth. in Appl. Mech. And Engrg, 121, (1995), 323-344.

[3] M. Petrangeli, V. Ciampi, *Equilibrium based iterative solutions for the non linear beam problem,* Int. J. Num. Meth. in Engrg, 40, (1997), 423-437.

[4] J. C. Simo, M. S. Rifai , *A class of assumed strain methods and the method of incompatible modes,* Int. J. for Num. Meth. in Engrg., 29, (1990), 1595-1638.

COMPUTATIONAL MECHANICS (Abstracts)
WCCM VI in conjunction with APCOM'04, *Sept. 5-10, 2004, Beijing, China*

Shape Identification for Navier-Stokes Equations with Unsteady Flow Using Bubble Function Element Stabilization Method

J. Matsumoto *

Research Institute for Computational Sciences, AIST (PRESTO, JST), Central-2, 1-1-1 Umezono, Tsukuba, Ibaraki, 305-8568, Japan
e-mail: junmcct@ni.aist.go.jp

Abstract

A new Petrov-Galerkin formulation of finite element method with bubble function is proposed for the incompressible Navier-Stokes equations [1]. Numerically, spatial discretization is applied to the mixed interpolation for the velocity and pressure fields by bubble element and linear element, respectively. Numerical solutions [2-5] for shape identification of flow past a circular cylinder are treated in this paper. The Sakawa-Shindo method [6] is employed as a minimization algorithm. The purpose of the study is to formulate and solve an analysis of shape identification for Navier-Stokes equations with unsteady flow. Generally, the state equations that govern incompressible viscous flow are written to include the Navier-Stokes equations. By described here in making the state equations a constraint of the performance function J, the Lagrange multiplier method is applied for shape identification [4-5]. To minimize the extended performance function J^*, it is necessary to solve the gradient of J^* with respect to the identified coordinates x_j. The necessary condition is that the stationary condition deriving from the first variation of J^* is set to zero. To improve efficiency, stability, and accuracy of the calculation, the mixed interpolation that uses an improved bubble function element for the state equations of incompressible viscous fluid is applied. To validate the present method, shape identification of flow past a circular cylinder with periodic flow is analyzed. The identified drag force results in periodic solution region are similar to the objective results.

REFERENCES

[1] J. Matsumoto, M. Kawahara, *Shape identification for fluid-structure interaction problem using improved bubble element*, Int. J. Comput. Fluid Dynamics, 15, (2001), 33-45.
[2] O. Pironneau, *On optimum profiles in Stokes flow*, J. Fluid Mech., 59(part 1), (1973), 117-128.
[3] R. K. Ganesh, *The minimun drag profile in laminar flow: a numerical*, Transaction of the ASME, J. Fluids Engng., 116, (1994), 456-462.
[4] H. Azegami, *Solution to boundary shape identification problem in elliptic boundary value problem using shape derivatives*, Inverse Problems in Engineering Mechanics II, (2000), pp. 277-284.
[5] Y. Ogawa, M. Kawahara, *Shape optimization of body located in incompressible viscous flow based on optimal control theory*, Int. J. Comput. Fluid Dynamics, 17, (2003), 243-251.
[6] Y. Sakawa, Y. Shindo, *On global convergence of an algorithm for optimal control*, Transactions on Automatic Control, IEEE, AC-25(6), (1980), 1149-1153.

COMPUTATIONAL MECHANICS (Abstracts)
WCCM VI in conjunction with APCOM'04, *Sept. 5-10, 2004, Beijing, China*

Efficient Dynamic Finite Element Method for 3D Objects with Uniform Cross-Sections

Yuki Onishi [1*], Masataka Urago [1], Kenji Amaya [2], Shigeru Aoki [3]

[1] *Department of Mechanical and Environmental Informatics, Tokyo Institute of Technology, Tokyo 152-8552, Japan*
[2] *Department of International Development Engineering, Tokyo Institute of Technology, Tokyo 152-8552, Japan*
[3] *Department of Computational Science and Engineering, Toyo University, Saitama 350-8585, Japan*

e-mail: yonishi@a.mei.titech.ac.jp, murago@ide.titech.ac.jp, kamaya@a.mei.titech.ac.jp, saoki@eng.toyo.ac.jp

Abstract

An efficient implicit dynamic finite element method(FEM) for elastic 3D objects with uniform cross-sections was developed. In this method, the finite element mesh is generated in such a way that the object to be analyzed is at first sliced into layers with the same thickness along the generatrix and then each layer is discretized into the same pattern. This way of discretization makes the mass, viscosity, and stiffness matrices into the repetitive block tridiagonal matrices. The repetitive block tridiagonal matrix has the characteristic, that the sequence of matrices which appears in the Gaussian elimination for the repetitive block tridiagonal matrix is a rapid convergent sequence. The process of the Gaussian elimination can be terminated when the sequence converges. The rest of the sequence is not necessary to be stored. The present method can save the computational time and memory by utilizing this characteristic of the repetitive block tridiagonal matrix. A few examples of analyses including whole Hopkinson-bar analysis were performed to demonstrate the effectiveness of the present method. The present method is applicable not only to the elasto-dynamics but also to many other problems, such as thermal and electrical problems.

REFERENCES

[1] C. Guo, P. Lancaster, *Iterative solution of two matrix equations*, Math. Comput, 68(228), (1999), 1589-1603.
[2] T. J. R. Hughes, *The finite element method: linear static and dynamic finite element analysis*, Dover Publications, (2000), 490-513.
[3] H. Kolsky, *Stress waves in solids*, Dover Publications, (1963), 87-91.

COMPUTATIONAL MECHANICS (Abstracts)
WCCM VI in conjunction with APCOM'04, *Sept. 5-10, 2004, Beijing, China*

A General Convergence Condition of the Newton-Raphson Algorithm Applied to Compressible Hyperelasticity

F. Peyraut [1*], Z. Q. Feng [2], N. Labed [1]

[1] *Laboratoire M3M, University of Technology of Belfort-Montbéliard, Rue du Château, Sévenans, 90010 Belfort cedex, France*

[2] *Laboratoire de Mécanique d'Evry, University of Evry-Val d'Essonne, 40, rue du Pelvoux, 91020 Evry cedex, France*

e-mail: francois.peyraut@utbm.fr, feng@iup.univ-evry.fr, nadia.labed@utbm.fr

Abstract

In the case of the Blatz-Ko hyperelastic compressible model [1], it is shown in [2] and [3] that the violation of orientation preservation can cause the Newton-Raphson algorithm to diverge. Within the framework of the finite element theory, we propose a new algorithm which provides a control of the orientation preservation in order to prevent Newton-Raphson divergence. Based on the eigenvalues of the deformation gradient matrix, this control is carried out at the Gauss integration points and within each Newton-Raphson iteration. We show that this control is very general because it works with any geometry and any load case. We also show that this control is optimal because we compute a loading step number which is small enough to minimize computing time but sufficiently large to ensure the Newton-Raphson algorithm to converge. The implementation in the finite element software FER is presented with a total Lagrangian formulation [4]. We have successfully tested it in two cases. The first one concerns the academic case of a hyperelastic cubo subjected to a compressive force in a rigid container [2]. The second case is more complex because it relates to frictional contact with large sliding between two hyperelastic blocks. The contact problem is solved by the bipotential method [5, 6].

REFERENCES

[1] P. J. Blatz, W. L. Ko, *Application of finite elastic theory to the deformation of rubbery materials*, Transactions of the Society of Rheology, 6, (1962), 223-251.

[2] F. Peyraut, N. Labed, *Préservation de l'orientation et convergence de Newton-Raphson avec le modèle hyperélastique compressible de Blatz-Ko*, Revue Européenne des Eléments Finis, 10, (2001), 595-605.

[3] F. Peyraut, *Orientation preservation and Newton-Raphson convergence in the case of an hyperelastic sphere subjected to an hydrostatic pressure*, Comput. Methods Appl. Mech. Engrg., 192, (2003), 1107–1117.

[4] R. D. Wood, O. C. Zienkiewicz, *Geometrically nonlinear finite element analysis of beams, frames, arches and axisymmetric shells*, Comput & Struct, 7, (1977), 725-735.

[5] G. De Saxcé, Z. Q. Feng, *The bipotential method: a constructive approach to design the complete contact law with friction, improved numerical algorithms*, Int. J. of Math. Comput. Model., Special Issue, *Recent Advances in Contact Mechanics*, 28(4–8), (1998), 225-245.

[6] Z. Q. Feng, F. Peyraut, N. Labed, *Solution of large deformation contact problems with friction between Blatz-Ko hyperelastic bodies*, Int. J. of Engineering Science, 41, (2003), 2213-2225.

COMPUTATIONAL MECHANICS (Abstracts)
WCCM VI in conjunction with APCOM'04, *Sept. 5-10, 2004, Beijing, China*

Special Hybrid Element with Drilling Degrees of Freedom

A. P. Wang*, Z. S. Tian

Department of Mechanics, Graduate School, Chinese Academy of Sciences, P. O. Box 2706, Beijing 100080, China
e-mail: zxq@mail.etp.ac.cn

Abstract

A special hybrid finite element with drilling degrees of freedom — which contains a traction-free circular boundary — has been developed based on the Hellinger-Reissner principle.
The assumed stress field satisfies the equilibrium equations in the elements as well as the traction-free conditions on assigned boundaries respectively. The assumed stress field also satisfies the compatibility equations when the element is degenerated into 2-dimensional case.
The combination of the special element with the ordinary assumed displacement elements can be efficiently used for analyzing the stress concentrations around some cutouts.
The stresses of plate with symmetric fillets— in different ratios of the height to the width — under tension or bending are analyzed to evaluate the following two cases: 1) Combining the special assumed stress hybrid elements with the ordinary isoparармetric elements which are derived by the conventional assumed displacement approach. 2) Using the conventional assumed displacement isoparametric elements everywhere.
Numerical results have demonstrated that the special element can provide much more accurate stress concentration factors and the distribution of circumferential stress σ_θ along the rim of the grooves than those obtained by using the conventional assumed displacement elements.

REFERENCES

[1] S. M. Yunus, *A study of different hybrid elements with and without rotational d.o.f. for plane stress and plane stain,* Comput. & Stuct., 30(5), (1988), 1127-1133.
[2] R. D. Cook, *A plane hybrid element with rotational d.o.f. and adjustable stiffness,* Int. J. Num. Meth. Engng, 24, (1987), 1499-1508.
[3] T. Y. Lee, Y. S. Kang, C. K. Choi, *Non-conforming membrane elements with drilling d.o.f.,* CD-ROM of WCCM'V, Vienna, Austria, (2002).

COMPUTATIONAL MECHANICS (Abstracts)
WCCM VI in conjunction with APCOM'04, *Sept. 5-10, 2004, Beijing, China*

The Finite Element Format of Large Increment Method in material nonlinearity problems

Kaijian Wang [1*], Xila Liu [2]

[1] *Department of Civil Engineering, University of Tsinghua, Beijing, 100084, China*
[2] *Department of Civil Engineering, University of Shanghai Jiaotong, Shanghai, 200030, China*
e-mail: wkj98@mails.tsinghua.edu.cn, xilaliu@tsinghua.edu.cn

Abstract

In the present paper the finite element formulation of Large Increment Method (LIM) for material nonlinear problems is given. The Large Increment Method (LIM) is a new iteration method [1][2], which is based on the force method and the generalized inverse matrix theory for solving material nonlinear problems, so LIM is also called the force method based on Generalized Inverse Matrix (GIM). Unlike the classical force method, LIM does not need to find a basic structure any more, so this method brings the force method a new light in the computer calculation field [3]. LIM doesn't need to solve the material nonlinearity problem step by step, as the classic method based on the displacement method does [4]. The system control equations are divided into a linear group, which includes the equilibrium equation and the compatibility equation, and a nonlinear group, which includes the constitutive equation [5]. The iteration procedure starts from a special solution of the equilibrium equation which is got by using the generalized inverse matrix theory. The whole iteration procedure can be divided into the global stage and the local stage. In the global stage, the linear group is used. In the local stage, the nonlinear group must be considered. Furthermore the Drucker postulate is considered in the local stage. With these general finite element formulations, this new method can not only solve the problems of truss system, but also the multi-dimensional problems. As the mathematical background of LIM, the generalized inverse matrix theory is also introduced. Besides, there are very strong parallel-calculating characteristics in LIM, which are different with the traditional sub-structural algorithm. Finally, an example of plane stress problem is given, which shows the efficiency and generality of this new method.

REFERENCES

[1] Xila Liu, Chunjun Zhang, *A large increment method based on the generalized inverse matrix*, Engineering Mechanics, supplement, (1994), pp. 21-35, (in Chinese).
[2] C. J. Zhang, X. L. Liu, *A large Increment Method for Material Nonlinearity Problems*, Advances in Structural Engineering, 1(2), (1997), 99-110.
[3] Xila Liu, *The Status and Prospects of Structural Engineering*, Beijing, People's Jiaotong Press, (1997), (in Chinese).
[4] Pierre Ladeveze, *Nonlinear Computational Structural Mechanics*, New York, Springer-Verlag, (1999).
[5] R. T. Fenner, *Finite Element Method for Engineers*, Macmillan, London, (1978).

COMPUTATIONAL MECHANICS (Abstracts)
WCCM VI in conjunction with APCOM'04, *Sept. 5-10, 2004, Beijing, China*

Finite Element of B-Spline Wavelet on the Interval and Its Engineering Application

Jiawei Xiang*, Xuefeng Chen, Bin Li, Zhengjia He

School of Mechanical Engineering, Xi'an Jiaotong University, Xi'an, 710049, China
e-mail: wxw8627@163.com, xjw27@163.com

Abstract

This paper discusses the up-to-date research for both B-Spline wavelets on the interval (BSWI) [1] and adaptive wavelet finite element method (FEM) [2-3]. FEM is a general-purpose numerical analysis tool, which has been playing an important role in many engineering fields. However, for many complex problems with high gradient or strong nonlinearities in engineering, traditional FEM has some disadvantages, such as low efficiency, insufficient accuracy, slow convergence to correct solutions etc. It is greatly significant not only in theory but also in practice that a new finite element method should be explored to overcome the above disadvantages and make it more efficient in engineering application. Wavelet finite element method has more advantages for high gradient problems and wavelet on the interval has many advantages on solving singular partial differential equations. BSWI have advantages on numerical approach and there are more nodes in each BSWI elements than traditional elements. So we can adopt fewer BSWI elements than traditional elements to get the same accuracy. A finite element method of B-Spline wavelet on the interval (FEM BSWI) is established by using the above advantages, and one-dimensional corresponding C0 and C1 elements are constructed. The temperature field and displacements of beam are calculated by using BSWI elements. The results coincide with corresponding exact solutions and numerical solutions obtained by using the commercial software ANSYS. What's more, the lifting algorithm of automatically adaptive wavelet FEM is present. The adaptive form is established based on the FEM BSWI. When the calculated accuracy of local areas is not satisfied, the wavelet terms are configured automatically to the corresponding elements so as to increase the exponent of the local areas and improve the calculative accuracy. And the stiffness matrix and transfer matrix on the previous scale are still useful. So we can adopt fewer wavelet basis functions to approximate actual solution according to automatically adaptive high exponent. The numerical examples testified the validity of automatically adaptive wavelet FEM in solving singular problems, such as the temperature field of office paper with high varying gradient. So this paper lays the foundation for solving singular problems in engineering. This work is supported by National Natural Science Foundation of China (No.50335030, 50175087).

REFERENCES

[1] Quak Ewald, Weyrich Norman, *Decomposition and reconstruction algorithms for spline wavelets on a bounded interval*, Applied and Computational Harmonic Analysis, 1(3), (1994), 217-231.

[2] C. Canuto, A. Tabacco, K. Urban, *The wavelet element method Part I: construction and analysis*, Applied and Computational Harmonic Analysis, 6, (1996), 1-52.

[3] A. Cohen, R. Masson, *Wavelet adaptive methods for elliptic problems-preconditioning and adaptivity*, SIAM J. Sci. Comp., 21, (1999), 1006-1026.

COMPUTATIONAL MECHANICS (Abstracts)
WCCM VI in conjunction with APCOM'04, *Sept. 5-10, 2004, Beijing, China*

Geometric Nonlinear Calculation of the Flexspline in Harmonic Drive

Wenbin Yao *, Qianjun Shao, Wei Zhang

Faculty of Engineering, Zhejiang Forestry University, Linan, Zhejiang, 311300, China
e-mail: wenbin925@sina.com

Abstract

Harmonic drive was called as revolution of machanic drive and flexspline is a key element in harmonic drive. Its service life decides lifetime of harmonic drive equipment. As flexspline will produce large deformation in assembling and driving, its stress level should be relatively low. Hence the problem of the flexspline calculation is typical geometric nonlinear problem.

The flexspline is a thin shell on which bending and twisting moments are exerted. The paper considers it as a large deflection problem in which flexspline is forced under non-axisymmetric loads. In order to calculate simply, the flexspline is considered as fixed and wave generator as initiative.The calculational cases are given.Calculational results indicate that dangerous section locates at the section of gear loop transition.The point agrees to practice.

It should be obvious that the finite element model is simple, direct and practical, by which the geometrical nonlinear questions in cylindrical shell and other shells having the complicated conditions of geometrical boundary and weighted conditions are easily handled. The research indicates that the computational data will be valuable if influence of large deflection is taken into account and the method can be satisfied with necessity in engineering.

COMPUTATIONAL MECHANICS (Abstracts)
WCCM VI in conjunction with APCOM'04, *Sept. 5-10, 2004, Beijing, China*

Exact Integration for Hypersingular Boundary Integral Equation of Two-Dimensional Potential Problems

Xin An, Xiaosong Zhang *

Department of Mechanics and Engineering Science, Shijiazhuang Railway Institute, Shijiazhuang, 050043, China
e-mail: tjanxin@yahoo.com.cn, xszhang@sjzri.edu.cn

Abstract

There are two ways to formulate the boundary value problem into boundary integral equation, i.e. the Cauchy singular boundary integral equation (BIE) and the hypersingular boundary integral equation (HBIE). The hypersingular boundary integral equation has been studied for symmetric Galerkin BEM, degenerate boundary problems and fracture mechanics. It also provided a natural way for error estimation in adaptive boundary element analysis.
Accurate and efficient evaluation of the integral in boundary element analysis is an important issue and has been studied intensively in past few decades. This paper presents the exact integration for hypersingular boundary integral equation of two-dimensional potential problems. The boundary of the domain under consideration is discretized with straight segments, while the primary variables within the elements are approximated by discontinuous linear elements. The integrals for the influence matrices in boundary element analysis are obtained in closed form. It is shown that no special treatment is needed for both singular in CPV and hypersingular integrals in HFP sense if the exact integration and discontinuous boundary elements are utilized. This greatly simplifies the computer codes for boundary element analysis.
Two numerical examples are considered. The first is a quadrilateral domain under prescribed potential boundary condition, while the second is Motz problem where the flux is singular. These problems are analyzed by both Cauchy singular boundary integral equation and hypersinular boundary integral equation to verify the correctness of the presented exact integration. The numerical results of the kind of boundary integral representations are in good agreement except for the singular points. This illustrates that the Cauchy singular and hypersingular boundary integral equation provided a natural way for the error indicator in adaptive boundary element method.

REFERENCES

[1] M. Fratantonio, J. J. Rencis, *Exact boundary element integration integrations for two-dimensional Laplace equation*, Eng Anal Bound Elem., 24, (2000), 325-42.

[2] X. Zhang, X. Zhang, Exact integration in boundary element method for two-dimensional elastostatics problems, Eng Anal Bound Elem., 27, (2003), 987-97.

[3] M. Guiggiani, Hypersingular boundary integral equations have an additional free term. Comput. Mech., 16, (1995), 245-48.

[4] J. F. Luo, Y. J. Liu, E. J. Berger, Analysis of two-dimensional thin structures (from micro- to nano-scales) using the boundary element method. Comput. Mech., 22, (1998), 404-12.

COMPUTATIONAL MECHANICS (Abstracts)
WCCM VI in conjunction with APCOM'04, *Sept. 5-10, 2004, Beijing, China*

A Simple Boundary Element Formulation for Shape Optimization of 2D Continuous Structures

Luciano M. Bezerra*, Jarbas C. Santos-Júnior

Department of Civil Engineering, University of Brasilia, Brasilia -DF, 70910-900, Brazil
E-mail: lmbz@unb.br

Abstract

A mathematical formulation based on the Boundary Element Method (BEM) and on deterministic optimization techniques is proposed in this paper. The optimal geometric configuration of Two-dimensional Continuous Structure (2DCS) is obtained with the minimization of an objective function. This function is written in terms of referential stresses located at points in the internal or at the boundary of the 2DCS. The process of geometric optimization is accomplished by reducing the difference between the referential stresses and the stresses calculated at the referential points. The 2DSC geometry is expressed in terms of design variables. Along the minimization process, the design variables are modified until the minimization of the objective function approximates the calculated stress values given at the reference points. The values of the design variables have to be within certain limits and, therefore, inside a feasible domain. The restrictions on the design variable values are enforced by the use of heuristic rules. The use of the BEM for problems of geometric optimization of 2DCS is quite attractive with respect to other methods that inevitably discretize the whole continuous, i.e., the Finite Elements and the Finite Difference Methods. When the body forces are not significant, the BEM discretizes just the boundary of the structure. During the minimization process of objective function, the geometry of the structure undergoes a sequence of geometric modifications which is accompanied with the appropriate discretization of the 2DCS. With the BEM such modifications are easy to perform while in formulations using the Finite Element or the Finite Difference methods such mesh modifications turn out to be very cumbersome. Several simple examples of the application of the proposed formulation are presented with some conclusions and suggestions for future work.

REFERENCES

[1] R. E. Ricketts, O. C. Zienkiewicz, *Shape optimization of continuum structures,* In E. Atrek, R. H. Gallagher, K. M. Ragsdeil, O. C. Zienkiewicz eds. *New directions in optimum structural design,* John Wiley & Sons Ltd, (1984).
[2] P. E. Gill, W. Murray, M. H. Wright, *Practical optimization,* London, Academic Press Inc., (1981).
[3] W. H. Press, B. P. Flannery, S. A. Teukolsky, W. T. Vetterling, *Numerical recipes,* New York, Cambridge University Press, (1986).
[4] R. H. Gallager, O. C. Zienkiewicz, *Optimum structural design: theory and applications.* London, John Willey, (1973).
[5] P. K., Banerjee, R. Butterfield, *Boundary element methods in engineering science,* London, McGraw-Hill, (1981).

COMPUTATIONAL MECHANICS (Abstracts)
WCCM VI in conjunction with APCOM'04, *Sept. 5-10, 2004, Beijing, China*

A Regularized Algorithm for the Calculation of Second Derivative Values in Potential BEM

H. B. Chen *

CAS Key Laboratory of Mechanical Behavior and Design of Materials, Department of Modern Mechanics, University of Science and Technology of China, Hefei, Anhui230026, China
e-mail: hbchen@ustc.edu.cn

Abstract

In the past decade, experiments have shown that materials display strong size effects when the characteristic length scale is on the order of microns. The conventional elasto -plasticity theory, however, can not predict this size dependence because its constitutive model possesses no internal length scale. So far, a number of models considering this length scale have been proposed, see e.g. [1], and numerical simulations have been made to this phenomenon through the finite element method (FEM) [2] and the boundary element method (BEM) [3]. In those numerical analyses, strain gradients were generally calculated in an indirect way, i.e. interpolating from the obtained strains or interpolating twice from the displacements. This procedure obviously induces significant numerical errors. Therefore, the aim of the present work is to develop a new algorithm to improve the numerical accuracy of the strain gradients. For the first step, this paper presents our new algorithm in the context of potential BEM to show how the potential second derivatives can be calculated in an agreeable accuracy.

The boundary integral equations (BIEs) of potential second derivatives are of third order singularities; here the strong singularity is referred to as the first order singularity and the hyper-singularity as the second order one. Obviously the direct calculation of the third order singular integrals is rather difficult, which has been seldom met in the BEM analysis so far. The idea of the present paper is to use a special indirect algorithm which is based on a regularized B IE formulation of the potential second derivatives, following the work of the present author and his coworkers [4]. In this algorithm, the singularities in the integrals are fully regularized, thus only a Gaussian quadrature is needed and the most common u sed C^0 continuous element can be used. Numerical examples show the accuracy and efficiency of the present algorithm.

REFERENCES

[1] S. H. Chen, T. C. Wang, *Interface crack problems with strain gradient effects*, Int. J. Fracture, 117, (2002), 25-37.

[2] Z. C. Xia, J. W. Hutchinson, *Crack tip fields in strain gradient plasticity*, J. Mech. Phys. Solids, 44, (1996), 1621-1648.

[3] A. Benallal, C. A. Fudoli, W. S. Venturini, *An implicit BEM formulation for gradient plasticity and localization phenomena*, Int. J. Numer. Meth. Engng, 53, (2002), 1853-1869.

[4] H. B. Chen, P. Lu, E. Schnack, *Effective algorithms for the calculation of values on and near boundaries in 2D elastic BEM*, Eng. Anal. Bound. Elem., 25, (2001), 851-876.

COMPUTATIONAL MECHANICS (Abstracts)
WCCM VI in conjunction with APCOM'04, *Sept. 5-10, 2004, Beijing, China*

Network Parallel BEM for Band Precision Rolling

Yiming Chen [1*], Yajuan Hao [1], Qingxue Huang [2], Guangxian Shen [3], Hong Xiao [3]

[1] *College of Science, Yanshan University, Qinhuangdao, 066004, China*
[2] *Department of Material Science and Engineering, Taiyuan Heavy Mechanical College, Taiyuan, 030024, China*
[3] *College of Mechanical Engineering, Yanshan Univer sity, Qinhuangdao, 066004, China*
e-mail: Chenym@ysu.edu.cn, moonhyj@sina.com.cn, HQX@tyhmi.edu.cn, {sgx35, Xhh}@ysu.edu.cn

Abstract

In this paper, through the network parallel computation, it is given of the three-dimensional deformation, stress field and surface force field and the distributions of surface force of the strip steel, under nine groups of rolling condition with the ratio of the width and thickness being 200, for the three-dimensional elasticity and elasto-plasticity frictional contact Boundary Element Method and the model for frictional rolling contact of strip steel precision rolling [1-3]. The rolling pressure, rolling direction and the landscape orientation friction are the true surface of the strip steel and roller's interface, which isn't the equivalent surface force given by FEM and not the assumed friction given by elementary difference method. It cost five days by the network parallel computing for the rolling with the ratio of the width and thickness is 200, which is shorten 6 times than 30 days by sequential computation.

This paper can be divided into the following four parts. 1) The micro landscape orientation deformation of strip rolling and three-dimensional rolling theory. 2) Roller deformation and three-dimensional coupling rolling theory. 3) Basic assumptions and the computational model: Basic assumptions for the contact of strip steel and roller; Computational flow chart; Parameters for the strip steel and roller, and nine groups of rolling conditions. 4) Distributions of three-dimensional rolling pressure and frictional force: Three-dimensional distributions of the rolling pressure, rolling and landscape orientation friction; Friction distributions on the surface of strip steel.

This project is supported by Hebei Province Natural Science Foundation (No. E2004000245).

REFERENCES

[1] N. Kamiya, H. Iwase, *Computational efficiency of parallel boundary element analysis via domain decomposition*, Proc.12 th Int. Conf. On BeTeQ, (1997), pp. 343-350.

[2] G.X. Shen, C.J. Hua, D.Y. Liu, *The parallel solution of 3-Delasto-plastic contact BEM based on network*, Proc.3rd Int. Conf. On BeTeQ, Beijing, 2002, Tsinghua University Press &Spring-Verlag, (2002), pp. 213-219.

[3] X. Yin, Z. H. Yao, J.G. Cheng, *A PC cluster system for large-scale BEM problem*, Mechanics and practice, (2), (2001), 43-45, (in Chinese).

COMPUTATIONAL MECHANICS (Abstracts)
WCCM VI in conjunction with APCOM'04, *Sept. 5-10, 2004, Beijing, China*

A-Posteriori Error Estimation via Regularized BIEs in Adaptive Plate Bending Analysis

D. J. Fu*, H. B. Chen, P. Q. Zhang

CAS Key Laboratory of Mechanical Behavior and Design of Materials, Department of Modern Mechanics, University of Science and Technology of China, Hefei, Anhui, 230026, China
e-mail: djfu@ ustc.edu, hbchen@ustc.edu.cn, pqzhang@ustc.edu.cn

Abstract

Error estimation and adaptive mesh refinement are important issues in the boundary element method (BEM). A number of researches have been made in the past two decades and some effective algorithms have been proposed [1]. However, comparing to the potential and elasticity problems, relative few work has been done to the fourth order plate bending problems which is also of engineering significance. In this paper, an overview is made to error estimation and adaptivity algorithms in plate bending BEM analysis and emphasis is put on the differences of the algorithms between plate bending and normally the potential and elasticity problems. Then a simple a-posteriori error estimation algorithm is proposed for plate bending BEM and it is further used to conduct an h-type adaptive mesh refinement procedure.

In the plate bending BEM analysis, boundary integral equations (BIEs) for normal and tangential slopes are both hyper-singular. Based on recent researches [2-3], the residual of the hyper-singular boundary integral equations (HBIEs) can be used as an a-posteriori error estimation for different kinds of problems, which is especially significant to adaptive scheme in BEM. In the present paper, the regularized hyper-singular plate bending BIEs are first investigated as an indirect calculation method of integrals. Rigid deflection and normal or tangential slope fields are introduced to regularize the singular or hyper-singular BIEs instead of placing the source boundary outside the actual boundary in some regularization methods. Two regularized equations for normal slope and tangential slope, which can greatly improve the accuracy of the numerical solutions in analysis, are obtained. The conventional plate bending BEM analysis and HBIEs are adopted simultaneously to present two solutions, and the measure of the difference of two solutions furnishes a new a-posteriori error indicator. Both Hermitian [4] and Lagrangian elements are used to test the efficiency of the proposed algorithm. Results of the typical numerical examples show that the new error indicator is simple and effective.

REFERENCES

[1] E. Kita, N. Kamiya, *Error estimation and adaptive mesh refinement in boundary element method, an overview,* Eng. Anal. Bound. Elem., 25, (2001), 479-495.

[2] H. B. Chen, P. Lu, E. Schnack, *Regularized algorithms for the calculation of values on and near boundaries in 2D elastic BEM,* Eng. Anal. Bound. Elem., 25, (2001), 851-876.

[3] H. B. Chen, D. H. Yu, E. Schnack, *A simple a-posteriori error estimation for adaptive BEM in elasticity,* Comput. Mech., 30, (2003), 343-354.

[4] Qinghua Du, Zhenhan Yao, Guoshu Song, *Solution of some plate bending problems using the boundary element method,* Appl. Math. Modeling, 8, (1984), 15-22.

COMPUTATIONAL MECHANICS (Abstracts)
WCCM VI in conjunction with APCOM'04, *Sept. 5-10, 2004, Beijing, China*

A Boundary Element Method for Analyzing the Dispersive Relation of Guided Waves in Bar with Rectangular Cross Section

Arief Gunawan*, Sohichi Hirose

Department of Mechanical and Environmental Informatics, Tokyo Institute of Technology, O-okayama, Meguro-ku, Tokyo 152-8552, Japan
e-mail: arief@qnde.mei.titech.ac.jp, shirose@cv.titech.ac.jp

Abstract

This study shows the application of boundary element method to find the dispersive relation of guided waves in a bar of a uniform rectangular cross section. The wave field **u** in the bar is assumed to be in the form of $\mathbf{u} = \hat{\mathbf{u}}(x_1, x_2)e^{i(\xi x_3 - \omega t)}$, where ω and ξ are the frequency and the wave number, and the x_3 axis is taken in the longitudinal direction. A boundary integral equation is derived with respect to the boundary of the cross section of the bar. Discretization of the boundary leads to a matrix equation $[\mathbf{T}]\{\mathbf{u}\} - [\mathbf{U}]\{\mathbf{t}\} = \{\mathbf{0}\}$, where $\{\mathbf{u}\}$ and $\{\mathbf{t}\}$ contain the displacements and tractions on the discretized boundaries, and [**U**] and [**T**] contain the integrals of the fundamental solutions derived by Li[1]. Taking into account the traction-free condition on the surface of the bar, det([**T**]) must be 0 to have a nontrivial solution of $\{\mathbf{u}\}$. det([**T**]) depends on ω, ξ, Poisson's ratio ν, and the shape of the cross section of the bar. When ν and the shape of the cross section are fixed, then the equation $\det([\mathbf{T}]) = 0$ gives the dispersive relation between ω and ξ. In this study, the equation $\det([\mathbf{T}]) = 0$ is solved numerically by using the Newton's method. It is found that the Newton's method is able to find ξ in both real values and complex values which satisfy $\det([\mathbf{T}]) = 0$ for a certain ω.
The boundary element method has the advantage that it can easily be applied to the other bars with complex cross sections. Furthermore, the boundary-type integral method has the computational efficiency because the discretization has to be performed only on the boundary. In this study, the dispersion curves of the longitudinal modes, the torsional modes, the screw 1 and 2 modes, and the flexural modes in a square steel bar with ν = 0.3 are shown. Comparison between our results with the results obtained by Tanaka [2] and Fraser [3], which apply semi-analytical approaches in finding the dispersion curves, shows a fairly good agreement. Our results also include the nonpropagating modes with complex wave numbers, which are not found in the other results.

REFERENCES

[1] Z. L. Li, J. D. Achenbach, I. Komsky, Y. C. Lee, *Reflection and Transmission of Obliquely Incident Surface Waves by an Edge of a Quarter Space: Theory and Experiment*, J. Applied Mechanics 59, (1992), pp.349-355.

[2] K. Tanaka, Y. Iwahashi, *Dispersion Relation of Elastic Waves in Bars of Rectangular Cross Section*, Bull. JSME 20(46), (1977), 922-929.

[3] W. B. Fraser, *Stress Wave Propagation in Rectangular Bars*, Int. J. Solids Structures 5, (1969), 379-397.

COMPUTATIONAL MECHANICS (Abstracts)
WCCM VI in conjunction with APCOM'04, *Sept. 5-10, 2004, Beijing, China*

Green's Functions and Boundary Element Method for an Infinite Magneto-Electro-Elastic Half-Plane

A. M. Jiang [1, 2*], H. J. Ding [1]

[1] *Department of Civil Engineering, Zhejiang University, Hangzhou, 310027, China*
[2] *West Branch of Zhejiang University of Technology, Quzhou, 324006, China*
e-mail: jam@vip.sina.com, hjding@mail.hz.zj.cn

Abstract

Fundamental solution or Green's function can not only be used to construct many important analytical solutions, but also be applied in dealing with defects and inclusions in materials. Moreover, Green's functions play an important role in the boundary element method.

In this paper, the general solution of infinite magneto-electro-elastic plane problems in the case of distinct eigenvalues is derived through four displacement functions and from the governing equations and the strict differential operator theorem. The displacements, electric potential, magnetic potential, stresses, electric displacements and magnetic inductions can all be expressed concisely in terms of the four displacement functions, all of which are harmonic.

On the basis of the general solution, the Green's functions for point forces, point charge and point current acting in the interior of an infinite magneto-electro-elastic half-plane are systematically studied. With the method of trial-and-error, we first assume the displacement functions for different point load cases, then, derive the necessary equations and calculate the undetermined constants of the displacement functions by considering the continuity condition of the generalized displacements and the whole equilibrium of a cut elastic strip. Further, setting one of the three loads(point force, point charge and point current) to be one unity ,and the others zero, respectively, we will derive the fundamental solutions for an infinite magneto -electro-elastic half-plane.

All the results are obtained in an exact and explicit way and hence are very convenient to be used in the associated computation with boundary element method. Finally, the boundary integral formulation is derived and the corresponding boundary element method program is implemented to perform numerical calculations. Numerical examples are also given.

REFERENCES

[1] H. J. Ding, G. Q. Wang, W. Q. Chen, *A boundary integral formulation and 2D fundamental solutions for piezoelectric media*, Comput. Meth. Appl. Mech. Engng., 158, (1998), 65-80.

[2] H. J. Ding, G. Q. Wang, W. Q. Chen, *Green's functions for a piezoelectric half plane*, Science in China (Series E), 41(1), (1998), 70-75.

[3] H. J. Ding, G. Q. Wang, W. Q. Chen, *Green's functions for a two-phase infinite piezoelectric plane*, Proc. R. Soc. Lond., 453A, (1998), 2241-2257.

[4] H. J. Ding, A. M. Jiang, *Fundamental solutions for transversely isotropic magneto-electro-elastic media*, Science in China (Series E), 46(6), (2003), 607-619.

[5] X. Wang, Y. P. Shen, *The general solution of three-dimensional problems in magneto- electro-elastic media*, Int. J. Engng. Sci., 40, (2002), 1069-1080.

[6] E. Pan, *Three-dimensional Green's functions in anisotropic magneto-electro-elastic bimaterials*, Z. Angrew. Math. Phys., 53, (2002), 815-838.

COMPUTATIONAL MECHANICS (Abstracts)
WCCM VI in conjunction with APCOM'04, *Sept. 5-10, 2004, Beijing, China*

A New Derivative Boundary Integral Equation in Two-Dimensional Potential Problems

Zhongrong Niu [1*], Huanlin Zhou [1], Xiuxi Wang [2]

[1] *Department of Engineering Mechanics, Hefei University of Technology, Hefei 230009, China*
[2] *Department of Mechanics, University of Science and Technology of China, Hefei 230026, China*
e-mail: niuzhong@mail.hf.ah.cn

Abstract

The boundary integral equations (BIE) of the primary fields have sufficient conditions to determine the solutions for the regular boundary problems, which are usually termed the conventional BIE. The conventional potential BIE can be applied to obtaining boundary potential function u and its normal derivative $\partial u/\partial n$. If the boundary potential derivative $\partial u/\partial x_i$ is required, one way is to differentiate the potential function along the boundary. However, it reduces the accuracy by one order. Another way seeks to establish the derivative boundary integral equation for the secondary field. Thus the hypersingular integral arises from the derivative BIE.

This paper deals with the derivative BIE in the two-dimensional potential problems. A series of transformations are manipulated on the conventional potential derivative BIE in order to eliminate the hyper-singularity. By partial integration, it makes that the sum of both the strongly singular integral and hypersingular integral in the derivative BIE is replaced by another strongly singular integral with the boundary variables $\partial u/\partial x_i$. It leads to a new natural BIE for the two-dimensional potential problems. The natural BIE also belongs to the derivative BIE, but only contains the strongly singular integral, which is easily evaluated by limit analysis. In the natural BIE, the boundary variables on a local boundary with the source point are the potential derivatives $\partial u/\partial x_i$, while the boundary variables on the remainder boundary are u and $\partial u/\partial n$.

As a result, another boundary element analysis according to the natural BIE can obtain more accurate potential derivatives on the boundary in comparison with the conventional BEM. It is interesting that the natural BEM can only be implemented in a local boundary if desirable. Moreover, the natural BIE can also calculate the potential derivatives of the interior points very close to the boundary instead of the conventional potential derivative BIE with the nearly hypersingular integrals. The technique is useful when the thin-wall structures are analyzed by the BEM.

REFERENCES

[1] H. Hong, J. Chen, *Derivations of integral equations for elasticity,* J. Eng. Mech., 114(6), (1988), 1028-1044.

[2] M. Guiggiani, G. Krishnasamy, T. J. Rudolphi, F. J. Rizzo, *A general algorithm for the numerical solution of hypersingular BEM,* J. Appl. Mech., 59(3), (1992), 604-614.

[3] Zhongrong Niu, Xiuxi Wang, Huanlin Zhou, Chenli Zhang, *A novel boundary integral equation method for linear elasticity — natural boundary integral equation*, ACTA Mechanica Solida Sinica, 14(1), (2001), 1-10.

COMPUTATIONAL MECHANICS (Abstracts)
WCCM VI in conjunction with APCOM'04, *Sept. 5-10, 2004, Beijing, China*

BEM Application to Calculation of Filtration through Membrane Considering Influence of Membrane Deformation

Mykola V. Polyakov [*], Dmytro V. Yevdokymov

Mechanical and Mathematical Faculty, Dniepropetrovsk National University, Nauchna str. 13, Dniepropetrovsk, 49050, Ukraine
e-mail: devd@mail.ru, devd@ukr.net

Abstract

Membrane technology is widespread in many brunches of modern industry. In mechanics a membrane is very thin body, which doesn't resist to bending. There are a lot of membrane kinds and it leads to great difficulties in the building of general mathematical model. The pressure difference must take place on the membrane to filtration process takes place. On the other hand the pressure difference leads to membrane deformation. However for many kinds of membranes, even relatively small deformation corresponds to sufficient porosity increasing an d, as a result, to sufficient increasing of filtration coefficient. The main aim of the present work is to build a family of mathematical models, which correctly phenomenologically describe the influence of membrane deformation on filtration coefficient. T he assumption, that the filtration coefficient is proportional to effective cross-section area of pores is used as a base of considering mathematical models. Such approach gives an opportunity to take into account an influence of membrane shape on filtrati on process velocity. Several mathematical models are proposed. The main differences between the models are change forms of initial cross-section of pores and forms of damage accumulation, corresponding to properties of different membrane kinds. In the any case the properties of the membrane are assumed changeable during filtration process. In the same time the filtration process can change in time too, for example, filtration from some closed volume, where the fluid is initially contained under high pressur e.

Since the geometrical shape of membrane can be very complex and membrane deflection must be substituted in quite complex formula for following integration, boundary element method is used for numerical calculation of the membrane deformation. From the m athematical point of view, it is boundary element method application to Poisson equation in some domain. In the case of constant pressure the solution of the mentioned problem is especially simple. If some fluid flow (as a rule, a Stokes flow) takes place due to filtration process, the linear mathematical model is used for the flow and boundary element method is applied for its calculation.

The problem of filtration through semipermeable membrane due to an osmotic pressure is considered as another example of developed approach. The osmotic pressure is proportional to concentration of solved substance, which is described by diffusion equation. Neglecting by influence of the membrane deflection on the concentration the concentration problem can be solved using correspondent Green's function for diffusion equation in the case half -space and using the boundary element method in general case. Euler time -stepping scheme is used for time discretization.

The developed approach is illustrated by several examples of nu merical calculations.

COMPUTATIONAL MECHANICS (Abstracts)
WCCM VI in conjunction with APCOM'04, *Sept. 5-10, 2004, Beijing, China*

Mode III Fracture Analysis of Piezoelectric Materials by Trefftz Boundary Element Method

Qinghua Qin *

Department of Engineering, Australian National University,Canberra, ACT 0200, Australian
e-mail: qinghua.qin@anu.edu.au

Abstract

Applications of the Trefftz boundary element method (BEM) to anti-plane electroelastic problems are presented in this paper. Both direct and indirect methods with domain decomposition are discussed in details. Each crack is treated as semi-infinite thin slit defined in a subregion, for which a particular solution of the anti-plane problem, satisfying exactly the crack-face condition, is derived. The stress intensity factors defined at each crack tip can be directly computed from the coefficients of the particular solution. The performance of the proposed formulation is assessed by an example and comparison is made with results obtained by other approaches. The Trefftz boundary element approach is demonstrated to be ideally suited for the analysis of the anti-plane problem of piezoelectric materials.

COMPUTATIONAL MECHANICS (Abstracts)
WCCM VI in conjunction with APCOM'04, *Sept. 5-10, 2004, Beijing, China*

Acceleration of Boundary Element Computations in Reanalysis of Problems in Elasticity

J. Trevelyan*, D. J. Scales, R. Morris, G. E. Bird

School of Engineering, University of Durham, South Road, Durham, DH1 3LE, United Kingdom
e-mail: Jon.Trevelyan@durham.ac.uk

Abstract

The objective of this work is to undertake a rapid boundary element re-analysis, i.e. an analysis of a problem similar in geometry to one already run, in elasticity. In particular, we seek to update stress contours in real time as a simple design geometry is perturbed interactively. The paper addresses the acceleration of computations in the integration phase and the solver, and goes on to describe how the internal point calculations may also be accelerated.

In order to address the integration problem, we define a set of look-up tables containing complete integrals (including the shape function terms as well as the fundamental solutions) for integrations over flat planar elements. We consider the optimal look-up table refinement and interpolation strategies, and show how an appropriate displacement smoothing strategy can be used to reduce the memory requirements.

The solver acceleration is brought about by re-use of the LU decomposition used to solve the original system as a preconditioner for an iterative GMRES solver for the perturbed matrix. We show how this greatly reduces the iteration count over both diagonal and ILU preconditioners, and explore the applicability of this technique to small and large geometric perturbations.

Internal point calculations can also benefit from re-use of information. It is possible to store the integrals resulting from collocation at each internal point to provide a set of coefficients to be multiplied by the new boundary solution. For the majority of internal points that may remain unperturbed in the design change, this can provide a very rapid internal stress/displacement computation.

COMPUTATIONAL MECHANICS (Abstracts)
WCCM VI in conjunction with APCOM'04, *Sept. 5-10, 2004, Beijing, China*

Analysis of 2-D Cracks Using Fast Mutipole BEM

Pengbo Wang *, Zhenhan Yao

Department of Engineering Mechanics, Tsinghua University, Beijing, 100084, China
e-mail: wangpengbo01@mails.tsinghua.edu.cn, demyzh@mail.tsinghua.edu.cn.

Abstract

In this paper, fast mutipole method (FMM) is applied to traction boundary integral equations for 2D elastic crack analysis.

Traction boundary integral equation is introduced to overcome the mathematical degeneration of the displacement boundary integral equation when the two surfaces of the same crack are co -planar.

The concept of finite-part integral is applied to deal with the hyper-singular integrals in traction boundary integral equation. Crack surfaces are discretized using discontinuous elements to satisfy continuity requirements associated with finite-part integral.

In order to enhance run-time and space efficiency, FMM is applied to Dual BEM. Due to the realization of mutipole expansion and local expansion in FMM, both computational complexity and memory requirement is reduced to *O(N)*, where *N* is the number of DOF.

In FMM, a new form of complex Taylor series is applied in mutipole expansion. This technique reduces the number of mutipole moments and local moments. Generalized minimum residual method (GMRES) is adopted as iterative solver. And sparse approximate inverse type is selected as the left preconditioner for GMRES. An effective sparsity pattern is developed to reduce the cost of preconditioner and improve convergence rate.

In numerical examples, the largest scale is more than 1,000,000 DOF. Numeric al results show that the presented fast mutipole Dual BEM scheme is greatly efficient for large-scale crack problems.

REFERENCES

[1] T. A. Cruse, *BIE fracture mechanics analysis 25 years of developments*, Comput. Mech., 18(1), (1996), 1-11.

[2] L. Greengard, V. Rokhlin, *A fast algorithm for particle simulation*, J. Comput. Phys., 135, (1997), 280-292.

[3] K. Yoshida, N. Nishimura, S. Kobayashi, *Application of new fast multipole boundary integral equation method to crack problems in 3D*, Engineering Analysis with Boundary Elements, 25, (2001), 239-247.

COMPUTATIONAL MECHANICS (Abstracts)
WCCM VI in conjunction with APCOM'04, *Sept. 5-10, 2004, Beijing, China*

Three-Dimensional Multipole BEM for Potential Problems and Computation of Temperature Field of Bearing Chock

Yinling Yan [1*], Guangxian Shen [1], Xiumin Chen [1], Chunxiao Yu [2]

[1] *College of Mechanical Engineering, Yanshan University, Qinhuangdao, 066004, China*
[2] *College of Science, Yanshan University, Qinhuangdao, 066004, China*
e-mail: sgx35@ysu.edu.cn, yucx77@88mail.ysu.edu.cn

Abstract

The discrete boundary integral equations give rise to dense algebraic problems when traditional BEM is used, so the operation scale is limited, for a large memory size is required and the solution procedure costs too long CPU time. Applications of Fast Multipole Method[1,2] (FMM) and Generalized Minimal Residual Method [3] (GMRES) to potential problems, combining with boundary integral equations, the computing structure of traditional 3-D potential BEM is updated, which makes the large-scale computing possible. FMM was employed to calculate the influence of elements located far from the node under consideration and applied to the fundamental solutions of integral equations directly. The discrete boundary integral equations of three-dimensional potential problems to a combination of FMM can be written as

$$c^i u^i(x^q) - \frac{1}{4\pi}\partial_m \sum_{k,l,s} \frac{1}{\left|x^q - y(\xi^s)\right|}\left\{u^{kl}\phi^l(\xi^s) n_m[y(\xi^s)] J[y(\xi^s)]\omega^s\right\}$$

$$-\frac{1}{4\pi}\sum_{k,l,s}\frac{1}{\left|x^q - y(\xi^s)\right|}\left\{q^{kl}\phi^l(\xi^s) J[y(\xi^s)]\omega^s\right\} = , \quad m = 1, 2, 3$$

Solving the discrete boundary integral equations by using GMRES without forming the coefficient matrix only produces a sequence of recursive vector, which can avoid the high memory demanding for larger scale problems, and reduce the amount of the computation largely. As a result, it makes the large-scale computing of subdivided elements possible.

In this paper, we developed 3-D potential problems Multipole BEM and source Programs, which are successfully used in solving the temperature distribution of bearing chock in 22.mm aluminium foil four-high mill. In which the degrees of freedom is 1294, in the condition of the same precision, with this new method, the need of computer memory size is only one twentieth of FEM.

This project is supported by the National Natural Science Foundation of China (No. 5.75.75) .

REFERENCES

[1] L. Greengard, V. Rokhlin, *A fast algorithms for particle simulations*, J. Comput. Phys., 73, (1987), 325-348.

[2] V. Rokhlin, *Rapid solution of integral equations of classical potential theory*, Journal of Computational Physics, 60, (1983), 87-207.

[3] Y. Saad, M. H. Schultz, *GMRES: a generalized minimal residual algorithm for solving nonsymmetric linear systems*, SIAM. J. Sci. Stat. Comput., 7, (1986), 856-869.

COMPUTATIONAL MECHANICS (Abstracts)
WCCM VI in conjunction with APCOM'04, *Sept. 5-10, 2004, Beijing, China*

Research for Lubricant Dynamic Characteristics of the Double Floating-Ring Bearing

Dequan Yang [1*], Tigui Fan [2], Xinyu Yang [3]

[1] *Inner Mongolia University for Nationalities, Tongliao, 028043, China*
[2] *Chifeng College, Chifeng, 024001, China*
[3] *Central University for Nationalities, Beijing, 100081, China*
e-mail: tlswz@sohu.com, fantigui@cfxy.cn, tlswz@soh u.com

Abstract

In this paper, hydrodynamics characteristics of double floating -ring bearing are studied by using Boundary Element Method [1, 2]. Change the satisfied differential equation of region into boundary integral equations, and then use the boundary finite element method to divide the boundary of the region into many small elements, using the known boundary conditions to change the integral equations into the algebraic equations. Solving the algebraic equations, we can obtain the unknown values of the points of boundary. And then, using the discrete forms of the integral equations we can obtain the numerical solution of the points of the region. In calculation, we obtained the curves of distribution of pressure on the surface of bearing -neck and bearing and the flow field of lubricant. All these results accord with Ref [3, 4].

We have compared the double floating-ring bearing with the single floating-ring bearing. As the result, the double floating-ring bearing when the relative rotational speed of i nner and outer floating-ring are fixed, the distribution of pressure on the surface of bearing -neck, bearing, inner and outer surfaces of the inner and outer floating -ring increase with the increasing of eccentric ratio. That manifests the load-bearing capacity of three lubricant films of the double floating -ring bearing increase with the increasing of eccentric ratio. From the distribution of the flow field we can see if there appears the vortex or not and the vortex is big or small in the lubricant relate to the eccentric ratio, and the domain of vortex increase with the increasing of the eccentric ratio. The load -bearing capacity of double floating-ring bearing is higher than single one. That is the rotational efficiency of double floating-ring bearing is higher.

REFERENCES

[1] C. A. Brebbia, *The Boundary Element Method for Engineers*, Penteen Press, (1978).
[2] Zhenhan Yao, M. H. Aliabadi, *Boundary Element Techniques*, Tsinghua University Press & Springer-Verlag, Beijing, (2002), pp. 95-100.
[3] Dequan Yang, *Boundary Element Method of Multi-lobe Hydrodynamic Lubrication Bearing for Numerical Analysis in Lubrication Theory*, International Journal for Numerical Methods in Engineering, 34(2), (1992), 667-673.
[4] Dequan Yang, Gouyou Zhang, Baolin Wu, Pengling Xing, *Boundary Element Analysis of Fluid Dynamics of Complex Shaped Bearings*, in Computational Mechanics, Vol.2, A. A. BALKEMA, Rotterdam, (1991), pp. 1469-1474.

COMPUTATIONAL MECHANICS (Abstracts)
WCCM VI in conjunction with APCOM'04, *Sept. 5-10, 2004, Beijing, China*

One Boundary Element Algorithm Application to Calculation of Motion near the Solid Boundary

Dmytro V. Yevdokymov [1*]

Mechanical and Mathematical Faculty, Dniepropetrovsk National University, Nauchna str. 13, Dniepropetrovsk, 49050, Ukraine
e-mail: devd@mail.ru, devd@ukr.net

Abstract

Motion of small objects near the solid boundary is fundamental proble m of applied hydrodynamics. The discrete vortex motion in fluid, small liquid drop motion in gas, small gas bubble motion in liquid, small solid particle motion in liquid and gas can be considered as examples of such problems. The main difficulty in this p roblem is necessity of very accurate velocity determination near the boundary. Traditional numerical methods (finite difference and finite element method) cannot provide desired accuracy. One of the most sufficient disadva ntages of the ordinary boundary element method is fast increasing of error near the boundary. This effect is especially strong on the distances less, than one half of boundary element length from the boundary. The aim of the present work is development of boundary element algorithm, which can overcome mentioned difficulties. As a base of such algorithm the functional equations of V. Kupradze [1] are used.

The error sources of boundary element method are considered in paper [2]. Comparing the accuracy of numerical calculations of test proble m for different domains it can be concluded, that the approximation of the boundary incomes the most contribution into the total error of method, at least near the boundary. Note that the approximation of the boundary is necessary for singular boundary element method, since otherwise it is difficult to evaluate singular integrals analytically. However for regular method the approximation of the boundary isn't obligatory, since all using integrals are regular. Excluding of the boundary approximation from the regular boundary element method, which is replacing of integration along approximating curve by integration along real boundary, is main idea of proposing algorithm.

Several problems with known analytical solutions were used as test problems, for example, external potential ideal fluid flow about the circle without circulation. Considered test problems confirmed high accuracy of the proposed algorithm near the boundary.

The proposed approach was applied to the spraying process in protective coating creatio n and to the problem of discrete vortex motion near the solid boundary in combined boundary element and discrete vortex method proposed in the paper [3].

REFERENCES

[1] V. D. Kupradze, *Potential method in elasticity theory,* Fizmatgiz, Moscow, (1963), (in Russian).

[2] D. V. Yevdokymov, *On one of regular boundary element method algorithm,* Visnyk Dniepropetrovsk University, Mechanics, 1(2), (1999), 150-156, (in Russian).

[3] D.V. Yevdokymov, *Boundary element and discrete vortices method for ideal fluid flow calculations,* in D. Durban and A. R. J. Pearson eds. *Non-linear singularities in deformation and flow.* Proceeding of IUTAM Symposium, Haifa, Israel, March 17-21, (1997), Kluwer Academic Publisher, pp. 217-230.

COMPUTATIONAL MECHANICS (Abstracts)
WCCM VI in conjunction with APCOM'04, *Sept. 5-10, 2004, Beijing, China*

Performance Research on FMM Based GMRES (m) Algorithm

Chunxiao Yu [1*], Guangxian Shen [2]

[1] *College of Science, Yanshan University, Qinhuangdao, 066004, China*
[2] *College of Mechanical Engineering, Yanshan University, Qinhuangdao, 066004, China*
e-mail: sgx35@ysu.edu.cn, yucx77@88mail.ysu.edu.cn

Abstract

In this paper, we studied the stability and convergen ce of GMRES (m) algorithm, which is often used in the solution of asymmetry positive definite linear system of equations. GMRES (m) algorithm is based on complete orthonormalization of Krylov vector, so an approximate minimal residual solution is obtained in Krylov subspace. GMRES (m) algorithm includes many merits such as less memory and computation, each step of iteration having optimality without break and so on. Now it has been successfully used in all kinds of engineering application fields [1].

However, current studies on GMRES (m) algorithm are limited to the quantitative analysis [2] of memory requirement and computation and the discussion how to reduce these costs, or to point out that each step of iteration has optimality. As for the problems why i terations will not break down and show excellent stability and convergence, no one has done related theoretical research about the performance yet. Therefore, we strictly demonstrate the essence of the above problems in mathematics and find out the reason why iterations will not break down, and prove the stability of GMRES (m) algorithm and the reliability of the convergent solution. For the convergence of GMRES (m) algorithm, FMM [3] is adopted to accelerate the iterations, for GMRES (m) algorithm and FMM are similar to each other ın the product operation of a matrix and a vector.

In the end, it is brought to light of the true reason why simulating computation shows stability and faster convergence through an example which is about the contact pressure analysis of interference fit of oil-film bearing sleeve and roll neck (from one of the Baogang projects. See fig.1 and fig.2.), which further shows excellent performance of FMM based GMRES (m) algorithm.

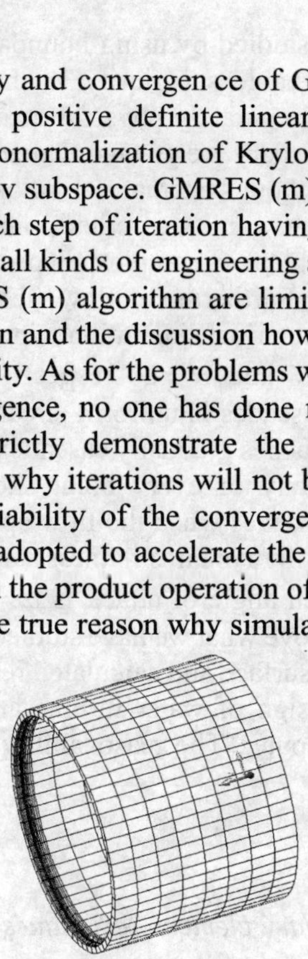

Fig.1 Model for the taper sleeve

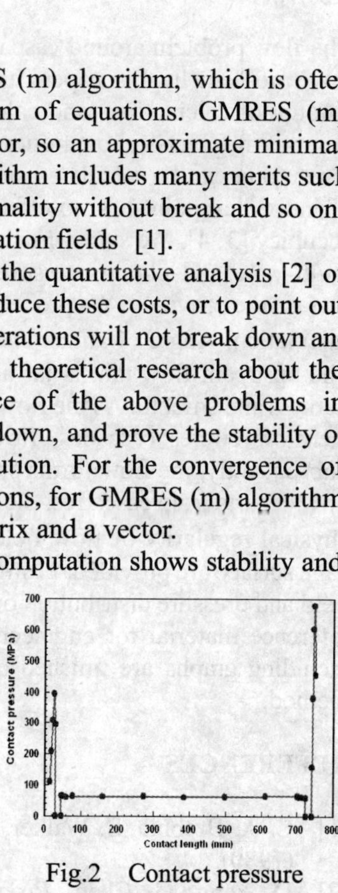

Fig.2 Contact pressure

This project is supported by the National Natural Science Foundation of China (No. 50075075).

REFERENCES

[1] C. X. Yu, G. X. Shen, D. Y. Liu, *Mathematical programming solution for the frictional contact multipole-BEM*, Tsinghua Science and Technology, 1, (2005).
[2] Y. Saad, M. H. Schultz, *GMRES: A generalized minimal residual algorithm for solving nonsymmetric linear systems*, SIAM J. SCI. STAT. COMPUT, 7, (1986), 856-869.
[3] V. Rokhlin, *Rapid solution of integral equations of classical potential theory*, Journal of Computational Physics, 60, (1983), 87-207.

COMPUTATIONAL MECHANICS (Abstracts)
WCCM VI in conjunction with APCOM'04, *Sept. 5-10, 2004, Beijing, China*

Boundary Element Method on Flow Problem around Cascades

Gaiping Zhao [*], Dequan Yang, Pengling Xing

Inner Mongolia University for Nationalities, Tongliao, 028043, China
e-mail: tlswz@sohu.com

Abstract

The flow problem around cascades is studied by using boundary element method [1, 2] of viscous flow problem in this paper. Its remarkable superiority is that reduces dimensionality, decreases amount of calculations and calculates separately physical quantities of inner points and boundary points. After physical quantities of boundary are given or calculated, physical quantities of arbitrary inner point can be calculated singly without systematic solution to interior of the domain. The error of individual point has no effect on the validity of other points' solution. The method has higher accuracy [3, 4]. We study the flow problem of complex shape body such as cascades that consist of multi-vane and obtain pressure field. Furthermore, flow field and pressure distribution of vane surface are provided. We proceed to numerical simulation with cascades that consist of four -vane in calculation. Angle of attack are $\alpha = 16^\circ$ and $\alpha = 26^\circ$ respectively. Results obtained are as follows: The upstream flow that is far away cascades is uniform stream. Moreover, the flow field exhibits flatness and smooth. As the flow approaches gradually cascade s, initial flow field is disturbed. When the flow circles cascades, flow velocity of every point shows regular changes in flow field. Subsequently, the flow forms small vortexes when the flow approaches the tail of cascades. As the flow departs from cascades, the flow still maintains flatness state. The flow's characteristics embody physical regularity of flow field. When angle of attack is $\alpha = 16^\circ$, pressure distribution curves on vane surface are provided. From the above what we have studied about cascades, the analysis of flow field and pressure distribution of vane surface and calculated results offer theoretical foundation and reference material for engineering design of water turbine and steam turbine. Calculated results including graphs are finished by computer. The reasonability of results shows the validity of the method.

REFERENCES

[1] C. A. Brebbia, S. Walker, *Boundary element Techniques in Engineering*, News-Butterworths, (1980).

[2] D. Yang, G. Chen, *Progress in Boundary Element Method of 2-D Viscous Fluid Flow*, Numerical Methods in Laminar and Turbulent Flow, 8(2), (1993), 1581-1586.

[3] Dequan Yang, *Boundary Element Method of Multi-lobe Hydrodynamic Lubrication Bearing for Numerical Analysis in Lubrication Theory*, International Journal for Numerical Methods in Engineering, 34(2), (1992), 667-673.

[4] Dequan Yang, Gouyou Zhang, Baolin Wu, Pengling Xing, *Boundary Element Analysis of Fluid Dynamics of Complex Shaped Bearings*, in Computational Mechanics, Vol.2: A. A. BALKEMA, Rotterdam, (1991), pp.1469-1474.

COMPUTATIONAL MECHANICS (Abstracts)
WCCM VI in conjunction with APCOM'04, *Sept. 5-10, 2004, Beijing, China*

Application of Fast Multipole BEM for 3D Elastostatics on Thin Shell Structures

Libin Zhao [1*], Zhenhan Yao [2]

[1] *School of Space Technology, BeiHang University, Beijing, 100083, China*
[2] *Department of Engineering Mechanics, Tsinghua University, Beijing, 100084, China*
e-mail: zhaolibin@tsinghua.org.cn, demyzh@tsinghua.edu.cn

Abstract

In this paper, the BEM based on the three dimensional elasticity, but not on the thin shell theory, is used to analysis thin shells structures. To enhance the computational efficiency, an O(*N*) Taylor series fast multipole boundary element method(FM-BEM) for three-dimensional elasticity problems is adopted. The basic idea of this FM-BEM can be summarized as: the boundary integral equation can be divided into two parts: the contributions from near-field via direct computation, and the contributions from far-field via approximately computation by the moment coefficients associated with far-field, which can be computed in advance. An iterative solver (GMRES) is used for obtaining the solution of the problem.

The study on the inner pressure cylindric shell shows that the BEM solution based on three-dimensional elasticity is in good agreement with the theory solution, and the difference between the FM-BEM and the theory solution is always smaller than 0.5%, which is acceptable in engineering practice. The further analysis on different sizes of the problem shows that the FM-BEM can improve the computing ability remarkably and the boundary element method based on the 3D elasticity has the expansive application prospect on the complex thin structures.

REFERENCES

[1] V. Rokhlin, *Rapid solution of integral equations of classical potential theory*, J. Comput. Phys, 60, (1985), 187-207.

[2] K. Hayami, S. A. Sauter, *Panel clustering for 3-D elastostatics using spherical harmonics, Boundary elements XX*, Computational Mechanics Publications, Southampton, England, (1998), pp. 289-298.

[3] V. Popow, H. Power, *An O(N) Taylor series multiple boundary element method for three-dimensional elasticity problems*, Eng. Analysis with Boundary Elem., 25, (2001), 7-18.

[4] L. B. Zhao, Z. H. Yao, *A study on fast multipole BEM for thin plates structures*, Proceeding of the Third International Conference on Boundary Element Techniques, Tsinghua University Press & Springer-Verlag. Sept. 10-12, (2002), pp. 89-94.

[5] L. B. Zhao, *Research of fast multipole BEM for 3D elastostatic problems and applications for thin structures*, Postdoctoral report, Tsinghua University, Beijing, China, (2003), (in Chinese).

COMPUTATIONAL MECHANICS (Abstracts)
WCCM VI in conjunction with APCOM'04, *Sept. 5-10, 2004, Beijing, China*

Analytical Integral Algorithm of Nearly Singular Integrals in the BEM of Two-Dimensional Thermoelasticity

Huanlin Zhou [1*], Zhongrong Niu [1], Changzheng Chen [1], Xiuxi Wang [2]

[1] *Department of Engineering Mechanics, Hefei University of Technology, Hefei, 230009, China*
[2] *Department of Mechanics, University of Science and Technology of China, Hefei, 230026, China*
e-mail: *zhouhl@ustc.edu

Abstract

There exist nearly singular integrals in the boundary element method, when a source point is very close to an integral element. The nearly singular integrals can 't be accurately calculated by conventional Gaussian quadrature. A new analytical integral algorithm is proposed to evaluate the nearly singular integrals in the BEM of two-dimensional steady-state uncoupled thermoelasticity problems. The discretized boundary integral equations include five integral forms whe n the boundary of the considered domain is discretized by a set of linear elements. New analytical formulas for calculating the five forms of nearly singular integrals are obtained. For an interior point, if it is very close to an integral element, the pre sent new analytical formulas are applied to evaluating the corresponding nearly singular integrals. Otherwise, the conventional Gaussian integral method is still used. Numerical results show that the calculation of thermal displacements at arbitrary points and thermal stress at interior points closer to the boundary is accurate with the present method. In other words, the nearly weakly and strongly singular integrals are eliminated overwhelmingly. At the same time, the singularity of the nearly hypersingula r integrals has been handled in a great degree.

The analytical algorithm is available for linear element. As to higher order element, after it is subdivided into several linear elements, the analytical integral algorithm can be easily implemented. What's more, the algorithm can be extended to analyze structures with thin bodies in surface and coating engineering.

REFERENCES

[1] M. Fratantonio, J. J. Rencis, *Exact boundary element integrations for two-dimensional Laplace equation*. Engng. Anal. Boundary Elements, 24, (2000), 325-342.

[2] G. S. Padhi, R. A. Shenoi, S. S. J. Moy, M. A. McCarthy, *Analytic integration of kernel shape function product integrals in the boundary element method*, Computers and structures, 79, (2001), 1325-1333.

[3] Zhongrong Niu, Xiuxi Wang, Huanlin Zhou, *A non-singular algorithm for the evaluation of the nearly singular integrals in boundary element methods*, Chinese Journal of Applied Mechanics, 18(4), (2001), 1-8, (in Chinese).

COMPUTATIONAL MECHANICS (Abstracts)
WCCM VI in conjunction with APCOM'04, *Sept. 5-10, 2004, Beijing, China*

Moving Structural Finite Elements Method

S. Hao *, W. K. Liu, T. Belytschko

Department of Mechanical Engineering, Northwestern University, Evanston, IL 60208, USA
e-mail: suhao@northwestern.edu, w-liu@northwestern.edu, tedbelytschko@northwestern.edu

Abstract

A recently proposed MPFEM approximation that combines a partition of unity with generalized shape functions is extended to multi-dimensions, e.g. thin plate and fluid dynamic (biofilm growth). It is shown that a crucial feature toward achieving computational efficiency in the use of an assumed strain method based on the Hu-Washizu principle. An approach applying "Partition of Unity" to solve the associated orthogonality have also been proposed, which provides additional conditions for determining (1) enriched strain; (2) the coefficient of B-bar matrix in a stabilized finite element scheme; (3) the window function in particle method.

REFERENCES

[1] J. C. Simo, T. J. R. Hughes, *On the variational foundations of assumed strain methods.* Journal of Applied Mechanics, Transactions of the ASME, 53(1), (1986), pp. 51-54.

[2] I. Babuska, J. M. Melenk, *The partition of unity method,* International Journal for Numerical Methods in Engineering, 40, (1997), pp. 727-758.

[3] O. C. Zienkiewicz, R. L. Taylor, *The finite element method*, vol. 2, McGraw-Hill, England, (1995).

[4] T. Belytschko, Y. Y. Lu, L. Gu, *Element-free Galerkin methods*, Int. J. Num. Meth. Engrg, 37, (1994), 229-256.

[5] W. K. Liu, S. Jun, Y. F. Zhang, *Reproducing kernel particle methods*, Int. J. Num. Meth. in Fluid, 20, (1995), 1081-1106.

[6] S. Hao, W. K. Liu, T. Belytschko, *Moving particle finite element method with global smoothness*, Int. J. Num. Meth. Engrg, 59, (2004), 1007-1020.

[7] S. Hao, W. K. Liu, *Moving particle finite element – interpolating using nodes and finite elements*, Proc. WCCM V, Vienna, (2002), (on-line publication).

COMPUTATIONAL MECHANICS (Abstracts)
WCCM VI in conjunction with APCOM'04, *Sept. 5-10, 2004, Beijing, China*

Sub-Structuring Techniques in EFGM

Ho-young Jo[1], Hae Sung Lee[2*]

[1] *Kunhwa Engineering Co., Ltd, 321-91 Sungsu 2-ga 1-dong, Sungdong-gu, Seoul 133-827, Korea*
[2] *School of Civil, Urban and Geosystem Engineering, Seoul National University, Sillim 9-dong, Gwanak-gu, Seoul 151-742, Korea*
e-mail: van96@freechal.com, chslee@plaza.snu.ac.kr

Abstract

The element free Galerkin method (EFGM) has been proven to be effective in the analysis of certain types of mechanical problems such as modeling of crack growth. However, the EFGM requires more computational time than usual finite element method. To overcome the drawback of the EFGM, the sub-domain technique is an attractive alternative. For an example of modeling crack growth, the computational cost would be reduced greatly if the domain in the vicinity of the crack path is modeled by the EFGM, and the other domains are discretized by the FEM. The sub-domain technique can be also applied to a parallel-processing algorithm to solve larger problems such as the analysis of 3-dimensional crack growth. Since integration cells are defined independently to the interpolation function of displacement, the sub-domain method used in the FEM can not be applied to the EFGM. This paper presents a sub-domain method based on the Lagrange multiplier approach for the application to the EFGM. The displacement field of each sub-domain is defined within a sub-domain without use of the information of the other sub-domain. The compatibility and equilibrium conditions are imposed on the interface between sub-domains using the Lagrange multiplier approach. Since both displacement and traction are unknown on the interface, a mixed formulation is derived in the proposed method. Instabilities of solutions may be triggered in mixed formulation unless the interpolation functions of the unknowns are carefully chosen. To avoid instabilities in mixed formulations, the continuous displacement field is used on the interfaces while the traction field is interpolated by discontinuous function. To demonstrate validity and efficiency of the proposed method, two numerical examples are presented. In the first example fatigue crack growth is simulated using a coupled FEM-EFGM based on the proposed sub-domain technique. The second example shows applicability of the proposed method to parallel algorithms. A 2-dimensional problem with 120,000 DOF is divided into 4 sub-domains, and solved on a cluster system with four 2.4 GHz Intel Zeon processors within 920 seconds. The proposed method yields very accurate results, and requires less computation time compared to the original EFGM.

REFERENCES

[1] T. Belytschko, Y. Krongauz, D. Organ, M. Fleming , P. Krysl, *Meshless methods: an overview and recent developments*, Computer Methods in Applied Mechanics and Engineering, 139, (1996), 3-47.

[2] O. C. Zienkiewicz, R. L. Taylor, *The finite element method*, 5th ed., Butterworth- Heinemann, (2000).

COMPUTATIONAL MECHANICS (Abstracts)
WCCM VI in conjunction with APCOM'04, *Sept. 5-10, 2004, Beijing, China*

A Meshless Approach to Aeroelastic Simulations with Large Control Surface Deflections

Michel Lesoinne*, Vivek Kaila

Department of Aerospace Engineering and Sciences and Center for Aerospace Structures, University of Colorado at Boulde, Boulder, CO80309-0429, USA
e-mail: Michel.Lesoinne@Colorado.edu

Abstract

One deficiency of most current aeroelastic codes is their inabilities to deal appropriately with control surface deployment. Though transpirations techniques. Can be used, they are only capable of handling very small deflection, as they ignore the actual non-linear changes of the geometry.

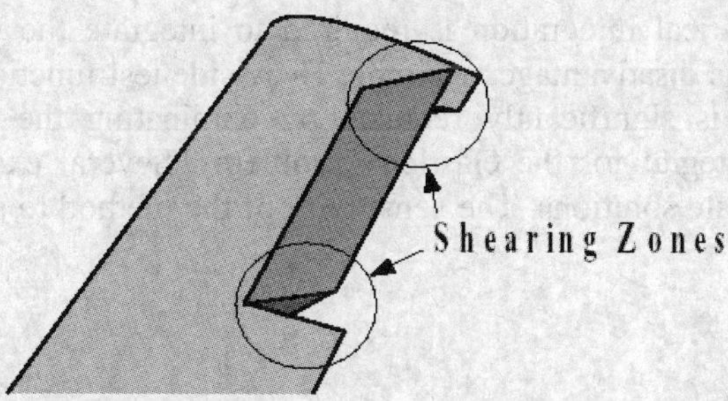

Fig. 1 Mesh shearing regions due to control surface deflections

The origin of the difficulties for the correct modeling of large control surface deflection is that they introduce shearing in the mesh at the place where the control surface meets the fixed part of the wing (seeFig.1). In this paper, we present a meshless approach to the aeroelastics immolation of aircraft involving large control surface deflections.

We make use of the Reproducing Kernel Particle Method combined with the Streamline-Upwind Petrov-Galerkin method to discretize the flow. We introduce a transition layer toward a projected solution to enforce the slip boundary conditions on the structural surface. By appropriately combining several transitions layers, our method allows us to deal with arbitrary relative motions of Structural elements, such as flaps, control surfaces, or a separating external missile or reservoir. As the structure moves, some nodes may penetrate or come out of the structure and we show how they can be dynamically activated or deactivated. We illustrate our approach with a two-dimensional simulation of an airfoil with a flap performing a separation motion.

COMPUTATIONAL MECHANICS (Abstracts)
WCCM VI in conjunction with APCOM'04, *Sept. 5-10, 2004, Beijing, China*

A Computationally Efficient Meshless Local Petrov-Galerkin Method for Elasticity Problems

Shuyao Long [*], De'an Hu

Department of Engineering Mechanics, Hunan University, Changsha, 410082, China
e-mail: sylong@hnu.cn , hudean@163.com

Abstract

The Meshless Local Petrov-Galerkin (MLPG) method is a promising numerical method to analyze potential and elasticity problems. In this method, a set of regularly or arbitrarily distributed nodes is used to interpolate the field variables. The method does not use either elements or a background mesh for integration and hence is truly meshless. The method, however, appears to be more computationally expensive than the finite element method. One of reasons for the high computational cost is that accurate numerical integration is required to integrate the weak form of governing equations. To overcome this disadvantage, a simple Heaviside test function is chosen in this paper. The computational effort is significantly reduced by eliminating the domain integral and by simplifying the domain integral for the elasticity problems. Several examples show the present method yielded very accurate solutions. The sensitivity of the method to several parameters is also studied.

REFERENCES

[1] Satya N. Atluri, Shenping Shen, *The meshless local Petrov-Galerkin (MLPG) method: a simple and less costly alternative to the finite element and boundary element methods,* CMES, 3(1), (2002), 11-51.

[2] I.S. Raju, T. Chen, *Meshless local petrov-galerkin method Applied to axisymmetric problems,* Proceedings of the ICES 2K Conference, Los Angeles, CA, August 21-25, (2000).

[3] S. N. Atluri, T. L. Zhu, *A new meshless local Petrov-Galerkin (MLPG) approach in computational mechanics,* Computational Mechanics, 22(2), (1998), 117-127.

[4] S. Y. Long, *A local Petrov-Galerkin method for the elasticity problem*, Acta. Mechanica Sinica, 33(4), (2001), 508-518.

COMPUTATIONAL MECHANICS (Abstracts)
WCCM VI in conjunction with APCOM'04, *Sept. 5-10, 2004, Beijing, China*

On the Analysis of the MEFG Method under Large Deformations

Rodrigo Rossi [1*], Marcelo K. Alves [2]

[1] *Departamento de Engenharia Mecânica, Universidade de Caxias do Sul, Cidade Universitária, Bloco D, Caxias do Sul, RS, 95070-560, Brazil*
[2] *Departamento de Engenharia Mecânica, Universidade Federal de Santa Catarina, Campus Trindade, Florianópolis, SC, 88010-970, Brazil*
e-mail: rrossi@ucs.br, krajnc@ufsc.br

Abstract

In this work we investigate a modified element-free Galerkin method when submitted to large deformation process. The modified element-free Galerkin method used in this work enables the direct imposition of the essential boundary conditions due to the Kronecker delta property of the final shape functions derived on the essential boundary. The model assumes the multiplicative decomposition of the deformation gradient into elastic and plastic parts and considers a J_2 elasto-plastic constitutive relationship with nonlinear isotropic hardening. The constitutive model is written in terms of the rotated Kirchoff stress and the logarithmic strain conjugate measures. A total Lagrangian formulation is considered. Some numerical results are presented, under axisymmetric and plane strain assumptions, in order to attest the performance of the proposed methodology. Aspects related to volumetric locking are investigated by considering a relaxed deformation gradie nt, which is constructed by the convex combination between the deformation gradient and the *Fbar* methodology.

REFERENCES

[1] A. L. Eterovic, K. J. Bathe, *A hyperelastic-based large strain elasto-plastic constitutive formulation with combined isotropic-kinematic hardening using the logarithmic stress and strain measures,* Int. J. Num. Meth. Eng., 30, (1990), 1099-1114.

[2] A. Srikanth, N. Zabaras, *An updated Lagrangian finite element sensitivity analysis of large deformations using quadrilateral elements,* Int. J. Num. Meth. Eng., 52, (2001), 1131-1163.

[3] E. A. Souza Neto, D. Peric, M. Dutko, D. R. J. Owen, *Design of simple low order finite elements for large strain analysis of nearly incompressible solids,* Internat. J. Solids and Structures, 33, (1996), 3277-3296.

[4] G. Weber, L. Anand, *Finite deformation constitutive equations and a time integration procedure for isotropic, hyperelastic-viscoplastic solids,* Computer Methods in Applied Mechanics and Engineering, 79, (1990), 173-202.

[5] J. C. Simo, F. Armero, *Geometrically non-linear enhanced strain mixed methods and the method of incompatible modes,* Int. J. Num. Meth. Eng., 33, (1992), 1413-1449.

[6] M. K.Alves, R. Rossi, *A modified element-free Galerkin method with essential boundary conditions, enforced by an extended partition of unity finite element weight function,* Int. J. Num. Meth. Eng., 57, (2003), 1523-1552.

COMPUTATIONAL MECHANICS (Abstracts)
WCCM VI in conjunction with APCOM'04, *Sept. 5-10, 2004, Beijing, China*

An Efficient Meshless Formulation for Thick Plate Analysis

Jurica Sorić[1*], Qiuzhan Li [2], Tomislav Jarak [1], Satya N. Atluri [2]

[1] *Faculty of Mechanical Engineering and Naval Architecture, University of Zagreb, Ivana Lučića 5, 10000 Zagreb, Croatia*
[2] *Center for Aerospace Research and Education, University of California Irvine, 5251 California Ave., Irvine, CA 92612, USA*
e-mail: jurica.soric@fsb.hr, qiuzhanl@uci.edu, tomislav.jarak@fsb.hr, satluri@uci.edu

Abstract

In this contribution, the Meshless Local Petrov -Galerkin (MLPG) method, originally proposed by Atluri and Zhu [1], is applied to the analysis of shear deformable thick plates. This efficient numerical approach is truly meshless because it requires no elements or background cells in either interpolation or integration. Using the kinematics of a three -dimensional continuum, the local symmetric weak form of the equilibrium equations is derived. All integrals are evaluated over the cylindrical local sub-domains, where the linear test function in the plate thickness direction is assumed. The essential boundary conditions are enforced by the penalty metho d. Discretization is performed employing the moving least squares approximation [2, 3] in the in-plane directions. Analogous to the finite element formulation in [4], the interpolation through the thickness is performed separately. The linear interpolation is used for the in-plane displacements, while the hierarchical quadratic interpolation over the thickness is adopted for the transversal displacement in order to avoid the undesired thickness locking effect. In the plate discretization, the nodal unknown variables are the three fictitious displacement components associated to the nodes on the upper and lower plate surfaces. An additional unknown variable due to the hierarchical quadratic interpolation appears, and it can be eliminated by means of the stati c condensation. The numerical efficiency of the proposed meshless formulation is illustrated by the numerical examples.

REFERENCES

[1] S.N. Atluri, T. Zhu, *A new meshless local Petrov-Galerkin (MLPG) approach in computational mechanics*, Comput. Mech., 22, (1998), 117-127.

[2] S. N. Atluri, S. Shen, *The meshless local Petrov-Galerkin (MLPG) method*, Tech. Science Press, Encino, USA, (2002).

[3] Q. Li, S. Shen, Z. D. Han, S. N. Atluri, *Application of Meshless Local Petrov-Galerkin (MLPG) to Problems with Singularities, and Material Discontinuities, in 3-D Elasticity*, CMES: Computer Modeling in Engineering & Sciences, 4, (2003), 571 -585.

[4] R. Hauptmann, K. Schweizerhof, *A systematic development of solid-shell element formulations for linear and non-linear analyses employing only displacement degrees of freedom*, Int. J. Numer. Meth. Engng, 42, (1998), 49 -69.

COMPUTATIONAL MECHANICS (Abstracts)
WCCM VI in conjunction with APCOM'04, *Sept. 5-10, 2004, Beijing, China*

A Comparison of Service Life Prediction of Concrete Structures Using Element-Free Galerkin (EFG) and Finite Element Method

H. Taghaddos*, F. Mahmoodzadeh, A. Noorzad, S. Mohammadi, A. Ansari

Department of Civil Engineering, University of Tehran, Tehran, Iran
e-mail: {taghados, mahmodzad, noorzad, smoham, ansari}@ut.ac.ir

Abstract

The reinforcement concrete structures exposed to sea environment suffer from corrosion of steel bars due to chloride ingress. It's generally assumed that the diffusion of chloride ions follows Fick's second low. Traditionally such partial differential equations are largely solved using numerical methods, such as Finite Element (FE) Method and Finite Difference (FD) Method. Although these methods are robust and efficient, but they exhibits some noises in the initial time steps results. In the present research, a detailed description of element-free Galerkin (EFG) method and its numerical implementation in one and two-dimensional models is presented. In this method, Moving Least Square (MLS) approximation is used to weak form the equation. So the independent variable, chloride content, and its gradient are continuous in the entire domain. By comparing EFG results with FE and FD and obtainable analytical solutions in special cases, it's seen some advantages in using this method such as fewer noises in the initial time steps results and smoother curves, increasing the rate of convergence. It also could be achieved to higher resolution in the interfaces such as coating on the surface of concrete.

COMPUTATIONAL MECHANICS (Abstracts)
WCCM VI in conjunction with APCOM'04, *Sept. 5-10, 2004, Beijing, China*

Evaluation of Stress Intensity Factor of Spot Welded Joints Using Enriched EFGM

Kazutoki Tanabe [1*], Noboru Tomioka [1], Akifumi Okabe [1], Masakazu Aoki [1], Xingzhu Pei [2]

[1] *College of Science & Technology, Nihon University, 1-8-14, Kanda-surugadai, Chiyodaku, Tokyo, 101-8308, Japan*

[2] *Civil Engineering and Mechanics Xi'an Jiao Tong University, 28 Xianning Xilu, Xi'an, 710049, China*

e-mail: {m053029, tomioka, okabe}@mech.cst.nihon-u.ac.jp, masakazu_0627@yahoo.co.jp, peixz@mail.xjtu.edu.en

Abstract

In the design of the automobile body structure, it is necessary to evaluate the fatigue strength of spot welding at the good accuracy. The evaluation by stress intensity factor K of the linear elastic fracture mechanics is carried out, because the contact surface of the joint by spot welding can be considered annular crack.

In this study, as the fundamental research of crack growth analysis in the thin plate spot welded joints, DC joint which was thin plate spot welded joint developed for multiaxial loading fatigue specimen was analyzed in Enriched Element-Free Galerkin Method (Enriched EFGM)[1], and the accuracy of the solution of the stress intensity factor was examined.

Main results are as follows:

1) The stress intensity factor of band plate with single -edge crack under uniform tensile stress was analyzed in Enriched EFGM. It was possible to obtain the good solution in Enriched EFGM, even if the node distribution is rough, because the analysis region is wide for the band plate with single-edge crack [2].

2) The analysis region is narrow for DC joint which is thin plate spot welded joint, because the boundary surface closes. Therefore, it is necessary to place the node distribution in the density in the whole analysis region in order to obtain the good solution of the stress intensity factor in Enrich ed EFGM.

3) The relationship between stress intensity factor and coupling region of the DC joint was examined. As outer radius in the coupling region field grows, the value of the stress intensity factor approaches a constant value gradually. The value of the stress intensity factor does not change and is almost constant even if inner radius in the coupling region field is changed.

4) The stress intensity factor obtained by Enriched EFGM can arrange the fatigue data of DC test specimens with different plate thickness in a narrow bandwidth.

REFERENCES

[1] T. Belytschko et al, *Enriched element-free Galerkin methods for crack tip fields*, International Journal For Numerical Methods In Engineering, 40, (1997), 1483-1504.

[2] Maejima, N. Tomioka, A. Okabe, *Evaluation of stress intensity factor of spot welded joint using enriched element-free Galerkin method*, Fifth World Congress on Computational Mechanics, July, (2002).

COMPUTATIONAL MECHANICS (Abstracts)
WCCM VI in conjunction with APCOM'04, *Sept. 5-10, 2004, Beijing, China*

A Gridless Solution Method for Two-Dimensional Unsteady Flow

Gang Wang *, Yingdan Sun, Zhengyin Ye

College of Aeronautics, Northwestern Polytechnical University, Xi'an, 710072, China
e-mail: wanggang@nwpu.edu.cn, yezy@nwpu.edu.cn

Abstract

In the past ten years, several gridless methods [1-3] have been developed for the solution of Euler and Navier-Stokes equations. These gridless solution methods were mainly concerned with steady flow. However, gridless methods have more potential capability for simulating unsteady flows since it is free of grid skewness and stretching distortions. The main purpose of this paper is to develop a gridless method for unsteady flow simulation. A quadrantal point infilling strategy is developed to generate point and combine clouds of points automatically. A point-moving algorithm is introduced to ensure the clouds of points following with movements of body boundaries. A dual time method for solving the two-dimensional Euler equations describing the compressible flow about bodies undergoing arbitrary motions is presented. In this approach, an implicit real-time discretization is used, and the equations are integrated in a fictitious pseudo time. Dual time method allows the real-time step to be chosen on the basis of accuracy rather than stability. It also permits the acceleration techniques, which are commonly used to speed up steady flow calculations, to be used when marching the equations in pseudo time. The spatial derivatives, which are used to estimating the inviscid flux, are directly approximated by using local least-squares curve method. An explicit multistage Runge-Kutta algorithm is used to advance the flow equations in pseudo time. In order to accelerate the solution to convergence, local time stepping technique and residual averaging are employed. A standard test case in which NACA0012 airfoil oscillates in transonic flow is computed and compared with experiment data to demonstrate the accuracy and practicability of the method. The gridless algorithm presented in this paper has produced solutions in good agreements with the experiment data of standard test case, so it is reliable and suitable for simulating unsteady flow problems with moving boundaries.

REFERENCES

[1] T. J. Barth, *A gridless Euler/Navier-Stokes solution algorithm for complex aircraft applications*, AIAA Paper 93-0333, (1993).

[2] K. Morinishi, *A gridless type solution for high Reynolds number multielement flow fields*, AIAA Paper 95-1856, (1995).

[3] J. L. Liu, S. Su, *A potentially gridless solution method for the compressible Euler/Navier-Stokes equations*, AIAA Paper 96-0526, (1996).

[4] E. Oñate, S. Idelsohn, *A mesh-free finite point method for advective-diffusive transport and fluid flow problems*, 21, (1998), 283-292.

[5] R. Lohner, C. Sacco, E. Oñate, S. Idelsohn, *A finite point method for compressible flows*, Int. J. Numer. Meth. Eng, 53, (2002), 1756-1799.

COMPUTATIONAL MECHANICS (Abstracts)
WCCM VI in conjunction with APCOM'04, *Sept. 5-10, 2004, Beijing, China*

Application of a Collocation Method Based on Moving Least Squares Interpolation in Structural Dynamic Analysis

Jieguang Wang *

Department of Civil engineering, Guilin Institute of Technology, No. 12 Jiangan Road, Guilin, 541004, China
e-mail: jieguangwang@163.com

Abstract

Structural dynamic problems are often analyzed with finite element method or Run ge-Kutta method. These methods are complicate and time-consuming. Another kind of analysis methods is direct integral methods, such as Wilson-θ's method and Newmark's method. These methods have the problem of stability and lower accuracy unless they take very small time-step.
In this paper the weighted residuals method based on moving least squares interpolation is applied in structural dynamic analysis. The length of time is divided into an appropriate number of segments, in which the moving equation of dynamic system can be transformed into the following forms :

$$(M / m_0)\ddot{u}(t) + 2n_0\dot{u}(t) + \omega_0^2 u(t) = q(T_1 t) \qquad (0 < t < 1),\ u(0) = a_0 / x_0 \ ,\quad \dot{u}(0) = \dot{a}_0 T_1 / x_0$$

A moving least squares interpolation $\mathbf{u}^{(1)}(t) = \sum_{i=1}^{n}\phi_i(t)w_i^{(1)}$ is taken as an approximate solution of the function $u(t)$ and $w_i^{(1)}$ as uncertain coefficients. Substituting $u^{(1)}(t)$ into (1) it become residuals equations. By minimizing the residuals in each segment with least squares point collocation method the coefficients $w_i^{(1)}$ can be obtained from the following equation:

$$\{w^{(1)}\} = [K]^{-1}\{P^{(1)}\} = [KV]\cdot\{P^{(1)}\}$$

Solutions for other segments can be given by a step-by-step formula as following:

$$\{w^{(k)}\} = [K]^{-1}\{P^{(k)}\} = [KV]\cdot\{P^{(k)}\} \qquad (k = 1,2,\cdots,L)$$

$$\{P^{(k)}\} = [P_n^{(k)}] + [P_0]\cdot\{w^{(k-1)}\} \qquad (k = 2,3,\cdots,L)$$

where $[K]$, $[P_0]$, $[P_n^{(1)}]$, $[P^{(1)}]$ are known when the number n and the length of the time -segment are given. Therefore the structural dynamic response is obtained.
The formula is stabilized and convergence when n>6 and the time segment is less than a per iod of vibration. Two examples are analyzed by this method, and the results are compared with theoretical results and Newmark's method, which have indicated that the weighted residuals method is not only very simple to use, time-saving but also has a high accuracy.

REFERENCES

[1] T. Belytschko, Y. Y. Lu, L. Gu, *Element-free Galerkin methods*, Int. J. Num. Meth. Eng., 37, (1994), 229-256.
[2] Jieguang Wang, *Application of the moving least squares interpolant in plate bending problems*, Mechanics in engineering, 24(5), (2002), 58-60.
[3] Jieguang Wang, *The application of weighted residuals with moving least square interpolating function to analysis of beams and plates on elastic foundation*, China Civil Engineering Journal, 36(12), (2003), 98-102.

COMPUTATIONAL MECHANICS (Abstracts)
WCCM VI in conjunction with APCOM'04, *Sept. 5-10, 2004, Beijing, China*

Application of MPM to Shock Wave Problems

Yuxin Wang *, Yuanxian Gu, Ming Sun

Department of Engineering Mechanics, Dalian University of Technology, Dalian, 116024, China
e-mail: wyxphd@sina.com, guyx96@dlut.edu.cn, cicd@sina.com

Abstract

FEM is main numerical simulation method in solid mechanics, but remeshing is difficult to solve problems in large deformation, fracture under impact loading. To accommodate the multi-scale discontinuity in structural responses under impact loading condition, a robust simulation tool without invoking fixed mesh connectivity should be applied [3]. As one of the innovative spatial discretization methods, the Material Point Method (MPM) is an extension of the Particle-in-Cell method (PIC)[2] to solve the problems in solid mechanics. The motivation of developing the MPM method is to simulate those problems with history-dependent internal state variables, such as contact/impact, penetration/perforation and solid-fluid interaction, and does not invoke master/slave nodes and remeshing. The essential idea of MPM is to take advantages of both the Eurlerian and Lagrangian methods; any continuum material body is represented by a finite collection of material points with lumped mass. The application of MPM method consists of three steps. First, as a map computation grid, fixed background mesh is divided in which the gradient and divergence terms are calculated by shape functions. Then, the continuum body is divided into N material points. In the last step, the governing equations are applied to calculate the numerical solution of each material point and background mesh. Although the MPM is still under development, some sample calculations have demonstrated the robustness and potential of this method for solving some challenging problems [2]. As one of the meshless methods, MPM has demonstrated many advantages in the numerical simulation of shock wave problems compared with the other methods existing in the literature. It is well suit to solve problems of impact load, shock wave, penetration and localized large deformation etc. In this paper, the theory of MPM, such as division of the background mesh and material points, shape function, map calculation between background mesh and material points, are described in detail [1]. The application of the MPM to shock wave problems is introduced. A numerical simulation program for MPM is also given by which shock wave propagation in bar under impact load is calculated and the result is shown in the graphs.

REFERENCES

[1] Z. Chen, W. Hu, L. Shen. *An evaluation of the MPM for simulating dynamic failure with damage diffusion*, Engineering Fracture Mechanics, 69, (2002), 1873-1877.

[2] D. Sulsky, Z. Chen, H. L. Schreyer, *A particle method for history-dependent materials*, Computer methods in applied mechanics and engineering, 118, (1994), 179-196.

[3] T. Belytschko, Y. Krongauz, D. Organ, *Meshless methods: An overview and recent developments*, Computer methods in applied mechanics and engineering, 139, (1996), 3-47.

COMPUTATIONAL MECHANICS (Abstracts)
WCCM VI in conjunction with APCOM'04, *Sept. 5-10, 2004, Beijing, China*

Solving Viscoelastic Problems with Cyclic Symmetry via a Precise Algorithm and EFGM

Haitian Yang*, Zhi Han, Ling Liu

State Key Lab of Structural Analysis of Industrial Equipment, Department of Engineering Mechanics, Dalian University of Technology, Dalian, 116024, China
e-mail: haitian@dlut.edu.cn

Abstract

The paper combines a self-adaptive precise algorithm in the time domain with the Meshless Element Free Galerkin Method (EFGM) for solving viscoelastic problems with rotationally periodic symmetry. By expanding variables in a discretized time interval, the variations of variables can be described more precisely, and iteration is not required for non-linear cases. A space-time domain coupled problem with initial and boundary values can be converted into a series of recursive boundary value problems, which are solved by a group theory based EFGM. It has been proved that the coefficient matrix of the global EFG equation for a rotationally periodic system is block-circulant so long as a kind of symmetry-adapted reference coordinate system is adopted, and then a partitioning algorithm, facilitating to parallel processing, was proposed via a completely orthogonal group transformation. Therefore instead of solving the original system, only a series of independent small sub-problems need to be solved, leading to computational convenience and a high computing efficiency. Numerical examples are given to illustrate the full advantages of proposed algorithm in the context.

REFERENCES

[1] D. L. Thomas, *Dynamics of rotationally periodic structure*, Int. J. Num. Meth. Eng., 14, (1979), 81-102.
[2] W. X. Zhong, C. H Qiu, *Analysis of symmetric or partially symmetric structures*, Comput. Meth. Appl. Mech. Eng., 38, (1983), 1-18.
[3] F. W. Williams, *Extract eigenvalue calculation for structures with rotationally periodic substructure*, Int. J. Num. Meth. Eng., 23, (1986), 695-706.
[4] G. F. Wu, H. T. Yang, *The use of cyclic symmetry in two-dimensional elastic stress analysis by BEM*, Int. J. Solids Struct., 31, (1994), 279-290.
[5] G. F. Wu, F. W. Williams, D. Kennedy, *Extract eigenvalue calculation for partially rotationally periodic structures,* Comput. Struct., 64, (1997), 275-284.
[6] T. Belytschko, Y. Y. Lu, L. Gu, *Element-free Galerkin methods*, *I*, Int. J. Num. Meth. Eng., 37, (1994), 229-256.
[7] T. Belytschko, Y. Krongauz, D. Organ, M. Fleming, P. Krysl, *Meshless methods: an overview and recent developments*, Comput.Meth. Appl. Mech. Eng., 139, (1996), 3-47.
[8] Y. Y. Lu, T. Belytschko, M. Tabbara, *Element-free Galerkin method for wave propagation and dynamic fracture*, Comput. Meth. Appl. Mech. Eng., 126, (1995), 131-153.

COMPUTATIONAL MECHANICS (Abstracts)
WCCM VI in conjunction with APCOM'04, *Sept. 5-10, 2004, Beijing, China*

Parallel Meshfree Computation for Finite-Strain Elasto-Plastic Problems with Cam-Clay Model

Jifa Zhang *, Yao Zheng

Center for Engineering and Scientific Computation, and College of Computer Science, Zhejiang University, Hangzhou, Zhejiang, 310027, PR China
e-mail: puppy@zju.edu.cn, yao.zheng@zju.edu.cn

Abstract

Owing to severe mesh distortion, standard finite element approaches are still ineffective in handling extreme material deformation, such as large deformations and moving discontinuities. In order to improve the effectivity, meshfree methods have been developed. Among these methods, Reproducing Kernel Particle Methods (RKPM) have demonstrated their greater suitability for structural analysis. After a review of meshfree methods and an introduction to the RKPM, applications of the RKPM to finite-strain elasto-plastic problems are presented. As an illustrative case, a slope stability problem with the finite-strain Cam-clay model for geomaterials is analyzed. Being different from using the classical additive decomposition, this finite-strain Cam-clay model is presented by using multiplicative elasto-plastic decomposition of the deformation gradients. The numerical simulations are carried out on an SGI Onyx 3900 supercomputer with 1, 4, 8, 16, 24 and 32 processors utilized, respectively. Under the same condition, the RKPM and the FEM are compared. The comparison shows that the RKPM is more suitable for the problems where there exists extreme large strain such as the slope sliding

REFERENCES

[1] W. K. Liu, S. Jun, S. Li, J. Adee, T. Belytschko, *Reproducing kernel particle methods for structural dynamics*, Int. J. Numer. Methods Eng., 38, (1995), 1655-1679.

[2] T. Belytschko, D. P. Flanagan, J. M. Kennedy, *Finite element methods with user-controlled meshes for fluid-structure interaction*, Comput. Methods Appl. Mech. Engrg., 33, (1982), 669-688.

[3] J. S. Chen, C. Pan, C. I. Wu, *Large deformation analysis of rubber based on a reproducing kernel particle method*, Comp. Mech., 19, (1997), 211-227.

[4] J. Donea, *Arbitrary Lagrangian-Eulerian finite element methods*, in T. Belytschko and T. Hughes, eds. *Computational methods for transient analysis*, Elsevier, Amsterdam, (1983), pp. 474-515.

[5] W. K. Liu, S. Li, T. Belytschko, *Moving least square reproducing kernel methods, I. methodology and convergence*, Comp. Methods Appl. Mech. Eng., 143, (1997), 113-154.

[6] K. Hashiguchi, *On the linear relations of v-lnp and lnv-lnp for isotropic consolidation of soils,* Int. J. Numer. Analyt. Methods Geomechanics, 19, (1995), 367-376.

[7] K. H. Roscoe, J. B. Burland, *On the generalized stress-strain behaviour of 'wet' clay*, Engineering Plasticity, Cambridge Univ. Press, Cambridge (1968), pp. 535-609.

[8] J. C. Simo, *Algorithms for static and dynamic multiplicative plasticity that preserve the classical return mapping schemes of the infinitesimal theory*, Comput. Methods Appl. Mech. Engrg., 99, (1992), 61-112.

COMPUTATIONAL MECHANICS (Abstracts)
WCCM VI in conjunction with APCOM'04, *Sept. 5-10, 2004, Beijing, China*

On h-Adaptivity Analysis Using Multiple Scale Reproducing Kernel Particle Method and Partition of Unity Quadrature

Zhiqian Zhang *, Jinxiong Zhou, Qiong Li, Wei Zou, Ling Zhang

School of Civil Engineering and Mechanics, Xi'an Jiaotong University, Xi'an, 710049, China
e-mail: zhang_zhiqian@mailst.xjtu.edu.cn, jxzhouxx@mail.xjtu.edu.cn

Abstract

The present work describes a framework of *h*-adaptivity analysis implemented by multiple scale reproducing kernel particle method (RKPM) [1] and the numerical integration technique based on the theory of partition of unity (PU). The multiple scale RKPM is briefly reviewed, and the penalty method for imposing the essential boundary conditions is employed subsequently [2]. In order to solve the derived integrals, the general formulas of PU quadrature are derived [3]. To perform the PU quadrature, two crucial aspects, generating the patches and to calculating the PU functions, are presented in details. The patches can be generated by zooming out or even just being the supports of the RKPM shape functions and the PU functions can be computed by Shepard method or be the RKPM shape functions [3]. A node refinement strategy, so-called local Delaunay triangulation (LDT) is introduced to refine the high gradient region detected by the edge detection technique (extreme detection with a threshold). With the proposed whole scheme of h-adaptivity analysis, two numerical examples are performed and the results demonstrate that the feasibility, convergence and performance of the h-adaptivity analysis scheme given in this paper.

REFERENCES

[1] W. K. Liu et al, *Generalized multiple scale reproducing kernel particle methods*, Comput. Mech., 18, (1996), 73-111.

[2] F. Bobaru, S. Mukherjee, *Shape sensitivity analysis and shape optimization in planar elasticity using the element-free Galerkin method*, Comput. Methods Appl. Mech. Engrg, 190, (2001), 4319-4337.

[3] Z. Q. Zhang, et al, *Investigations on reproducing kernel particle method enriched by partition of unity and visibility criterion*, Compt. Mech., (accepted).

COMPUTATIONAL MECHANICS (Abstracts)
WCCM VI in conjunction with APCOM'04, *Sept. 5-10, 2004, Beijing, China*

Modelling of Forging Processes, Including Ductile Plastic Damage, Using a Meshless Method

Cai Zheng*, J. M. A. Cesar de Sa

IDMEC-Institute for Mechanical Engineering, Faculty of Engineering, University of Porto, Rua Dr. Roberto Frias, s/n, 4200-465 Porto, Portugal
e-mail: cai@fe.up.pt, cesarsa@fe.up.pt

Abstract

For analysis of metal forming processes with large deformation the conventional finite element method usually requires several remeshing operations due to severe mesh distortion. The new class of computational methods developed in the recent years, usually designated by meshless methods, offers an attractive approach to avoid those time-consuming remeshing efforts. These new methods use a set of points to represent the problem domain without need of an additional mesh.
A meshless approach based on RKPM (Reproducing Kernel Particle Method) has been applied to simulate metal forming processes with the purpose of predicting defects in formed parts. The plastic deformation model is based on the "flow formulation" and includes ductile damage effects by coupling the rigid-plastic constitutive law with the damage evolution equations proposed by Lemaitre. The velocity-base shape functions are constructed by reproducing kernel approximations. A collocation formulation is used in the discretization of essential boundary integral using a penalty method. Contact detection is implemented by a variable time integration method. Special emphasis is placed on the treatment of frictional effects along the contact interface. The importance of the fundamental parameter such as the size of support domain and the choice of weight function is investigated and the results are compared with those obtained by the well-established finite element method.
Several numerical examples have been carried out to verify the validity of the proposed meshless method on the simulation of metal forming processes and damage detection.

REFERENCES

[1] W. K. Liu, S. Jun, Y. F. Zhang, *Reproducing kernel particle methods*, Int. J. Num. Meth. Fluids, 20, (1995), 1081-1106.
[2] J. S. Chen, C. Pan, C. T. Wu, *Reproducing kernel particle methods for large deformation analysis of nonlinear structures*, Comput. Meth. Appl. Mech. Eng., 139, (1996), 195-229.
[3] F. M. Andrade Pires, J. M. A. César de Sá, L. Costa Sousa, R. M. Natal Jorge, *Numerical modelling of ductile plastic damage in bulk metal forming*, International Journal of Mechanical Sciences, 45, (2003), 273-294.
[4] J. Lemaitre, *A course on damage mechanics*, Berlin, Heidelberg, New York: Springer, (1996).

COMPUTATIONAL MECHANICS (Abstracts)
WCCM VI in conjunction with APCOM'04, *Sept. 5-10, 2004, Beijing, China*

A Mathematical Study of Meshfree Methods by Use of Non-Standard Analysis (NSA)

Z. Zong, K. Y. Lam, C. Lu *

Institute of High Performance Computing, 1 Science Park Road, #01-01 The Capricorn, Singapore Science Park II, Singapore 117528
e-mail: zongzhi@ihpc.a-star.edu.sg, lam_khin_yong@ipc.a-star.edu.sg, luc@ihpc.a-star.edu.sg

Abstract

The technique that a function is approximated (either through interpolation or fitting) on overlapping domains is the character of most meshless methods. It frees us from the workload of meshing, but brings in the problem of non-uniqueness, that is, the function under cons ideration resumes different values if approximations on different domains are used. Overlapping domain approximation results in a multi-valued function. In this paper, non-standard analysis (NSA) created by Robinson in 1966 is used to analyze multi-valued function.
NSA extends the real analysis to hyperreal number analysis by introducing infinitesimals and infinite numbers into the real system. It is capable of describing multi-valued function in the limit sense. Using the tool, we prove that a locally goo d enough approximation is also globally good enough, a very useful conclusion for construction of local approximations.

REFERENCES

[1] T. Belytschko, Y. Krongauz, D. Organ et al, *Meshless methods: An overview and recent developments*, Comput. Methods Appl. Mech., 139, (1996), 3-47.
[2] B. Nayroles, G. Touzot, P. Villon *Generalizing the finite element method: diffuse approximation and diffuse elements*, Comput. Mech., 10, (1992), 307-318.
[3] T. Belytschko, Y. Y. Lu, L. Gu, *Element-free Galerkin methods*. Int. J. Numer. Meth. Engng, 37, (1994), 229-256.
[4] W. K. Liu, S. Jun, Y. F. Zhang, *Reproducing kernel particle methods*, Int. J. Numer. Meth. Fluids, 20, (1995), 1081-1106.
[5] T. J. Liszka, C. A. M. Duarte, W. W. Tworzydlo, *Hp-meshless cloud method*, Comput. Meth. Appl. Mech. Engng, 139, (1996), 263-288.
[6] S. N. Atluri, T. Zhu, *A new meshless local Petrov-Galerkin approach in computational mechanics*, Comput. Mech., 22, (1998), 117-127.
[7] G. R. Liu, Y. T. Gu, *A local point interpolation method for stress analysis of two-dimensional solids*. Struct. Engng and Mech., 11, (2001), 221-236.
[8] N. R. Aluru, G. Li, *Finite cloud method: A true meshless technique based on a fixed reproducing kernel approximation*, Int. J. Numer. Meth. Engng, 50, (2001), 2373-2410.
[9] C. Franke, R. Schaback, *Solving partial differential equations by collocation using radial basis functions*, Appl. Math. Comput., 93, (1998), 73-82.
[10] A. Robinson, *Non-Standard Analysis,* North-Holland, Amsterdam, (1966).

COMPUTATIONAL MECHANICS (Abstracts)
WCCM VI in conjunction with APCOM'04, *Sept. 5-10, 2004, Beijing, China*

Soft Computing Methods for Estimation of Microplane Model Parameters

Anna Kučerová *, Matěj Lepš, Jan Zeman

Department of Structural Mechanics, Faculty of Civil Engineering, Czech Technical University in Prague, Thákurova 7, Prague 6, 166 29, Czech Republic
e-mail: anicka@cml.fsv.cvut.cz, leps@cml.fsv.cvut.cz, zemanj@cml.fsv.cvut.cz

Abstract

Concrete as a man-made heterogeneous material shows very complex non-linear behavior which is extremely difficult to model both theoretically and numerically. The microplane model [1] is a fully three-dimensional material law that includes tensional and compressive softening, damage of the material, different combinations of loading, unloading and cyclic loading along with the development of damage-induced anisotropy of the material. As a result, it is fully capable of predicting behavior of real-world concrete structures [2] once provided with proper input data.
The major disadvantages of this model are, however, the high computational cost associated with structural analysis even in parallel implementation [2] and a large number of phenomenological material parameters. Although the authors of the model proposed a heuristic calibration procedure [1], it is based on the trial-and-error method and provides guide to determination of selected material parameters. Therefore, a reliable and inexpensive procedure for parameters identification of this model is on demand.
In the view of potential improvements proposed in a recent work on soft computing methods [3], a new procedure based on artificial neural networks for the microplane parameter identification is proposed in the present paper. Novelties are usage of the Latin Hypercube Sampling method for generation of training sets, sensitivity analysis and genetic algorithm based training of a neural network by the evolutionary algorithm SADE [4]. In this contribution, advantages and disadvantages of this approach together with possible extensions are thoroughly discussed and analyzed.

REFERENCES

[1] Z. P. Bazant, F. C. Caner, I. Carol, M. D. Adley, S. A. Akers, *Microplane model M4 for concrete, part I: formulation with work-conjugate deviatoric stress, part II: algorithm and calibration*, Journal of Engineering Mechanics, ASCE, 126, (2000), 944-953, 954-961.

[2] J. Němeček, B. Patzák, D. Rypl, Z. Bittnar, *Microplane models: computational aspects and proposed parallel algorithm,* Computers and Structures, 80, (2002), 2099-2108.

[3] B. Pichler, R. Lackner, H. A. Mang, *Back analysis of model parameters in geotechnical engineering by means of soft computing,* International Journal for Numerical Methods in Engineering, 57, (2003), 1943-1978.

[4] O. Hrstka, A. Kučerová. *Improvements of the different types of binary and real coded genetic algorithms preventing the premature convergence,* Advances in Engineering Software, (accepted).

COMPUTATIONAL MECHANICS (Abstracts)
WCCM VI in conjunction with APCOM'04, *Sept. 5-10, 2004, Beijing, China*

Super Efficient Neural Network Based Simulation of Highly Skewed Non-Gaussian Stochastic Fields

N. D. Lagaros *, G. Stefanou , M. Papadrakakis

Civil Engineering Department, National Technical University of Athens, Zografou Campus, 15780 Athens, Greece
e-mail: {nlagaros, stegesa, mpapadra}@central.ntua.gr

Abstract

The problem of simulating non-Gaussian stochastic processes and fields has gained considerable interest lately in the field of stochastic mechanics. This is due to the fact that several quantities arising in practical engineering problems (e.g. material and geometric properties of structural systems, soil properties in geotechnical engineering applications) exhibit non-Gaussian probabilistic characteristics. Especially the simulation of highly skewed narrow-banded fields is well recognized today as a benchmark that reveals the limitations of the existing simulation methods [1-2].
In this paper, a super efficient procedure for the simulation of homogeneous non-Gaussian stochastic fields with prescribed target marginal distribution and spectral density function is presented. This procedure is based on the method developed by Deodatis and Micaletti [2], which constitutes an improved version of the algorithm proposed by Yamazaki and Shinozuka [3]. Both approaches are related to the translation field concept [1] and are spectral representation based in the sense that they use the spectral representation method in order to produce sample functions of the underlying Gaussian field.
The procedure described in this work takes advantage of the improvements developed in [2] but further reduces the computational effort of the simulation by predicting the unknown Gaussian spectrum using a neural network approach. This approach is based on a two-layered feedforward architecture, which is used to approximate the Gaussian spectrum. The proposed method can be successfully applied for simulating narrow-banded fields with very large skewness. Its efficiency is demonstrated in a numerical example involving a field that follows the lognormal distribution.

REFERENCES

[1] M. Grigoriu, *Simulation of stationary non-Gaussian translation processes,* J. of Engrg. Mech., 124(2), (1998), 121-126.
[2] G. Deodatis, R. C. Micaletti, *Simulation of highly skewed non-Gaussian stochastic processes,* J. of Engrg. Mech., 127(12), (2001), 1284-1295.
[3] F. Yamazaki, M. Shinozuka, *Digital generation of non-Gaussian stochastic fields,* J. of Engrg. Mech., 114(7), (1988), 1183-1197.

COMPUTATIONAL MECHANICS (Abstracts)
WCCM VI in conjunction with APCOM'04, *Sept. 5-10, 2004, Beijing, China*

SCNN and GA Solver for Hybrid Stiffness Matrix with Neural Network Elements

Shuixiang Li *, Yanxiang Ren, Shuli Sun

Department of Mechanics and Engineering Science, Peking University, Beijing 100871, China
e-mail: lsx@pku.edu.cn, 790105312@163.com, sunsl@mech.pku.edu.cn

Abstract

This paper presents SCNN (structure controllable multi-layer forward neural network) and GA (genetic algorithm) solver for hybrid stiffness matrix in finite element analysis. The hybrid stiffness matrix model [1] which developed from Takeuchi&Kosugi [2] approach is presented to simulate the structural analysis with unknown components. The hybrid stiffness matrix is a global stiffness matrix which combines both conventional finite elements and neural network elements. Because of the neural networks element, the hybrid stiffness matrix is the implicit representation that elements of the matrix are unknown. Conversional direct or iterative methods which concerns matrix decomposition are no longer adoptable. The SCNN approach, which computes with only matrix multiplication, is improved and applied to solve linear equations and eigenvalues of the hybrid stiffness matrix. Another consequence benefit of the SCNN approach is the selection of training set of neural networks. Only n+1 mode should be matched, where n is number of freedoms of the element. The advantages of SCNN solver for hybrid stiffness matrix are summarized as follows: 1) Flexible frame for solving various matrix algebra problems; 2) Performing only matrix multiplication but no decomposition; 3) Input is fixed to base vectors, so the neural network element gives high precision and the solver is stable in all cases; 4) Suitable for parallel computation.

In the SCNN approach, a three-layer SCNN is constructed to solve inverse and eigenvalues/eigenvectors of the hybrid stiffness matrix. Since the input of SCNN is fixed to base vectors, the SCNN solver is stable and gives high precision results compared with previous approaches. Improvement of standard SCNN approach is also presented by adding a smoothing term. Example shows significant increase of convergence rate has been achieved. In the GA approach, an improved and optimized GA is employed to solve linear systems of equations of the hybrid stiffness matrix. A local searching algorithm is presented to overcome the early convergence of standard GA. Additionally, new genetic operations are introduced to increase the convergence rate of standard GA. Example gives a satisfied result.

REFERENCES

[1] S. X. Li, *Global flexibility simulation and element stiffness simulation in finite element analysis with neural networks*, Computer Methods in Applied Mechanics and Engineering, 186(1), (2000), 101-108.

[2] J. Takeuchi, Y. Kosugi, *Neural network representation of finite element method*, Neural Networks, 7(2), (1994), 389-395.

[3] Jiao Licheng, *Neurocomputation*, Xidian University Press: Xian, (1993).

COMPUTATIONAL MECHANICS (Abstracts)
WCCM VI in conjunction with APCOM'04, *Sept. 5-10, 2004, Beijing, China*

Fault Identification in Rotating Machinery Using Artificial Neural Networks

H. Nahvi *

Mechanical Engineering Department, Isfahan University of Technology, Isfahan, Iran
e-mail: hnahvi@cc.iut.ac.ir

Abstract

Vibration analysis has long been used for the detection of machine fault conditions. Vibration analysis of machines is a nondestructive technique that may be performed during the operation of machines. The specific characteristics of the vibration frequency spectrum that are related to the common fault conditions are well known. The fault identification is based on comparison of frequency spectrum of different points with the reference frequency spectrum. Due to the measurement errors and special techniques needed in some situations, a suitable data structure is required.

In this paper, an artificial neural network system is designed and employed for fault prediction of rotating machinery systems. An artificial neural network is a circuit composed of interconnected simple circuit elements called neurons [1]. Each neuron represents a map typically with multiple inputs and a single output. Multi-layer feed-forward networks, constituted of nonlinear neurons, have been employed. In a feed-forward neural network, the neurons are interconnected in layers, so that the data flow only in one direction. Normalization scheme is implemented on the input and output vectors. The performance of the expert structure is optimized to encounter input data with different intensities and non-regular data.

More than forty rotating machinery faults [2] are introduced to the algorithm. To train the network, the data of the vibration identification chart consisting of vibration signals of common rotating machinery faults is used. A computer software is developed to detect machinery faults by using above techniques and is validated for fault detection of different machinery systems. In this work no attempt has been made on quantification of the fault, once it is identified. The focus was to generate data for faulty rotor-systems and develop a neural network diagnosis structure. It is found that the designed network is capable to identify unknown faults in rotating machinery. The effectiveness of the proposed neural network algorithm is displayed by an example.

REFERENCES

[1] J. Stuart, P. Norving, *Artificial intelligence*, Hall Inc., New Jersey (1995).
[2] H. P. Bloch, F. K. Gaitner, *Practical machinery management for process plant*, *vol. 2: machinery failure analysis and troubleshooting*, Gulf Pub. Company, (1990).

COMPUTATIONAL MECHANICS (Abstracts)
WCCM VI in conjunction with APCOM'04, *Sept. 5-10, 2004, Beijing, China*

Joint Stiffness Estimation of Thin-walled Structure using Neural Network

Yosuke Sato*, Akifumi Okabe, Noboru Tomioka, Takashi Hirose

College of Science & Technology, Nihon University, 1-8-14, Kanda-surugadai, Chiyodaku, Tokyo, 101-8308, Japan
e-mail: m053017@edu.cst.nihon-u.ac.jp

Abstract

The connections, which consist of thin-walled box beams with complex sectional contour like the automobile body structure, are not rigid perfectly. This flexible characteristic (called joint stiffness after this) contributes one of the key factors dominating the following basic characteristics, strength, a vibration, and a stiffness of a whole body structure. Thus it must be required to estimate the value of joint stiffness quantitatively in the stage of designing body structure.

However, in general, it is hard to obtain the joint stiffness quantitatively. Because the jointed parts have many spot welded parts, service holes, and the box section is complex. If we will try to obtain the value of joint stiffness as exactly as possible, the detailed analysis by the experiments and finite element method is necessary. If the method that is quickly and accurately able to obtain the joint structure from the joint stiffness is developed, it must be useful in designing an automobile body structure.

In this study, the joint stiffness is represented by joint stiffness matrix that can represent a size and a property of the joint stiffness very well. [1] And we propose about the new method to estimate the joint stiffness by inputting the design parameters (thickness, shape of cross section, length of partition, position of flange etc.) to the trained neural network. [2] This method was applied to the Y shape joint structure composed of three-beams as an automobile body structure.

At first, some learning data are prepared by changing the design parameters. Next, the relation between joint stiffness and the design parameters are constructed by using neural network. Once the trained neural network is given, the values of joint stiffness can be estimated by only inputting the design parameters into the trained neural network.

It was shown that the joint stiffness was quickly and accurately obtained from the design parameters by using the trained neural network.

REFERENCES

[1] K. Shimomaki et al, *Joint stiffness of body structure, part 1: its evaluation method*, Transactions of JSAE, 43, (1990), 138-142, (in Japanese).

[2] Y. Sato, A. Okabe, N. Tomioka, Estimating each element of joint stiffness matrix using neural network, applied to a joint structure composed of three beams, The Japan Society of Mechanical Engineers, 31(2), (2003), 1-2, (Japanese).

COMPUTATIONAL MECHANICS (Abstracts)
WCCM VI in conjunction with APCOM'04, *Sept. 5-10, 2004, Beijing, China*

Study on Active Control in Centrifuge-Vibration Compound Environment by Genetic Algorithm-Neural Network System

Yan Song *, Yongtao Wei, Jianhua Yu

Department of Civil Engineering & Applied Mechanics, Sichuan University, Chengdu 610065, China
e-mail: wyt2119@sina.com

Abstract

Working under the vibration-centrifugal compound environment, the components of the vibrator are unbalanced because of the centrifugal force, which prevents the vibrator from working properly. So the active control on the vibrator is necessary. Because the stiffness of the vibrator differs from the rotation speed, and the different test subjects also bring out the changes of stiffness and damp of the system, the system of the vibrator under the vibration-centrifugal compound environment is highly nonlinear, whose parameters alter sharply [1]. Obviously the traditional control method is hardly to work. The computational intelligence may be a good choice.

This paper aims at controlling the acceleration of the vibrator board under the vibration-centrifugal compound environment by means of the computational intelligence. On the basis of the governing motion equation of the vibrator, a new active control method is proposed, which uses the genetic algorithm and neural network system. In the active control system, the neural network is used as the emulator to simulate the movement of the vibrator [2], which is embedded into the simple genetic algorithm as the evaluation function to produce the fitness of the every individual in the population. The genetic algorithm is applied to determine the optimal control force.

In this paper a neural network with four neural cells for the input and output is set, the first three of them correspond to three input-values and the last one cell is related to the output-value. For the three input-values, the first cell is set for the vibration force at the current moment, and the second cell is for control force at the same time, and the last one is related to the acceleration of the vibrator board. For the output-value, one neural cell is used to export the acceleration of the vibrator board at the next moment. After training the neural network, it can work as the emulator to forecast the acceleration of the vibrator board at the next moment. In the simple genetic algorithm, the binary codes in the population are set to express the control forces. By decoding the binary codes into float numbers and inputting the float numbers into the neural network, the neural network produces the fitness of the every individual in the population. Roulette wheel selection operator, crossover operator and mutation operator act on the initial population to bring out the new population. Through the fitness of the every individual of the population, we can find out the optimal control force that can control the acceleration in the advisable range. At last, we give some numerical examples, which prove the effectiveness of the proposed method.

This research is founded by Chinese natural science foundation item (No. 10276030).

REFERENCES

[1] Tao Lin, *Numerical simulation of centrifugal force-vibration compound environment*, Master Degree Thesis, Sichuan University, (2003), (in Chinese).

[2] Shuangwu Xu, *The design and simulation of active structure control based on neural network*, Master Degree Thesis, Sichuan University, (1999), (in Chinese).

COMPUTATIONAL MECHANICS (Abstracts)
WCCM VI in conjunction with APCOM'04, *Sept. 5-10, 2004, Beijing, China*

A Triangular Stress/Strain Element Based on Yield-Line Finite Element Theory

M. Gohnert*

School of Civil and Environmental Engineering, University of the Witwatersrand, Private Bag x3, P.O. Wits, 2050, South Africa
e-mail: gohnert@civil.wits.ac.za

Abstract

There are several triangular plane stress/strain element formulations in existence. The displacement element is by far the most accepted formulation. In the elastic range, the element performs well if the mesh is refined. However, the transition to a plastic analysis is cumbersome. Yield-line elements are a new type of element which incorporates virtual deformation theory commonly used in plastic analysis. These virtual work equations are equilibrium relationships that are equally valid in the elastic and plastic range. Thus, a smooth or natural transition to a plastic analysis is possible. The element described in this paper is limited to an elastic analysis. However, a plastic analysis is possible as demonstrated by other elements derived by this method. The element also lends itself to heat transfer problems, which was the original motivation behind the development of this element. The triangular element described is based on yield-line element theory. This element is part of a collection of elements using a similar process of formulation and is unique in a sense that equilibrium is established by forming a yield-line pattern in the element. Although yield-line theory is used almost exclusively to determine the collapse load of plates and slabs, the method is equally useful to establish equilibrium equations between elements connected to a common node. Compatibility, however, is established by formulating linear displacement equations that are equated at common nodes. This method also differs from traditional methods of achieving compatibility between elements. Although shape functions are similarly derived, the necessity of using complete polynomials is not as critical as it is with the displacement method. The only criterion is that the shape function must have sufficient and appropriate terms to represent the intended deformation. In stress/strain elements, the function is linear which significantly simplifies the compatibility equations. The final criterion is the formulation of constitutive relationships. This is achieved by combining the equilibrium and compatibility equations, which is easily achieved since unknowns are common to both.

REFERENCES

[1] M. Gohnert, *Yield-line elements for elastic bending of plates and slabs*, Engineering Structures, 17(2), (1995), 87-94.
[2] M. Gohnert, *Collapse load analysis for yield-line elements*, Engineering Structures, 22(8), (2000), 1048-1054.
[3] A. R. Kemp, *A consistent mixed approach to computer analysis of frames*, The Civil Engineer in SA, 30(7), (1986), 317-322.

COMPUTATIONAL MECHANICS (Abstracts)
WCCM VI in conjunction with APCOM'04, *Sept. 5-10, 2004, Beijing, China*

Numerical Scheme for Continuum Mechanics Using C_1-Continuity Helmholtz-Element

Junya Imamura [1*], Takahiko Tanahashi [2]

[1] *IMI Computational Engineering Laboratory, Honcho 31-9-803, Wako-shi, 351-0114, Japan*
[2] *Department of Mechanical Engineering, Keio University, Hiyoshi 3-14-1, Kohoku-ku, Yokohama, 223-8552, Japan*
e-mail: jimamura@ra2.so-net.ne.jp, taka@mech.keio.ne.jp

Abstract
The primitive variable method, which uses velocities and pressure elements, is the most popular scheme in CFD as well as solid analysis. This method can be referred to as a hybrid application of the Helmholtz's decomposition: Instead of linear combination of differentials of vector potentials, the pressure is calculated without boundary condition and participates in the method as a pseudo-potential in some independent velocity elements. The ψ-ω method is also a hybrid application of the Helmholtz's decomposition, in which the basis function of elements is expressed with other differentials of vector potentials. We have proposed a CFD scheme in 2D analysis using Helmholtz-element, which is based on the Helmholtz's theorem and satisfies the continuity condition for incompressible fluid as strong solution. The compatible element, however, requires C_1-continuity. In addition, Coulomb's gauge needs to be satisfied in 3D analysis as an extra constraint for the better analysis. In this report, we propose a method under Coulomb's gauge in 3D, and examine it numerically in cavity flow that represents benchmark test problem. Our method utilizes the Helmholtz decomposition that expresses the displacement vector in solid analysis with 6 components, potentially bypassing the locking problem imposed on near incompressible materials. We also perform a numerical examination of our method in terms of this problem by showing a simple example in 2D (Poisson's ratio 0.475).

REFERENCES

[1] J. Imamura, T. Tanahashi, *Numerical scheme for conservation of vorticity using Helmholtz-element and for conservation of irrotational component*, 17th Nat. Sym. on wind Eng., Tokyo, 226-230, (2002).

[2] J. Imamura, T. Tanahashi, *Voxel-analysis using C1-continuity Helmholtz-element and its boundary-fitting technique,* Transact. JSCES, Paper No.20030020, (2003).

COMPUTATIONAL MECHANICS (Abstracts)
WCCM VI in conjunction with APCOM'04, *Sept. 5-10, 2004, Beijing, China*

A Continuum-Based Discrete Element Method for Continuous Deformation and Failure Process

Shihai Li [1*], Manhong Zhao [1], Yuannian Wang [1], J.G. Wang [2]

[1] *Institute of Mechanics, Chinese Academy of Science, Beijing 10080, China*
[2] *Tropical Marine Science Institute, National University of Singapore, Singapore 119260*
e-mail: mh_zhao@21cn.com, shli@imech.ac.cn, deepen_eye@yahoo.com.cn, tmswjg@nus.edu.sg

Abstract

This paper presents a novel continuum -based discrete element method (called CDEM). This CDEM combines finite element method (FEM) and particle -spring model of discrete element method (DEM). This CDEM treats the 8-node solid isoparametric element in FEM as a fundamental system composed of 8 particles and interconnecting springs. The particle mass is determined by material density and its spatial distribution. The stiffness matrix of the fundamental system is the same as the element stiffness matrix of 8-node solid isoparametric element. This stiffness matrix is divided into 3×3 stiffness submatrix to determine the spring stiffness. The eigenvalues and eigenvectors of the 3×3 stiffness submatrix determine the magnitude and direction of springs from particle to particle. If three eigenvalues are all real, the two particles are connected by three springs, and the magnitude of springs is equal to the eigenvalu es and the directions of the springs are along the eigenvectors. If there are two conjugate eigenvalues, the elastic body between two particles is presented by five instead of three springs in three directions, among which two should have Poisson effect. I t is noted that whether three-springs or five-springs, the stiffness matrix of each fundamental system is the same as element stiffness in FEM if the material is continuum. Furthermore, the breakage of springs is introduced to describe the evolution from c ontinuous deformation to discontinuous deformation. As examples, uniaxial loading test and cantilever beam subjected to point load are simulated by the current method. The results are fairly good agreement with theoretical solution as well as FEM in continuum. The simulation is further extended up to failure process. The simulated failure pattern is consistent with experimental observation.
As a conclusion, this CDEM is the combination of FEM which is based on continuum with particle-spring model of DEM whi ch is based on discontinuous deformation. It can describe whole process from continuous to discontinuous deformations with the same method. In continuum, the current CDEM obtains the same results as FEM. In discontinuum, the principle of CDEM is the same as DEM except the connecting pattern of springs including their number, stiffness coefficient, and direction, but the CDEM is more capable in the simulation for failure process.

REFERENCES

[1] C. H. Dowding, T. B. Belytschko, H. J. Yen. *Dynamic computational analysis of opening in joint rock,* J of Geotechnical Engineering, ASCE, 109(12), (1983), 1551-1556.
[2] L. Jing, J. A. Hudson, *Numerical methods in rock mechanics,* Int. J of Rock Mechanics and Mining Sciences, 39(4), (2002), 409-427.

COMPUTATIONAL MECHANICS (Abstracts)
WCCM VI in conjunction with APCOM'04, *Sept. 5-10, 2004, Beijing, China*

Numerical Analysis for Structural Dynamic Responses Using a Highly Accurate Differential Quadrature Method

Junping Pu *

College of Architecture & Civil Engineering, Zhejiang University of Technology, Hangzhou 310014, China
e-mail: pujunping@tsinghua.org.cn

Abstract

The differential quadrature method was proposed by Bellman and Casti in 1971 as an analogous extension of the quadrature for integrals [1]. Its essence is to be expressed the values of the derivatives at each grid point as weighted linear sums approximately of the function values at all grid points within the domain under consideration. As a distinct numerical solution technique for the initial and/or boundary value problems of engineering and physical sciences, the advantage of producing highly accurate solutions with minimal computational effort of the DQM is gradually emerging [2] compared with the conventional numerical solution techniques such as the finite difference, the finite element methods. Currently, the research areas in which the application of the DQM include fluid mechanics, static and dynamic structural mechanics, heat transfer, biosciences, transport processes, static aeroelasticity, lubrication mechanics and some problems deal with petrochemical engineering. Bert et al first applied the DQM to solve the structural mechanics in 1987; the algorithm was successfully applied to the static and dynamic structural analysis for the structural components soon afterwards. In this paper, a kind of unconditionally stable higher-order accurate time step integration algorithms [3] based on the DQM for second-order initial value problems has been applied. The dynamic responses have been computed numerically for a double-degree-of-freedom systems and a beam forced by a changed load. Numerical results changing is not distinct for the algorithm with the time step Δt changing, the solving procession is stable. The time step can be taken much large more then other algorithms and the solving efficiency is remarkable. The results show that the algorithm has the capability of producing highly accurate solutions with minimal time consumption comparing with the conventional numerical method, and the energy is perfectly conserved for dynamic response analysis by the algorithm.

REFERENCES

[1] R. Bellman, J. Casti, *Differential quadrature and long-term integration, Journal of Mathematical Analysis and Applications*, 34, (1971), 235-238.

[2] C. W. Bert, M. Malik, *Differential quadrature method in computational mechanics: a review*, Applied Mechanics Reviews, 49, (1996), 1 -28.

[3] T. C. Fung, *Solving initial value problems by differential quadrature method- Part 2: second-and higher-order equations*, International Journal for Numerical Methods in Engineering, 50, (2001), 1429-1454.

COMPUTATIONAL MECHANICS (Abstracts)
WCCM VI in conjunction with APCOM'04, *Sept. 5-10, 2004, Beijing, China*

A Fully Discrete Difference Approximation for the Time Fractional Diffusion Equation

Shujun Shen [1*], Fawang Liu [1, 2]

[1] *School of Mathematical Sciences, Xiamen University, Xiamen 361005, China*
[2] *School of Mathematical Sciences, Queensland University of Technology, GPO Box 2434, Brisbane, Qld. 4001, Australia*
e-mail: shshj07@tom.com, fwliu@xmu.edu.cn

Abstract

The time fractional diffusion equation (TFDE) is obtained from the standard diffusion equation by replacing the first-order time derivative with a fractional derivative of order $\alpha \in (0,1)$. From a physical view-point this generalized diffusion equation is obtained from a fractional Fick law which describes transport processes with long memory. TFDE has been investigated by several authors for different purposes [1-3]. Wyss [1] considered the time fractional diffusion equation and the solution is given in closed form in terms of H-function. Gorenflo et al. [2] used the similarity method and the method of Laplace transform to obtain the scale-invariant solution of time-fractional diffusion-wave equation in terms of the Wright function. We note that these special functions are not suitable for a numerical evaluation. Therefore a new numerical strategy is important in solving these equations. By adopting a suitable finite-difference scheme of solution, Gorenflo et al. [3] generated models of random walk suitable for simulating random variables whose spatial probability density evolves in time according to this fractional diffusion equation. Numerical methods and theorem analysis of fractional differential equations are very difficult tasks [4]. In this paper, TFDE in a bounded domain is considered. The relationship between Caputo, Riemann-Liouville, and Grünwald-Letnikov fractional derivatives [5] are used. A fully explicit finite difference approximation for TFDE is proposed. The stability and convergence of the fully discrete difference approximation are proved. Finally, some numerical examples are presented to show the application of the present technique. Results show that for time fractional derivatives of order of $\alpha \in (0,1)$, the system exhibits diffusion behaviours.

REFERENCES

[1] W. Wyss, *The fractional diffusion equation*, J. Math. Phys., 7, (1986), 2782-2785.

[2] R. Gorenflo, Yu. Luchko, F. Mainardi, *Wright function as scale-invariant solutions of the diffusion-wave equation*, J. Comp. Appl. Math., 118, (2000), 175-191.

[3] R. Gorenflo, F. Mainardi, D. Moretti, P. Paradisi, *Time Fractional Diffusion: A Discrete Random Walk Approach*, Nonlinear Dynamics, 29, (2000), 129-143.

[4] F. Liu, V. Anh, I. Turner, *Numerical solution of space fractional Fokker-Planck equation*, J. Comp. and Appl. Math., 164, (2004), (in press).

[5] I. Podlubny, *Fractional Differential Equations*, Academic, Press, New York, (1999).

COMPUTATIONAL MECHANICS (Abstracts)
WCCM VI in conjunction with APCOM'04, *Sept. 5-10, 2004, Beijing, China*

A New Version of Differential Quadrature and Its Applications to Static and Free Vibration Analyses of Thin Sectorial Plates

Xinwei Wang *, Yongliang Wang, Feng Liu

College of Aerospace Engineering, NUAA, Nanjing 210016, China
e-mail: wangx@nuaa.edu.cn, wyl646@sohu.com, lhjlf999@sohu.com

Abstract

As is known that in many cases, computational effort can be alleviated by Differential Quadrature (DQ) method, introduced by Bellman and Casti in 1971. Bert and his coworkers first used DQ method to solve problems in structural mechanics in 1988. Since then, a lot of developments have been made on DQ method and the method has been applied successfully to a variety of problems in structural mechanics. It is noticed, however, that difficulty arises for applying the multi-boundary conditions when the ordinary DQ method is used to solve fourth-order or even higher-order differential equations. Although the so-called δ method and method similar to the mixed collocation method can give satisfactory results for one-dimensional problems, but they are either not accurate enough or cannot easily treat the multi-boundary conditions at corner points. While the method with multi-degree of freedoms at the boundary points can yield accurate results as well as treat the multi-boundary conditions at corner points without any difficulty, but Hermit-type interpolation functions, instead of Lagrange interpolation functions used in the ordinary DQ method, are used. Earlier a new way to apply the multi-boundary conditions to increase the solution accuracy has been introduced by the senior author for the DQ method to make the method more attractive in engineering practice. The essential idea of the method is that the multi-boundary conditions are applied during the formulations of the weighting coefficients for higher-order derivatives. However, the proposed method cannot be used for all boundary conditions. Later Malik and Bert extend its application range to all boundary conditions but still have to resort the δ method for some combinations of boundary conditions. Therefore, the main disadvantage existing earlier has not been overcome satisfactory. A novel way to apply the multi-boundary conditions, named the new version of Differential Quadrature method herein, is proposed by the authors to solve the above-mentioned problem satisfactory. The essential idea of the method is that the weighting coefficients of higher order derivatives, e.g., the third and fourth order derivatives if a fourth order differential equation is involved, are computed slight differently form the ordinary DQ method with the help of the introduction of multi-degrees of freedom at the boundary points. The method is described first in one dimension for clarity and then extended to analyze the static and free vibration of thin sectorial plates with various sector angles and combinations of boundary conditions. Detailed formulations are given. Convergence of the method is studied and several examples are solved. Numerical results are compared with existing analytical and/or numerical data to demonstrate the accuracy and applicability of the DQ method. Finally, some conclusions are drawn based on the results presented herein.

COMPUTATIONAL MECHANICS(Abstracts)
WCCM VI in conjunction with APCOM'04, *Sept. 5-10, 2004, Beijing, China*

Differential Cubature Solution for Helmholtz Equation

Lanhe Wu [1*] , Zhiqing Jiang [2]

[1] *Department of Mechanics and Engineering Science, Shijiazhuang Railway Institute, Shijiazhuang 050043, China*
[2] *Department of Mechanics and Engineering Science, Shijiazhuang Railway Institute, Shijiazhuang 050043, China*
e-mail: lxx_of@sohu.com

Abstract

In this paper, a novel numerical solution technique, the differential cubature method is employed to solve two-dimensional Helmholtz equations, which are frequently encountered in various fields of engineering and physics. Differential cubature method is a direct discretization method, which approximates a linear operation such as any order of partial derivative of a function or a combination of them by means of polynomials, which expressed as a weighted linear sum of function values at the grid points in the overall physical domain. Since the grid points are arranged arbitrarily in position, this method is capable of solving boundary value problems in various shaped domains. By using this method, Helmholtz equation is transformed into sets of linear algebraic equations about the potential functions at each discrete grid point. Boundary conditions are implemented through discrete grid points on boundary by constraining the values of potential function or the values of derivatives of potential function. Solving this set of algebraic equations, numerical values at each discrete point can be obtained. The convergence, accuracy and applicability of this method are all demonstrated through solving some sample problems of different wave numbers, which have exact solutions. It is concluded that the differential cubature method converged monotonically and very rapidly with increasing the number of grid points for solving Helmholtz equations with various number of waves. By comparison study, we find that the accuracy of present differential cubature method is 32 times greater than the least squares finite element method when the computing scale is the same. It was found that this method is more suitable for solving the Helmholtz equations of lower wave numbers. For Helmholtz equation of higher wave number, much more discrete points are needed to achieve higher accurate solutions. We also find that the accuracy of the present method cannot be improved again when the number of grid points is above a value. This is due to the intrinsic property of the differential cubature method by which the matrix for determining the weighting coefficients will be singular when too much grid points are used, since it is based on polynomials.

COMPUTATIONAL MECHANICS (Abstracts)
WCCM VI in conjunction with APCOM'04, *Sept. 5-10, 2004, Beijing, China*

Mixed Model for Partitioned Interface Stress Element Method with Arbitrary Shape of Discrete Element-Finite Element Method

Qing Zhang *, Jiashou Zhuo

Department of Engineering Mechanics, Hohai University, No.1 Xikang Roads, Nanjingy, 210098, China
e-mail: lxzhangqing@hhu.edu.cn, hhudem@hhu.edu.cn

Abstract

On the base of the rigid body-spring element model which solves the homogeneity elastic problems by professor Kawai[1], we found the ISEM (interface stress element model) of the discontinuous mechanics[2]. The paper[3] has also established the ISEM in which the block element can be arbitrary shape, making the geometric simulation of complicated structure much more convenient. Compared with the traditional finite element method (FEM), the ISEM is more accurate and applicable. But generally, the total number of freedom degree of ISEM in dealing with three-dimensional problems is higher than that of FEM, which often brings about the negative affects on efficiency of calculation. By making use of the good compatibility of ISEM, this paper introduces the concept of transitional interface element, and proposes a mixed model of ISEM-FEM. Assume s_j is the cross-section between ISE and FE, the relative displacement of an arbitrary point on s_j along the local coordinate is the sum of the deformation accumulated on s_j by the ISE and the displacement induced by the FE. So we can get the relative displacement of an arbitrary point on s_j, which relates to the centroid displacement of block element in ISEM and node displacement of element in FEM. Then the interface stress of s_j can be obtained by the relative deformation of the adjacent element and the material properties. By incomplete generalized variation method which loosens the displacement compatibility conditions on interface, we can establish the control equation of ISEM-FEM. After solving the control equation, we can get the centroid displacement of block element in ISEM and node displacement of element in FEM, and obtain the stress of ISE, transitional interface element and FE by applying the stress formula of them respectively.

The paper also gives two numerical examples. The calculate results proves that the accuracy of transitional interface element is lower than that of ISEM but higher than that of FEM. The mixed model proposed in this paper can solve in a large degree the contractions between accuracy and efficiency of calculation.

REFERENCES

[1] T. T. Kawai, *A new discrete model for analysis of solid mechanics problem*, Seisan Kenkyn, 29, (1977), 204-207.

[2] J. S. Zhuo, Q. Zhang, *Interface stress element method of discontinuous mechanics*, Beijing, China Science Press, (2000), (in Chinese).

[3] Y. L. Fang, J. S. Zhuo, Q. Zhang, *Interface element method with arbitrary shape of discrete element*, Engineering Mechanics, 15, (1998), 27-37, (in Chinese).

COMPUTATIONAL MECHANICS (Abstracts)
WCCM VI in conjunction with APCOM'04, *Sept. 5-10, 2004, Beijing, China*

A Novel Numerical Method for the Vibration Analysis of Plates

L. Zhou *, W. X. Zheng

School of Quantitative Methods and Mathematical Sciences, University of Western Sydney, Penrith South DC, NSW 1797, Australia
e-mail: l.zhou@uws.edu.au, w.zheng@uws.edu.au

Abstract

This paper presents a novel numerical method, the moving least square Ritz method (MLS-Ritz), to analyse the vibration behaviour of classical thin plates. A set of points is pre-selected on a plate domain and boundaries and the Ritz trial function for the transverse displacement of the plate is established through the moving least square technique [1]. The weight function in the moving least square technique is chosen to be $\varpi(\mathbf{r}) = (1-\mathbf{r}^2)^n$, where $\mathbf{r} = \sqrt{(x-x_i)^2 + (y-y_i)^2}\,/\,d$, d is the radius of support and n is an positive integer. The boundary conditions of the plate are satisfied by forcing the boundary points to meet the geometric boundary conditions of the plate via a point substitution technique. Table 1 presents the convergence and comparison studies for a simply supported plate and a clamped square plate. It shows that the MLS-Ritz method can obtain accurate frequency parameters for the considered cases. Comparing to the existing Ritz methods, the proposed MLS-Ritz method is highly stable and accurate and is extremely flexible for dealing with plates of arbitrary shapes, different internal point and line supports and boundary conditions. Several selected cases for plates of various shapes and support conditions are also presented in the full paper to demonstrate the versatility and accuracy of the proposed MLS-Ritz method. The authors believe that the MLS-Ritz method can be further developed to be a powerful numerical tool in computational mechanics.

Table 1. Convergence and comparison study of frequency parameters $\lambda = (\omega \mathbf{b}^2/\pi^2)\sqrt{\rho h/D}$ for simply supported and clamped square plates

Grid Point Size	Simply Supported			Clamped		
	Mode 1	Mode 2	Mode 3	Mode 1	Mode 2	Mode3
5x5	6060.9	53051	55369	8.4900	20.261	20.263
9x9	2.0019	5.0357	5.0357	3.6521	7.4610	7.4610
15x15	2.0000	5.0001	5.0001	3.6461	7.4364	7.4364
Ref. [2]	2.0000	5.0000	5.0000	3.6468	7.4383	7.3483

REFERENCES

[1] T. Belytschko, Y. Krongauz, D. Organ, M. Fleming, P. Krysl. *Meshless methods: an overview and recent developments*, Computer Methods in Applied Mechanics and Engineering, 139, (1996), 3-47.

[2] A. W. Leissa, *The free vibration of rectangular plates,* Journal of Sound and Vibration, 31, (1973), 257-293.

COMPUTATIONAL MECHANICS (Abstracts)
WCCM VI in conjunction with APCOM'04, *Sept. 5-10, 2004, Beijing, China*

Numerical Simulation of Fracture Process in Anisotropic Brittle Material

Fusheng Zhu [1*], Dehai Zhang [1], Jibo Xing [2]

[1] *Department of Civil Engineering, Northeastern University, Shenyang, 110004 China*
[2] *Department of Civil Engineering, Yantai University, Yantai, 264005 China*
e-mail: neuzfs@etang.com, dehaizhang@sohu.com, jbxing@ytu.edu.cn

Abstract

In the paper, a numerical model BPM ^{2D} (beam-particle model of two dimensions) is developed to simulate the mechanical properties and fracture process in anisotropic brittle material. According to the correlative literature information, the discrete element met hod (DEM) and the lattice model have their own advantages and weaknesses. The DEM was developed for calculating the failure problems accompanying with the transition from continuum to non -continuum media, but the calculating for continuum media is not accu rate. It is fitting that the lattice model forecasts the failure area of the material, but it can' t calculate the failure process in brittle media. Therefore, to combine the two methods to set up a compromising model becomes a good choice. The BPM ^{2D} is presented on the basis of the DEM and the lattice model. Material is schematized at mesoscale level as a two dimensional assembly of particle elements in the BPM ^{2D}, and the nearest neighboring particles are connected through a lattice of elastic beam elements . Compact and anisotropic synthetic media are generated automatically and are used to investigate the mechanical behaviour of the low-porosity material. According to the property of particles linked by beams, the beams have three types: powered beam, conne cting beam and surface beam. Every type of beam has different property. For example, the strength of powered beam is the highest and the strength of connecting beam is the lowest. The random assignments of three types beam in the BPM ^{2D} simulate the propert y of spatial variation in anisotropic brittle material. The mechanical property of every type beam is randomly allocated according to Weibull distribution to reflect the initial heterogeneity of modulus and strength at mesoscale level. Cracks in material are simulated through breaking beam elements.

The uniaxial compression tests of rock and concrete specimens are considered as examples of application. The numerical simulation results indicate that the failure modes are different according to parameters of Weibull distribution. It has been revealed by the comparison of numerical simulation that failure modes of the specimen vary with the initial distribution heterogeneity of the material properties at meso-level, and the initiation, propagation and linking o f crack in the specimen at mesoscale level produce the macro-failure of the material. The comparison of uniaxial compression tests of rock with failure modes of pillars in underground mines proves the BPM^{2D} is highly appropriate and efficient for simulatin g fracture propagation of anisotropic brittle material.

REFERENCES

[1] C. S. Chang, T. K. Wang, L. J. Sluys, et al, *Fracture modeling using a microstructural mechanics approach*, Engineering Fracture Mechanics, 69 , (2002), 1959-1976.

[2] Alain Sellier, Benoît Bary, *Coupled damage tensors and weakest link theory for the description of crack induced anisotropy in concrete*, Engineering Fracture Mechanics, 69, (2002), 1925-1939.

COMPUTATIONAL MECHANICS (Abstracts)
WCCM VI in conjunction with APCOM'04, *Sept. 5-10, 2004, Beijing, China*

Fully Vectorial Algorithm for PIM Simulation with Mixture Theory

Z. Cheng [1*], T. Barriere [2], B. Liu [1], J. C. Gelin [2]

[1] *Department of Applied Mechanics, Southwest Jiaotong University, 610031 Chengdu, China*
[2] *Applied Mechanics Laboratory, ENSMM/CNRS, 24 Chemin de l'épitaphe, 25000 Besancon, France*
e-mail: thierry.barriere@univ-fcomte.fr, zqcheng@netease.com, bsliu@home.swjtu.edu.cn, jean-claude.gelin@univ-fcomte.fr

Abstract

The simulation of Powder Injection Molding (PIM) requires the prediction of powder segregation in injection stage. The application of mixture theory is necessary, which induces the modeling of injection flow by a bi-phasic model. In such a case, two coupled Navier-Stokes equation should be solved for filling process. The computation becomes an extremely heavy charge for the traditional algorithms, which prevents severely its real application. To achieve the efficient PIM simulation, the authors proposed previously a fully vectorial explicit algorithm for simulation of the filling process. The important extension to bi-phasic simulation for the prediction of powder segregation is proposed to realize a fully explicit and vectorial solver. In the new solver for simulation of the bi-phasic injection model, there is no more global solution or the construction of any global matrix except the small operations at element level. In global sense, the computation is carried out only by the vectorial operations. So the computation cost increases only linearly proportional to the degree of freedom number. Moreover, the new solver employs the elements of equal-order interpolations instead of the MINI elements. Incompressibility of the mixture is retained by a corrective feedback operation, with a procedure of systematic smoothing to avoid the possibility of mesh locking. Another smoothing method is proposed to avoid instability at the starting stage of filling simulation because of the sudden change of filled zone cause by discretization of the element model. The developed new solver shows the proper advantages for simulation of the industrial problems in large scale. Furthermore, the software is easy to be parallelized on a high performance system with multi-clusters because of its vectorial feature. The comparison of numerical results with previous simulations proves the validity and efficiency of new algorith m.

REFERENCES

[1] P. M. Gresho, S. T. Chan, R. L. Lee, C. D. Upson, *A modified finite element method for solving the time dependent incompressible Navier-Stokes equations*, *part 1 and part2*, Int. J. for Num. Methods in Fluids, 4, (1984), 557-598.

[2] R. W. Lewis, A. S. Usmani, T. J. Cross, *Efficient mould filling simulation in castings by an explicit finite element method*, Int. J. for Num. Methods in Fluids, 20, (1995), 493-506.

[3] T. Barriere, B. Liu, J. C. Gelin, *Analysis of phase segregation effects arising in fluid-particle flows during metal injection molding*, Int. J. of Forming Process, 4, (2001), 199-216.

[4] Z. Cheng, T. Barriere, B. Liu, J. C. Gelin, *A fully explicit vectorial algorithm for 3D simulation of the metal injection moulding*, 6th ESAFORM, Mini-symposia 16: New Advanced and Numerical Strategies Informing Process Simulation, Salerno, Italy, (2003), pp. 643-646.

COMPUTATIONAL MECHANICS (Abstracts)
WCCM VI in conjunction with APCOM'04, *Sept. 5-10, 2004, Beijing, China*

Metric Map and Anisotropic Mesh Adaptation for Static and Moving Surfaces

T. Coupez *, C. Gruau, C. Pecquet, J. Bruchon

CEMEF-Ecole des Mines de Paris-umr CNRS 7635 BP 207-06904 Sophia Antipolis Cedex, France
e-mail: Thierry.Coupez@ensmp.fr

Abstract

The proposed paper is concerned with the use of mesh adaptation of anisotropic type as a possible technique to capture and calculate accurately the interfaces between different domains : solid/solid, solid/fluid, fluid/fluid. The case of moving free surfaces and complex fluid flow calculation is also investigated and numerous application examples will be pr oposed.
We first address the 3D mesh generation of mixed thin/thick geometry. The natural metric is introduced to drive an anisotropic mesher [3]. The metric construction is done by calculating a conformation tensor which contains the statistical distribut ion of length in any direction. From any mesh conforming a given boundary an adaptive procedure leads to a more useful mesh (enough elements in the thickness). The next step is to relax on the boundary constraint and multidomain mesh generation is revisite d. Here again a metric map containing a geometrical information is passed to the mesher. Interfaces between subdomains are then captured within a thin layer of stretched elements. Several iterations are needed to fully complete the mesh.
Dynamic is investi gated by considering complex moving free surfaces and moving interfaces in continuation of our previous work [1, 2]. Interfaces motion are assumed to be solution of a transport equation (Level set, VOF, pseudo concentration methods for instance). The appro ximation error results in numerical diffusion which depends on the local element size crossed by interfaces. It can be drastically reduced by using the proposed anisotropic adaptive procedure.
The proposed approach does not rely on a particular mesh adapta tion technique and example will be run using a genuine 3D anisotropic mesh generator as well as in conjunction with a simple moving node technique. Different kind of 3D applications will be presented including 3D mould filling, fluid structure interaction, interacting bubble growth in a liquid [4].

REFERENCES

[1] E. Pichelin, T. Coupez, *Finite element solution of the 3d mold filling problem for viscous incompressible fluid*, Comput. Methods Appl. Mech. Engrg., 163, (1998), 359-371.

[2] E. Bigot, T. Coupez, *Capture of 3D moving free surfaces and material interfaces by mesh deformation*, European Congress on Computational Methods in Applied Sciences and Engineering, ECCOMAS 2000, Barcelona, (2000), (CD-ROM).

[3] C. Gruau, T. Coupez, *3D tetrahedral, unstructured and anisotropic mesh generation with adaptation to natural and multidomain metric*, Comput. Methods Appl. Mech. Engrg, (submitted).

[4] J. Bruchon, T. Coupez, *The foam structure prediction by the 3D calculation of the bubble growth in a molten polymer*, J. of Non Newtonian Fluid Mech., (submitted).

COMPUTATIONAL MECHANICS (Abstracts)
WCCM VI in conjunction with APCOM'04, *Sept. 5-10, 2004, Beijing, China*

Numerical-Value Perturbation Method with Application of Solving Convective-Diffusion Integral Equation

Zhi Gao [*], Guowei Yang, Yiqing Shen

Institute of Mechanics, Chinese Academy of Sciences, Beijing, 100080, China
e-mail: yqshen@imech.ac.cn

Abstract

The convective-diffusion equation is of primary importance in such fields as fluid dynamics and heat transfer. In the numerical methods solving the convective-diffusion equation, the finite volume method can use conveniently diversified grids (structured and unstructured grids) and is suitable for very complex geometry. The disadvantage of FV methods compared to the finite difference method is that FV-methods of order higher than second are more difficult to develop in three-dimensional cases. The second-order central scheme (2cs) offers a good compromise among accuracy, simplicity and efficiency, however, it will produce oscillatory solutions when the grid Reynolds numbers are large and then very fine grids are required to obtain accurate solution. The simplest first-order upwind (1UW) scheme satisfies the convective boundedness criteria; however, its numerical diffusion is large. The power-law scheme, QUICK and second-order upwind (2UW) schemes are also often used in some commercial codes. Their numerical accurate are roughly consistent with that of 2CS. Therefore, it is meaningful to offer higher-accurate three point FV scheme.

In this paper, the numerical-value perturbational method suggested by Zhi Gao is used to develop an upwind and mixed FV scheme using any higher-order interpolation and second-order integration approximations, which is called perturbational finite volume (PFV) scheme. The PFV scheme uses the least nodes similar to the standard three-point schemes, namely, the number of the nodes needed equals to unity plus the face-number of the control volume. For instance, in the two-dimensional (2-D) case, only four nodes for the triangle grids and five nodes for the Cartesian grids are utilized, respectively. The PFV scheme is applied on a number of 1-D problems, 2-D and 3-D flow model equations. Comparing with other standard three-point schemes, The PFV scheme has much smaller numerical diffusion than the first-order upwind(1UW) scheme, its numerical accuracy are also higher than the second-order central scheme(2CS), the power-law scheme (PLS), the QUICK scheme and the second-order upwind(2UW) scheme.

REFERENCES

[1] J. H. Ferizger, M. Peric, *Computational methods for fluid dynamics* (third edition), Springer, (2002).

[2] K.A. Hoffma, *Advanced computational fluid dynamics*, AIAA distance learning, July 23, (2003).

[3] M. S. Darwith, *A new high-resolution scheme based on the normalized variable formulation*, Numer. Heat Transfer, Part B, 24, (1993), 353-371.

[4] D. B. Spalding, *A novel finite-difference formulation for differential expressions involving both first and second derivatives*, Int. J. Numer. Methods Engrg., 4, (1972), 551-559.

[5] W. Q. Tao, *Numerical heat transfer*, 2nd edition, Xi'an Jiaotong University Press, (2001).

COMPUTATIONAL MECHANICS (Abstracts)
WCCM VI in conjunction with APCOM'04, *Sept. 5-10, 2004, Beijing, China*

The Numerical Analysis of Oscillatory Characteristic in 3D Supersonic Open Cavity

Zhongxi Hou*, Gang Xia, Chengyao Wang

Institute of Aerospace and Material technology, NUDT, Hunan, Changsha, 410073, China
e-mail: cn_hzx@sina.com

Abstract

The unsteady oscillatory characteristic in supersonic cavity was numerical studied at steady inflow condition. 5 order WENO scheme and 3 order Runge-Kutta algorithm were used. The B-L turbulent model was adopted. Through numerical calculated and analyzed the pressure changing procedure, sound pressure level, oscillatory frequency in cavity, it shows that there is typi cal pressure oscillatory in cavity. The oscillatory has approximate cycle, but the rule of cycle is very complex. 3D cavity has little effect on main oscillatory frequency, but the oscillatory intensity is more different in 2D and 3D. The 3D impact is important at outer of the cavity.

REFERENCES

[1] C. J. Tam, *Algebraic turbulence model simulations of supersonic open-cavity flow physics*, AIAA Journal, 34(11), (1996).
[2] R.L. Sarno, M. E. Franke, *Suppression of flow-induced pressure oscillations in cavities*, Journal of Aircraft, 31(1), (1994), 90-96.
[3] S. H. Shih, A. Hamed, *Unsteady supersonic cavity flow simulations using coupled k-ε and Navier-Stokes equations*, AIAA Journal, 32(10), (1994).
[4] C. M. Robert, *Characteristics of compressible shear layer over a cavity*, AIAA Journal, 39(3), (2001).
[5] A. D. Gorski, *Influence of the trailing edge of a cavity on the flow fluctuation int ensity*, Fluid Dynamics, 36(3), (2001).
[6] B.Y. Adela, K.H. Ronald, *Cavity flame-holders for ignition and flame stabilization in scramjets: an overview*, Journal of Propulsion and Power, 17 (4), (2001).
[7] O.H. Unalmis et al, *Planar laser Imaging of high speed cavi ty flow dynamics*, AIAA Paper 98-0776, (1998).
[8] S. Y. Lin, J. J. Hu, *Parameter study of weighted essentially non-oscillatory schemes for computational aeroacoustics*. AIAA Journal, 39, (2001).
[9] S. Lawrence et.al, Suppression of pressure loads in cavity flows, AIAA Journal, (1), (2004).
[10] R. A. Braurle et al, Hybrid simulation approach for cavity flows: blending, algorithm, and boundary treatment issues, AIAA Journal, 41(8), (2003).
[11] J. Poggie et al, *Large-scale structures in a compressible Mixing layer over a cavity*, AIAA J., 41(12), (2003).

COMPUTATIONAL MECHANICS (Abstracts)
WCCM VI in conjunction with APCOM'04, *Sept. 5-10, 2004, Beijing, China*

A Computational Investigation of the Stationary-State Progression of a Semi-Infinite Bubble through a Corrugated Flow Chamber as a Model of Airway Reopening

Anne-Marie Jacob *, Donald P. Gaver III

Department of Biomedical Engineering, Tulane University, Boggs Suite 500, New Orleans, LA, 70118, USA
e-mail: dpg@tulane.edu, ajacob@tulane.edu

Abstract

This computational investigation examines the local mechanical environment of the lung's lining epithelium in a 2-D model of airway reopening. To this end, the boundary element method (BEM), in conjunction with lubrication theory, is implemented to assess the stationary-state behavior of a flow-driven semi-infinite bubble traveling through a horizontal, liquid-occluded parallel plate flow chamber whose walls are sinusoidally corrugated by a rigid epithelial cell layer. The Newtonian fluid occlusion is characterized by a constant viscosity, μ, density, ρ, and surface tension, γ. Slow, viscous fluid flow is assumed, and the system's behavior is governed by the Stokes and continuity equations. Our results indicate that the interactions between the microgeometry of the walls and the interfacial kinematics surrounding the bubble's tip result in a complex, spatially- and temporally-dependent stress distribution. The walls' non-planar topography magnifies the normal and shear stresses and stress gradients. In addition, we find that decreasing the bubble's speed serves to increase the maximum normal stress and stress gradient while decreasing the maximum shear stress and stress gradient. Our results give credence to the recently-proposed pressuregradient-induced epithelial cell damage theory.[1] Not only may the normal stress gradient likely be responsible for epithelial damage in this model of airway reopening, its magnitude may be even more severe than had been originally predicted.

REFERENCES

[1] A. M. Bilek, K. C. Dee, D. P. Gaver III, *Mechanisms of surface-tension-induced epithelial cell damage in a model of pulmonary airway reopening*, J. Appl. Physiol., 94, (2003), 770-783.

COMPUTATIONAL MECHANICS (Abstracts)
WCCM VI in conjunction with APCOM'04, *Sept. 5-10, 2004, Beijing, China*

Numerical Simulation of the Fine Particles Separation in a Rotational Tube Separator

Jinyu Jiao [1*], Ying Zheng [1], Guogang Sun [2]

[1] *Department of Chemical Engineering, University of New Brunswick, 15 Dineen Drive, P.O. Box 4400, Fredrecton, N.B., E3B 5A3, Canada*
[2] *Department of Chemical Engineering, University of Petroleum, Beijing, 102249, China*
e-mail: jinyu.jiao@unb.ca, yzheng@unb.ca, ggsun@bj163.com

Abstract

This paper presents a numerical analysis of gas -solid separation in a rotational tube separator. This tube separator composes of a bundle of cylindrical tubes with 5 mm in inner diameter and 360 mm in length. The tubes rotate around their common rotating axis. The tubes that have the shortest distance of 130 mm to the common rotating axis are selected to study in this work. Two solid particles with different densities (600kg/m^3 and 2320 kg/m^3) were used as the discrete phase. The separation efficiency and critical particle diameter of the separator were studied using Fluent. It is assumed that solid particles are separated from gas stream once they hit on the wall of the tube separator. As the rotational speed reaches up to 3000 rpm, complete separation can go as low as 2 μm in diameter for both solid particles. The inlet gas ve locity is a key factor of the separation efficiency. Low inlet gas velocity allows laminar gas flow presenting in the rotational tube, which therefore enhances high separation efficiency. Turbulence occurs in the tube separator when the inlet gas velocity is high enough. The separation efficiency can be significantly decreased due to the presence of turbulence. The simulation also showed that the Saffman force has little impact on the separation efficiency of this separator. Furthermore, the critical partic le diameter of this tube separator was calculated theoretically without considering the Saffman force. The simulation predictions on the critical particle diameter agree well with the theory calculations.

REFERENCES

[1] N. Huber, M. Sommerfeld, *Modelling and numerical calculation of dilute-phase pneumatic conveying in pipe systems,* Powder Technol., 99, (1998), 90-101.

[2] David I. Graham, Rana A. Moyeed, *How many particles for my Lagrangian simulations?* Powder Technol., 125, (2002), 179-186.

[3] A. V. Nguyen, C. A. J. Fletcher, *Particle interaction with the wall surface in two-phase gas-solid particle flow,* Int. J. Multiph., Flow, 25, (1999), 139-154.

[4] Y. S. Cheng, Y. Zhou, B. T. Chen, *Particle deposition in a cast of human oral airways,* Aerosol Sci. Technol., 31, (1999), 286-300.

COMPUTATIONAL MECHANICS (Abstracts)
WCCM VI in conjunction with APCOM'04, *Sept. 5-10, 2004, Beijing, China*

Numerical Simulation of Unsteady Shock Waves Using HGA Method

Tomohiko Jimbo *, Takahiko Tanahashi

Department of Mechanical Engineering, Keio University, 3-14-1, Hiyoshi, Kohoku-ku, Yokohama, Kanagawa, 223-8522, Japan
e-mail: jinbo@tana.mech.keio.ac.jp, taka@mech.keio.ac.jp

Abstract

In recent large-scale numerical simulations, efficiency of grid generation and efficient use of limited computer resources are demanded. Various solution-algorithms have been proposed to solve them. In problems of especially unsteady high-speed compressible flows, discontinuity of physical quantity like Shock Waves is dealt with, changing in the location at every moment. In order to obtain higher resolutions, high-resolution schemes and fine grids are needed. As refining whole domain by fine grids, much computer memory is needed and it causes computational speed to be late. Therefore, in order to use limited computer resources efficiently, AMR (Adaptive Mesh Refinement) Method [1] is used, which obtains higher resolutions with minimum requirement computer memory by collecting fine grids only in the location where the change of the physical quantity is intense. As analysis objects are complex in recent numerical simulations, efficiency of grid generation is demanded. Therefore, it is very important to generate grids easily constructed by cells that are not distortion around complex objects.

In this study, the some kinds of cell are paid attention to, which have great effect on grid generation, analysis accurate and computational efficiency. In numerical simulations, fully Triangular or Quadrilateral in two-dimensional problems, Tetrahedron or Hexahedron in three-dimensional problems is usually used. Triangular and Tetrahedron have advantage of shape flexibility. Quadrilateral and Hexahedron can analyze flow problem efficiently. In three-dimensional problems Prism and Pyramid are used, too. Prism has the advantage of both shape flexibility and analysis accurate because the base of Prism is Triangular and it is possible to generate grids which intersect objects at right angles. And Pyramid is used as transition grids. Therefore, using Hybrid (Triangular/ Quadrilateral, Tetrahedron/ Pyramid/ Prism/ Hexahedron) grids makes it possible to take advantage of both shape flexibility and computational efficiency. In order to analyze them more efficiently, HGA (Hybrid Grid Adaptation) Method, included AMR Method based on Hybrid grid is used. Effectiveness of HGA Method is verified by analyzing benchmark problems for compressible flows using Hybrid grid, the following conclusions are obtained. The numerical results of benchmark problems by using the present scheme agree with other numerical results and experimental results. Making comparison of the numerical results by using between HGA Method with Hybrid gird and AMR Method with fully Triangular or Tetrahedron, the former obtain higher resolutions than the latter. Therefore it is verified that HGA Method takes advantage both shape flexibility and computational efficiency.

REFERENCES

[1] R. Löhner, *An adaptive finite element scheme for transient problems in CFD*, Comput. Meth. Appl. Mech. and Eng., North-Holland, 61, (1987), 323-338.

COMPUTATIONAL MECHANICS (Abstracts)
WCCM VI in conjunction with APCOM'04, *Sept. 5-10, 2004, Beijing, China*

Hall Current Effect on MHD Mixed Convection Flow Past a Semi Infinite Vertical Plate with Thermal Stratification

R. Kandasamy*

Department of Mathematics, Institute of Road and Transport Technology, Erode-638 316, India
e-mail: kandan_kkk@yahoo.co.in

Abstract

The study of magnetohydrodynamic mixed convection through a viscous fluid past a heated semi-infinite plate is considered very essential to understand the behavior of the performance of fluid motion in several applications. Since no attempt has been made to analyze the Hall current effect on magnetohydrodynamic mixed-convection flow past a semi-infinite vertical plate with thermal stratification and heat transfer under the influence of uniform magnetic field, we have investigated the same in this article. The similarity transformation has been utilized to convert the governing nonlinear partial differential equations into nonlinear ordinary differential equations and then the numerical solution of the problem is drawn using R.K.Gill method. Numerical calculations were carried out for different values of dimensionless parameters of the problem under consideration for the purpose of illustrating the results graphically. Examination of such flow models reveals the influence of magnetic field on velocity and temperature profiles. The analysis of the results obtained shows that the flow field is influenced appreciably by the presence of Hall current, thermal stratification and magnetic effects.

It is hoped that the present investigation of the study of physics of flow over a vertical surface can be utilized as the basis for many scientific and engineering applications and for studying more complex vertical problems involving the flow of electrically conducting flows. The findings may be useful for the study of movement of oil or gas and water through the reservoir of an oil or gas field, in the migration of underground water and in the filtration and water purification processes. The results of the problem are also of great interest in geophysics in the study of interaction of the geomagnetic field with the fluid in the geothermal region.

REFERENCES

[1] E. M. Aboeldahab, E. M. E. Elbarbary, *Hall current effect on MHD free convection flow past a semi infinite vertical plate with mass transfer,* Int. J. Engng. Sci., 39, (2001), 1641- 1653.

[2] S. P. Anjali Devi, R. Kandasamy, *Thermal stratification effects on heat transfer over a wedge with suction or injection*, Mechanics Res. communications, 28, (2001), 349-354.

[3] E. M. Elabashbeshy, *Heat and mass transfer along a vertical plate with variable surface tension and concentration in the presence of magnetic field,* Int. J. Engng. Sci., 34, (1997), 515-522.

COMPUTATIONAL MECHANICS (Abstracts)
WCCM VI in conjunction with APCOM'04, *Sept. 5-10, 2004, Beijing, China*

Low-Dimensional Description of Nonlinear Transient Thin-Film Flow

Roger E. Khayat *

Department of Mechanical and Materials Engineering, the University of Western Ontario, London, Ontario, N6A 5B9, Canada
e-mail: rkhayat@eng.uwo.ca

Abstract

The two-dimensional steady and transient nonlinear flow of thin films is examined in this study. The character of the nonlinearities is both geometrical, resulting from the presence of free surface, and convective, in the form of inertia. The fluid is assumed to emerge from a narrow channel to flow on a flat or curved stationary solid boundary (substrate). The flow is induced by a constant pressure gradient maintained inside the channel. The governing equations are established using the boundary-layer approximation. The equations are solved by expanding the flow field in terms of an orthonormal polynomial basis in the depthwise direction, using the Galerkin projection, combined with a time-stepping implicit scheme, and integration along the flow direction using a variable-step finite-difference scheme for the resulting two-point boundary-value problem. The accuracy of the method as well as convergence are established by comparing the solution to limit flow solutions. It is found that only a few modes are required to achieve acceptable convergence. Given the hyperbolic nature of the governing equations, initial conditions play a dominant role in determining the nature of the transient flow. For some class of initial conditions the film exhibits a steep wave train, and the steady state is unstable.

COMPUTATIONAL MECHANICS (Abstracts)
WCCM VI in conjunction with APCOM'04, *Sept. 5-10, 2004, Beijing, China*

A CFD Study of the Luminal Surface Concentration of Low Density Lipoproteins in Coronary Arteries: the Effects of Wall Motion

M. K. Kolandavel [1], E. T. Fruend [2], E. M. Pedersen [2], S. Ringgaard [2], P. G. Walker [1*]

[1] *Centre for Biomechanics and Medical Engineering, School of Mechanical Engineering, University of Leeds, Leeds, LS2 9JT, United Kingdom*
[2] *MR-Centre, Institute of Experimental Clinical Research, Skejby Sygehus, Denmark*
e-mail: {men1kkm, p.g.walker} @leeds.ac.uk, etf@mr.au.dk, erik.morre@iekf.au.dk, steffen@mr.au.dk

Abstract

The transport of low density lipoproteins (LDLs) from the blood to the arterial wall and their accumulation within the arterial intima are key steps in atherogenesis. In coronary arteries these may be affected by the local haemodynamic factors including their physiological motion, whose effect on LDL transport has not been studied elsewhere. Therefore, this study was initiated to investigate the transport of LDL to the endothelium (a semi permeable membrane) in a moving 3-D model of the Left Anterior Descending (LAD) coronary artery segment by using the commercial finite volume CFD solver, FLUENT (v6.1.18, Fluent Inc). A time varying inlet parabolic velocity [1] and planar vessel wall motion waveforms [2-3] obtained from literature were imposed as boundary conditions. The results showed the occurrence of flow-dependent concentration polarization (the accumulation and elevation of luminal surface concentration of LDL due to water filtration through the semi permeable endothelium) of LDL [4] at the endothelial surface in varying degrees in pulsatile flow (static case) and pulsatile flow with wall motion (dynamic case). In both the cases, this effect was more pronounced at the inner wall of bend where wall shear stress (WSS) was lower. The motion induced elevation of WSS and stronger secondary flows seem to promote better cross mixing, resulting in a lower concentration of LDL, especially at the inner wall when compared to the pulsatile flow simulations. Taken together, these results suggest that the vessel wall motion may have a dominant influence on LDL transport. However the degree of influence would also depend on the individual motion patterns and the reaction mechanisms used. This warrants future studies on more physiologic models to gain a fundamental understanding on LDL transport and consequently on conditions such as atherosclerosis.

REFERENCES

[1] J. T. Marcus, H. G. Smeenk, J. P. A. Kuijer, R. J. Van der Geest, R. M. Heethaar, A. C. Van Rossum, *Flow profiles in the left anterior descending and the right coronary artery assessed by MR velocity quantification: effects of through-plane and in-plane motion of the heart*, J. Comp. Assist. Tomogr., 23, (1999), 567-576.

[2] S. A. Mackay, M. J. Potel, J. M. Rubin, *Graphics method for tracking three-dimensional heart wall motion*, Comp. and Biomed. Res., 15, (1982), 455-473.

[3] S. Kozerke, M. B. Scheidegger, E. M. Pederson, P. Boesiger, *Heart motion adapted cine phase-contrast flow measurements through the aortic valve*, MRM, 42, (1999), 970-978.

[4] S. Wada, T. Karino, *Theoretical prediction of low-density lipoproteins concentration at the luminal surface of an artery with a multiple bend*, Ann. Biomed. Eng., 30, (2002), 778-791.

COMPUTATIONAL MECHANICS (Abstracts)
WCCM VI in conjunction with APCOM'04, *Sept. 5-10, 2004, Beijing, China*

QUICK Scheme on Unstructured Grids

Xijun Lai [1*], Zhidong Wang [2], Deguan Wang [3]

[1] *Nanjing Institute of Geography & Limnology, CAS, Nanjing, 210008, China*
[2] *Instituite of East-China Ship-building, Zhenjiang, 212000, China*
[3] *College of Environmental Science and Engineering, Hohai University, Nanjing, 210098, China*
e-mail: laixijin@163.com, ljin1977@hotmail.com, cywzd@sohu.com

Abstract

When a convective and diffusive equation is solved using numerical methods, the discretization of convective term is one of the main difficulties due to its special characteristic. The studies on it have become the key point and the hot topic in computational fluid dynamics. The choice of discretization method for convective term has a direct effect o n the stability of numerical algorithm, the accuracy of numerical results and the economy of numerical simulation. So the success of simulation mainly depends on the convective approximation. On structured grids, many numerical schemes are successfully applied, e.g. QUICK-like schemes. These schemes have low numerical diffusion. Using them, high accuracy can generally be obtained and stability can be kept.
To meet the requirement of simulating flow field in complex geometry, researchers developed many numerical methods on non-orthogonal curvilinear grids, block-structured grids and unstructured grids for solving Navier-Stokes equations. Due to the flexibility of grid generation, the generality and extensibility of program, the collocated finite volume method based on unstructured grids is favored. However, the high-order accuracy scheme is difficult to implement on unstructured grids because of it always involves the indirect neighboring points. Ferziger & Peric thought that the second-order scheme offered the best compromise among accuracy, generality and simplicity and high-order accuracy may be obtained by shape function.
But high-order schemes with high performance are still desired and some high-order schemes can simply be conducted by re-constructed techniques. In the paper, simplified QUICK (Quadratic upwind interpolation of convective kinematics) scheme on unstructured grids was implemented by linear re-construction. The general QUICK scheme on one-dimension inhomogeneous grids was firstly written as general formula for different upwind direction. And then substituted the linear re-constructed expression of variables' values at "upwind cell" of upwind cell into general formula, QUICK scheme on unstructured (UQUICK) was obtained. The implementation metho d of UQUICK proposed by Hayase was adopted.
To validate performance and efficiency of the scheme, three numerical experiments on unstructured grids were carried out. First, numerical results of skewed driven cavity flow with 30 degree inclined angle were compared among different schemes and benchmark solutions. Then, Flows past cylinders in low Reynolds number was studied. Finally, double diffusive natural convection was simulated at different buoyant ratio.

REFERENCES

[1] Peric M. Ferziger, *Computational methods for fluids dynamics*, Springer, Germany, (1999).
[2] T. J. Barth, D. C. Jespersen, *The design and application of upwind schemes on unstructured meshes*, AIAA Paper 89-0366, (1989).
[3] T. Hayase, J. A. C. Humphrey, R. A. Grif, J. Comput. Phys., 98, (1992), 108-118.

COMPUTATIONAL MECHANICS (Abstracts)
WCCM VI in conjunction with APCOM'04, *Sept. 5-10, 2004, Beijing, China*

Finite Element Numerical Simulation of Estuary Tides and Tidal Current Process

Dongfeng Li [1, 2*], Bing Zou [2], Huijun Hu [2], Hongwu Zhang [3], Guangqian Wang [3], Zhiyong Lu [1]

[1] *Beijing University of Aeronautics and Astronautics, Beijing, 100083, China*
[2] *Department of Hydraulic Engineering, Zhejiang Water Conservancy and Hydropower College, Hangzhou, 310018, China*
[3] *Department of Hydraulic Engineering, Tsinghua University, Beijing, 100084, China*
e-mail: lidongfeng998@163.com

Abstract

The Yellow River is famous for it's plenty of sediment and the most height-concentration. Because of much limit which it is not possible to get much field data and remote sensing data of the tides and tidal current flow .So it is difficult to make known and disclose the features of estuary tides rising and ebb, tidal current flow, sediment diffusion and transport and bed deformation. But these features are important for the estuary and coast harness especially when the local economic has been developed. Comparing by numerical simulating based on the mathematical model with the getting field data, the former has much more virtues than the latter and by the former these problems can be solved.
Based on the established and verified Yellow River estuary and ocean two dimensional finite element numerical model [1], tides and tidal current flow are numeric simulated, from the calculation results, the typical and enough tidal current vectors charts, velocity contour, tidal stage distribution diagrams are given. These diagrams indicate that the flow is coastwise and there are two flood and ebb tides in a day and night in this the Yellow River channel estuary and the Laizhou gulf region. The simulated results also make known that in this region the tide and tidal current features show great variations greatly during a day resulting from the half-closed Bohai Sea coastlines and the wave from the East China Sea. These results are also good agreement with the field data, satellite remote-sensing data and their analysis [2-4]. These show that these laws could be disclosed by the numerical simulation method based on the sea-dynamic two-dimensional finite element mathematical model.
This work was supported by Zhejiang Provincial Natural Science Foundation (ZJNSF-M503254); Zhejiang Provincial Education Foundation (20031064); Foundation for the Author of National Excellent Doctoral Dissertation of PR China (FANEDD-No.199935).

REFERENCES

[1] Dongfeng Li, Y. J. Cheng, B. Zou, et al, *Numeical two dimensioal finite element model and application of the Yellow River estuary sediment,* Journal Advances in Marine Science, 22(1), (2004), (in Chinese).
[2] P. Y. Li, S. Y. Wu, Q. Y. Zang, *Tidal flat topographic and its erosion and deposition variation in the Yellow River harbour region,* Acta Oceanologica Sinica, 14(6), (1992), 74-84, (in Chinese).
[3] Q. Y. Zang, *The Yellow River delta sediment,* Beijing, Ocean Press, (1996), 1-152, (in Chinese).
[4] D. Y. Xu, *Remote-sensing research on the Yellow River estuary,* meteorological press, (1990), 139-161, (in Chinese).

COMPUTATIONAL MECHANICS (Abstracts)
WCCM VI in conjunction with APCOM'04, *Sept. 5-10, 2004, Beijing, China*

2D and 3D Numerical Simulation of Aerodynamic Performance for Unmanned Air Vehicle

H. Y. Li*, J. J. Xiong, X. H. Li, N. Chen

Aircraft Department, Beihang University, Beijing, 100083, China
E-mail: lida1026@sina.com, lixiaohui_2003@tom.com, xiongjj2003@yahoo.com.cn

Abstract

This paper seeks to explore the influences of mesh distribution and turbulence model on flow fields of unmanned air vehicle with low velocity. 2D and 3D physical models are established to simulate laminar and turbulent flow fields of airfoil of unmanned air vehicle. Using the structured and unstructured meshes, the mesh configuration models of fuselage and complete vehicle are obtained for aerodynamic computation respectively. Then 2D and 3D, laminar and turbulent flow fields are numerically simulated by means of the CFD methodology at two kinds of Mach numbers in low velocity range respectively. The pressure distributions and lift and drag coefficients of wing surfaces, fuselage and complete vehicle are also calculated. Finally, contrast analysis is conducted between the simulated flow fields in order to examine the presented simulation procedure, demonstrating the practical and economic use of the proposed technique.

COMPUTATIONAL MECHANICS (Abstracts)
WCCM VI in conjunction with APCOM'04, *Sept. 5-10, 2004, Beijing, China*

High Accurate Combined Compact Upwind Difference Schemes for the Incompressible Flow Simulations

Xian Liang [1*], Zhenfu Tian [1, 2]

[1] *Institute of Applied Mathematics and Engineering Mechanics, University of Ningxia, Yinchuan, Ningxia, 750021, China*

[2] *Shanghai Institute of Applied Mathematics and Mechanics, University of Shanghai, Shanghai, 200072, China*

e-mail: liangx_yan@263.net , zftian@nxu.edu.cn

Abstract

The aim of this paper is to develop a high-order compact finite difference methods for solving the Navier-Stokes (NS) equations based on projection method [1]. The present methods are constructed on staggered grid system and have fourth-order accuracy. Discretization of the convection terms in the NS equations is the key step. We employed a kind of combined compact upwind schemes to complete this goal in our paper. Firstly, a new compact upwind scheme was developed, at the same time a existed fifth-order upwind scheme was quoted [2]. Like all upwind schemes, both of them have bigger errors in short wave range. Secondly, we combined these two schemes with certain weight in order to decrease such errors. Fourier accuracy analysis shows that such combination can obtain better performance. The viscous terms are discretized with fourth-order compact symmetrical finite difference schemes, which will lead to solve the triple diagonal systems. We design special fourth-order finite difference scheme to approximate the pressure Poisson equation and pressure gradient by employing projection techniques.

Furthermore, we list two numerical tests results [2-3], one is Taylor's vortex array problem with analysis solution, and another is a doubly periodic double shear layer flow problem [3] with big gradient and high Reynolds number, up to 10000. The results show that the present methods have desired accuracy and robustness, which are fit for directly numerical simulation complex fluids flow problem.

This work is supported by the National Natural Science Foundation of China under Grant 19702008 and 19902010 and the Teaching and Research Award Program for Outstanding Young Teachers in Higher Education Institutions of MOE.

REFERENCES

[1] Z. Tian, Y. Li, *Numerical solution of the incompressible Navier-Stokes equations with a three-point fourth-order upwind compact difference scheme*, ICNM-IV, Shanghai, (2002), pp. 942-946.

[2] Yanwen Ma, Dexun Fu, T. Kobarash, N. Taniguchi, *Numerical solution of the incompressible Navier-Stokes equations with an upwind compact difference scheme*, Int. J. Numer. Meth. Fluids, 30, (1999), 509-521.

[3] J. B. Bell, P. Colella, *A second-order projection method for the incompressible Navier-Stokes equations*, J. Comp. Phys., 85, (1989), 257-283.

COMPUTATIONAL MECHANICS (Abstracts)
WCCM VI in conjunction with APCOM'04, *Sept. 5-10, 2004, Beijing, China*

Numerical Simulation on Cavitating Flows

Chuanjing Lu *, Lei Wu, Leiping Xue

Department of Engineering Mechanics, Shanghai Jiaotong University, Shanghai, 200030, China
e-mail: cjlu@mail.sjtu.edu.cn

Abstract

Cavitation is a natural phenomenon especially existing in flowing liquids. It happens in flow field where the local pressure is lower than the critical pressure of cavitation. Along with great improvement of speed for kinds of vehicle both on and under water, the cavitation phenomenon is inevitable and it takes a dominant position for cavitation damage and noise, as well as hydrodynamic characteristic and stability controlling of high-speed bodies. Up to now, cavitation mechanism and unstable characteristic of cavitating flows are still the important topics in fluid dynamics and studies conducted are of great significance both in theory and engineering applications. This work is concentrated on the study of reasonable cavitation models and numerical calculation methods suitable for two-dimensional unsteady cavitating flows. Researches are mainly developed to the contents as follows.

A new two-phase cavitating flow model, which involves viscosity and multi-phase effects, is presented. Based on the physical behaviors of water/vapor mixture and sound velocity characteristics, a function between density and pressure is adopted to account for the two-phase flow of water/vapor and the transition from one phase to the other.

A numerical calculation method for three-dimensional unsteady cavitating flows is developed. A algorithm for calculating density variable N-S equations of cavitating flow problem is put forward based on the Navier-Stokes solver. The typical two-dimensional numerical examples presented indicate that the cavity model is rational and the algorithm is applicable for steady and unsteady cavitating flows effectively.

The inner flow of cavity for two-dimensional hydrofoils is studied to analyze the cavity shape and the water/vapor structure. The mechanism of vortex that causes turbulence in the rear part of cavity and the effects of pressure in cavity on numerical results are also investigated. It is found that more accurate results can be obtained by using the modified cavitation number defined according to the actual pressure in the cavity.

The numerical simulation on the unsteady cavitating flows passing two-dimensional hydrofoils under periodic income flow condition shows that when income flow takes periodic swing under certain range of cavity number and angle of attack, the cavity behaves corresponding periodically. Break-off phenomenon takes place at the rear part of cavity under the condition of small cavity number or larger swing amplitude. Thereafter, the change of cavity is affected.

Two-dimensional unsteady cavitating flows under uniform-flow condition are simulated. The periodic evolution, which includes inception, development, break-off, shedding of cavitating vortices and finally shedding cavity collapsing at the high-pressure region downstream, is obtained. The calculation results of shedding frequency, break-off instantaneousness and shedding velocity match up to experimental data quite well. It is found that re-entrant jet is not absolutely constituted of liquid, but vapor/liquid mixture and the break-off process is directly induced by reverse flow closely along the wall. The shedding cavity moves downstream and the cavity near the leading edge shrinks with the reverse velocity. As the shedding cavity collapsing downstream, a new cavity incepts near the leading edge.

COMPUTATIONAL MECHANICS (Abstracts)
WCCM VI in conjunction with APCOM'04, *Sept. 5-10, 2004, Beijing, China*

A Numeric Method for Hydrodynamic Performance Computation of POD Propulsion Systems

Cheng Ma [1*], Xu Zhang [1], Zhengfang Qian [1], Du Du [1], Chenjun Yang [2]

[1] *Naval Research Center, Beijing, 100073, China*
[2] *Department of Naval Architecture & Ocean Eng., Shanghai Jiaotong University, Shanghai, 200030, China*

e-mail: dudu0736@yahoo.com.cn

Abstract

POD propulsion is the up-to-date development trend of electric propulsion for ships. Based on potential flow theory, a numerical method is proposed in this paper for predicting the steady hydrodynamic performance of podded propulsors. Podded propulsors can be divided into three types, which are tractor-, pusher-, and tandem-type propulsors. As the first and third types are applied more often and the calculation method of the second type is similar with the first, the emphasis is put on introduction of calculation methods of tractor- and tandem-type propulsors. The propeller blades are calculated with a vortex lattice method, while the pod including its strut with a boundary element method for non-lifting bodies. Hydrodynamic interactions between the propeller and the pod are treated via iterative calculations. For the propeller, the linear system of equation for solution of load distribution on the blades can be obtained by satisfaction of all object plane boundary conditions on all control points on the camber plane of blades. For the pod, the linear system of equation for solution of source sink intensity of each unit can also be obtained by satisfaction of all object plane boundary conditions on center of each unit (control point) on its surface. For the towed propeller and the pusher-type one, due to the different positions of these propellers, the influence of pod on the fore propeller vortex has to be considered for the towed one; for the pusher-type one, it is similar to that without pod. The other calculation and processing are the same. For the twin-screw POD propulsion, not only the influence of pod on the wake of fore propeller but also the mutual influence among the fore propeller, the pod and the aft propeller have been considered in the iterative calculation. Then a real case computation has been carried out and the results obtained agree well with those obtained from the experiments carried out in a cavitation tunnel.

REFERENCES

[1] Cheng Ma, Zhengfang Qian, Xu Zhang, *Research on performances of POD propulsion and its military application*, in Proc. of the 6th National Academic Conference on Hydrodynamics. Hong Kong, 2003, Ocean Press, Beijing, (2003), (in Chinese).

[2] P. Lohi, J. Savikurki, *Azimuth thrusters in icebreakers: a revolutionary step ahead*, in Proc. of Int. Shipbuilding Conf., 1994, St. Petersburg, (1994).

[3] The Propulsion Committee, *Final report and recommendations to the 23rd ITTC*, Proceeding of 23rd ITTC, (2002).

[4] D. S. Greeley, J. E. Kerwin, *Numerical methods for propeller design and analysis in steady flow*, Trans. SNAME, 90, (1982), 415-453.

[5] Chenjun Yang, Zhengfang Qian, Cheng Ma, *Influences of pod on hydrodynamic performances of propeller*, Journal of Shanghai Jiaotong University, 37(4), (2003), 449-453, (in Chinese).

COMPUTATIONAL MECHANICS (Abstracts)
WCCM VI in conjunction with APCOM'04, *Sept. 5-10, 2004, Beijing, China*

The Finite Element Method for MHD Flow at High Hartmann Number

Ali I. Nesliturk *

Department of Mathematics, Izmir Institute of Technology, Urla, Izmir, 35430, Turkey
e-mail: alinesliturk@iyte.edu.tr

Abstract

A stabilized finite element method using the residual -free bubble functions (RFB) is proposed for solving the governing equations of steady magn etohydrodynamic duct flow. A distinguished feature of the RFB method is the resolving capability of high gradients near the layer regions without refining mesh. We show that the RFB method is stable by proving that the numerical method is coercive even not only at low values but also at moderate and high values of the Hartmann number. Numerical results confirming theoretical findings are presented for several configurations of interest. The approximate solution obtained by the RFB method is also compared wi th the analytical solution of Shercliff's problem.

REFERENCES

[1] F. Brezzi, A. Russo, *Choosing bubbles for advection-diffusion problems,* Math. Models Meth. Appl. Sci., 4, (1994), 571-587.
[2] L. P. Franca, A. Nesliturk, M. Stynes, *On the stability of residual-free bubbles for convection-diffusion problems and their approximation by a two-level finite element method,* Comput. Methods Appl. Mech. Engrg., 166, (1998), 35-49.
[3] B. Singh, J. Lal, *FEM for MHD channel flow with arbitrary wall conductivity,* J. Math. Phys. Sci., 18, (1984), 501-516.
[4] M. Tezer-Sezgin, S. Koksal, *FEM for solving MHD flow in a rectangular duct,* Int. J. Num. Meth. Eng., 28, (1989), 445-459.
[5] Z. Demendy, T. Nagy, *A new algorithm for solution of equations of MHD channel flows at moderate Hartmann numbers,* Acta Mechanica, 123, (1997), 135-149.

COMPUTATIONAL MECHANICS (Abstracts)
WCCM VI in conjunction with APCOM'04, *Sept. 5-10, 2004, Beijing, China*

Parallelization of Lattice Boltzmann model in Thermohydrodynamics Problems and Simulating Natural Convection in Cavity

A. Niavarani-Kheiri [*], M. Darbandi

Department of Aerospace Engineering, Sharif University of Technology, Azadi Ave., P.O. Box 11365-8639, Tehran, Iran
e-mail: niavarani_a@hotmail.com, darbandi@sharif.edu

Abstract

The main objective of the current work is to develop a Lattice -Boltzmann approach on parallel computers in order to treat thermobuoyant flows. The Lattice -Boltzmann method is a relatively new computational tool which is based on mesoscopic kinetic equations. Indeed, this app roach is a bridge between the microscopic phenomena and the continuum macroscopic behavior. The continuum equations can be treated using a number of different discretization schemes with a second order of accuracy in time and space. Since the macroscopic b ehavior of the system is computed with respect to the kinetic equation and distribution functions, the macroscopic averaged properties follow the Navier-Stokes equations. However, the Lattice-Boltzmann approach can be modified to treat thermo-hydrodynamics problems if the buoyant force terms are included in the Boltzmann equation. In this regard, specific distribution functions and equilibrium distribution functions are required which are consequently derived. The discrete velocity set for the second equation is not the same as the Boltzmann equation which can neglect the effect of viscous and dissipation terms. From parallel processing point of view, the Lattice-Boltzmann method essentially needs only the nearest neighbor information, hence, the algorithm i s superior for parallel computations. Therefore, the current parallel code has been written in C and uses MPI (Massage Passing Interface) libraries for portability. The computational domain is primarily divided into blocks. The experience shows that the no n-parallelizable part of the computational time is between 0.7 to 3 percent of the total computational time. This is because we have tried to reduce the Communication time between the blocks as much as possible. This is done by choosing different geometries for the sub-domains and changing their directions. The results show that the speedup of the system is excellent.

REFERENCES

[1] D. Kandhai, A. Koponen, A. Hoekstra, M. Kataja, J. Timonen, P. Sloot, *Lattice-Boltzmann hydrodynamics on parallel systems*, Comput. Phys. Commun., 111, (1998), 14-26.

[2] J. Onishi, Y. Chen, H, Ohashi, *Lattice Boltzmann simulation of natural convection in square cavity*, JSME International Journal, Series B, 44 (1), (2001).

COMPUTATIONAL MECHANICS (Abstracts)
WCCM VI in conjunction with APCOM'04, *Sept. 5-10, 2004, Beijing, China*

Mesh Generation for Wind Flow Analysis in City

Kazuya Nojima [*], Mutsuto Kawahara

Department of Civil Engineering, Chuo University, Kasuga 1-13-27, Bunkyo-ku, Tokyo, 112-8551, Japan
e-mail: noji@kc.chuo-u.ac.jp, kawa@civil.chuo-u.ac.jp

Abstract

In the field of civil engineering, it is important to estimate the wind flow before the construction of buildings. The high-rise buildings changes wind field and makes many problems. A strong wind sometimes has a bad influence on life of human. For instance, a wind circulation behind the buildings raises dust. The strong wind makes noise. In Japan, houses and office building concentrate in small space. Therefore, the change of the wind caused by high-rise building makes a lot of troubles.
In this paper, development of mesh generation algorithm for fluid flow analysis with finite element method is proposed. The unstructured mesh generation, which is based on Delaunay triangulation, is used because it is difficult to make the structural mesh of the buildings. There are several algorithms used to construct the Delaunay triangulation. In this research, the algorithm proposed by Bowyer [1] is used. Because of the computation of wind flow in city, the Delaunay triangulation extended to three dimensions. The configuration of building is made using some copy points [2]. However the Delaunay tetrahedrarization has some problems. The generation of negative volume elements and the disappearance of the nodes are the some of them. In this paper, some techniques are proposed to fix these problems. Additionally, the improvements of the Delaunay method for non-convex domain are proposed.
The high-resolution finite element mesh is required for wind analysis. Since the huge number of nodes is required for the analysis, improvements in the speed and the robustness of mesh generation are necessary. The radiation datum line method, which is an algorithm of distribution of nodal points and can distribute the nodal points fast, is proposed.
For the verification of the finite element mesh, a computation is carried out. In this paper, the flow is assumed as incompressible viscous flow. The bubble function element [3] is used as stabilization scheme. For the large-scale computation, the parallel computing that is based on domain decomposition method is employed.

REFERENCES

[1] A. Bowyer, *Computing dirichlet tessellations,* The Computer Journal, 24, (1981), 162-166.
[2] K. Nojima, M. Kawahara, *An analysis of wind environment around buildings with unstructured mesh generation technique,* Computational Fluid and Solid Mechanics, (1), (2003), 1066-1071.
[3] J. Matsumoto, T. Umetsu, M. Kawahara, *Shape identification for fluid-structure interaction problem using improved bubble element*, IJCFD, 15, (2001), 33-45.

COMPUTATIONAL MECHANICS (Abstracts)
WCCM VI in conjunction with APCOM'04, *Sept. 5-10, 2004, Beijing, China*

A Multiscale-Domain Decomposition Method for High Reynolds Number Flows

Yiqing Shen *, Guowei Yang, Zhi Gao

Institute of Mechanics, Chinese Academy of Sciences, Beijing, 100080, China
e-mail: yqshen@imech.ac.cn

Abstract

The high Reynolds number flow contains a wide range of length and time scales, and the flow domain can be divided into several sub-domains with different characteristic scales. In some sub-domains, the viscosity dissipation scale can only be considered in a certain direction; in some sub-domains, the viscosity dissipation scales need to be considered in all directions; in some sub-domains, the viscosity dissipation scales are unnecessary to be considered at all.

For laminar boundary layer region, the characteristic length scales in the streamwise and normal directions are L and $L\,\mathrm{Re}^{-1/2}$, respectively. The characteristic length scale and the velocity scale in the outer region of the boundary layer are L and U, respectively. In the neighborhood region of the separated point, the length scale $l \ll L$, the velocity scale $u \ll U$, and $ul/\nu = O(1)$. In the light of this reason there exists a great disparity of the grid Reynolds numbers $R_{\Delta x_i}$ between different cells in Navier-Stokes (NS) equations computations for high Reynolds number flows, an idea of solving the conservation equations for discrete cells was proposed and named the discrete fluid dynamics (DFD) algorithm. Analysis shows that the basic conservative equations for discrete cells are the Euler equations, NS- and diffusion parabolized (DP) NS equations.

In this paper, a new multiscale-domain decomposition method is developed for the high Reynolds number flow. First, the whole domain is decomposed to different sub-domains with the different characteristic scales. Then the different dominant equation of all sub-domains is defined according to the diffusion parabolized (DP) theory of viscous flow. Finally these different equations are solved simultaneously in whole computational region.

For numerical tests of high Reynolds numerical flows, two -dimensional supersonic flows over rearward and frontward steps as well as an interaction flow between shock wave and boundary layer were solved numerically. The pressure distributions and local coefficients of skin friction on the wall are given. The numerical results obtained b y the multiscale-domain decomposition algorithm are well agreement with those by NS equations. Comparing with the usual method of solving the Navier-Stokes equations in the whole flow, under the same numerical accuracy, the present multiscale domain decomposition method decreases CPU consuming about 20% and reflects the physical mechanism of practical flow more accurately.

REFERENCES

[1] F. G. Zhuang, D. L. Zhang, *Significance of diffusion-parabolized NS equations and its application to computational,* Acta Aerodynamica Sinica, 21, (2003), 1-10.

[2] Z. Gao, *Hierachical structure of the simplified Navier-Stokes equations (SNSE) and its mechanical connotation and application,* Science in China (Ser A.), 32, (1989), 168.

[3] Z. Gao, Y. Q. Shen, *New strategy for finite difference computing of viscous fluid dynamics*, Sciences in China (Ser. A), 29, (1999), 433-443.

COMPUTATIONAL MECHANICS (Abstracts)
WCCM VI in conjunction with APCOM'04, *Sept. 5-10, 2004, Beijing, China*

Numerical Manipulations in Measurement of the Dynamic Evolution of Surface Tension of Oscillating Jet

Guowen Tan[1*], Stephen E. Bechtel[2], Kurt W. Koelling[3]

[1] *Department of Engineering Mechanics, Tsinghua University, Beijing, 100084, China*
[2] *Department of Mechanical Engineering, The Ohio State University, Columbus, Ohio, 43210, U.S.A.*
[3] *Department of Chemical Engineering, The Ohio State University, Columbus, Ohio, 43210, U.S.A.*
E-mail: tanguowen@tsinghua.org.cn, bechtel.3@osu.edu, koelling.1@osu.edu

Abstract

Due to the presence of surfactant additives, the surface tension in air of some solutions, such a s pesticides, decays significantly with surface age in the first several milliseconds of existence of the free surface, often by a factor of 2 or more, before it stabilizes to its static, equilibrium value. Since drop formation and atomization occur within this interval of rapidly changing surface tension before equilibrium is reached, and surface tension dominates the hydrodynamic instability leading to droplets, accurate methods for the determination of the evolution of surface tension of new surfaces on the millisecond time-scale are necessary to understand and model spray formation.

To exploit the oscillating jet phenomenon to determine the dynamic surface tension, the measurements from the oscillating jet are combined with an analytical model which rela tes these measurements to the surface tension. The dynamic surface tension is deduced from the experimental free surface measurements employing systems of governing equation [1, 2]. In the formulation, the measured free surface profiles, their spatial deri vatives, axial velocities, and velocity gradients are required. To obtain these necessary functions a numerical manipulation is applied in this paper: filtering of the primitively measured free surface profile together with dis crete numerical differentiation. This method is superior to polynomial curve fitting of the primitively measured free surface profile together with continuous analytical differentiation [3, 4].

Then, the forward problem, in which the surface tension of the jet is specified and the an alytical model predicts the resulting jet profile, is inverted to deduce the surface tension of the interface from physical measurements of the free surface profile of a jet of a test fluid, with the measured free surface profiles, their spatial derivative s, axial velocities, and velocity gradients known.

REFERENCES

[1] S. E. Bechtel, J. A. Cooper, M. G. Forest, N. A. Petersson, D. L. Reichard, A. Saleh, V. Venkataramanan, *A new model to determine dynamic surface tension and elongational viscosity using oscillating jet measurements,* J. Fluid Mech., 293, (1995), 379-403.

[2] S. E. Bechtel, M. G. Forest, N. Youssef, H. Zhou, *The effect of dynamic surface tension on the oscillation of slender elliptical Newtonian jets*, J. Applied Mech., 65, (1998), 694.

[3] Guowen Tan, *Material Characterization of Polymer Solutions and Surfactant Systems Using Free Surface Measurements*, PhD Thesis, The Ohio State University, USA, (2002).

[4] W. H. Nguyen, *Measurement of dynamic surface tension on the submillisecond timescale*, M.S. Thesis, The Ohio State University, USA, (1999).

COMPUTATIONAL MECHANICS (Abstracts)
WCCM VI in conjunction with APCOM'04, *Sept. 5-10, 2004, Beijing, China*

Numerical Investigation on Mode-Transition of Laminar Natural Convection in Inclined Enclosures with the Higher-Order Compact FDM

Zhenfu Tian [1*], Bo Hong [2]

[1] *Shanghai Institute of Applied Mathematics and Mechanics, Shanghai University, Shanghai , 200072, China*
[2] *LED, South China Sea Institute of Oceanology, China Academy of Science, Guangzhou , 510301, China*
e-mail: zftian@263.net, hongbo@scsio.ac.cn

Abstract

Using the technique described by first author [1], a fourth-order compact finite difference algorithm for the solution of two-dimensional incompressible laminar thermal flows using the vorticity-stream function formulation was developed in [2]. In the present work, this algorithm is applied to report a numerical study of natural convection and the associated mode-transition in a two-dimensional differentially heated inclined enclosure. To check the validity of the present high-order algorithm, buoyancy-driven flows in a square cavity with the benchmark solutions [3] were solved. The angle of inclination of cavity was varied from 0° to 90°. Enclosure is assumed to fill with a Boussinesq fluid with a Prandtl number of 0.71. For Rayleigh numbers from 10^3 to 10^5, a detailed numerical analysis is made for enclosures with overall aspect ratios 4. Air-filled ($Pr = 0.71$) enclosures with aspect ratios $As = 4$ is chosen as the flow model to examine the influences of the inclination at various Rayleigh numbers. The present numerical results on coarser mesh size are compared with ones reported on finer mesh size [4]. The comparison shows quite good agreement in either maximum velocities or mean Nusselt numbers.

REFERENCES

[1] Z. Tian, Y. Ge, *A fourth-order compact finite difference scheme for the steady streamfunction-vorticity formulation of the Navier-Stokes/Boussinesq equations*, Int. J. Numer. Meth. Fluids, 41, (2003), 495-518.

[2] B. Hong, *High-order compact finite difference algorithms of Navier-Stokes/Boussinesq equations and its Application*, Master Thesis, Ningxia University, China, (2003), (in Chinese).

[3] G. De Vahl Davis, *Natural convection of air in a square cavity: a benchmark numerical solution*, Int. J. Numer. Meth. Fluids, 3, (1983), 249-264.

[4] C. Y. Soong, P. Y. Tzeng, D. C. Chiang, C. D. Hsieh, *Numerical study on mode- transition of natural convection in differentially heated inclined enclosures*, Int. J. Heat Mass Transfer, 39, (1996), 2869-2882.

COMPUTATIONAL MECHANICS (Abstracts)
WCCM VI in conjunction with APCOM'04, *Sept. 5-10, 2004, Beijing, China*

Direct Simulations of Aerodynamic Sound by the Finite Difference Lattice Boltzmann Method

Michihisa Tsutahara*, Takeshi Kataoka, Kenji Shikata

Graduate School of Science and Technology, Kobe University, 1-1 Rokkodai-machi, Nada, Kobe, 657-8501, Japan
e-mail: tutahara@mech.kobe-u.ac.jp, kataoka@ mech.kobe-u.ac.jp, shikata@mi-1.scitec.kobe-u.ac.jp

Abstract

Conventional lattice BGK model for compressible fluids are usually unstable in computations, but employing stable schemes we can make the computations stable, which is called finite difference lattice Boltzmann method. The authors present in this study some examples of direct simulations of Aerodynamic sounds, for example the Aeolian tones and edge tones.
The lattice BGK equation added by the negative viscosity term is given by

$$\frac{\partial f_i(t,\mathbf{r})}{\partial t} + c_{i\alpha}\frac{\partial f_i(t,\mathbf{r})}{\partial r_\alpha} - \frac{A c_{i\alpha}}{\phi}\frac{\partial\left(f_i - f_i^{(0)}\right)}{\partial r_\alpha} = -\frac{1}{\phi}\left[f_i(t,\mathbf{r}) - f_i^{(0)}(t,\mathbf{r})\right]$$

In this equation, f_i is the particle distribution function, $f_i^{(0)}$ on the RHS refers to its local equilibrium state, c_i is the particle velocity, and ϕ is called the relaxation time factor. The term on RHS represents the collision. The third term on the LHS is an additional term that represents negative viscosity, by which we can choose appropriate time increment for high Reynolds number flows.
We show the sound pressure fields for Aeolian tone emitted by a circular cylinder and edge tone.

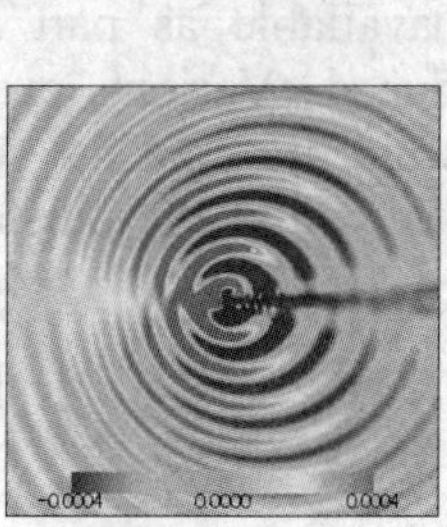

Fig.1 Aeolian tone (Re=200, Ma=0.27, γ =1.4)

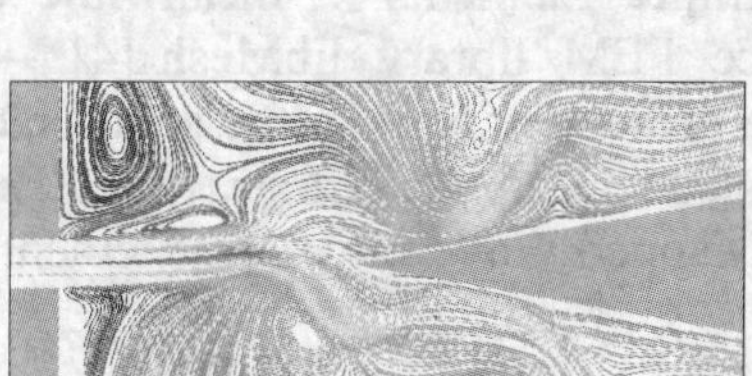

Fig.2 Edge tone (Streamlines and Sound pressure field)

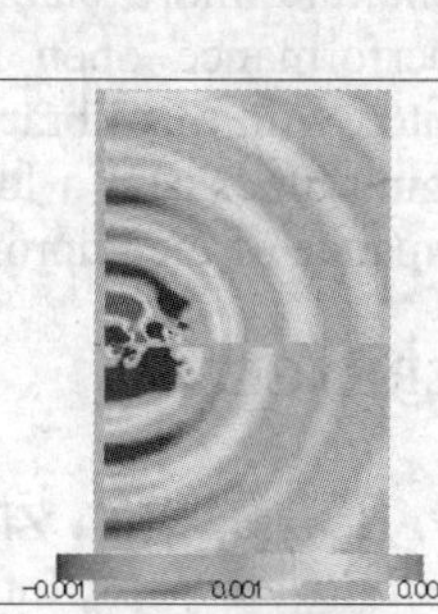

REFERENCES

[1] S. Succi, *The lattice Boltzmann equation for fluid dynamics and beyond*, Oxford, (2001), pp. 233-253.

[2] N. S. Cao, S. Chen, S. Jin, D. Martinez, *Physical symmetry and lattice symmetry in the lattice Boltzmann method*, Phys. Rev., E, 55, (1997), R21-R24.

[3] M. E. Goldstein, *Aeroacoustics*, McGraw-Hill, (1976), p.114, 157.

COMPUTATIONAL MECHANICS (Abstracts)
WCCM VI in conjunction with APCOM'04, *Sept. 5-10, 2004, Beijing, China*

An Efficient Finite Element-Infinite Element Method for Exterior Acoustics Simulation

O. von Estorff[1*], D. Dreyer[2], S. Petersen[1]

[1] *Mechanics and Ocean Engineering, Technical University Hamburg-Harburg, D-21071 Hamburg, Germany*
[2] *Audi AG, D-85045 Ingolstadt, Germany*
e-mail: estorff@tu-harburg.de

Abstract

For the simulation of exterior acoustics using the Finite Element Method (FEM), appropriate measures are needed in order to represent the radiation of acoustic energy into the far field. Among several alternatives, Infinite Elements have been proven to be a rather interesting approach. In fact, first formulations originate from the 70's already [1]. In recent years, several formulations of Infinite Elements have been developed. Especially, an optimized variant of the Mapped Wave Envelope Element concept [2] has been formulated recently by the authors [3], which is currently one of the most advanced methodologies for time-harmonic as well as selected transient acoustic simulations. The optimized formulation is based on Jacobi polynomials for the interpolation in radial (infinite) direction, enabling an efficient solution of large scaled exterior acoustic simulations using iterative solution techniques, such as Krylov subspace solvers. Before, these solvers were prone to failure due to disadvantageous eigenvalue distributions stemming from the Infinite Elements in particular.

In the present paper, the new Infinite Element formulation is summarized, critically reviewed, and its applicability to real life acoustical problems is discussed. In particular, a whole family of efficient and more stable Infinite Elements — meanwhile generally available as part of the high-performance open source FEM library libMesh [4] — is introduced. Various examples computed with this library are presented, namely the sound radiation from a tire on a flat surface, an engine block and a long, slender, submerged structure. The examples exhibit the outstanding performance of the improved Mapped Wave Envelope Elements.

REFERENCES

[1] P. Bettess, O. C. Zienkiewicz, *Diffraction and refraction of surface waves using finite and infinite elements*, Int. J. Numer. Meth. Engng, 11, (1977), 1271-1290.

[2] R. J. Astley, G. J. Macaulay, J.-P. Coyette, L. Cremers, *Three-dimensional wave-envelope elements of variable order for acoustic radiation and scattering*, *part I. formulation in the frequency domain*, Journal of the Acoustical Society of America, 103, (1998), 49-63.

[3] D. Dreyer, O. von Estorff, *Improved conditioning of infinite elements for Helmholtz problems*, Int. J. Numer. Meth. Engng, 58, (2003), 933-953.

[4] Kirk, B. S., Peterson, J. W., http://libmesh.sourceforge.net, (2004).

COMPUTATIONAL MECHANICS (Abstracts)
WCCM VI in conjunction with APCOM'04, *Sept. 5-10, 2004, Beijing, China*

Liquid/Solid Wall Slip Analysis Using Complementary Algorithm

Chengwei Wu *

State Key Laboratory of Structural Analysis for Industrial Equipment, Department of Engineering Mechanics, Dalian University of Technology, Dalian 116024, China
e-mail: cwwu@dlut.edu.cn

Abstract

The velocity boundary condition at a liquid/solid interface is an old but challenging problem for over one hundred years. The classical fluid mechanics assumes that no wall slip occurs at the liquid/solid interfaces, i.e., the so called no-slip boundary condition. The validity of this assumption has been doubted for over a century. Early experimental measurements observed that wall slips occur for mercury flowing in very narrow, fused quartz capillaries, as well as for water flowing at a glass surface treated with dimethyldichlorosilane. During the last few years, both experimental observation [1] and molecular dynamics simulation [2] found that wall slips occur not only on poorly wetting surfaces but also on wetting surfaces.
In the present paper boundary slip for a fluid flow at a smooth solid surface is numerically analyzed using a limiting shear stress model and complementary algorithm. The physical model is a smooth sphere approaching a flat plane. The wall slip velocity is controlled by the liquid/solid interface limiting shear stress, which is different from the so-called slip length model [3] where slip length controls the wall slip velocity. In our numerical methods, no iterative process is needed. The fluid pressure, boundary slip velocity and boundary shear stress are obtained at the same time. It is found that almost perfect agreements exist between the present theoretical predictions and the existing experimental observations for wall slip. It is found that the predicted hydrodynamic force in the case of wall slip is one order smaller than that predicted by the classical fluid hydrodynamic theory. This work not only provides a new model for the scientific research of wall slips occurring in both a simple flow and complex flow of fluids, but also gives a reliable numerical method for fluid hydrodynamics simulation considering wall slips. The present slip model gives a new way to search the wall slip mechanisms and let us think again the possibility of the so-called slip length model.

REFERENCES

[1] Y. X. Zhu, S. Granick, Phys. Rev. Lett., 87(9), (2001), 096105.
[2] P. A. Thompson and S. M. Troian, Nature, 389, (1997), 360.
[3] O. I. Vinogradova, Langmuir, 11, (1995), 2213.

COMPUTATIONAL MECHANICS (Abstracts)
WCCM VI in conjunction with APCOM'04, *Sept. 5-10, 2004, Beijing, China*

CFD Modeling of Gas-Liquid Flow and Mass Transfer in Rotating Packed Beds

Ming Xu *, Jianwen Zhang, Jianfeng Chen, Jin Zhao

Research Center for High Gravity Engineering and Technology, College of Chemical Engineering, Beijing University of Chemical Technology, Beijing, 100029, China
e-mail: xu_ming1979@hotmail.com, {zhangjw, chenjf}@mail.buct.edu.cn, zhao_jin1979@yahoo.com.cn

Abstract

As a novel multiphase contactor, Rotating Packed Bed (RPB) can greatly intensify the processes of micromixing and mass transfer. Recently, RPB has been successfully applied to manufacture nanomaterials by reactive precipitation and for separation in process industry [1]. There have been a number of experimental studies concerning the phenomena of fluid flow and mass transfer within RPB [2,3,4], most of which ended up in obtaining empirical correlations. In this paper, the gas flow and the dispersed liquid flow in RPB are modeled using a Eulerian-Lagrangian approach. Based on appropriate simplifications on the dispersed liquid flow and the construction of net packing, the gas flow and the trajectory of the liquid droplets in RPB are simulated by SIMPLE algorithm and particle trajectory model respectively with coalescence and breakage of the liquid droplets being considered. Upon the numerical results of the gas phase and the dispersed liquid flow, the mass transfer coefficient of liquid phase is calculated. The model results are in fair agreement with the experimental data of desorption of O_2 dissolved in water by N_2 giving an average relative error of 7.6%, which verifies the methodology and the applicability of present model in predicting the fluid flow and the mass transfer in RPB. Present numerical simulation illustrates the development of gas velocity mainly depends on geometric structure of RPB and rotating speed of packing, which is especially different from the conventional packed beds. It is depicted that the droplets in the voids of the rotating packing flow outward in different directions, which deviate from the incident direction. Further, the influences of some key operation parameters such as the rotating speed, the flow rates of the liquid phase and the gas phase as well as the inner diameter of the packing on the mass transfer coefficient are also discussed. It is found that the coefficient of mass transfer increases greatly as the flow rate of liquid and the inner diameter of packing rises. According to the model, the intensification on the mass transfer is primarily contributed to the liquid atomization by the rotating packing.

REFERENCES

[1] J. F. Chen, Y. H. Wang, F. Guo, X. M. Wang, C. Zheng, *Synthesis of nanoparticles with novel technology : high-gravity reactive precipitation*, Ind. Eng. Chem. Res., 39, (2000), 948-954.

[2] J. R. Burns, C. Ramshaw, *Process intensification: Visual study of liquid maldistribution in rotating packed beds*, Chem. Eng. Sci., 51, (1996), 1347-1352.

[3] F. Guo, C. Zheng, K. Guo, Y. D. Feng, N. C. Gardner, *Hydrodynamics and mass transfer in cross-flow in rotating packed bed*, Chem. Eng. Sci., 52, (1997), 3853-3859.

[4] K. Guo, F. Guo, Y. D. Feng, J. F. Chen, C. Zheng, *Synchronous visual and RTD study on liquid flow in rotating packed-bed contactor*, Chem. Eng. Sci., 55(2000), 1699-1706.

COMPUTATIONAL MECHANICS (Abstracts)
WCCM VI in conjunction with APCOM'04, *Sept. 5-10, 2004, Beijing, China*

A Boundary Condition for Numerical Simulation of Multi-body Movement in Viscous Incompressible Flow

J. Yu [1,2*], Y. Y. Nie [1,2]

[1] *Shenyang Institute of Automation, Academia Sinica, Nanta Street 114, Shenyang, 110016, China*
[2] *School of Graduate, Academia Sinica, Beijing, 10039, China*
e-mail: yujun@sia.cn, yynie@sia.cn

Abstract

An accurate, efficient method is developed for simulating 2D unsteady viscous incompressible flows around multiple moving rigid objects in another paper. This method employs dynamical difference mesh to solve the Navier-Stokes (N-S) equations written in terms of primitive variabl e. There are two problems to solve: one is that the boundary conditions on given curves are imposed to computational grid points which are not coinciding with the boundary geometry; the other is that some grid points' initial variables of every time step n ear the objects boundary are required. In [3], a ghost cell method was used to enforce immersed boundary condition in complex geometry, which can treat both Dirichlet and Neumann boundary conditions while preserving the overall second-order accuracy of the base solver. In this paper, this ghost cell method is extended to the dynamic mesh for dealing with the internal moving boundaries. Through these approaches, the boundary conditions on an arbitrary curve are distributed to regular difference mesh points, and the initial variables can be provide d to the points changing fr om rigid objects domain to fluid domain following time. The flow equations are solved on a difference mesh. Thus the standard N -S equations' solvers (for example, SIMPLE method) can be empl oyed through a few modifications. So the advantages and efficiency of regular solvers are retained. The method is validated using flow past a cylinder. The flows around two cylinders which are moving relative to each other are calculated.

REFERENCES

[1] E. A. Fadlun, R. Verzicco, P. Orlandi, J. Mohd-Yusof, *Combined immersed-boundary finite-difference methods for three-dimensional complex flow simulations*, J. Comput. Phys, 161, (2000), 35-60.

[2] David Russell, Z. Jane Wang, *A cartesian grid method for modeling mu ltiple moving objects in 2D incompressible viscous flow*, J. Comput. Phys, 191, (2003), 177-205.

[3] Yu-Heng Tseng, Joel H. Ferziger, *A ghost-cell immersed boundary method for flow in complex geometry*, J. Comput. Phys, 192, (2003), 593-623.

[4] R. P. Fedkiw, T. Aslam, B. Merriman, S. Osher, *A non-oscillatory Eulerian approach to interfaces in multimaterial flows (the ghost fluid method)*, J. Comput. Phys, 152, (1999), 457-492.

[5] Wenqun Tao, *Numerical heat transfer*, Xi'an Jiaotong University Press, (2001).

COMPUTATIONAL MECHANICS (Abstracts)
WCCM VI in conjunction with APCOM'04, *Sept. 5-10, 2004, Beijing, China*

A Finite Volume Scheme for Incompressible Fluids

Sebastien Zimmermann *

Department of Applied Mathematics, University of Centrale Lyon, 36 avenue Guy de Collongues, Ecully Cedex, 69134, France
e-mail: SebastienZimmermann69@hotmail.com

Abstract

Incompressible fluids can be found in numerous real -life situations, including meteorology problems for example.
In many situations, it is useful to have a scheme that allows controlling the numerical diffusion introduced. Also, in case of discontinuous features in the flow, it may be useful to have a numerical method that is able to track them.
Finite volume methods combine these two features. An example of computations done with such a method can be found in [1].
In this talk, we will discuss the numerical analysis of one of these methods with techniques introduced in [2] and [3]. The method uses a Chorin-Temam fractional-step method for the computation of pressure. Depending on the discrete operators used, different regularity must be assumed on the mesh. A few numerical results will then validate the method.

REFERENCES

[1] D. Kim, H. Choi, *A second-order time-accurate finite volume method for unsteady incompressible flow on hybrid unstructured grids,* J. Comp. Phys., 162, (2000), 411-428.
[2] R. Eymard, T. Gallouet, R. Herbin, *The finite volume method,* in Ph. Ciarlet J. L. Lions eds. *Handbook for Numerical Analysis,* North Holland, (2000).
[3] R. Temam, *Navier-Stokes equations: theory and numerical analysis,* North Holland, (1984).

COMPUTATIONAL MECHANICS (Abstracts)
WCCM VI in conjunction with APCOM'04, *Sept. 5-10, 2004, Beijing, China*

Cartesian Grid Method for Free Surface Flows with Arbitrary Immersed Boundaries

M. H. Chung *

Sinotech Engineering Consultants Inc., Taipei, Taiwan, China
e-mail: mhsuan@sinotech.org.tw

Abstract

The major advantages of the Cartesian grid method over the body-fitted grid method include the simplicity of grid generation and neat discretization equations. Particularly, a prominent advantage appears when the immersed boundary is moving arbitrarily. In this paper, a Cartesian grid method has been developed to simulate two dimensional unsteady viscous incompressible free surface flows with arbitrary immersed boundaries. The level-set approach is adopted to capture the free surface motion and the cut cell approach to treat complex geometry.

A collocated finite volume method with nominally second-order accurate schemes in space and time is used for discretization. The solution node is defined as the cell center located in the fluid region. The cut cells whose centers lie outside the boundary must be merged with neighbouring cut or non-cut cells with solution nodes to ensure global mass/momentum conservation and to avoid exceptionally small size of control volume which causes impractical time step interval. First, the intersection points between the immersed boundary and the cell faces are located and recorded. The center, length, and normal of the immersed boundary segment in each cut cell can then be derived from these information. The no-slip condition is implemented directly by assigning a prescribed velocity to the segment center. Along the normal line emanating from the segment center, two appropriately separated points are established in order. The solution variables at these points are bilinearly interpolated from the neighbouring solution nodes. Then the pressure at the segment center is extrapolated from these two points and the normal diffusive flux obtained by the finite difference between the first point and the segment center. Second, the values and fluxes on the cell faces near the boundary segment are approximated by the polynomial fit to the neighbouring solution nodes or segment centers. The interpolation stencil is comprised of 4 and 6 solution nodes (or segment centers) for cut and non-cut cell faces respectively. The whole procedure is similar to that used by [1]. The difference between the two includes: 1) in the present work, all the discretization terms related to the irregular geometry, except the time-derivative term, are treated explicitly for the sake of simplicity and not deviating severely the coefficient matrix of the discretized equations; 2) the solution node stencil used for polynomial interpolation is simpler and more systematic while the spatial accuracy remains second-order in the present approach.

REFERENCES

[1] T. Ye, R. Mittal, H. S. Udaykumar, W. Shyy, *An accurate Cartesian grid method for viscous incompressible flows with complex immersed boundaries,* Journal of Computational Physics, 156, (1999), 209-240.

COMPUTATIONAL MECHANICS (Abstracts)
WCCM VI in conjunction with APCOM'04, *Sept. 5-10, 2004, Beijing, China*

Assessment of Turbulence Model for Ship Flow at Full Scale

G. B. Deng *, R. Duvigneau, P. Queutey, M. Visonneau

Laboratoire de Mécanique des Fluides, Ecole Centrale de Nantes,44321,France
e-mail: Ganbo.deng@ec-nantes.fr

Abstract

Considerable progresses have been made during recent years in viscous flows computation around a ship. Due to careful refinement of turbulence models validated by detailed experimental measurement as well as remarkable progress made in numerical algorithm and computer hardware, CFD tools for ship flow applications have acquired a certain maturity. It is generally accepted that turbulence modelization is of crucial importance in numerical simulation of an aft -body flow. Reliable prediction can be obtained when advanced turbulence models such as Reynolds stress transport model is employed [1]. However, nearly all experiments and numerical validations have been performed at model scale. The reliability of the CFD tools at full scale is unclear due to the fact that few measurements have been performed at full scale, and to author's knowledge, no systematical comparison between numerical simulation and measurement for viscous flow at full scale has been conducted. To fill this gap, an European cooperative project known as EFFORT (European Full -scale Flow Research and Technology) has been carried out since 2002 [2]. Part of the work pe rformed in this project devoted to turbulence model assessment is presented in this paper. Free surface is not taken into account in the present study. A double model is employed. Numerical computations have been performed for the *Navigator* at full scale at a constant advance speed of 13 knots giving a Reynolds number of about 3.10^8 to evaluate the robustness and the performance of different turbulence models. Three different models have been selected, namely a Reynolds stress transport model, an explicit algebraic stress model, and the k-ω SST model. All of them use low Reynolds number formulation. The EASM model includes a rotation correction. Turbulence modelization has been found to play an important role in flow prediction. Due to its capability to repr esent normal stress anisotropy as well as transport effect, the Reynolds stress model predicts a much stronger bilge vortex than the SST model. The EASM model also improves the prediction of the bilge vortex, although the intensity of the vortex is less st rong compared with the results obtained with the Reynolds stress model. The EASM model may be a good alternative to the Reynolds stress model which is much more expensive and sometimes unstable for ship flow computation. The robustness of different models is confirmed in a computation including the rudder with an unstructured grid containing about 4 millions grid nodes. The reliability of numerical computation using wall function at full scale has also been addressed. Computation performed with the SST mode l shows that satisfactory results can be obtained as long as the first grid points are located inside the log law region, that is, below $y^+=1000$ for the Navigator at full scale.

REFERENCES

[1] L. Larsson, F. Stern, V. Bertram, *Benchmarking of Computational Fluid Dynamics for Ship Flows*, The Gothenburg 2000 Workshop, J. of Ship Research, 47, (2003), 63-81.

[2] http://www.marin.nl/JIP.asp

COMPUTATIONAL MECHANICS (Abstracts)
WCCM VI in conjunction with APCOM'04, *Sept. 5-10, 2004, Beijing, China*

Collision in an Incompressible Fluid

E. Dimnet*

Laboratoire Central des Ponts et Chaussées, 58, boulevard Lefèbre, 75732 Paris cedex 15, France
e-mail: dimnet@lcpc.fr

Abstract

When a collision occurs between deformable solids and/or fluids, in a first step, very large forces and stresses are active during a very short time, in order to make the diffe rent velocities compatible. After this, the collision may still last (the colliding bodies keep in contact) but the evolution is smoother and can be modelised by regular equations.

The first step, called instantaneous collision, is described by an instant aneous discontinuity of the velocities, and the efforts are concentrated in time and modelised by percussions. Thus, for an incompressible fluid, the equations of an instantaneous collision are

$$\rho(V^+ - V^-) = div\Sigma, \quad \Sigma = 2\gamma D\left(\frac{V^+ + V^-}{2}\right) - pI, \quad div\left(\frac{V^+ + V^-}{2}\right) = 0.$$

V^+ is the right and V^- the left limit of the velocity V, Σ is the percussion stress tensor and p the Lagrange multiplier associated to the incompressibility constrain $div\left[\left(V^+ + V^-\right)/2\right] = 0$.

This theory is exposed in the paper and examples of numerical simulations are presented.

REFERENCES

[1] E. Bonetti, M. Frémond, *Collisions and Fractures*, Proc. 4th ECCOMAS Conf., Jyväskylä, Finland, 24 - 28 July, (2004).

[2] E. Dimnet, *Instantaneous collisions in incompressible fluids: theory and applications*, Proc. 4th ECCOMAS Conf., Jyväskylä, Finland, 24 - 28 July (2004).

COMPUTATIONAL MECHANICS (Abstracts)
WCCM VI in conjunction with APCOM'04, *Sept. 5-10, 2004, Beijing, China*

Numerical Simulation of Unstationary Separation Process for the Space Transport System

L. Jiang[*], C. Breitsamter, B. Laschka

Mechanical Department, Technical University of Munich, Boltzmannstr. 15, Garching, 85748, Germany
e-mail: jiang_lei@hotmail.com, Breitsamter@flm.mw.tum.de, Laschka@flm.mw.tum.de

Abstract

To depict the separation process of an idealized two-stage vehicle for the orbit transport system, pitching movement, translation and the manoeuvre shown in Fig. 1 with different reduced frequencies k_{red} (Fig. 2) as well as different angles of attack are numerically simulated. Here a detailed 3-D orbital stage is used while the carrier stage is idealized by a flat plate.
The software FLMNAV is suitable for simulating the Euler and Navier-Stokes equations (laminar flow) about the flow physics associated with the separation process [1-2]. For the orbital stage in the beginning of the separation manoeuvre, the lift- and pitching moment coefficients may be negative. The drag coefficient may increases with increasing the relative distance from the carrier stage. This has to be carefully considered in the flight control system.

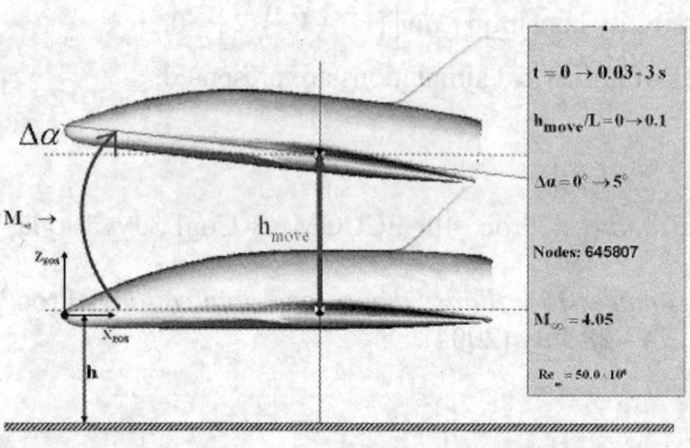

Fig. 1 Sketch of the unsteady separation process (L: length of orbital stage)

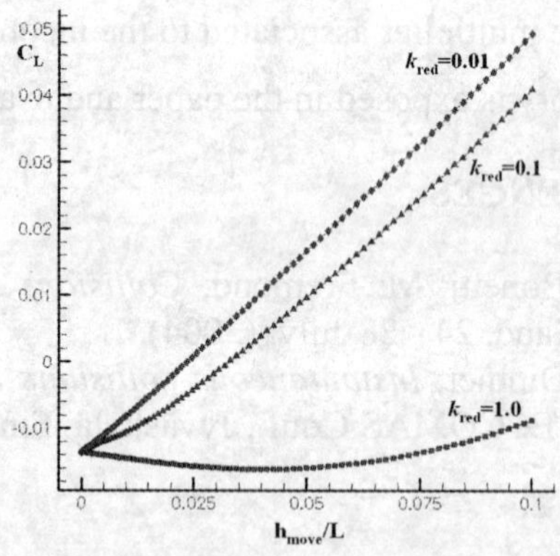

Fig. 2 Lift coefficients C_L as function of h_{move}/L for unsteady separation process

REFERENCES

[1] L. Jiang, C. Breitsamter, B. Laschka, *Numerical simulation of the stage separation flow physics of an idealized two stage space transport vehicle*, Jahrbuch der Deutschen Gessellschaft für Luft- und Raumfahrt–Lilienthal-Oberth e.V., DGLR-JT2003-131, Deutscher Luft- und Raumfahrtkongreß / DGLR- Jahrestagung, München, Nov. 17-20, (2003).

[2] T. Cvrlje, C. Breitsamter, C. Weishäupl, B. Laschka, *Euler and Navier-Stokes simulations of two-stage hypersonic vehicle longitudinal motions*, Journal of Spacecraft and Rockets, 37(2), (2000), 242-251.

COMPUTATIONAL MECHANICS (Abstracts)
WCCM VI in conjunction with APCOM'04, *Sept. 5-10, 2004, Beijing, China*

Advances in the Flow-Condition-Based Interpolation Procedure for Advection-Diffusion Problems

H. Kohno*, K.J. Bathe

Department of Mechanical Engineering, Massachusetts Institute of Technology, 77 Massachusetts Avenue, Cambridge, MA, 02139-4307, USA
e-mail: kohno@mit.edu, kjb@mit.edu

Abstract

The flow-condition-based interpolation (FCBI) finite element procedure is improved for the solution of two-dimensional advection-diffusion problems. Novel numerical schemes are developed by introducing, in one scheme, the link-cutting bubbles [1] and, in another scheme, the general solution of the advection-diffusion equation into the trial functions for the advection term. In the proposed algorithms, as in the conventional FCBI procedure [2,3], no artificial parameters are included to reach stability for high Peclet number flow. The governing equation is discretized by employing a Petrov-Galerkin formulation, in which step functions are used as weight functions to satisfy local flux conservation in selected control volumes of the finite element discretization. Similarities and relations among the FCBI procedure, the link-cutting bubbles and the streamline-upwind Petrov-Galerkin (SUPG) formulations [4] are established considering the solution of the one-dimensional advection-diffusion problem. Then, the proposed numerical schemes are applied to two-dimensional thermal flows over a square domain. Various problems are considered with various mesh topologies. In this way the capabilities and the effectiveness of the new methods are identified and demonstrated for a wide range of Peclet numbers. The research accomplishments will be published in a full paper [5].

REFERENCES

[1] F. Brezzi, G. Hauke, L.D. Marini, G. Sangalli, *Link-cutting bubbles for the stabilization of convection-diffusion-reaction problems*, Math. Models Meth. Appl. Sci., 13, (2003), 445-461.

[2] K. J. Bathe, H. Zhang, *A flow-condition-based interpolation finite element procedure for incompressible fluid flows*, Comput. Struc., 80, (2002), 1267-1277.

[3] K. J. Bathe, H. Zhang, *Finite element developments for general fluid flows with structural interactions*, Int. J. Numer. Meth. Engng., (in press).

[4] A. N. Brooks, T. J. R. Hughes, *Streamline upwind/Petrov-Galerkin formulations for convection dominated flows with particular emphasis on the incompressible Navier-Stokes equations*, Comput. Meth. Appl. Mech. Engng., 32, (1982), 199-259.

[5] H. Kohno, K. J. Bathe, *Advances in the flow-condition-based interpolation procedure for advection-diffusion problems*, (in preparation).

COMPUTATIONAL MECHANICS (Abstracts)
WCCM VI in conjunction with APCOM'04, *Sept. 5-10, 2004, Beijing, China*

Effect of RANS Turbulence Models on Computation of Separated Flows over a Wing-Body Configuration

Zhong Lei *

Advanced Aircraft Technology Center, Japan Aerospace Exploration Agency (JAXA), 7-44-1 Jindaiji-higashi, Chofu, Tokyo 182-8522, Japan
e-mail: lei@chofu.jaxa.jp

Abstract

The objective of this paper is to investigate the effect of turbulence mo dels on the numerical simulation of the separated flow over a wing-body configuration at low speed and high angles of attack. At high angles of attack, the aerodynamics of a typical delta wing-body configuration is dominated by the roll-up structure of the concentrated vortices, which are separated from the body, the outboard/inboard leading-edge and the wing tip. This kind of flow is so complex that numerical simulation is very difficult in predicting separation, reattachment, vortex behaviors, and so on. Computational results may be strongly dependent on the turbulence models used.

In this study, Reynolds-averaged Navier-Stokes (RANS) equations were solved by finite difference method. A diagonalized ADI method was applied for implicit time integration. The third order scheme of Chakravarthy-Osher's TVD was used for explicit convection terms, and the second order central differential scheme was used for viscous terms. Turbulence models are employed to simulate turbulence flow and include three popular turbulence models, Spalart-Allmaras model, Menter's shear stress transport model and Launder-Sharma $k-\varepsilon$ model. Computational results are compared with experiments. The Spalart-Allmaras model shows too large dissipation in the vortical flow away from the surface. The Menter model predicts reasonable agreement with the experiment in details and capture vortex structures well. The Launder-Sharma $k-\varepsilon$ performs as well as the Menter model does in most part of the flowfield at high angles of attack. Although details of the flowfield may have large differences with different turbulence models, the choice of turbulence models is not important for predicting aerodynamic forces.

REFERENCES

[1] D. Kwak, K. Miyata, M. Noguchi, Y. Sunada, K. Rinoie, *Experimental investigation of high lift device for SST*, NAL TR-1450, Japan Aerospace Exploration Agency, (2002).

[2] P. R. Spalart, S. R., Allmaras, *A one-equation turbulence model aerodynamic flows*, AIAA paper 92-0439, (1992).

[3] F. R. Menter, *Zonal two equation $k-\omega$ turbulence models for aerodynamic flows*, AIAA paper 93-2906, (1993).

[4] B. E. Launder, B. I. Sharma, *Application of the energy dissipation model of turbulence to the calculation of flow near a spinning disc*, Letters in Heat and Mass Transfer, 1, (1974), 131-138.

COMPUTATIONAL MECHANICS (Abstracts)
WCCM VI in conjunction with APCOM'04, *Sept. 5-10, 2004, Beijing, China*

Numerical Simulation on Aerodynamic Interference Fields of Parallel Twin Blades at Low Velocity

X. H. Li, J. J. Xiong *

Aircraft Department, Beihang University, Beijing, 100083, China
e-mail: lixiaohui_2003@tom.com, xiongjj2003@yahoo.com.cn

Abstract

This paper seeks to explore the influences of level interval, attack angle and Mach number on aerodynamic interference fields of parallel twin blades. Based on the NACA0012 aerofoil, a physical model is established for the aerodynamic computation of level parallel twin blades. Using the structured C-shape, partition mixed and unstructured triangle meshes, the mesh models are produced for aerodynamic computation of single-blade and twin blades respectively. Then the aerodynamic interference fields of between twin blades are numerically simulated by means of the CFD methodology at two kinds of Mach numbers in low velocity range, three kinds of attack angles and eleven kinds of level intervals respectively. The pressure distributions and lift and drag coefficients of wing surfaces are also calculated. Finally, contrast analysis is conducted between the simulated flow fields of single-blade and twin blades in order to examine the presented simulation procedure, demonstrating the practical and economic use of the proposed technique. It is clear from the results that the variation trends of lift and drag forces on twin blades at different attack angles are consistent with that on single-blade. The lift coefficients increase with increasing of attack angle, and then decrease after stalled attack angle of aerofoil, while the drag coefficients always increase with increasing of attack angle. In the case of the attack angles of 4° and 10°, the interference effect between twin blades becomes severer and severer with increasing of attack angle before stall occurs on the upper parting surface, while in the case of the attack angle of 17°, it has a severe influence on the lift forces of both blades only after stall occurs on the upper parting surface of front wing.

REFERENCES

[1] N. Ng, R. Hillier, *Numerical investigation of the transonic blade-vortex interaction,* AIAA Paper, The 28th AIAA Fluid Dynamics Conference, Snowmass Village, CO, (1997), pp. 97-1846.

[2] G. R. Srinivasan, W. J. McCroskey, *Navier-Stokes calculations of hovering rotor flowfields,* Journal of Aircraft, 25(10), (1998), 865-874.

[3] Du Yunle, Lin Guohua, et al, *On numerical simulation for twin blades interaction,* Journal of Aerodynamics, 2003, (accept), (in Chinese).

COMPUTATIONAL MECHANICS (Abstracts)
WCCM VI in conjunction with APCOM'04, *Sept. 5-10, 2004, Beijing, China*

The Explicit and Implicit Finite Difference Approximations for Space Fractional Advection-Dispersion Equation

Xuanzhu Lu [1*], Fawang Liu [2, 3]

[1] *School of Mathematical and Computer Sciences, Fuzhou University, Fuzhou 350002, China*
[2] *School of Mathematical Sciences, Xiamen University, Xiamen 361005, China*
[3] *School of Mathematical Sciences, Queensland University of Technology, GPO Box 2434, Brisbane, Qld. 4001, Australia*
e-mail: lxz541226@sina.com, fwliu@xmu.cdu.cn

Abstract

Fractional derivatives have recently been applied to many problems in physics, finance, and hydrology [1-6]. Fractional derivatives play a key role in modeling particle transport in anomalous diffusion or dispersion, where a particle plume spreads at a rate inconsistent with the classical Brownian motion model. When a fractional derivative replaces the second derivative in a diffusion or dispersion model, it leads to enhanced diffusion [1]. A fractional diffusion equation is obtained from generalized Fick's laws which account for spatial and/or temporal non-locality [2]. In practical applications, it is important to establish suitable numerical methods for solving these equations. However, numerical methods and theoretical analysis of fractional differential equations are very difficult tasks [3]. Recently, the numerical methods have be considered by several authors for fractional diffusion equations [1-2]. Fractional advection-dispersion equations are used in groundwater hydrology to model the transport of passive tracers carried by fluid in a porous medium [1, 5, 6]. In this paper, a space fractional advection-dispersion mixed problem is considered in a bounded space domain. The equation obtained is from the standard advection-dispersion equation with the second order derivative in space replaced by a fractional derivative in space of order α $(1<\alpha\leq 2)$. The explicit and implicit finite difference approximations for a space fractional advection-dispersion equation with variable coefficients are presented, and its discretization error is estimated. The stability and convergence of the finite difference approximations are proven.

REFERENCES

[1] Mark M. Meerschaert, Charles Tadjeran, *Finite difference approximations for fractional advection-dispersion flow equations*, J. Comp. and Appl. Math., (2004), (in press).

[2] R. Gorenflo, F. Mainardi, D. Moretti et al, *Discrete random walk models for space-time fractional diffusion*, Chemical Physics, 284, (2002), 521-541.

[3] F. Liu, V. Anh, I. Turner, *Numerical solution of space fractional Fokker-Planck equation*, J. Comp. and Appl. Math., 166, (2004), 209-219.

[4] I. Podlubny, *Fractional Differential Equations*, Academic Press, New York, (1999).

[5] Rina Schumer, David A. Benson, et al, *Eulerian derivation of the fractional advection-dispersion equation*, Journal of Contaminant, 48, (2001), 69-88.

[6] D. A. Benson, *The fractional advection-dispersion equation*, Ph. D. Thesis, University of Nevada, Reno, USA, (1998).

COMPUTATIONAL MECHANICS (Abstracts)
WCCM VI in conjunction with APCOM'04, *Sept. 5-10, 2004, Beijing, China*

Moving Least Square Approximation in the Fitting of Characteristics Curves of Reversible Pump Turbine

W. Y. Shao *

Department of Civil Engineering, Zhejiang University, Hangzhou, 310027, China
e-mail: shaowy@zju.edu.cn

Abstract

The complete characteristic curves of reversible pump turbine, which plays a key role in the transient simulation in pumped storage plant because the serious safety problem in the plants are usually induced by the overpressure in water conveyance system after a full-load rejection during abnormal conditions, has been approximated by variety of approaches such as linear interpolation and least square surface fitting [1]. However, the least square surface fitting is mesh-dependence which requires regular distribution of the data that the scattered testing data of complete characteristic curves can not meet. Therefore, the mesh free method, moving least square (MLS) approximation, is introduced in this paper to approximate the curves transferred from the new transformation method [2] for the first time. Then, the effects of different basis functions and weight functions, weight function parameters and scaling parameters on the error norms of MLS approximation of the characteristic curves are discussed one by one while the total number of nodes within the domain of influence of the fixed node is initially fixed and the radius of support domain are adjusted all the time to satisfy the requirement. Finally, based on the comparison of the fitting accuracy of the characteristic curves, quadratic basis function and Gaussian weight function surpass other functions and are chosen for MLS approximation of the curves. Gaussian weight function parameter, which determines the shape of the weight function, and scaling parameter, which determines the influence domain of the weight function, are recommended to be within the range of 3.25 to 4.25 and the range of 1.0 to 1.5 respectively, because the overall error norm [3] are achieved their minimum value within those ranges.

REFERENCES

[1] L. S. Wang, L. S. Suo, D. Y. Liu, *A new transformation method for the complete characteristic of reversible pump turbines*, J. of Hydroelectric Eng., (3), (2003), 68-74.
[2] W. Y. Shao, G. H. Mao, G. H. Liu, *New transformation method for the complete characteristic curves of reversible pump turbine based on surface fitting*, J. of Zhejiang University (Science), (2), (2004), 40- 44.
[3] Y. X. Mukherjee, S. Mukherjee, *On boundary condition in the element-free Galerkin method*, Comp. Mech., 19, (1997), 264-270.

COMPUTATIONAL MECHANICS (Abstracts)
WCCM VI in conjunction with APCOM'04, *Sept. 5-10, 2004, Beijing, China*

Parallel Lattice BGK Simulations of Three-Dimensional Lid-Driven Cavity Flow

Baochang Shi [*], Haifeng Han

Department of Mathematics, Huazhong University of Science and Technology, Wuhan, 430074, China
e-mail: {sbchust0382, haifenghan}@sina.com

Abstract

Over the last two decades, the lattice Boltzmann method (LBM) has emerged as a powerful technique for computational fluid dynamics (CFD). It is a relatively new approach that uses simple microscopic kinetic models to simulate complicated macroscopic behaviour of transport phenomena. Compared with conventional CFD approach based on discretizations of macroscopic continuum equations, the LBM algorithm is simple, fast and intrinsically parallel. It is also easy to approximate fluid properties at the micro-kinetic scale and can handle complex boundary conditions in a relatively straightforward way. In recent years, the LBM has achieved great success in simulations of fluid flows and modeling physics in fluids [1]. However, the LBM is still computationally very expensive when simulating real systems. As a result, parallel computing is a prerequisite for most LB simulations. Because of its parallelism in nature the LBM is an ideal candidate to take advantage of parallel computing resources. The most popular LB model is the so-called lattice BGK (LBGK) model.

Recently, we developed an efficient incompressible LBGK model [2] and simulated the three-dimensional lid-driven cavity flow for *Re*=3200 and span aspect ratio SAR=1, 2 and 3 on $48\times48\times(48\times\mathrm{SAR})$ uniform lattice in detail by using the model on a personal computer. The test results agree well with those in precious works. This model can simulate the one- to three-dimensional incompressible flow at larger *Re* on coarser lattice accurately and efficiently. In this work, we implemented the parallel version of the model using the Message Passing Interface (MPI). We simulated the same problem in [2] at a higher resolution on our high performance cluster Lenovo DeepComp 1800 by using the model. With the grid refinement, much more accurate results were yielded, and some new phenomena were found, one of which was predicted in [2].

REFERENCES

[1] S. Y. Chen, G. D. Doolen, *Lattice Boltzmann method for fluid flows*, Ann. Rev. Fluid Mech., 30, (1998), 329-364.

[2] N. Z. He, N. C. Wang, B. C. Shi, Z. L. Guo, *A unified incompressible lattice BGK model and its application to three-dimensional lid-driven cavity flow*, Chin Phys, 13(1), (2004), 40-46.

COMPUTATIONAL MECHANICS (Abstracts)
WCCM VI in conjunction with APCOM'04, *Sept. 5-10, 2004, Beijing, China*

Efficient Surface Grid Generation Based on Remeshing

D.Wang*, O. Hassan, K.Morgan, N.P.Weatherill

Department of Computational and Civil Engineering, University of Wales Swansea, Singleton Park, Swansea SA2 8PP, U.K.
e-mail: {desheng.wang, o.hassan, k.morgan, n.p.weatherill}@swansea.ac.uk

Abstract

Surface mesh generation is still one of the most challenging and difficult prerequisites for various numerical simulations, and especially for large-scale volumetric mesh generation. Existing approaches can be essentially categorized into two kinds: direct and indirect [1]. In the direct approach, mesh is generated directly on the surface in 3D such as the Octree-based method and the Advancing Front method. The indirect approach relies on the generation in the parametric domain using Advancing Front method, Delaunay-based technique, or any 2D mesh generation technique [1]. The mesh is then mapped onto the surface using the reverse mapping. The direct methods obviously have the difficulty in mesh validity checking, while the indirect methods are facing the trouble to control the sizing and shapes of the elements to be generated in the 2D region. Accordingly, it is still very time-consuming for generating large-scale surface meshes of sufficient quality.

In this paper a new surface mesh generation method is described for efficient automatic generation of high-quality surface grids based on remeshing technique [2]. Supported geometry import formats include composite parametric surface with analytic definition or discrete composite parametric Ferguson curve and surfaces. Initially, for each independent surface patch, a boundary triangulation is generated in the parametric domain, and those patch meshes are mapped onto the surface and assembled into the beginning surface mesh conformingly. Next, remeshing techniques including edges splitting and contraction are utilized for the refinement or coarsening of this initial surface mesh. Metric controlled edge length is computed with respect to surface curvature or coupled with user-specified spacing, both for isotropic or anisotropic cases. Grid points are added accurately through edges splitting using the surface definition; edges contractions are performed robustly with help of determinants computation of triangles in the parametric domain related to the contracted edge. After the refinement, the surface mesh is optimized through edges swapping and points relocation. All the operations are locally performed and high efficiency is resulted. Above one million elements with high quality can be generated in less than one minute in a PC with reasonable CPU speed. Furthermore, smal l-angle elements due to neighboring patches interface can be avoided by the remeshing techniques naturally. Also, the existing method can be used directly for remeshing surfaces with SGL, STL, IGES formats as geometry input. At the end of the paper, various complex geometry configurations, such as fighter-type aircraft, human organs such as heart and kidney are tested as examples to demonstrate the capabilities of this approach.

REFERENCES

[1] J. F. Thompson, B. K. Soni, N. P. Weatherill, *Handbook of grid generation*, CRC Press LLC, (1999).
[2] P. J. Frey, *About surface remeshing*, 9th International Meshing Roundtable, (2000), 123-156.

COMPUTATIONAL MECHANICS (Abstracts)
WCCM VI in conjunction with APCOM'04, *Sept. 5-10, 2004, Beijing, China*

A General Method for Mesh Generation from Medical Images

D. Wang, O. Hassan*, K. Morgan, N. P. Weatherill

Department of Computational and Civil Engineering, University of Wales Swansea, Singleton Park, Swansea SA2 8PP, U.K.
e-mail: {desheng.wang, o.hassan, k.morgan, n.p.weatherill}@swansea.ac.uk

Abstract

The use of numerical methods for biomedical simulation is becoming increasingly popular. It provides promising ways for understanding, predicting and providing control of biomechanic features of human body and organs. The key to a successful interactive computation and prediction for patient-specific surgical analysis and planning is a rapid geometry modeling and grid generation from medical images which are generated through CT, MRI or other methods. The existing approaches for this end consists of several parts: image processing for cross sectional contours extraction, surface reconstruction from the sectional contours and grid generation for the constructed surface model . One kind of methods generate three dimensional meshes between any two neighboring sectional slices[1] with several meshing operators. This obviously lacks the flexibility to treat the issues of curvature-based surface meshing and good model approximation. Furthermore, if the number of slices is too large, the resulted elements number is formidable. Another category generates meshes applying existing mature grid generation tools[4], but also lacks generality and reasonable efficiency for practical usage.

In this paper, we propose a general method for automatic mesh generation from medical images, which is modular-oriented, systematic and efficient enough for prompt interactive simulation. For cross sectional contours extraction, we either use various image processing methods or directly apply existing mature commercial softwares; for surface reconstruction or modeling, we first apply a modified version of the robust and efficient Delaunay-based method[3] to obtain an initial surface triangulation, then surface remeshing including edges splitting , contractions and optimizations are performed to improve the surface mesh with respect to both elements sizing, shape and geometry smoothness[2]; finally, for the volumetric meshing, the improved Delaunay tetrahedral mesh generator for any complex geometry is used, with robust boundary recovery, efficient field points generation and insertion, and effective optimizations. Using the developed software, the whole process from receiving the medical images to the generation of a CFD mesh, takes less than one hour. Various examples are presented in the paper for demonstrating the effectiveness of the method.

REFERENCES

[1] G. Kwon, S. Chae, K. Lee, *Automatic generation of tetrahedral meshes from medical images*, Computers and Structures, 81, (2003), 765-775.

[2] P. J. Frey, *About surface remeshing*, 9th International Meshing Roundtable, (2000), 123-156.

[3] J. D. Boissonnat, *Shape reconstruction from planar cross sections*, Comput Vision, Graph Image Proc, 44(1), (1988), 1-29.

[4] J. R. Cebaral, R. Lohner, *From medical images to CFD meshes*, 8th International Meshing Roundtable, (1999).

COMPUTATIONAL MECHANICS (Abstracts)
WCCM VI in conjunction with APCOM'04, *Sept. 5-10, 2004, Beijing, China*

Surface Grid Generation for Complex Geometry with 'Dirty' Boundary Representation

D. Wang, O. Hassan, K. Morgan*, N. P. Weatherill

Department of Computational and Civil Engineering, University of Wales Swansea, Singleton Park, Swansea SA2 8PP, U.K.
e-mail: {desheng.wang, o.hassan, k.morgan, n.p.weatherill}@swansea.ac.uk

Abstract

In computational fluid dynamics unstructured grids are widely used and its popularity is due to the flexibility in tackling complicated geometries and adapting the computational grids in line with the flow features. Traditional volumetric meshing approaches such as AFT, Delaunay-based methods require a 'water-tight' boundary surface grid as an input. If a complex geometry with 'non-water-tight' boundary representation (BREP) which is composed of surface triangles is given, the grid generation can become very time consuming. The repair of the geometry such as cleaning of cracks, overlaps or invalid manifolds is time consuming and in many cases may not be practical. A Cartesian grid generation approach can be utilized [3] for this kind of 'Dirty' geometry. However the proposed method does not provide high quality grids near intersection curves and such curves can be completely omitted in some cases.

In this paper a new surface mesh generation method is described for this 'Dirty' geometry. Initially, an adaptive Cartesian grid is generated for the approximation of the geometry boundary. Projections of the points forming the staircase which surrounds the boundary onto the closest boundary surface is then performed. Next, we perform constrained refinement to capture the intersection curves and geometric features, and a water-tight coarse surface triangulation is resulted. Remeshing techniques including edges splitting and contraction are utilized for the refinement or coarsening of this initial surface mesh [2]. Metric controlled edge length is computed with respect to surface curvature or coupled with user-specified spacing, both for isotropic or anisotropic cases. Grid points are added through edges splitting using the G1 interpolations; edges contractions are performed robustly with help of determinants computation of triangles in the local parametric domain and dihedral angles checking with respect to the contracted edge. Finally, the surface mesh is optimized through edges swapping and points relocation. At the end of the paper, various complex geometry configurations with 'Dirty' STL boundary representations, such as a complete tank, a representation of the circulatory system and a representation of the urinary system are tested as examples to demonstrate the capabilities of this approach.

REFERENCES

[1] J. F. Thompson, B. K. Soni, N. P. Weatherill, *Handbook of Grid Generation*, CRC Press LLC, (1999).
[2] P. J. Frey, *About surface remeshing*, 9th International Meshing Roundtable, (2000), 123-156.
[3] Z. J. Wang, K. Srinivasan, *An adaptive cartesian grid generation method for 'Dirty' geometry*, International Journal For Numerical Methods in Fluids, 39, (2002), 703-717.

COMPUTATIONAL MECHANICS (Abstracts)
WCCM VI in conjunction with APCOM'04, *Sept. 5-10, 2004, Beijing, China*

Numerical Simulation of Interaction between Blood Flow and Elastic Arteries in Aneurysm and Stenosis

Feng Gao [1*], Teruo Matsuzawa [1], Aike Qiao [2]

[1] *Center of Information Science, Japan Advanced Institute of Science and Technology, 923-1292, Japan*

[2] *Beijing University of Technology, Beijing, 100022, China*

e-mail: feng-g@jaist.ac.jp, matuzawa@jaist.ac.jp, qak@bjpu.edu.cn

Abstract

The fully coupled solution of fluid flows with structural interactions, a rapidly evolving discipline, represents the natural next step in simulating mechanical systems. At present, one application area of fluid-flow analysis is that incompressible-flow analyses are performed on medical field. The analysis of blood flow in a diseased artery is a biomechanical application of great interest. In this paper, author created aneurysm vessel and stenosis vessel model to simulate the interaction between blood flow and elastic arteries. The Arbitrary Lagrangian-Eulerian (ALE) technique has been employed. In an ALE framework, the coordinates are not attached to the material nor are they fixed in space. The finite element mesh can be changed according to the user's needs. In aneurysm, the velocity of the blood flow decreases and the back part of aneurysm wall had a larger deformation. In stenosis, the velocity of the blood flow increases and the front part of stenosis wall had a larger deformation. The simulation results indicated that: (1) the ALE technique can describe fluid-structure interaction; (2) this finding could help to better understand the effects of aneurysm and stenosis in atherosclerosis and the effects that disabilities, surgical procedures, and other physical changes have on them.

REFERENCES

[1] S. A. Berger, L-D Jou, *Flows in stenotic vessels*, Annual Review of Fluid Mechanics, 32, (2000), 347-382.

[2] S. Glagov, *Intimal hyperplasia*, Vascular Modeling and the Restenosis Problem, Circu lation, 89, (1994), 2888-2891.

[3] Karl A. Illig, Arvind Soni, Jacky Williams, K. Cynthia Shortell, Richard M. Green, *Clinical review: irradiation for lower extremity arterial occlusive disease*, Cardiovascular Radiation Medicine, 1(3), (1999), 288-296.

[4] Jr., J. E. Moore, C. Xu, S. Glagov, C. K. Zarins, D. N. Ku, *Fluid wall shear stress measurements in a model of the human abdominal aorta; oscillatory behavior and relationship to atherosclerosis*, Atherosclerosis, 110, (1994), 225-240.

[5] C. A. Taylor, J. R. Thomas, T. H. Hughes, C. K. Zarins, *Finite element modeling of blood flow in arteries*, Computer Methods in Applied Mechanics and Engineering, 158, (1998), 156-196.

[6] T. H. Hughes, C. Taylor, C. Zarins. *Finite element modeling of blood flow in arteries*, Comp. Mech. Eng., 158, (1998), 155-196.

[7] A. Quarteroni, M. Tuveri, A. Veneziani, *Computational vascular fluid dynamics: problems, models and methods*, Computing and Visualisation in Science, 2, (2000), 163-197.

[8] A. Tozeren, *Elastic properties of arteries and their influen ce on the cardiovascular system*, ASME J. Biomech. Eng., 106, (1984), 182-185.

COMPUTATIONAL MECHANICS (Abstracts)
WCCM VI in conjunction with APCOM'04, *Sept. 5-10, 2004, Beijing, China*

Numerical Analysis of Hemodynamics in the Operation for Cerebral Aneurysm Using Coil, Catheter, Stent and Baloon

H. Iwase [1*], R. Himeno [1], K. Fukasaku [2]

[1] *Advanced Center for Computing and Communication, RIKEN, 351-0198, Wako-city Hirosawa 2-1, Saitama, Japan*
[2] *Neuro Surgery, Kusakabe Central Hospital, 344-0063, Kasukabe-city, Midori-cho, 5-9-4, Saitama, Japan*
e-mail: iwase@postman.riken.go.jp, himeno@postman.riken.go.jp, Fukasaku@med.Nagoya-u.ac.jp

Abstract

In the treatment of cerebral aneurysms without opening of skull, doctors use some operation equipments such as coil, catheter, stent and baloon, which have the complexity geometry. We want to predict the hemodynamics around operation equipments in the therapy of cerebral aneurysm using computational fluid dynamics (CFD). The studies [1-2] of hemodynamics simulation by CFD in a cerebral aneurysm have been performed in the case without therapeutic devices in the aneuryms.
The simulation model, called as solid model, of a brain artery network like Willis ring based on medical surgery images and operation equipments are constructed using CAD modeling software by means of our proposed modeling method. Furthermore, the simulation solid models are converted to voxel models for numerical simulation using V-CAD developed by RIKEN [3], which can attain to take a very short time for making voxel model.
Finite difference method with rectangular coordinate is applied for numerical simulation. The model of operation equipments are coil, catheter, stent, and baloon model. The results show that coil and stent control the generation of vorteies in a cerebral aneurysm.

REFERENCES

[1] M. Oshima et al, *Finite element simulation of blood flow in the cerebral artery*, Comp. Methods Appl. Mech. Engrg., 191, (2001), 661-667.
[2] K. Yokoi et al, *Numerical method for blood flow in a cerebral artery with multiple aneurysms*, RIKEN Symposium, (2003), 23-24.
[3] K. Kase, Y. Teshima, S. Usami, H. Ohmori, C. Teodoosiu, A. Mkinouchi, *Volume CAD*, Internation Workshop on Volume Graphics (VG03), (IEEE TCVG, Eurographics & ACM SIGGRAPH), Tokyo, July, (2003).

WCCM VI in conjunction with APCOM'04, *Sept. 5-10, 2004, Beijing, China*

Influences of Graft Diameter on the Hemodynamics of Femoral 2-Way Bypassing Models

Aike Qiao [1,2,*], Teruo Matszawa [1]

[1] *Center of Information Science, Japan Advanced Institute of Science and Technology, 1-1, Asahidai, Tatsunokuchi, Ishikawa 923-1292 Japan*

[2] *College of Electromechanical Engineering, Beijing University of Technology, Pingleyuan 100, Chaoyang, Beijing, 100022, China*

e-mail: a-qiao@jaist.ac.jp, matuzawa@jaist.ac.jp

Abstract

The geometry configuration of anastomosis has profound effects on the hemodynamics of femoral bypass graft. Numerical simulation methods have been extensively applied to study the relationship between the hemodynamics and the geometry configuration. In order to clarify the influence of graft diameter on the hemodynamics of femoral 2-way bypassing surgery, the physiological blood flows in two femoral bypassing models were simulated with computational fluid dynamics approaches. Unlike the conventional 1-way bypass graft procedure, the present models were featured with symmetrically implanted 2-way bypass grafts. These models were constructed with identical geometry parameters except the different diameters of grafts. Two models with small and large grafts were studied. The flow domain were subdivided into multiple control volumes in order to discretize the models with different meshing density according to the variation of geometry and the complexity of flow patterns. The boundary conditions for the simulations of blood flow were constant for both models. The blood flow was assumed to be Newtonian transient laminar flow. The maximum Reynolds number was 832.8, and the Womersley number was 6.14. The emphases of results were on the comparison analysis of flow fields in the vicinity of the distal anastomosis. The temporal-spatial distributions of velocity vectors, pressure drop between the proximal and distal toes, wall shear stresses, wall shear stress gradients and oscillating shear index were analyzed and compared. The present study indicated that femoral artery bypassed with a large graft demonstrated relatively uniform wall shear stresses, lower pressure and less refluences, whereas it does not have advantages in the distributions of secondary flow and the oscillating shear index. Although there are advantages and disadvantages in both the small and the large graft models, the large model exhibits better and more regular hemodynamic phenomena near the vessel wall and may be effective in preventing the initiation and development of postoperative intimal hyperplasia and restenosis.

REFERENCES

[1] J. S. Cole, J. K. Watterson, M. J. G. O'Reilly, *Numerical investigation of the haemodynamics at a patched arterial bypass anastomosis*, Medical Engineering & Physics, 24, (2002), 393-401.

[2] M. Lei, C. Kleinstreuer, J. Jr. Archie, *Geometric design improvements for femoral graft–artery junctions mitigating restenosis*, J. Biomech., 29, (1996), 1605-1614.

[3] P. E. Hughes, T. V. How, *Effects of geometry and flow division on flow structures in models of the distal end-to-side anastomosis*, J. Biomech., 29, (1996), 866-872.

COMPUTATIONAL MECHANICS (Abstracts)
WCCM VI in conjunction with APCOM'04, *Sept. 5-10, 2004, Beijing, China*

Adaptive Optimal Control of the Swimming Rule of a Fish

Liang Wang [1, 3*], Chuijie Wu [1, 2, 3]

[1] *Institute of Hydraulics and Hydroelectrics, Hohai University, Nanjing, 210098, China*
[2] *LNM, Institute of Mechanics, Chinese Academy of Sciences, Beijing, 100080, China*
[3] *Research Center for Fluid Dynamics, PLA University of Science and Technology, Nanjing, 211101, China*
e-mail: cjwu@mech.pku.edu.cn

Abstract

In this paper, combined with the moving boundary CFD method, we developed a new real-time optimal control method which can be used to adaptively optimize the swimming rule of fish. The results show that there is a 54.0% increase in the thrust obtained by the unsteady optimal swimming rule. Fig .1 and Fig. 2 are the vorticity field and vector field, from which the averaged flow has been removed, before and after optimization respectively. It can be seen that the reverse Kármán vertex street is strengthened after optimization. In addition, to reduce the cost of computation and deepen the understanding of the swimming rule, the maximum velocity at the tip of fish tail is taken as the objective functional, with which the thrust is increased by 22.9%.

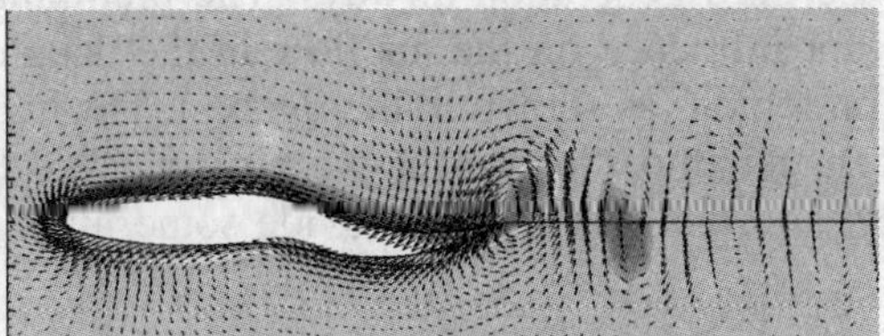

Fig. 1 Before optimization

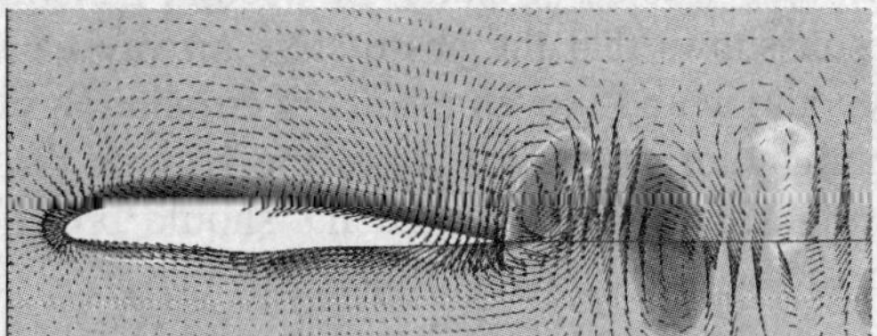

Fig. 2 After optimization

REFERENCES

[1] J. H. Ferziger, M. Peric, *Computational method for fluid dynamics,* Spinger, (1999).
[2] E. K. Kaufman, D. J. Leeming, G. D. Taylor, *An ODE-based approach to nonlinearly constrained minimax problems,* Num. Alg., 9, (1995), 25-37.
[3] O. K. Rediniotis, D. C. Lagoudas, T. Mashio, L. J. Garner, M. A. Qidwai, *Theoretical and experimental investigations of an active hydrofoil with SMA actuators,* Internet.

COMPUTATIONAL MECHANICS (Abstracts)
WCCM VI in conjunction with APCOM'04, *Sept. 5-10, 2004, Beijing, China*

Simulation of Effect of COPD on Respiratory Flow

X. L. Yang, Y. Liu *

Department of Mechanical Engineering, The Hong Kong Polytechnic University, Hung Hom, Kowloon, Hong Kong, China
e-mail: mmyliu@polyu.edu.hk

Abstract

The study of bifurcation flow in human lung airways is important to better predict the particle deposition in drug therapy and inhalat ion toxicology. Chronic Obstructive Pulmonary Diseases (COPD), e.g. bronchitis and asthma, is one of the most common diseases in human community. The COPD always results in inflammation that leads to narrowing and obstruction of the airways. The obstructive airways have serious effect on respiratory flow.

In order to understand the flow phenomenon in such obstructive airways, four different three - dimensional three–generation models based on the 23 generations model of Weibel [1] are considered, respectively. The fully three-dimensional incompressible laminar Navier-Stokes equations have been solved using CFD solver on unstructured tetrahedral meshes. A symmetric three-generation airway model is served as the reference, and the other three models are conside red to be obstructed at either the second generation or the third generation airways. The preliminary result shows that the obstructive airway has significant influence on the air flow in the daughter generation and it even results in circulation in juncti on region (as shown in the figures). The higher momentum of the air flow, the less of the flow can get into the obstructive airway. This has explained why COPD patients should breathe gently, and this might also provide some hints for medicine powder deposition.

Fig. 1 Flow Vectors in the Bifurcation Plane

REFERENCES

[1] E. R. Weibel, *Morphometry of the human lung*, Springer, Academic Press, Berlin, New York, (1963).

COMPUTATIONAL MECHANICS (Abstracts)
WCCM VI in conjunction with APCOM'04, *Sept. 5-10, 2004, Beijing, China*

Vortex Formation in Spatially Modulated Taylor-Vortex Flow

Zhenyu Li[*], Roger E. Khayat, Robert J. Martinuzzi

Department of Mechanical and Materials Engineering, University of Western Ontario, London, Ontario, N6A 5B9, Canada
e-mail: zli2@engga.uwo.ca, rkhayat@eng.uwo.ca, rmartinuzzi@eng.uwo.ca

Abstract

Low-inertia vortex formation and pattern selection are examined for axisymmetric Taylor vortex flow with spatially modulated cylinders. The forcing is arbitrary but remains periodic. The modulation amplitude is assumed to be small, and a regular perturbation expansion is used to determine the flow field at small to moderately large Taylor numbers (below the critical threshold). It is found that the presence of a weak modulation leads unambiguously to the emergence of steady Taylor-vortex flow even at vanishingly small Taylor number. This situation is closely reminiscent of the effect of end plates, and the consequent onset of imperfect bifurcation. The vortex structure is found to have the same periodicity as the forcing when only one of the cylinders is modulated, or when the modulations are commensurate. For incommensurate modulations, the vortex pattern is quasi-periodic, with regions of almost purely azimuthal flow. When the counter-rotation speed of the outer cylinder increases, the original vortices are gradually replaced by new ones that end up spanning the entire gap width, and in turn break up into two vortices resulting in two rows of vortices commensurate with each cylinder modulation. The results show good agreement with experiments.

COMPUTATIONAL MECHANICS (Abstracts)
WCCM VI in conjunction with APCOM'04, *Sept. 5-10, 2004, Beijing, China*

Multiscale Method for Turbulent-Flow Computations

Yiqing Shen [1*], Zhi Gao [1], Fenggan Zhuang [2]

[1] *Institute of Mechanics, Chinese Academy of Sciences, Beijing, 100080, China*
[2] *China Aerospace Science Technology Corporation, Beijing, 100830, China*
e-mail: yqshen@imech.ac.cn

Abstract

A main method of predicting turbulent flows is to solve LES equations, which was called traditional LES method. The traditional LES method solves the motions of large eddies of size larger than filtering scale Δ_n while modeling unresolved scales less than Δ_n. Hughes et al argued that many shortcomings of the traditional LES approaches were associated with their inabilities to successfully differentiate between large and small scales. One may guess that a priori scale-separation would be better, because it can predict scale-interaction well compared with posteriori scale-separation. To this end, a multi-scale method was suggested to perform scale-separation computation. The primary contents of the multiscale method are: 1) A space average is used to differentiate scale. 2) The basic equations include the large scale equations and fluctuation equations. 3) The large-scale equations and fluctuation equations are coupled through turbulent stress terms. We use the multiscale equations of n=2, i.e., the large and small scale (LSS) equations, to simulate 3-D evolutions of a channel flow and a planar mixing layer flow. Some interesting results are given.

REFERENCES

[1] Z. Gao, F. G. Zhuang, *Time-space scale effects in computing flow and a new approach to flow numerical simulation,* Lecture Notes in Physics, 453, (1995), 256.

[2] T. J. R. Hughes, L. Mazzei, A. A. Oberal, A. A. Wray, *The multiscale formulation of large eddy simulation: Decay of homogeneous isotropic turbulence,* Phys. Fluids, 13(2), (2001), 505.

[3] Y. Zhou, C. G. Speziale, *Advances in the fundamental aspects of turbulence: energy transfer, interacting scales and self-preservation inisotropic decay,* Appl. Mec. Rev., 51(4), (1998), 267-301.

[4] J. A. Domarazki, E. M. Saiki, *Subgrid-scale model based on the estimation of unresolved scales of turbulence,* Phys. Fluids, 9, (1997), 2148.

[5] W. G. Wang, Z. Gao, F. G. Zhuang, *A numerical comparison of the large and small scale (LSS) equations with the Navier-Stokes equations: the three-dimensional evolution of a planar mixing layer flow,* in Proc. of 7th International Symposium on Computational Fluid Dynamics, International Academic Publishers, Beijing, (1997), pp. 484-490.

[6] J. Jimenez, *Transition to turbulence in 2-D Poiseuille flow,* J. Fluid. Mech., 218, (1990), 265-297.

[7] K. D. Moser, M. M. Rogers, The three-dimensional temporal evolution of a planar mixing layer flow, J. Fluid. Mech., 273, (1993), 31-60.

COMPUTATIONAL MECHANICS (Abstracts)
WCCM VI in conjunction with APCOM'04, *Sept. 5-10, 2004, Beijing, China*

Numerical Investigation of Unsteady Flow around Existing Airship Body at Angle Incidence by Vortex Method

Subagyo*, Kyoji Kamemoto

Department of Mechanical Engineering and Material Science, Yokohama National University, 79-5 Tokiwadai Hodogayaku, Yokohama, 240-8501, Japan
e-mail: d01sb193@ynu.ac.jp

Abstract

The grid-less Vortex Method was developed as an analysis tool for two-dimension (2D) and three-dimension (3D) complex flow in various engineering applications. In the present study, unsteady flow around existing airship body at angle incidence $\alpha=10^{o}$, 20°, and 30° were investigated by vortex method for high Reynolds number $Re=3.9x10^{5}$. Numerical simulation for high Reynolds number in the engineering point of view can be simplified as simulation of the large-scale flow and Vortex Method can treat this flow easily. In the present study, the introduction of vortex blob shedding from the body surface is very important. Due to that problem the nascent vortex element was replaced by vortex blob at a certain distance from the body surface with an arrangement related to the value of Reynolds number. In this stu dy vortex method can be considered as Lagrangian direct simulation.

The numerical calculation results are flow -pattern, velocity field, coefficient pressure distribution on the airship body surface, and coefficient aerodynamics. The numerical calculation r esults compared with the experimental and other simulations show the good agreement and also show this method is quite well for predicting characteristic aerodynamics flow around 3D airship body at angle of incidence. The numerical calculation result indic ated that the complex 3D separation flow behind airship body could be simulated successfully on this study.

The purpose of this paper gives a perspective on the Grid -less Vortex Method as computational fluid dynamics tool with some advantages and strengthe ns, and the reasonable on the cost computational on the Personal Computer (PC) based without any losses in the accuracy. Considering the applicability of this algorithm, the future work will be expected concerning analysis of the complex unsteady flow around airship geometry.

REFERENCES

[1] E. Achenbach, *Vortex shedding from spheres, part 2*, J. Fluid Mech., 62, (1974), 209-221.
[2] A. J. Chorin, *Numerical study of slightly viscous flow,* J. Fluid Mech., 57, (1973), 785-796.
[3] A. Leonard, *Vortex methods for flow simulations,* J. Comp. Phys., 37, (1980), 289-335.
[4] C. E. Costis, N. T. Hoang, D. P. Telionis, *Laminar separation flow over prolate spheroid,* Journal of aircraft, 26(9), (1989), 810-816.
[5] C. Greengard, J. Comp. Phys., 6l, (1985), 345-348.
[6] W. H. Su, B. Tao, L. Xu, AIAA J., 3l, (1993), 2175-2176.
[7] J. C. Wu, J. F. Thompson, Computers & Fluids, l, (1973), 197- 215.
[8] C. J. Chesnakas, R. L. Simpson, AIAA J., 35(6), (1997), 990-999.
[9] T. G. Wetzel, R. L. Simpson, C. J. Chesnakas, AIAA J., 36(4), (1998), 557-564.

COMPUTATIONAL MECHANICS (Abstracts)
WCCM VI in conjunction with APCOM'04, *Sept. 5-10, 2004, Beijing, China*

Effects of Finite Difference Errors on the Accuracy of Subgrid-Scale Models for Large Eddy Simulation

Makoto Tsubokura*, Shigeaki Kuroda

Dept. of Mechanical Engineering and Intelligent Systems, University of Electro-Communications, Chofugaoka 1-5-1, Chofu-shi, Tokyo 182-8585, Japan
e-mail: tsubo@mce.uec.ac.jp, kuroda@mce.uec.ac.jp

Abstract

The difficulty of Large Eddy Simulation (LES) lies in the fact that not only the subgrid-scale (SGS) turbulence models but also the discretization method strongly affects the reproduced flow field. Especially in the lower-order Finite Difference Method (FDM) commonly used in the engineering problems, the obtained flow field possibly is contaminated severely by a numerical error especially at a higher wavenumber region close to a grid width.

That contamination near the cutoff wave number becomes more serious when we use the dynamic procedure of Germano [1], because the grid-scale (GS) turbulence near the cutoff region is utilized to dynamically estimate the unknown model coefficient by explicit filtering. In fact, the original procedure was found to be seriously degraded when we applied lower-order FDM which predicted excessively large mean velocity at log-law region in wall-turbulence, contrary to the spectral method.

In the previous study, we proposed a new eddy viscosity model [2] which was developed for the dynamic procedure using FDM by considering the consistency of the numerical error in the procedure and showed validity compared with the traditional Smagorinsky's model.

The objective of this study is to further estimate the proposed model as well as traditional models in simple wall-turbulence (pipe flow) in which various reliable experimental and numerical data are available. We intensively study the dependence of the SGS models on the FDM errors by changing the accuracy of FDM from 2nd up to 10th order (which is almost comparable to the spectral method) at various grid resolutions. For this purpose we have adopted the recently developed fully conservative FDM scheme of Morinishi [3] for cylindrical coordinates, in which not only momentum but also kinetic energy is well conserved in a discretized space.

REFERENCES

[1] M. Germano et al, *A dynamic subgrid-scale eddy viscosity model*, Phys. Fluids, A3, (1991), 1760-1765.

[2] M. Tsubokura, *Proper representation of the subgrid-scale eddy viscosity for the dynamic procedure in large eddy simulation using finite difference method*, Phys. Fluids, 13, (2001), 500-504.

[3] Y. Morinishi et al, *Fully conservative finite difference scheme in cylindrical coordinates for incompressible flow simulations*, J. Comp. Phys., (2004), (to appear).

COMPUTATIONAL MECHANICS (Abstracts)
WCCM VI in conjunction with APCOM'04, *Sept. 5-10, 2004, Beijing, China*

Nonlinear FE Analysis of Aseismic Retrofitting of Reinforced Concrete Frame Structure with Steel Braced under Cyclic Loads

Surong Fan [1*], Weiqing Liu [1], Yantao Xue [2]

[1] *College of Civil Engineering, PO Box 32, Nanjing University of Technology, Nanjing, 210009, China*
[2] *Institute of Earthquake Engineering, China Academy of Building Research, China Academy of Building Research, 100013,Beijing, China*
e-mail:srfan@sohu.com, wqliu@njut.edu.cn

Abstract

In this paper, two kinds of seismic retrofitting of two 2 -span 1-story reinforced concrete frames with steel bracing systems have been tested under reversed cy clic load. Based on the results of the model experiments, the mechanic characteristics, failure modes, ductility and energy dissipating ability of these kinds of frames structures are studied, the nonlinear finite element analysis for the hysteresis behavior of the model frame under cyclic loads is investigated by employing the command stream edited with the ANSYS Parametric Design Language, and the calculated result are much coincided with the test results, which proved that the design method for the aseis mic retrofitting of reinforced concrete frame structure with steel braced is feasible and practical.

REFERENCES

[1] M. H. Nariman, R. Shepherd, *Inelastic behaviour of X-bracing in plane frames,* ASCE, Journal of Structural Engineering, 112(4), (1986), 764-780.
[2] T. D. Bush, *Behaviour of RC frame strengthened using structural steel bracing*, ASCE, Journal of Structural Engineering, l17(4), (1991), 1115-1126.
[3] Badoux Marc Eric, *Seismic retrofitting of reinforced concrete structures with steel bracing systems*. UMI Dissertation Information Service, (1987), 25-31.

COMPUTATIONAL MECHANICS (Abstracts)
WCCM VI in conjunction with APCOM'04, *Sept. 5-10, 2004, Beijing, China*

Analysis of Stresses in Welded Structures subjected to Cyclic Loading

Grzegorz Glinka [1], Jin Qian [2*], Mohammad El-Zein [2]

[1] *University of Waterloo, Department of Mechanical Engineering, Waterloo, Ontario, N2L 3G1 Canada*
[2] *Deere & Co., Technical Center, Moline, IL61265, USA*
e-mail: gggreg@mecheng1.uwaterloo.ca, {Qianjin, El-ZeinmohamadS}@Johndeere.com

Abstract

Fatigue evaluations of welded structures require detail analysis of stresses in the weld toe region. Due to complex geometry of welded joints characterized by var iety of global and local geometrical features analytical solutions are seldom feasible and the Finite Element (FE) method 3 -dimensional models require prohibitively large number of small elements. In the case of cyclic loading, inducing localized plastic yielding; such FE models and analyses are impractical in daily engineering practice. Therefore simpler and more robust approaches to stress analysis are needed when dealing with fatigue and durability evaluations of welded structures.

The proposed method is based on the shell Finite Element modeling of welded structures. This approach does not require modeling the small local geometrical features of weldments, such as the weld toe radius, weld angle, etc. In fact the welds are not modeled at all. However, th e shell membrane and bending stresses determined from the shell Finite Element model of given welded structure form the base for the calculation of the stress concentration at the weld toe and the through-thickness stress distribution in the critical cross section. Those stresses can be subsequently used to predict the fatigue crack initiation life at the weld toe and subsequent through thickness fatigue crack growth. The advantage of using the proposed method lies in the fact that complex welded structures can be analyzed using relatively coarse shell Finite Element mesh models without losing the accuracy of required stress quantities such as the stress concentration factor and the stress gradient in critical cross sections of analyzed objects.

Both the stress concentrations and the through thickness stress distributions are determined using the shell Finite Element stress data. Detail discussion of available shell Finite Element models appropriate for welded structures and the analytical aspects of the meth od are also discussed including the limitations concerning the finite element size and the accuracy of predicted stresses. The stresses predicted using the proposed methodology was verified by comparing with results obtained from detail 3-dimensional finite element models using multi-nodal brick elements.

COMPUTATIONAL MECHANICS (Abstracts)
WCCM VI in conjunction with APCOM'04, *Sept. 5-10, 2004, Beijing, China*

Comparative and Parametric Studies of Slab-Edge Column Connections with Varying Locations of Openings

H. Guan *

School of Engineering, Griffith University, PMB 50 Gold Coast Mail Centre, Queensland 9726, Australia
e-mail: h.guan@griffith.edu.au

Abstract

Reinforced concrete flat plate structures are advantageous in construction because of flat soffits, reduced storey heights, simplified formwork and ease of reinforcement placement. Although being economical and efficient, the flat plate structures are prone to localized, brittle and disastrous punching shear failure. For flat slabs with openings in th e vicinity of edge columns, the punching shear strength capacity and ductility of the slab-column connections are reduced and the failure mechanism around the connections becomes more complicated. To strengthen such connections and to provide adequate duc tility, properly placed and anchored shear stud reinforcement (SSR) is often utilized.

Two series of experimental testing [1 -2] was undertaken at the University of Waterloo on the influence of the locations of openings and SSR on the punching shear behaviour of slab-edge column connections. Specimens in series I have no SSR, whereas specimens in series II, being the counterparts of those in series I, contain SSR. In this study, six test specimens (XXX, SF0, SF1 and SF2 without SSR and XXX-R and SF0-R with SSR) representing varying locations of openings are analysed. An extended nonlinear layered finite element method (LFEM) [3], which is capable of predicting the punching shear strength, load -deflection response and crack patterns for the slab -edge column connections, is used as the analysis tool. The numerical results are compared with the experimental findings and good agreement has been achieved. Further parametric study has also been carried out to determine the influence of the locations of openings in slab-edge column connections with SSR (SF1 -R and SF2-R), where no experimental data is available. The geometric configuration, boundary condition, material setting and loading type of SF1 -R and SF2-R are identical to SF1 and SF2, respectively. The re sults of parametric study confirm the general understanding of the punching shear behaviour of slab -edge column connections with varying locations of openings and SSR.

The comparative and parametric studies also confirm the applicability, accuracy and effe ctiveness of the LFEM in the analysis of slab -edge column connections when both openings and SSR are present.

REFERENCES

[1] E. F. El-Salakawy, M. A. Polak, M. H. Soliman, *Reinforced concrete slab-column edge connections with openings*, ACI Struct. J., 96(1), (1999), 79-87.

[2] E. F. El-Salakawy, M. A. Polak, M. H. Soliman, *Reinforced concrete slab-column edge connections with shear studs*, Can. J. Civ. Eng., 27, (2000), 338 -348.

[3] H. Guan, Y. C. Loo, *Layered finite element method in cracking and failure analysis of RC beams and beam-column-slab connections*, Struct. Eng. Mech. - An Int. J., 5(5), (1997), 645 -662.

COMPUTATIONAL MECHANICS (Abstracts)
WCCM VI in conjunction with APCOM'04, *Sept. 5-10, 2004, Beijing, China*

Alternative Reliability Indices in Structural Mechanics Using Support Vector Machines

Jorge E. Hurtado *

Department of Civil Engineering, National University of Colombia, Manizales, Colombia
e-mail: jhurtado14@epm.net.co

Abstract

A common situation in mechanical safety analysis is to assess the reliability of a system under scarce statistical information on the input variables. The widely used geometric index, however, has the shortcoming of not possessing the desirable ordering property, as it is weakly related to the limit state function. On the other hand, the generalized index, which overcomes this deficiency, requires the computation of a multidimensional integral which makes difficult its calculation. Also, it is defined as a quantile, with the consequence that it can be negative. In this paper, a pattern recognition approach to this problem is adopted. This consists in regarding the safe and failure domains as statistical classes, thus allowing using the sign of the performance mechanical function as the identifying label. It is shown that the use of an advanced pattern recognition method, namely the support vector machine, allows the development of at least two alternatives for reliability indices: 1) an index derived from a secant hyper plane that overcomes the main limitation of the popular geometric index. 2) A more sophisticated index that operates upon the support vector nonlinear mapping of the limit state function on a high dimensional space. The index is thus defined as the size of a hypercube equivalent to the volume defined by the hyper plane. Both indices exhibit the desirable positivity, ordering, formulation invariance and dimension invariance properties. Their practical use is demonstrated with some examples.

COMPUTATIONAL MECHANICS (Abstracts)
WCCM VI in conjunction with APCOM'04, *Sept. 5-10, 2004, Beijing, China*

Target Displacement in Pushover Analysis for Eccentric Building Structures Considering Torsion Effect

Gang Li [1*], Yong Liu [2]

[1] *Department of Engineering Mechanics, State Key Laboratory of Structural Analysis of Industrial Equipment, Dalian University of Technology, Dalian, 116023, China*
[2] *Department of Civil Engineering, Hong Kong University of Science and Technology, Clear Water bay, Kowloon, Hong Kong, China*
e-mail: ligang@dlut.edu.cn, liuyong@ust.hk

Abstract

The concept of performance based seismic design was proposed in the early 1990s in the USA, and has been gradually accepted by the earthquake engineering profession recently. Compared with the current seismic design codes, performance based seismic design takes cost-effectiveness criterion as its important principle, and pays more attention to the structural performance in the inelastic stage. The static pushover analysis is becoming popular in seismic performance evaluation due to its simplicity compared with the inelastic time-history analysis. Pushover analysis can provide the information on the global nonlinear deformation, local hinge progres s, weak story and irregular strength distribution, etc. In the recent two decades, especially the past several years, the pushover analysis method has undergone a lot of developments in researches and applications, and it is suggested as a nonlinear analys is method of the structural deformation under severe earthquakes in the new seismic design code in China

Structural irregularity is one factor that restricts the use of pushover analysis. Eccentricity is one kind of irregularity and the torsion effect of e ccentric building structures may affects the performance of the building considerably. In the practical engineering, eccentric buildings are sometimes built due to the limit of space or the aesthetic requirements. Therefore, it is necessary to study the pu shover analysis of eccentric building structures considering torsion effect.

The determination of target displacement is an important step in pushover analysis, in which the selection of shape vector is a key problem. The existing methods of the selection of shape vector do not account for the torsion effect on the target displacement of the eccentric building structures. In this paper, a new method to construct the shape vector is proposed based on the existing methods, which combines the first vibration mode by dynamic analysis and the nonlinear deformation of the MDOF system by pushover analysis. The combined shape vector is then employed to transform the MDOF system into the equivalent SDOF system, and then the SDOF system is used to calculate the target displacement.

The results of the numerical example show that the proposed method can account for the torsion effect of eccentric building structures.

COMPUTATIONAL MECHANICS (Abstracts)
WCCM VI in conjunction with APCOM'04, Sept. 5-10, 2004, Beijing, China

Inverse Iteration Method in Form-Finding of Cable Structure

Guoqiang Li *, Liyuan Shen, Yongfeng Luo

Department of Structure Engineering, University of Tongji, 1239 Siping Road, Shanghai, 200092, China
e-mail: gqli@mail.tongji.edu.cn, shenly77@sina.com, yfluo93@mail.tongji.edu.cn

Abstract

Form-finding process of cable structures is a significant problem for the structure. Unlike normal structures three are three states in tension structures: 1) initial state; 2) pre -stress state; 3) load state. For the design of tension structures including cable structures and membrane structures, the form-finding process that determinates the pre -stress distribution and corresponding geometry of the structure is a key problem. The problem is classified into three kinds: 1) The geometry is known but the pre-stress distribution is unknown; 2) The geometry is unknown but the pre-stress distribution is known; 3) Both the geometry and the pre -stress distribution are unknown.
For the problem of the second and the third kinds, many methods are developed such as the finite element method, the dynamic relaxation, the force dens ity method etc. A successful method for the first kind of the problem of the cable structures is the inverse iteration method. The principal of the inverse iteration method is stated and the method is applied to form -finding process. According the geometry of the pre-stress state of the structure, the inverse iteration method is used to determinate the initial geometry of the structures and the pres -tress distribution when the structure is deformed to the pre-stress state from the initial state in which mem bers of the structures have initial lengths. Numerical examples show the high accuracy and the universality of this method.

COMPUTATIONAL MECHANICS (Abstracts)
WCCM VI in conjunction with APCOM'04, *Sept. 5-10, 2004, Beijing, China*

Shooting Method for Thermal Post-Buckling Analysis of Extensible Timoshenko Beams

Shirong Li [1,2], Xi Song [*1], Romesh C. Batra [2]

[1] *Department of Engineering Mechanics, Lanzhou University of Technology, Lanzhou, Gansu 730050, China*
[2] *Department of Engineering Science and Mechanics, Virginia Polytechnic Institute and State University, Blacksburg, VA 24061, USA*
e-mail address: lisr@lut.cn, songx@lut.cn , rbatrathird@vt.edu

Abstract

Geometrically nonlinear deformation of axially extensional Timoshenko beams subjected to mechanical as well as thermal loadings are characterized by a system of 7 coupled and highly nonlinear ordinary differential equations, which results in a complicated two -point boundary-value problem. They are accurate formulat ions of elastic Timoshenko beams in the sense of the first order shearing theory, in which the shearing angle of the cross -section and the stretching of the central axis of the beam are expressed in terms of the components of the internal resultant forces, rotational angle of the cross -section, and the temperature rise parameter. By using shooting method a numerical computing procedure for solving the highly nonlinear boundary -value problem is established, by which numerical solution to the problem can be o btained efficiently. Based on the above-mentioned mathematical formulation and numerical procedure, thermal post -buckling analysis of an elastic Timoshenko beam, subjected to a transversely non -uniform temperature rise and with immovably pinned -pinned as well as fixed-fixed ends, is presented. Characteristic curves showing the relationships between the beam deformation and temperature rise are illustrated. Especially, the effects of shear deformation on the post-buckling response are quantitatively investigated. The numerical results show, as we know, that shear deformation effects become significant with the decrease of the slenderness and with the increase of the shear flexibility. It is also found that the non-dimensional critical temperature remains the same magnitude for different values of the ratio of the length with the height of the beam. Finally, equilibrium paths of the beam subjected transversely temperature rise are given. These curves are similar to those of an axial compressed beam associated w ith transversely loads.

REFERENCES

[1] S. R. Li, C. J. Cheng, *Analysis of thermal post-buckling of heated elastic rods*, Applied Mathematics and Mechanics, 21, (2000), 133 -140.

[2] S. R. Li, Y. H. Zhou, X. J. Zheng, *Thermal post-buckling of a heated elastic rod with pinned-fixed ends*, Journal of Thermal Stresses, 25, (2002), 45-56.

[3] A. A. Khdeir, *Thermal buckling of cross-ply laminated composite beams*, Acta Mechanica, 149, (2001), 1949-1510.

[4] S. R. Li, *Geometrically nonlinear analysis of Timoshenko beams under thermomechanical loadings*, Journal of Thermal Stresses, 26(9), (2003), 861-872.

COMPUTATIONAL MECHANICS (Abstracts)
WCCM VI in conjunction with APCOM'04, *Sept. 5-10, 2004, Beijing, China*

Flexibility-Based Formulation for the Nonlinear Frame Element with Lateral Deformable Supports

Suchart Limkatanyu [1*], Enrico Spacone [2, 3]

[1] *Department of Civil Engineering, Faculty of Engineering, Prince of Songkla University, Hadyai, Songkla, 90110, Thailand*
[2] *Department PRICOS, University of Chieti, viale Pindaro 42, 65127, Pescara, Italy*
[3] *Department of Civil Engineering, University of Colorado, Boulder, CO, 80309-0428, USA*
e-mail: lsuchart@ratree.psu.ac.th

Abstract

This paper presents the theory and the numerical validation of new proposed nonlinear frame element with nonlinear lateral deformable supports. The governing differential equations (strong form) of the problem are derived first and the flexibility-based finite element formulation (weak form) is then presented. The flexibility-based model is derived based on the virtual force principle. Tonti's diagrams are used to conveniently represent the equations governing both the strong and the weak forms of the problem. The general matrix equations of the flexibility-based formulation are presented, with some details on the issues regarding the elements' implementations in a general-purpose finite element program. The convergence, accuracy, and computational times of the proposed model are studied through a numerical example. The distinctive element characteristics in terms of force and deformation discontinuities between adjacent elements are discussed. The capability of the proposed frame model to trace the softening response due to softening of the foundation is also investigated. Overall, the force-based model is much more accurate than the *traditional* displacement-based model and requires very few elements to reach the converged solution. The nonlinear frame model proposed in this paper has practical applications in modeling the soil-pile structural system, geosynthetics/fiber-glass reinforcement of foundation soils, beam on deformable foundations, etc.

REFERENCES

[1] S. Limkatanyu, E. Spacone, *R/C frame element with bond interfaces: formulations and numerical validations*, ASCE Journal of Structural Engineering, 128(3), (2002), 346-364.
[2] M. R. Salari, E. Spacone, *Analysis of steel-concrete composite frames with bond-slip*, ASCE Journal of Structural Engineering, 127(11), (2001), 1243-1250.
[3] E. Spacone, F. C. Filippou, F. F. Taucer, *Fiber beam-column model for nonlinear analysis of R/C frames: formulation and applications*, Earthquake Engineering and Structural Dynamics, 25(7), (1996), 711-742.
[4] E. Winkler, *Theory of elasticity and strength*, Dominicus, Czechoslovakia, (1867).
[5] S. Limkatanyu, E. Spacone, *Frame element with lateral deformable supports: formulations and numerical validation*, International Journal for Numerical Methods in Engineering, (under review).

COMPUTATIONAL MECHANICS (Abstracts)
WCCM VI in conjunction with APCOM'04, *Sept. 5-10, 2004, Beijing, China*

Orthogonal Complex Base Spline and Its Application in Structure Analysis

Fuyong Lin *

Huaqiao University, Quanzhou, Fujian, 362011, China
e-mail: fylin@hqu.edu.cn

Abstract

For isotropic space and time there are limitless symmetric axles, using these symmetric transform group one can obtain a serial of symmetric function bases, any function defined in isotropic space can be approached by these symmetric function bases. The Abel group is the simplest of approaches of the symmetric transform group of isotropic space. The orthogonal complex bases of Abel group's space is

$Q_i = \sum_{k=0}^{N-1} e^{\frac{j2ik\pi}{N}} C_N^k, i = 0,1,2\cdots, N$, where C_N^k is shift operator that: $C_N^k f(x) = f(x - \frac{kL}{N})$.

Acting these orthogonal complex bases of Abel group's space on spline bases $N_m(x)$. One can obtain the orthogonal bases of spline space:

$$Q_i(x) = \sum_{k=0}^{N-1} e^{\frac{j2ik\pi}{N}} N_m(x - \frac{kL}{N}).$$

These orthogonal function bases are orthogonal to each other in periodic zone $[0, L]$. Any periodic function can be approached by the constructed function bases, functions other than periodic one can be expanded to periodic one by adding some functions, and therefore can be approached by these orthogonal function bases.

The method is applied to the structure analysis, using energy method we get a new method to get the solution of bending problem based on orthogonal complex function bases of spline. Concrete equation is presented in the paper. Because the proposed function is orthogonal to each other, the calculation is greatly reduced, by using Fast Fourier transform and other fast calculation the computation is further reduced, the computation is O($N \log N$) by Fast Fourier transform method for periodic plate and beam system, the structure other than the periodic system can be expanded to a periodic system by adding additional forces on the boundary or outside the defined areas, the computation is similar to boundary element method. The accuracy of this method is the same as finite element. An example is present in the paper, the simple example show the method have good accuracy.

REFERENCES

[1] Lin Fuyong, *Orthogonal continuous segmentation function*, Applied Mathematics and Computation, (in press).
[2] Lin Fuyong, *Orthogonal base of B-spline and its application*, Mechanical Science and Technology, 22, (2003), 123-125.

COMPUTATIONAL MECHANICS (Abstracts)
WCCM VI in conjunction with APCOM'04, *Sept. 5-10, 2004, Beijing, China*

Structural Internal Force Prediction of Long Suspension Bridges Due to Buffeting

Gao Liu [1*], Ledong Zhu [1], Jiahao Lin [2], Youlin Xu [3]

[1] *Department of Bridge Engineering, Tongji University, 1239 Siping Road, Shanghai, 200092, China*
[2] *Department of Engineering Mechanics, Dalian University of Technology, Dalian, 116023, China*
[3] *Department of Civil and Structural Engineering, The Hong Kong Polytechnic University, Hung Hom, Kowloon, Hong Kong, China*
e-mail: liugao@dlut.edu.cn, jhlin@ dlut.edu.cn, ledong@mail.tongji.deu.cn, ceylxu@polyu.edu.hk

Abstract

Buffeting of long span bridges by storm winds is a random vibration issue in nature. In current design practice, only the displacement and acceleration responses of long span bridges due to buffeting are analyzed by the random vibration method directly, while the structural internal force responses are generally estimated by means of the static analysis procedure in terms of the equivalent static buffeting loading (ESBL) method. The motivation of this study is to investigate the precision of the ESBL method comparing with the random vibration method. Firstly, the mechanism of the ESBL method was studied based on the principle of structural dynamics. Then, an improved random vibration method for internal force response of long span bridges due to buffeting was presented, which is featured mainly by the combination of finite element approach and pseudo-excitation method in the frequency-domain. Finally, both the ESBL method and the improved random vibration method are applied to the Tsing Ma suspension Bridge in Hong Kong. Traditionally, the ESBL method only includes several fundamental modes of vibration, the effect of inter-modal coupling is omitted, and the ESBL on the bridge deck is considered in computation only. The computational results show that the ESBL method can give the same prediction in structural internal force responses comparing with the random vibration method for one mode of vibration when the ESBL on the bridge deck, main cables and bridge towers were all included. However, the structural internal force responses would be underestimated as much as 35% when the ESBL on the bridge deck was considered only. Furthermore, the contribution of inter-modal coupling and some high order modes of vibration had important influence on structural internal force responses in some cases.

REFERENCES

[1] A. G. Davenport, *Buffeting of a suspension bridge by storm winds*, J. Struct. Div., ASCE, 88 (ST3), (1962), 233-268.

[2] N. P. Jones, R. H. Scanlan, A. Jain, H. Katsuchi, *Advanced (and challenges) in the prediction of long-span bridge response to wind*, in A. Larsen, S. Esdahl eds. *Bridge aerodynamics*, Balkema, Rotterdam, The Netherlands, (1998), pp. 59-85.

[3] J. H. Lin, *A fast CQC algorithm of PSD matrices for random seismic responses*, Computers & Structures, 44(3), (1992), 683- 687.

COMPUTATIONAL MECHANICS (Abstracts)
WCCM VI in conjunction with APCOM'04, *Sept. 5-10, 2004, Beijing, China*

Solution of Linear Function Pressures on a Hollow Cylinder and the Limit Solution when $l \to \infty$

Wen Liu *, Rui Shan

College of Science, Yanshan University, Qinhuangdao, 066004, China
e-mail: lw@ysu.edu.cn

Abstract

The problem of linear function pressures on a hollow cylindrical is solved with a new stress function, which provides the basis for the solution of space symmetrical deformation of a hollow cylinder. Examples of elastic deformation produced in the engineering are widespread .Liquid press working urn or squeezing cylinder is a practical example [1]. It is symmetry problem of three-dimension space. But only the solution uniformly distributed pressures on a hollow cylinder is obtained by transforming the problem in three-dimension space into one in two-dimension space in many books up to date. It can result in very big margin of error. In fact the pressure is not certainly distributed uniformly and the sketch of pressure on a hollow cylinder can be described with multiplication. An analytical solution for liner function pressures on a hollow cylinder with new stress function satisfying both the biharmonic equation [2-3] and boundary conditions is presented here. It is very significant to get the analytical solution to theory and practice. So it is very significant to get the solution of arbitrarily distributed pressure that can be expressed with constants, quadratic functions, cubic functions, sine functions, hyperbolic functions and so on.

This paper can be divided into the following four parts : 1) Basic assumptions and the computational model; 2) Establishing a new stress function ; 3) The solution of linear function distribution pressures ; 4) When $l \to \infty$, the hollow cylinder is infinitely long, the pressure can be regarded as invariable in the axial direction. We can get the well-known Lamè principle.

REFERENCES

[1] Z. B. Liu, *New technology plastic deforming and its mechanical principles,* Beijing China Machine Press, (1995), (in Chinese).

[2] Z. L. Xu, *Applied elasticity,* Higher Education Press Wiley Eastern Limited, (1992).

[3] Rui Shan, Zhubai Liu, Wen Liu, *Solution of sine pressures on a hollow cylinder and the limit when* $k \to 0$, Chinese Journal of Mechanical Engineering, 40, (2004), 73 -77, (in Chinese).

COMPUTATIONAL MECHANICS (Abstracts)
WCCM VI in conjunction with APCOM'04, *Sept. 5-10, 2004, Beijing, China*

The Elastic and Plastic Analysis for the Linearly Elastic and Power Hardening Structure of the Pin-Jointed Bars

Xiequan Liu *, Xinhua Ni

Mechanical Engineering College, Shijiazhuang, 050003, China
e-mail:lxqnxh@163.com

Abstract

A structure of pin-jointed bars is considered. For *i*-th bar, the length is l_i, the cross-sectional area is A_i. The oriented angles of the unit vector $\boldsymbol{r}_i$ along *i*-th bar about three coordinate axes x, y, z are α_i, β_i, γ_i respectively. Applied forces acted on the joint are P_x, P_y, P_z. The constitutive relation can be assumed

$$\sigma = \begin{cases} E_1\varepsilon & |\varepsilon| \le \varepsilon_s \\ \operatorname{sgn}(\varepsilon)\left[\sigma_s + E_2\left(|\varepsilon| - \varepsilon_s\right)^n\right] & |\varepsilon| > \varepsilon_s \end{cases}$$

If the displacements of the joint along three coordinate axes are *u*, *v* and *w*, deformation of *i*-th bar is can be expressed out as

$$\Delta l_i = u\cos\alpha_i + v\cos\beta_i + w\cos\gamma_i\,)$$

Thus the strain, stress and axial force of the bar can be obtained.
Equilibrium equations of the joint can be determined from static

$$\sum N_i \cos\alpha_i - P_x = 0, \sum N_i \cos\beta_i - P_y = 0, \sum N_i \cos\gamma_i - P_z = 0$$

We know that the expressions are a group of nonlinear equations of *u*, *v*, *w*. *u*, *v*, *w* can be figured out, then the strain, stress and axial force of *i*-th bar will be obtained.
Above equations may be transformed to

$$f_1 = \sum N_i \cos\alpha_i - P_x = 0, f_2 = \sum N_i \cos\beta_i - P_y = 0, f_3 = \sum N_i \cos\gamma_i - P_z = 0$$

The nonlinear equations may be solved by descending method.
Define the aim function: $F(u,v,w) = f_1^2(u,v,w) + f_2^2(u,v,w) + f_3^2(u,v,w)$

Evidently, if $F(u^*, v^*, w^*) \leqslant \xi$, (u^*, v^*, w^*) will be thought as a group of solutions that are satisfied for precision ξ.
It is the more general term that constitutive relation of materials was referred to as linearly elastic and power hardening formulation. Example, if $\sigma_s=\infty$, it will be transformed to linearly elastic model; If $\sigma_s=0$, it will be transformed to power hardening function. If $n=1$, it will be linearly hardening elastic-plastic model. If $E_2=0$, it will be perfectly elastic-plastic model.

REFERENCES

[1] Liu Xiequan et al., *Stress-strain analysis of nonlinearly elastic structure of bars*, Mechanics and Practice, (3), (1999), (in Chinese).

[2] Liu Xiequan et al., *Numerical solution of the elastic and plastic analysis for the linearly elastic and power hardening structure of the bars*, Journal of Computational Mechanics, (2), (2000), (in Chinese).

COMPUTATIONAL MECHANICS (Abstracts)
WCCM VI in conjunction with APCOM'04, *Sept. 5-10, 2004, Beijing, China*

The Stress-Strain Analysis for Strength Difference Plane Structure of the Pin-jointed Bars

Xinhua Ni*, Xiequan Liu

Mechanical Engineering College, Shijiazhuang, 050003, China
e-mail: lxqnxh@163.com

Abstract

A plane structure of the pin-jointed bars is considered. For i-th bar, the length is l_i the cross-sectional area is A_i and the orented angle is α_i. The applied forces are P_x and P_y. The constitutive relation can be written out.

$$\sigma = \begin{cases} C_L \varepsilon^{n1} & \varepsilon \geq 0 \\ -C_Y |\varepsilon|^{n2} & \varepsilon < 0 \end{cases}$$

Where $n1$ and $n2$ are the hardening coefficients in tension and compression respectively. C_L and C_Y are constants of material in tension and compression.
If the horizontal and the vertical displacements of the joint are u and v respectively, deformations of i-th bar can be expressed out as: $\Delta l_i = -u\cos\alpha_i - v\sin\alpha_i$
Thus the strain, stress and axial force of the bar are obtained
Equilibrium equations at the joint can be determined from static

$$\sum N_i \cos\alpha_i + P_x = 0, \quad \sum N_i \sin\alpha_i + P_y = 0$$

We know that the expressions are a group of nonlinear equations of u and v. u and v can be figured out, then the strain and axial force of i-th bar will be obtained
Above equations may be transformed to

$$f_1(u,v) = \sum N_i \cos\alpha_i + P_x = 0, \quad f_2(u,v) = \sum N_i \sin\alpha_i + P_y = 0$$

The nonlinear equations may be solved by descending method.
Define the aim function: $F(u,v) = f_1^2(u,v) + f_2^2(u,v)$
Evidently, if $F(u^*,v^*) \leqslant \xi$, u^* and v^* will be thought of as a group of solutions of the equations that are satisfied for precision.
The constitutive relation of materials was referred to as strength difference and nonlinearly elastic formulation, it is the more general term. Example, if C_L=C_Y and $n1$=$n2$ it will be transformed to strength same and nonlinearly elastic. If $n1$=$n2$=1, it will be transformed to elastic modulus difference and linearly elastic model. If C_L=C_Y and $n1$=$n2$=1, it will be transformed to strength same and linearly elastic model.

REFERENCES

[1] Xiequan Liu et al, *Stress-strain analysis of nonlinearly elastic structure of bars*, Mechanics and Practice. (3), (1999), (in Chinese).
[2] Xiequan Liu et al, *Numerical solution of the elastic and plastic analysis for the linearly elastic and power hardening structure of the bars*, Journal of Computational Mechanics, (2), (2000), (in Chinese).

COMPUTATIONAL MECHANICS (Abstracts)
WCCM VI in conjunction with APCOM'04, *Sept. 5-10, 2004, Beijing, China*

Salini Method 2004

Reus Salini *

Salini Engineering, R. Heitor Luz, 225/709, Centro, Florianópolis, 88015-500, Brazil
e-mail: r.salini@gmx.net

Abstract

The Salini Method was published in 1999 as a simple but scientific way to calculate the strains in the last fiber in the asphalt layer from deflections measured with Benkelman Beam, FWD/HWD or any similar equipment. These strains are critical to estimate the service-life for new and existent asphalt pavements. As the pavement works under flexion, with a compressed and a tractioned area, to improve the accuracy the "neutral line position" is now introduced as variable and the method called "Salini Method 2004". The method is available as scientific publication and freeware software.

REFERENCES

[1] Reus Salini, *Avaliação da Capacidade Estrutural dos Pavimentos a Partir dos Dados Deflectométricos - Proposição de Metodologia*, In: 3º Simpósio Internacional de Avaliação de Pavimentos e Projectos de Reforço - 3º SINAPPRE, (1999), Belém, Brazil, (in Portuguese).

COMPUTATIONAL MECHANICS (Abstracts)
WCCM VI in conjunction with APCOM'04, *Sept. 5-10, 2004, Beijing, China*

Weak Forms of Generalized Governing Equations in Elasticity and Their Applications in Computational Structural Mechanics

G. Shi [1*], L. Tang [2]

[1] *School of Material Engineering, Suzhou University, Suzhou, 215021, China*
[2] *Department of Engineering Mechanics, Dalian University of Technology, Dalian, 116024, China*
e-mail: shiguangyu@suda.edu.cn

Abstract

This paper presents the derivation of the generalized governing equations in elasticity, their weak forms and the some applications in the numerical analysis of structural mechanics. Unlike the classical elasticity theory, the generalized equations of the equilibrium and compatibility equations presented here are in the integral forms, and the generalized equilibrium equations contain the classical differential equations and the boundary conditions in one form. As a result, not only the finite element method but also the finite difference method can possess the natural boundary conditions. By using appropriate test functions, the weak forms of these generalized governing equations can be achieved. It can be shown that various variational principles are merely the special cases of these weak forms of generalized governing equations in elasticity.

Some of the previous works are reviewed and examined by the newly established equations. It indicates that the so-called discrete operator equations [1] can be deduced from the weak forms of generalized equations by using local function in them; the powerful quasi-conforming element method [2] is the exact solutions of the generalized compatibility conditions and it satisfies the weak form continuity requirement naturally; and the theoretical foundations of some assumed stress beam [3] and plate [4] elements are the same as the present weak forms of the generalized governing equations. In fact, this new integral theory of elasticity widens greatly the space of trial functions in seeking approximate solutions. Therefore, the weak forms of generalized governing equations in elasticity provide a powerful modeling tool in the computational structural mechanics.

REFERENCES

[1] L. Tang et al, *Establishment and applications of week form of generalized basic equations in elasticity*, J. Dalian Univ. Tech., 41, (2001), 1-8, (in Chinese).

[2] L. Tang, W. Chen, Y. Liu, *String net function approximation and quasi-conforming technique*, in S. N. Atluri, R. H. Gallagher, O. C. Zienkiewich eds. *Hybrid and mixed finite element methods*, Wiley, New York, (1983).

[3] G. Shi, S. N. Atluri, *Elasto-plastic large deformation analysis of space-frames: a plastic-hinge and stress-based explicit derivation of tangent stiffnesses*, Int. J. Num. Meth. Eng., 26, (1988), 589-615.

[4] G. Shi, P. Tong, *Assumed stress C^0 quadrilateral/triangular plate elements by interrelated edge displacements*, Int. J. Num. Meth. Eng., 39 (1996), 1041-1051.

COMPUTATIONAL MECHANICS (Abstracts)
WCCM VI in conjunction with APCOM'04, *Sept. 5-10, 2004, Beijing, China*

A New Model on the Buckling of a Rod in Tubing

Chao Sun *, Stanislaw Lukasiewicz

Department of Mechanical & Manufacturing Engineering, University of Calgary, 2500 University Drive NW, Calgary, Alberta, T2N 1N4, Canada
e-mail: csun@ucalgary.ca , lukasiew@ucalgary.ca

Abstract

The sucker rod pumping system is widely applied in oil production. The pump and the plunger are mounted on the bottom end of the tubing and the rod respectively. The rod moves the plunger of the pump up and down to lift the crude oil from the reserv oir to the surface of ground. The bottom part of the rod is under oscillating tensile and compressive forces. In its downward running, the rod buckles first laterally and next helically due to the fluid pressure acting on its bottom end. The buckling of the rod reduces the axial translation of the plunger dramatically. Consequently the effectiveness of the pumping system and the producing ability of the oil well are decreased greatly. In some extreme case, the plunger even has a zero axial translation while the rod keeps oscillating and no oil is lifted to the surface. So it is of great importance to determine and predict the critical buckling forces and the post-buckling configuration. Past studies [1,2] were based on the helix model, which focused only on the helical buckling part of the rod instead of the whole rod.

In fact, the length of helix increases gradually from 0 to almost the whole length of the rod with the increase of the axial load and the development of helical buckling. The helix model is acc urate enough when the length of helix is close to the length of the rod, and will leads to greater errors when the length of helix is quite short. Thus a new model is presented in this paper, which models the whole rod, including the helix and non-helix part of the rod. The principle of virtue works is employed to simulate analytically the buckling of a rod in tubing. The principle requires first modeling the geometry of the well and the deformed rod, next calculating the changes of the strain energy of the rod during buckling and works done by external forces and finally minimizing the total potential energy. The difficulty here is the accurate description of the whole deformed rod. The indoor experiments are also carried out to verify and improve the analy tical simulations.

The successful simulations of buckling of a rod in tubing offer an effective tool to predict the critical buckling forces and the post-buckling configuration of the rod in tubing and thus enable us to increase the effectiveness of the p umping system and the producing ability of oil wells. It also lay a base for searching the measures of increasing the effectiveness of the sucker rod pumping and evaluating the fatigue strength of the rod.

REFERENCES

[1] A. Lubinski, *Developments in Petroleum Engineering*, vol.1, Gulf Publishing Company, Houston, Texas, USA, (1987).

[2] N. C. Huang, P. D. Pattillo, *Helical buckling of a tube in an inclined well bore*. International Journal of Non-Linear Mechanics. 35, (2000), 911-923.

COMPUTATIONAL MECHANICS (Abstracts)
WCCM VI in conjunction with APCOM'04, *Sept. 5-10, 2004, Beijing, China*

New Fusion Pattern of Analytic Method and Numerical Method in Analysis of Thin Plates: Composite Fourier Series Method

Weiming Sun [1*], Guangsong Yang [2], Zimao Zhang [1]

[1] *Institute of Engineering Mechanics, Beijing Jiaotong University, Beijing, 100044, China*
[2] *Mailbox Zi-7, Qinhe Building, Beijing, 100085, China*
e-mail: weimingsun-1974@263.net, warm_maxwell@elong.com, zmzhang@center.njtu.edu.cn

Abstract

Fusion of the analytic method, the numeric method and the experimental method is an important issue in the research of the analysis techniques of modern mechanics. Nowadays, attentive efforts on the fusion research of the analytic method and the numeric method have proven to be effective and a successful fusion pattern has been established [1 -3], where the numeric method is the fusion matrix and the analytic method is the fusion reinforcer. And accordingly the newly proposed numeric analysis techniques gain desired properties not only of the traditional analytic method, but also of the traditional numeric method. This demonstrates the advantage for fusion strategy of methods of different kinds. By employing a new fusion pattern, where the analytic method is the fusion matrix and the numeric method is the fusion reinforcer, the composite Fourier series method is presented for the thin plate bending problem. This new type analytic method solves a specific prob lem in two steps: the construction of the approximate solution and its fusion with an analytic solution in Fourier series. The first step is intended for determination of the major deformation pattern of thin plates, by which the original problem changes into the analysis of the plate under quasi zero load and quasi homogeneous boundary conditions. Therefore the approximate solution must be precise to a certain degree and allow of further adjustments in expression. There exists no limitation on the source o f the approximate solution, and in this paper the approximate solution in polynomial is established with the aid of the finite element method and the weighted residual method. And as to the second step, some little revision of the usually used Fourier seri es analytic solution of the thin plates is needed. Because of the elimination of the main influence of loads and the approximately prior satisfaction of the boundary conditions, the fusion solution only comprises the general solution and the special solution and has good convergence. The composite Fourier series method is a new development of the fusion research of methods of different kinds and lays a solid foundation of the improved understanding and wide application of the traditional analytic method.

REFERENCES

[1] Z. Y. Cao, Y. K. Cheung, *Semi-Analytical and Semi-Numerical Method*, National Defence Industry Press, Beijing, China, (1992), (in Chinese).

[2] W. M. Sun, G. S. Yang, *Rational Finite Element Method for Elastic Bending of Reissner Plates*, Applied Mathematics and Mechanics, 20(2), (1999), 193-199.

[3] P. Zeng, *New Numerical Method in Computational Mechanics*: *Composite Element Method*, Science in China (Series E), 43(2), (2000), 120-128.

COMPUTATIONAL MECHANICS (Abstracts)
WCCM VI in conjunction with APCOM'04, *Sept. 5-10, 2004, Beijing, China*

Adaptive Analysis of RC Frames with Microplane Joints

Ladislav Svoboda *, Bořek Patzák, Zdeněk Bittnar

Department of Structural Mechanics, Czech Technical University, Thákurova 7, Prague, 166 29, Czech Republic
e-mail: ladislav.svoboda@fsv.cvut.cz, borek.patzak@fsv.cvut.cz, bittnar@fsv.cvut.cz

Abstract

In this paper the use of the microplane model in an analysis of reinforced concrete frames is discussed. The microplane model is a constitutive model for concrete. It is a fully three-dimensional model capable to describe concrete in its very complex behaviour. The classical approach to the constitutive modeling is based on a direct relationship between strain tensor and stress tensor and their invariants. In contrary to it, the constitutive relations of microplane model for concrete are formulated in terms of strain and stress components on planes of ar bitrary spatial orientation, so called microplanes [1]. This approach excels in conceptual simplicity and allows straightforward modelling of anisotropy and other processes connected with planes with different orientations. The penalty to be paid is a grea t increase of computational effort.
For its qualities the microplane model is used for simulations of experimental results, for example the behaviour of reinforced concrete columns [2]. However, wider usage is not possible because of high computational cost. That is why we proposed adaptive analysis of RC frames. Due to nonlinearity of microplane model, the loading is applied in an incremental manner.
In first loading steps the whole frame is created only by beam elements with nonlinear behaviour, which is achieved by "fibered" cross-sections. It means a cross-section on each beam element is meshed into a number of fibres with different materials. For fibres at place of reinforcement a common bilinear material model is used. For other fibres at place of conc rete a special material model is used. This two-dimensional model should simulate behaviour of microplane model fully in tension and partly in pressure.
In next loading steps, when some concrete fibres reach highly nonlinear part of strain -stress diagram, appropriate beam elements are replaced by fully three-dimensional model of beam with microplane material. Attachments between beam elements and 3D elements are realised by rigid arms.
At the moment of collapse only in some sections of the frame (especially close to joints) the microplane model is used, so computational effort is lower against completely 3D frame. Finally, we conclude our paper by some numerical examples.

REFERENCES

[1] Z. P. Bažant, F. C. Caner, I. Carol, M. D. Adlez, S. A. Akers, *Microplane model M4 for concrete*, Journal of Engineering Mechanics, (1999).
[2] J. Němeček, B. Patzák, D. Rypl, Z. Bittnar, *Microplane models: computational aspects and proposed parallel algorithm*, Computers and Structures, 80, (2002), 2099-2108.

COMPUTATIONAL MECHANICS (Abstracts)
WCCM VI in conjunction with APCOM'04, *Sept. 5-10, 2004, Beijing, China*

A Two-dimensional Analysis of Surface Acoustic Waves in Finite Solids

Ji Wang [1*], Yongping Wan [1, 2], Jingbo Lin [1], Zheng Zhong [2]

[1] *Department of Engineering Mechanics and Material Science, School of Engineering, Ningbo University, 1211 Zhenning Xilu, Ningbo, Zhejiang 315211, China*

[2] *Department of Engineering Mechanics, School of Science, Tongji University, 1239 Siping Road, Shanghai 20092, China*

E-mail: wangji@nbu.edu.cn

Abstract

The analysis of surface acoustic wave propagation in finite elastic solids is of essential importance in applications concerning the modeling and design of various SAW resonat or devices, and tremendous efforts have been made lately in employing both analytical and numerical methods for accurate solutions of the phase velocity and displacements. The theoretical foundation of these analyses has only limited choices, and the appr oximations are made at the expenses of accuracy on many aspects including the oversimplified one-dimensional model and the three-dimensional analysis of a single period of the large periodic structure. Of course, these analyses are justified for their spec ific applications with satisfactory solutions. In case of situations requiring a full analysis, three - dimensional finite element analysis has been studied with special elements and algorithms. However, differences based on the choices of models and algor ithms demonstrate the need for more efficient methods.

To overcome the current difficulties in the models and computational intensity, a two -dimensional theory similar to Mindlin and Lee plate theories has been established for the surface acoustic wave propagation in finite elastic solids. A systematic simplification of the three -dimensional equations resulted in a two-dimensional theory with accuracy and potential to solve surface acoustic wave problems in solids in a more efficient manner. The applicat ion of the two-dimensional theory will reduce the problem size of a surface acoustic wave resonator analysis significantly when the thickness is no longer discretized. This will make the full analysis of SAW resonators more accurate without spending too much on the computation. In this study, we shall demonstrate the finite element application of the two-dimensional theory with accurate results and widely available computing resources. We expect the two -dimensional theory will be accepted as a standard m ethod for surface acoustic wave resonator design just as Mindlin plate theory for bulk acoustic wave analysis.

REFERENCES

[1] Ji Wang, Ken-ya Hashimoto, *A two-dimensional theory for the analysis of surface acoustic waves in anisotropic elastic solids*, Proc. of 2003 IEEE Intl. Ultrasonics Sym., pp. 637-640.

[2] Ji Wang, Jingbo Lin, *A two-dimensional theory for surface acoustic wave analysis in finite piezoelectric solids*, *Journal of Intelligent Material Systems and Structures*, (submitted).

[3] Ji Wang, Yongping Wan, *The mass effect of electrodes on the surface acoustic wave propagation in elastic solids*, Proc. 18th European Frequency and Time Forum, Guilford, England, April 5-7, (2004).

COMPUTATIONAL MECHANICS (Abstracts)
WCCM VI in conjunction with APCOM'04, *Sept. 5-10, 2004, Beijing, China*

Complex Base Orthogonal Continuous Segmentation Function and Its Application in Structure Analysis

Taiyong Wang [1*], Fuyong Lin [2], Xuda Qin [1]

[1] *Department of Mechanical Engineering, Tianjin University, Tianjin, 200072, China*
[2] *Department of Mechanical Engineering, Huaqiao University, 362011, China*
e-mail: fylin@edu.hqu.cn

Abstract

For isotropic space and time there are limitless number of symmetric axes, therefore there exist transform group to represent the symmetry. Using eigen-value equation of the symmetric transform group's one can obtain a serial of orthogonal bases of the group space, and future act on local defined function, one can get orthogonal function bases. Any function in isotropic space can be approached by these function bases. The Abel group is the simplest of approach of the symmetric transform group of isotropic space. The orthogonal complex bases of Abel group's space is

$Q_i = \sum_{k=0}^{N-1} e^{\frac{j2ik\pi}{N}} C_N^k, i = 0,1,2\cdots,N$, where C_N^k is shift operator: $C_N^k f(x) = f(x - \frac{kL}{N})$.

Acting these orthogonal complex bases of Abel group's space on Hermite bases $\mathbf{N}(x) = [N_1(x), N_2(x)]^T$.

One can obtains the orthogonal bases of Hermite space:

$$\mathbf{Q}_i(x) = \sum_{k=0}^{N-1} e^{\frac{j2ik\pi}{N}} \mathbf{N}(x - \frac{kL}{N}) .$$

These orthogonal function bases are orthogonal to each other in periodic zone [0, L]. Any periodic function can be approached by the constructed function bases, functions other than periodic one can be expanded to periodic one by adding some functions, and therefore can be approached by these orthogonal function bases.

The method is applied to the structure analysis, using energy method we get a new method to get the solution of bending problem based on orthogonal complex function bases of Hermite segmentation function. Because the proposed function is orthogonal to each other, the calculation is greatly reduced, by using Fast Fourier transform and other fast calculation the computation is further reduced, the computation is $O(N \log N)$ by Fast Fourier transform method for periodic plate and beam system, the structure other than the periodic system can be expanded to a periodic system by adding additional forces on the boundary or outside the defined areas, the computation is similar to boundary element method. The accuracy of this method is the same as finite element. An example is present in the paper, the simple example show the method have good accuracy.

REFERENCES

[1] Lin Fuyong, *Orthogonal continuous segmentation function*, Applied Mathematics and Computation, (in press).

[2] Lin Fuyong, *Orthogonal base of B-spline and its application*, Mechanical Science and Technology, 22, (2003), 123-125.

COMPUTATIONAL MECHANICS (Abstracts)
WCCM VI in conjunction with APCOM'04, *Sept. 5-10, 2004, Beijing, China*

A New Family of Generalized-α Time Integration Algorithms for Structural Dynamics with Nonlinear Stiffness

Kaiping Yu [1], Jie Zhao [1], Deguang Liu [2], Jingxiang Zou [1], Shiwei Pang [1*]

[1] *Dept. of Astronautics and Mechanics, Harbin Institute of Technology, Harbin, 15001, China*
[2] *School of Aeronautic Science and Engineering, Beijing University of Aeronautics and Astronautics, Beijing, 100083, China*
e-mail: hityukp@sohu.com, jiezhao1978@263.net

Abstract

The new numerical algorithm, which is called the new Generalized -α method, is presented for solving structural dynamics problems with nonlinear stiffness. The traditional Generalized-α (G-α) method [1] has the optimal and continuously controllable numerical dissipation in the high-frequency for the linear problems, and is stable in an energy sense for nonlinear dynamic problems, but this algorithms inherit undesired overshoot properties of a class of α- method. The algorithmic overshoot behavior leads often to convergence problems of the employed Newton-Raphson algorithm and overshoot is clearly exhibited by displacement and velocity even in the case of convergent Newton-Raphson iterations. The main purpose of this paper is to eliminate these inherited overshoot behaviors in the class of the G -α algorithm. Seven free parameters are introduced into the single-step three-stage algorithm formulations and the nonlinear internal force was approximated by means of the generalized trapezoidal rule. Then an analysis of the stability, accuracy, energy and overshoot properties of the proposed scheme was performed in the nonlinear regime. A value or range of value of the seven free parameters is given sequentially in the analysis process. The displacement, velocity and acceleration evaluated with the new algorithm was fully consistent and of order two. Significantly, new schemes eliminate inherent overshoot behavior of a class of traditional G-α method. The presented algorithms are zero-stable and when the algorithmic parameter $\rho = 1$, the presented methods can guarantees energy conservation in the high-frequency range and are stable in an energy sense. Whereas, when the algorithmic parameter $\rho \in [0, 1)$, presented methods are stable in an energy sense and guarantees energy decay as long as the potential energy is negligible compared to the remaining term of the energy expression.

REFERENCES

[1] S. Erlicher, L. Bonaventura, O. S. Bursi, *The analysis of the Generalized-α method for non-linear dynamic problems*. Computational Mechanics, 28, (2002), 83-104.

[2] J. Chung, G. M. Hulbert. *A time integration algorithm for structural dynamics with improved numerical dissipation: the generalized-α method*. Transactions of the ASME, Journal of Applied Mechanics, 60, (1993), 371-375.

[3] J. Chung, G. M. Hulbert, *A family of single-step Houbolt time integration algorithms for structural dynamics*. Computer Methods in Applied Mechanics and Engineering, 118, (1994), 1-11.

[4] T. J. R. Hughes. *The finite element method: linear static and dynamic finite element analysis*. Prentice-Hall, Englewood Cliffs, New Jersey, (1987), pp. 459-552.

COMPUTATIONAL MECHANICS (Abstracts)
WCCM VI in conjunction with APCOM'04, *Sept. 5-10, 2004, Beijing, China*

Temperature Caused by Hydration Heat in Massive Concrete Structures

Guorong Chen [1*], Yan Zhang [1], Ziming Zhang [2]

[1] *Department of Material Science, Hohai University, Nanjing, 210098, China*
[2] *Department of Engineering Mechanics, Hohai University, Nanjing, 210098, China*
e-mail: ziming58@163.com, yearover@163.com

Abstract

After development of the theory, this paper analyzes the temperature field caused by hydration heat in massive concrete wall. The analytical iteration solution of nonlinear heat conduction equation on this problem is presented. The analysis takes into account the fact that the adiabatic temperature rise is a function of equivalent time based on Arrhenius theory, which leads to a nonlinear differential equation. The systems of up to 180 nonlinear equations are solved in this paper by the iteration method. The following conclusions are drawn: 1) the placement temperature has a severe effect on the maximum temperature rise in massive concrete wall, the more the placement temperature, the more the maximum temperature difference between the inside and the surface of the concrete wall, 2) due to the low thermal conductivity of concrete, the center will hotter than the surface, resulting in different rates of hydration heat and different equivalent time at different location in the cross section, 3) the rate of hydration heat increases with temperature, consequently, the rate of the concrete stiffening is accelerated, its initial and finial setting time are reduced, which may be concluded that in hot weather conditions, the use of low heat cement is to be preferred, 4) the temperature inside the concrete is generally higher than the temperature of the concrete surface, resulting in different rate of hydration heat and different equivalent time at different location in the crosses section, which will increase the complexity of analysis.

REFERENCES

[1] F. De Larrard, P. Acker, R. Le. Roy, Shrinkage creep and thermal properties, chapter 3, high performance concrete: properties and applications, edited by S. P. Shah, S. H. Ahmad, McGraw Hill Inc, (1994), pp. 65-114.

[2] Z. P. Bazant, M. F. Kaplan, Concrete at high temperature: material properties and mathematical models (monograph and reference volume), Longman (Addison-Wesley), London, (1996).

[3] *Standard practice for estimating concrete strength by maturity method*, (ASTM C 1074-93), 1996 Annual Book of ASTM Standards, Vol. 04.02, ASTM, West Conshchocken, (1996), pp. 529-535.

[4] Ziming Zhang, Zuoxin Fu, Analytical solution of temperature field simulating construction of RCC dams, Hongshui River (CODEN HNGSEK), 15(5), (1996), (in Chinese).

[5] Ziming Zhang, V. K. Garga, *Temperature and Temperature Induced Stresses for RCC Dams*, Dam Engineering, 7(2), (1996).

[6] Ziming Zhang, V. K. Garga, *State of Temperature and Thermal Stresses in Mass Concrete Structure Subjected to Thermal Shock*, Dam Engineering, 7(4), (1996).

COMPUTATIONAL MECHANICS (Abstracts)
WCCM VI in conjunction with APCOM'04, *Sept. 5-10, 2004, Beijing, China*

Modeling of Adiabatic Temperature Rise of Concrete

Ziming Zhang [1*], Yan Zhang [2]

[1] *Department of Engineering Mechanics, Hohai University, Nanjing, 210098, China*
[2] *Department of Material Science, Hohai University, Nanjing, 210098, China*
e-mail: ziming58@163.com, yearover@163.com

Abstract

This paper presents an experimental study on the adiabatic temperature rise under different curing temperature. The form of the rate constant versus temperature function is needed to describe the combined effects of time and temperature on adiabatic temperature rise. This study investigates the relationship between the rate constant and curing temperature. Based on adiabatic temperature rise data for concrete cured at 4.4, 23.3 and 40.0 ^{o}C (40, 74 and 104 ^{o}F), the following conclusions are drawn: (1) the rate of hydration increases with temperature, consequently bring about the stiffening of the mix at a correspondingly higher rate, (2) adiabatic temperature rise can be represented by a three parameter hyperbolic function; (3) the rate constant is a nonlinear function of curing temperature, and Arrhenius function or a simple exponential function can be used to describe this relationship; (4) tests of appropriate concrete specimens provide the information needed to predict the temperature field of concrete structure, (5) the initial temperature does affect the limiting adiabatic temperature rise and there is a unique adiabatic temperature rise-equivalent time curve that is described by the rate constant model. The two numerical techniques (in elapsed time and in equivalent time) with which the nonlinear heat conduction equation based on equivalent time can be solved are provided.

REFERENCES

[1] Ziming Zhang, V. K. Garga, *Temperature and Temperature Induced Stresses for RCC Dams*, Dam Engineering, 7(2), (1996).

[2] Ziming Zhang, V. K. Garga, *State of Temperature and Thermal Stresses in Mass Concrete Structure Subjected to Thermal Shock*, Dam Engineering, 7(4), (1996).

[3] Ziming Zhang, Zuoxin Fu, Analytical solution of temperature field simulating construction of RCC dams, Hongshui River (CODEN HNGSEK), 15(5), (1996), (in Chinese).

[4] F. De Larrard, P. Acker, R. Le. Roy, Shrinkage creep and thermal properties, chapter 3, high performance concrete: properties and applications, edited by S. P. Shah, S. H. Ahmad, McGraw Hill Inc, (1994), pp. 65-114.

[5] L. E. Copeland, D. L. Kantro, G. Verbeck, *Chemistry of hydration of portland cement*, Proceedings of Fourth International Symposium on the Chemistry of Cement, Washington, D.C., (1960), National Bureau of Standards Monograph, 43, Paper 3, pp.429-465.

COMPUTATIONAL MECHANICS (Abstracts)
WCCM VI in conjunction with APCOM'04, *Sept. 5-10, 2004, Beijing, China*

Some Thermo-Mechanical Problems of Deformations and Vibrations of Plate and Shell Structures under Thermal Loading

P. Biswas*

Head/Vibration Research Group, Von Karman Society for Advanced Study and Research in Mathematical and Social Sciences, Old Polic Line, Jalpaiguri-735101, West Bengal, India
e-mail: biswas_paritosh@yahoo.com

Abstract

In many fields of engineering structures, such as high-speed spacecrafts, nuclear power plants, chemical plants, offshore and ship-building structures, storage and building structures, plates and shells of different shapes find enormous applications as integral structural components. Such structural components are likely to be subjected to various kinds of static loads or excitations such as mechanical, seismic, blast, hydrodynamic, aerodynamic, with or without thermal loading. Engineers and scientists all over the world are exerting relentless efforts to design and construct economic, efficient and smart structures with very low failure probability. In view of this, for proper modeling, analyzing and designing, structural engineers should be well-acquainted with the static and dynamic behaviour of such structural elements with different boundary and loading conditions.

In the present paper extensive survey of the bibliography of previous works done in the above field have been made and presented in the Reference Section for use of future Researchers. For precise analysis of such problems one needs to consider middle surface strains leading to nonlinear equations. Compared to linear analysis, the literature on large (nonlinear) thermal deflections, buckling and vibrations of elastic plates and shells is somewhat scanty due to high nonlinearity of the classical von Karman field equations extended to thermal loading for static and dynamic cases. Since classical von Karman equations are highly nonlinear in the coupled form without any possibility of having closed-form solutions besides being difficult to deal with, H.M. Berger (in 1955) proposed a set of quasi-linear partial differential equations in the decoupled form. These equations were extended to the static and dynamic cases by many authors with the inclusion of thermal loading. Since Berger's equations faced severe criticisms owing to absurdity of results in several cases such as for movable edge boundary conditions, a separate set of differential equations, also in the decoupled form, were proposed by some authors as a new approach demanding elimination of the earlier constraints contained in the Berger's method. A number of problems were however, solved giving satisfactory results.

In this paper four problems have been considered based on the above three methods: Case I. Nonlinear vibrations of inhomogeneous elastic shells under severe thermal loading; Case II. Nonlinear dynamic analysis of shell structures under thermal loading (in the von Karman sense); Case III. Thermal stresses and deformations of shallow shell panel. Case IV. Basic governing equations for thermal vibrations have been derived using the "new approach" and solved for some plate and shell structures. A comparative discussion has been presented on the three methods of analysis.

COMPUTATIONAL MECHANICS (Abstracts)
WCCM VI in conjunction with APCOM'04, *Sept. 5-10, 2004, Beijing, China*

Automatic Generation of Nonlinear 3D-Shell Finite Element Code by Symbolic Manipulations

B. Brank [1*], J. Korelc[1], U. Bohinc[2]

[1] *Faculty of Civil and Geodetic Engineering, University of Ljubljana, p.p. 3422, Ljubljana, Slovenia*
[2] *Slovenian National Building and Civil Engineering Institute, Dimičeva 12, Ljubljana, Slovenia*
e-mail: bbrank@ikpir.fgg.uni-lj.si, jkorelc@fgg.uni-lj.si, uros.bohinc@zag.si

Abstract

The nonlinear analysis of shells has mainly been restricted to "classical" shell model with "Reissner-Mindlin" kinematics. Even in such relatively simplified setting the numerical implementation issues have been finally settled fairly recently; mostly due to complexity of nonlinear kinematics of large rotations. The recent research on the nonlinear shell problem, e.g. [1] and references therein, is attempting to close the gap between the "classical" shell model (which is basically 2D) and the 3D solid model by trying to produce a 3D-shell model which fits in between. Such a 3D-shell model should possibly retain the "classical" shell model computational efficiency and be able to predict the 3D stress field in thicker shells that would be close to the "true" 3D solid stress field. A task needed to be addressed is to identify an optimal 3D-shell model and its optimal finite element implementation among a vast set of possible 3D-shell models and implementation procedures. This can be achieved by developing a model error estimator in a sense of [2], which would indicate an error of a 3D-shell formulation with respect to the 3D solid formulation for a problem at hand. In order to be able to study performance of different 3D-shell finite element formulations, one needs to produce their finite element codes in an efficient way. Symbolic manipulations, more precisely, automatic generator of numerical codes AceGen [3], has been used to that end. The generator is a Mathematica package able to resolve the mayor problem of symbolic generation of nonlinear finite element arrays (e.g. stiffness matrix, etc.), which is an exponential growth of size of the derived expressions. An approach, implemented in AceGen, avoids this problem by combining: symbolic and algebraic capabilities of Mathematica, automatic differentiation technique, simultaneous optimization of expressions and theorem proving. The generator translates final symbolic formulas in a compiled language and incorporates the code into a nonlinear finite element analysis environment (e.g. Feap). The first aim of this paper is to present how the above mentioned automatic generator can produce finite element codes of complex 3-D shell finite rotation formulations efficiently. The second aim is to present the initial results of our work on model error estimation for shell formulations.

REFERENCES

[1] B. Brank, J. Korelc, A. Ibrahimbegovic, *Nonlinear shell problem formulation accounting for through-the-thickness stretching and its finite element implementation*, Computers and Structures, 80, (2002), 699-717.

[2] E. Stein, S. Ohnimus, *Coupled model- and solution-adaptivity in the finite-element method*, Comput. Methods Appl. Mech. Engrg., 150, (1997), 327-350.

[3] J. Korelc, *Automatic generation of numerical codes with introduction to AceGen 4.0 symbolic code generator*, University of Ljubljana, available from www.fgg.uni-lj.si/Symech/, (2003).

COMPUTATIONAL MECHANICS (Abstracts)
WCCM VI in conjunction with APCOM'04, *Sept. 5-10, 2004, Beijing, China*

On the Membrane Locking in the Computation of Thin Elastic Shells, Some New Tests and Benchmarks

Daniel Choi *

Laboratoire de Math′ematiques Nicolas Oresme, Universit′e de Caen, Bld Mar′echal Juin, 14032 Caen, France
e-mail: choi@meca.unicaen.fr

Abstract

In the framework of the linear theory of thin shell (Koiter or Naghdi Model) the asymptotic analysis shows two distinct behaviors of the shell depending on whether the middle surface of the shell is geometrically rigid or not, that is to say, whether it admits or not displacement leaving the intrinsic metric of the middle surface unchanged (in linearized sense). This analysis enlights numerical problems encountered in the computation of very thin shells, see [1]. It is emphasized that the geometrical rigidity is only driven by the geometry and the boundary conditions, not the material.
A thin shell can then be classified as [with pure flexion] inhibited or not inhibited according the terminology of E. Sanchez-Palencia; it is also read as flexural and membrane dominated [4].
Actually, the asymptotic analysis, when thickness $\to 0$, shows that the limit solution is also solution of a limit problem, we call bending limit problem, which stands in space of admissible inextensional displacements whenever, it is not reduced to set $\{0\}$. As it has been shown in [3], the non-robustness of a finite element method scheme, giving the membrane locking is due to the inability of the polynomials of the finite element method to perform such inextensional displacements.
Although pessimistic, for irreductible formulation, the last assertion is to be nuanced as the membrane locking can appear significant only for very thin elastic shell, maybe physically unrealistic.
As a benchmark for the membrane locking, we use the solution of bending limit problem on specific examples. Following a rotational point of view of the theory of inextensional displacement as in [2], these solutions are obtained thanks to a new expression of the flexion bilinear energy form, based on cartesian expressions instead of covariant.
We propose here some new numerical tests and compare Ganev-Argyris, DKT and MITC schemes on cylindrical and hyperbolic non-inhibited shell to the solution of the limit problem, showing membrane locking appearing at different levels.

REFERENCES

[1] D. Chapelle, K. J. Bathe, *Fundamental considerations for the finite element analysis of shell structures*, Computers and Structures, 66(1), (1998), 19-36.

[2] D. Choi, *Computations on thin non-inhibited hyperbolic elastic shells. benchmarks for membrane locking*, Journal Mathematical Methods Applie Sciences, 22(15), (1999), 1293-1321.

[3] D. Choi, F. J. Palma, E. Sanchez-Palencia, M. A. Vilariño, *Remarks on membrane locking in the finite element computation of very thin elastic shells*, Mod′elisation Math′ematique et Analyse Num′erique, 32(2), (1998), 131-152.

[4] P. G. Ciarlet, *Mathematical elasticity volume III : theory of shells*, Studies in Mathematics and its Applications, North Holland, (2000).

COMPUTATIONAL MECHANICS (Abstracts)
WCCM VI in conjunction with APCOM'04, *Sept. 5-10, 2004, Beijing, China*

Numerical and Experimental Study of Dent Resistance of Thermoplastic Composite Shells

Z. Cui [*], G. Moltschaniwskyj, D. Bhattacharyya

Centre for Advanced Composite Materials, Department of Mechanical Engineering The University of Auckland, New Zealand
e-mail: zcui001@ec.auckland.ac.nz, {g.moltschaniwskyj, d.bhattacharyya}@auckland.ac.nz

Abstract

The paper aims at describing the large deformation behaviour of composite shells under concentrated transverse loading quasi-statically. Thermoplastics reinforced with woven and knitted fabrics have been investigated. The composite shells with radius to thickness ratio of 22~53 have been examined both numerically and experimentally. Non-linear stress analysis technique using Finite Element Method has been adopted to predict the large deformation responses including buckling. The ways of simulating bifurcation buckling have been explored either in a numerical disturbance manner or as an initial imperfection input. Lost structural stability and delamination are two major modes of failure. Panels with single curvature would collapse while the doubly-curved shells would still take up load after initial buckling. Rolling-plastic-hinge post-buckling responses are evident experimentally for doubly-curved shells, which is still a subject of further simulation work. Rupture or severe localized plastic deformation would occur eventually for the shallow domes at the contact spot. Some interesting buckling modes due to material orthotropy have been presented as well. Experimental results have well agreed the predicted behaviour across the whole deformation region

REFERENCES

[1] K.-J. Bathe, *Finite element procedures*, Upper Saddle River, Prentice Hall, (1996).
[2] D. O. Brush, B. O. Almroth, *Buckling of bars, plates, and shells*, McGraw-hill, Kogakusha, (1975).
[3] B. L. Wardle, P. A. Lagace, *Behavior of composite shells under transverse impact and quasi-static loading,* AIAA, 36(6), (1998), 1065-1073.
[4] K. Kim, G. Z. Voyiadjis, *Non-linear Finite Element Analysis of Composite Panels,* Composites Part B: Engineering, 30, (1999), 365-381.
[5] Z. Cui, D. Bhattacharyya, G. Moltschaniwskyj, *Limit point buckling of cylindrical composite shells compressed transversely between two rigid platens*, in ICCE10, New Orleans, USA, (2003).
[6] Z. Cui, D. Bhattacharyya, G. Moltschaniwskyj, *Buckling response of composite shells compressed between rigid platens,* Composites Part B: Engineering, (submitted).
[7] Z. Cui, G. Moltschaniwskyj, D. Bhattacharyya. *Large deformation behaviour of composite domes compressed between rigid platens*, in ICCS-12, Monash University, Melbourne, Australia, (2003).
[8] Z. Cui, G. Moltschaniwskyj, D. Bhattacharyya, *Non-linear response of composite domes compressed between rigid platens,* International Journal on composite structures, (accepted).

COMPUTATIONAL MECHANICS (Abstracts)
WCCM VI in conjunction with APCOM'04, *Sept. 5-10, 2004, Beijing, China*

An Advanced Finite Element for Nonlinear Analyses of Shell Structures

M. Duan *

School of Engineering, James Cook University, Queensland 4811, Australia
e-mail: mei.duan@jcu.edu.au

Abstract

At present available general-purpose finite element analysis packages have increased considerably to apply in engineering analysis. Numcrous significant gains have been made in various fields. The finite element analysis of shell structures has been a very active research field for two decades. So far, much effort has been implemented for shell analyses and many different shell elements have been proposed. However, it is not still enough success in describing the feature of practical projects. The progress of many engineering analyses is still limited in several cases because of the shortcoming of the classical finite element formulation.

The main objective of this paper is to develop and evaluate an advanced nonlinear shell element based on variational theory. The formulation and characteristics of the element are shown and examined. The numerical analyses of shell structures will be efficiently implemented and compared with previously presented elements [1-3]. The reliability and effectiveness of the presented element are assessed. The high accuracy and effective convergence of the element are revealed by theoretical and numerical analyses. The practical analysis results will indicate that the presented shell element will have a good application forecast in structural engineering analysis.

REFERENCES

[1] K. J. Bathe, E. N. Dvorkin, *A formulation of general shell elements – the use of mixed interpolation of tensorial components*, Int. J. Num. Meth. Eng., 22, (1986), 697-722.

[2] T. H. H. Pian, *State-of-the-art development of hybrid/mixed finite element method*, Finite Elem. Anal. Design, 21, (1995), 5-20.

[3] O. C. Zienkiewicz, D. Lefebvre, *A robust triangular plate bending element of the Reissner-Mindlin type*, Int. J. Num. Meth. Eng., 26, (1988), 1169-1184.

COMPUTATIONAL MECHANICS (Abstracts)
WCCM VI in conjunction with APCOM'04, *Sept. 5-10, 2004, Beijing, China*

Analysis and Approximation of a New Model Arising in Nonlinear Shell Theory

Liliana Gratie *

Liu Bie Ju Centre for Mathematical Sciences, City University of Hong Kong, 83 Tat Chee Avenue, Kowloon, Hong Kong, China
e-mail: mcgratie@cityu.edu.hk

Abstract

Equations that generalize the classical Marguerre-von Kármán equations for a nonlinearly elastic shallow shell have been recently identified (see [3]) by means of techniques from asymptotic analysis. The main feature of this model is that it allows more realistic boundary conditions that may change their type along the lateral face of the shell. Yet the boundary conditions on the Airy function can still be determined from the data of the problem.
These more general equations form a semi-linear elliptic system, which can be reduced to a single "cubic" operator equation whole sole unknown is the vertical displacement of the shell. This equation generalizes a cubic operator equation introduced by M.S. Berger and P. Fife for analyzing the von Kármán equations for a nonlinearly elastic plate (see [1]).
The main difficulty in handling such an equation lies in the lack of strict positivity in the leading "cubic" term, by contrast with the classical Marguerre-von Kármán equations. Nevertheless, an elegant compactness method due to J.L. Lions (see [4]) allows us to establish both the existence of a solution to this operator equation (see [2]) and the convergence of its finite element approximation.

REFERENCES

[1] P. G. Ciarlet, *Mathematical Elasticity*, *Vol. II: Theory of Plates*, North-Holland, Amsterdam, (1997).
[2] P. G. Ciarlet, L.Gratie, *On the existence of solutions to the generalized Marguerre-von Kármán equations*, (to appear).
[3] L. Gratie, *Generalized Marguerre-von Kármán equations of a nonlinearly elastic shallow shell*, Applicable Analysis, 81, (2002), 1107-1126.
[4] J. L. Lions, *Quelques Méthodes de Résolution de Problèmes aux Limites Non Linéaires*, Dunod, Paris, (1969).

COMPUTATIONAL MECHANICS (Abstracts)
WCCM VI in conjunction with APCOM'04, *Sept. 5-10, 2004, Beijing, China*

A Unified Stress Update Algorithm for Explicit Transient Shell Dynamics

Jason Har*

Korea Institute of Aerospace Technology, Korean Air, 461-1, Junmin, Yusung, Taejon, Korea
e-mail: mephdjh@krpost.net

Abstract

This paper presents applications of the objective stress rates of the Kirchhoff or Cauchy stress to a unified stress update algorithm for transient shell dynamic analysis within the context of explicit time integration. The Eulerian rate constitutive equations of hypo elasto-plastic materials are based on the additive decomposition of the rate of decomposition tensor and the assumption that its elastic part is characterized to be hypoelastic with grade zero, i.e., the fourth-rank constant isotropic tensor. Several objective stress rates are derived through the Lie derivative of the Kirchhoff or Cauchy stress tensor. Those stress rates cover, for instance, the Jaumann, Green-Naghdi, Truesdell, Oldroyd, Cotter-Rivlin stress rates. Further a unified stress update algorithm of those stress rates, based on the radial return method and the phenomenological plasticity theory with combined isotropic and kinematic hardenings, is presented, and their results are respectively compared through some transient shell dynamics examples. In the implementation of the unified stress update algorithm, the Belytschko-Lin-Tsay shell theory is employed. For the Kirchhoff stress formulation, the Jacobian determination algorithm is newly presented for the shell theory. A modified secant algorithm for the zero transverse normal stress of the shell is also proposed. Several numerical examples are demonstrated including non-contact transient shell dynamics examples and pipe whip problems along with the master-slave slideline contact search algorithm, by which proposed algorithms are compared with respect to the accuracy and effectiveness.

REFERENCES

[1] C. Truesdell, *Hypo-elsticity,* J. Rat'l Mech. Anal., 4, (1955), 83-133.

[2] J. K. Dienes, *On the analysis of rotation and stress rate in deformation bodies,* Acta Mechanica, 32, (1979), 217-232.

[3] L. M. Taylor, D. P. Flanagan, *PRONTO 3D: a three-dimensional transient solid dynamics problem,* Sandia National Laboratories, SAND87-1912, (1989).

[4] C. ruesdell, *Corrections and additions to: the mechanics foundations of elasticity and fluid dynamics,* J. Rat'l Mech. Anal., 2, (1953), 593-616.

[5] J. C. Simo, K. S. Pister, *Remarks on rate constitutive equations for finite deformation problems: computational implications,* Comp. Meth. Appl. Mech. Engrg., 46, (1984), 201-215.

[6] C. Truesdell, W. Noll, *The non-linear field theories of mechanics,* 2nd ed., Springer-Verlag, Berlin, (1992).

[7] J. Har, R. E. Fulton, *A parallel finite element procedure for contact-impact problems,* Engineering with Computers, 19, (2003), 67-84.

COMPUTATIONAL MECHANICS (Abstracts)
WCCM VI in conjunction with APCOM'04, *Sept. 5-10, 2004, Beijing, China*

Dynamic Response Analysis of R/C Cooling Tower Shell

T. Hara [1*]

[1]*Department of Civil Engineering and Architecture, Tokuyama College of Technology, 3538 Kume Takajo, Shunan, 745-8585, Japan*
e-mail: t-hara@tokuyama.ac.jp

Abstract

R/C cooling towers are used for many kinds of industrial and power plants. These are huge structures and also show thin shell structures. R/C cooling structures are subjected to its self weight and the dynamic load such as an earthquake motion and a wind effects. Especially, dynamic analyses of these structures are important factor to design R/C cooling tower structures. However, it is difficult to solve these problems because R/C structure composes of a concrete and reinforcements and both material shows nonlinear behavior. In particular, nonlinear behavior of concrete is complicate to represent mathematically. The dynamic responses of these materials are complicate as well. Therefore, a quasi-static loading is applied to design an R/C cooling tower instead of dynamic response analysis. In addition, these structures have huge surfaces of concrete with increasing its constructional height and R/C shell structure is usually placed on the supporting columns to take a cold air into it. R/C cooling tower represents the combinations of R/C shell and R/C column structures.

Consequently, we must evaluate the huge R/C shell structural behavior with local discrete elements. To solve such problems, many numerical schemes have been proposed combining the axisymmetric analyses or modal analyses [1, 2]. However, it is difficult to represent the local behavior of these structures by such numerical scheme. To overcome these problems, the finite element method (FEM) is one of the useful schemes. On the other hand, applying FEM to such problems, we must spend much numerical effort to solve these structures under a dynamic loading. Particularly, dynamic analyses are quite time consuming problems.

In this paper, dynamic behavior of R/C cooling tower shell under a earthquake loading are analyzed by use of FEM. To reduce the computational efforts to solve such problems, the technique for a parallel computing is applied. To solve the nonlinear dynamic response of a huge cooling tower with local deviation, the element by element (EBE) parallel approach is adopted by using of PC cluster.

From the numerical analyses, the nonlinear dynamic response of R/C cooling tower is examined and the effects of the combination of the shell and column structural behavior are calculated. Also, the applicability of EBE approach to the huge R/C structures is discussed.

REFERENCES

[1] T. Hara, P. L. Gould, *Local-global analysis of cooling tower with cutouts*, Computers & Structures, 80, (2002), 2157-2166.

[2] C. Lang, R. Meiswinkel, U. Wittek, *Anwendung von Schalenringelementen zur nichtlinearren dynamischen Berechnung von Stahlbeton-Rotationsschalen*, Beton- und Stahlbetonbau, 98, (2003), 123-134, (in German).

COMPUTATIONAL MECHANICS (Abstracts)
WCCM VI in conjunction with APCOM'04, *Sept. 5-10, 2004, Beijing, China*

Non-Linear Stability of Cylindrical Shells with Stiffened Cutouts Subjected to Axial Compression

M. Z. Kabir [*], A. Rismandoust

Dept. of Civil and Environmental Engineering, Amir Kabir University of Technology (Polytechnic of Tehran), Tehran, Iran
e-mail: R8024406@cic.aut.ac.ir

Abstract

Many structures in aerospace and civil engineering are made of cyl indrical shells, such as vessels and fuselage of airplanes and missiles. Some of them are subjected to axial compression in some cases of loading in their service life. On the other hand, some causes such as accessibility necessity (or facilities) dictate existence of one or more cutouts on the structure. These cutouts cause relatively much decrease of bearing load of structure dependent on size of cutouts. Thus stiffening of cutouts is an interesting scheme to restoration of bearing of structure.
An experimental and numerical analysis investigation is performed on the stability of cylindrical shells with cutouts. Both linear and nonlinear analysis is done by use of the finite element method. The analysis is based on the computer program ANSYS using shell an d beam elements and the effect of geometrical nonlinearity is also considered. In primary step, the classical and numerical buckling load of a perfect cylinder is calculated and compared with together. It shows very good agreement between them with maximum 0.2 percent of error ratio. The sensitivity analysis on the number of elements is carried on build up the numerical model leading to find out the optimum mesh size. Two opposite cutouts are assigned due to the symmetry of structure and omitting the unwelc ome effects, that only the influence of cutouts properties is considered in the analysis. At next step, analysis is done by taking into account the nonlinearity of material and geometry to simulate the actual structure much carefully. Then the variation of applied loads versus size and shape of cutouts is investigated. The most important step is the analysis of reinforced cutouts that contains two types of opening frame and stringer stiffeners. Results indicate that stringer stiffeners enhance both membrane and flextural rigidity around cutouts and is more effective reinforcements. Conversely, the opening frame stiffeners that stiffen only the membrane rigidity of cutouts region are not much effective on load carrying capacity of structure. Then sensitivity analysis is performed versus dimensions and thickness of stiffeners in order to find optimum shape. The main aspect in selecting of optimal shape of stiffeners is reducing of stress concentration and increasing the load carrying of cylinder. Due to verifying the numerical analysis with experiments, twelve cylinders containing unstiffened and stiffened cutouts are manufactured. Comparing between analysis and experiment indicates good agreement.

COMPUTATIONAL MECHANICS (Abstracts)
WCCM VI in conjunction with APCOM'04, *Sept. 5-10, 2004, Beijing, China*

Four-Field Mixed Formulation for a Solid-Shell Element

S. Klinkel [1*], F. Gruttmann [2], W. Wagner [1]

[1] *Institut f. Baustatik, Universitaet Karlsruhe (TH), Kaiserstr. 12, Karlsruhe, 76131, Germany*
[2] *Institut f. Werkstoffe u. Mechanik im Bauwesen, Technische Universitaet Darmstadt, Alexanderstr. 7, Darmstadt, 64283, Germany*
e-mail: sk@bs.uni-karlsruhe.de, gruttmann@iwmb.tu-darmstadt.de, ww@bs.uni-karlsruhe.de

Abstract

The paper is concerned with a geometrically non-linear solid shell finite element formulation, which is based on a variational principle of the Hu-Washizu type. The novel aspect is that four independent fields are considered in the variational principle, which reads

$$\Pi\left(\boldsymbol{u}, \tilde{\boldsymbol{S}}, \tilde{\boldsymbol{E}}, \hat{\boldsymbol{E}}\right) = \int_{B_0} \left(W_0(\tilde{\boldsymbol{E}}, \hat{\boldsymbol{E}}) + \tilde{\boldsymbol{S}} : (\boldsymbol{E} - \tilde{\boldsymbol{E}}) - \tilde{\boldsymbol{S}} : \hat{\boldsymbol{E}} \right) \mathrm{d}V + \Pi^{ext}(\boldsymbol{u}) .$$

Here $\boldsymbol{u}$, $\tilde{\boldsymbol{S}}$, $\tilde{\boldsymbol{E}}$, $\hat{\boldsymbol{E}}$ are the displacements, the stress field, the independent strain field and the enhanced assumed strain (EAS) field, respectively. The strain energy W_0 is a function of $\tilde{\boldsymbol{E}}$, $\hat{\boldsymbol{E}}$ and not of the Green-Lagrangean strains $\boldsymbol{E}$. In accordance with [1] it is supposed that the stress and the enhanced assumed strain interpolations are orthogonal, but in contrast to [1] this leads not to an elimination of the stress field from the finite element equations. Here, the orthogonality condition determines the interpolations for the EAS field and the stresses on element level. The independent strain field is approximated with the same interpolations as the stress field. For the displacements tri-linear shape functions are considered within the isoparametric concept. Shear locking is treated by using the assumed natural strain (ANS) interpolations for the transverse shear strains as proposed by [2]. To circumvent locking due to artificial thickness strains especially for thin shell structures ANS interpolations for the thickness strains are applied, which were introduced in [3]. The static condensation on element level leads to a simple low order hexahedral solid shell element formulation with 8-nodes and three displacement degrees of freedom per node.
Some numerical tests show that only three EAS parameters are necessary to improve the Poisson-thickness locking and to pass the membrane and the out-of-plane bending patch tests. The good behavior in membrane and plate bending problems also with distorted meshes is demonstrated. One of the most striking features of the model is the robustness of the formulation with respect to the size of the load step. Even for huge load steps the element formulation needs only a few equilibrium iterations to obtain convergence.

REFERENCES

[1] J. C. Simo, M. S. Rifai, *A class of mixed assumed strain methods and the method of incompatible modes*, Int. J. Num. Meth. Eng., 29, (1990), 1595-1638.
[2] E. N. Dvorkin, K.-J. Bathe, *Continuum mechanics based four-node shell element for general non-linear analysis*, Eng. Comput., 1, (1984), 77-88.
[3] P. Betsch, E. Stein, *An assumed strain approach avoiding artificial thickness straining for a non-linear 4-nod shell element*, Commun. Numer. Methods Eng., 11, (1995), 899-909.

COMPUTATIONAL MECHANICS (Abstracts)
WCCM VI in conjunction with APCOM'04, *Sept. 5-10, 2004, Beijing, China*

Matrix Method in the Contact Problems for Anisotropic Cylindrical Shells

A. Maksymuk [1], N. Shcherbyna [1*], B. Kopey [2], V. Kopey [2], V. Rozhoniuk [3]

[1] *Pidstryhach Institute of Applied Problems of Mechanics and Mathematics, NAS Ukraine, Naukova 3b, 79601 Lviv, Ukraine*
[2] *National Technical University of Oil and Gas, Karpatska, 15, 76019 Ivano -Frankivsk, Ukraine*
[3] *National Stock Joint Company "NaftaGas of Ukraine", Khmelnytsky, 6, 03000 Kyiv, Ukraine*
e-mail: khaldabd@yandcx.ru

Abstract

The efficient engineering application of structural joint thin -walled elements working in complicated conditions of contact l oading is associated with the solution of a contact problem. It is notice that, the contact problem for thin -walled elements (shells, plates) made from composite materials have specific character of statement and solution. The difficulty of this problem co nsists in determining the contact stresses.
The present paper is devoted the authors' new investigations on mechanics of contact interaction of anisotropic cylindrical shells including the mathematical models and methods for realized. In this study, the contact problem for shells is given in a refined statement. A mathematical is based on the generalized Timoshenko-Reissner theory of anisotropic shells. Choice of this model includes its solvability requirements in engineering calculations and correspondence of the results obtained to experimental data. The present model allows taking account of specific properties of mechanical behaviour for composite shells (the effect of transverse shear deformation, normal stretch and etc.). In particular, proposed model allows considering different kinds of contact interaction (friction, slipping, wear and others) and, finally, to obtain non -contradictory solutions.
To determine the contact stiffness and contact stresses we developed a matrix method. Using the numerical algorithm we obtain the approximate solution of the contact problem. The effectiveness and robustness of proposed approach is illustrated on several examples. This technique provides the high accuracy of results acceptable for engineering practice. Correspo nding solution within the framework of Timoshenko type shear model and classical Kirchhoff model are obtained as particular cases.
The effect of geometric and physical -mechanical parameters of the cylindrical shell subjected to contact loading (axisymmetri c and plane cases) on the contact pressure and displacements distribution is investigated in detail.
The numerical results obtained for the case of finite -length transversely loading cylindrical shell of fiber-reinforced composite materials are analyzed a pplying to concrete technique objects such as the tubes and its joints.

COMPUTATIONAL MECHANICS (Abstracts)
WCCM VI in conjunction with APCOM'04, *Sept. 5-10, 2004, Beijing, China*

Mass Matrix Lumping in Transient and Vibration Analysis of Shells

Elias A. Paraskevopoulos *, Demosthenes G. Talaslidis

Department of Civil Engineering, Aristotle University Thessaloniki, Institute of Structural Analysis, Mail Stop 502, Thessaloniki, GR-54124, Greece
e-mail: talaslid@civil.auth.gr

Abstract

There is a series of benefits associated with lumped (diagonal) mass matrices in transient or vibration analysis of shells. Most of the existing methods are based on ad hoc techniques or intuition, others devoid of physical meaning. They may yield mass matrices with zero or negative diagonal elements, or lead to inconsistencies in case of degrees of freedom with different physical meaning (e.g., displacements and rotations, as it is the case in shells). Furthermore, methodological issues related to the selection of appropriate shape functions are lacking. The foregoing discussion reveals the need for devising a systematic and mathematically sound procedure.

In this paper a variationally consistent methodology is presented, which yields simple and reliable elemental formulations and diagonal mass matrices for quadrilateral flat shell elements. A modified Hu-Washizu principle in elastodynamics is employed. The formulation incorporates the effects of transverse shear. Attention is being given to reducing the various types of excessive energy (locking). The displacements, rotations, stresses, and strains are treated as independent variables. Furthermore, the translational and angular velocities are also considered as independent variables. The relationships between the weak velocities and the time derivatives of the displacements and rotations are introduced via Lagrange multipliers (momentum type variables) into the functional. The kinetic energy is expressed in terms of the weak velocities. In choosing appropriate approximations for the aforementioned variables, we follow a systematic procedure and utilize our previous experience with elemental approximations based on the generalized Hu-Washizu principle (see, e.g., [1-4]). The velocities, rotations and mass density are assumed to have piecewise constant approximations within sub domains of the element. The momentum type variables are approximated by employing a combination of piecewise constants and Dirac distributions. Finally, it is proved that the proposed approach ensures consistency and stability.

REFERENCES

[1] G. Wempner, D. Talaslidis, C.-M. Hwang, *A simple and efficient approximation of shells via quadrilateral elements,* ASME, J. of Appl. Mech., 49, (1982), 115-120.
[2] G. Wempner, D. Talaslidis, *Mechanics of solids and shells: theories and approximations,* CRC-Press, Boca Raton, USA, (2003).
[3] E. Paraskevopoulos, D. Talaslidis, *Reduction of excessive energy in the four-noded membrane quadrilateral element,* Comput. Methods Appl. Mech. Engrg, (submitted).
[4] E. Paraskevopoulos, D. Talaslidis, *A rational approach to mass matrix diagonalization in 2D elastodynamics,* Internat. J. Numer. Methods Engrg, (accepted).

COMPUTATIONAL MECHANICS (Abstracts)
WCCM VI in conjunction with APCOM'04, *Sept. 5-10, 2004, Beijing, China*

Stress Analysis of Pressure Shaft Bifurcation Liner with Stiffeners for Pressure Loading

C. V. Ramakrishnan [1], B. P. B. Somashekar [1], M. R. Joshi [1*], P. K. Yadav [2]

[1] *Department of Applied Mechanics, Indian Institute of Technology, New Delhi, 110 016, India*
[2] *WAPCOS, New Delhi, 110066, India*
e-mail: cvrama@am.iitd.ernet.in

Abstract

The bifurcation for hydraulic tunnel liner pipe is analysed for stresses. The unsymmetric bifurcation presents complexities in both modeling and stress analysis, apart from the difficulties in fabrication. The hydraulic heads and flows are quite high hence very high level of reliability needs to be established with the design. The finite element model consists of intersecting inlet pipe and outlet pipes joined together with ring beams and a sickle plate. These components are modeled as assemblage of shell elements [3] capable of exhibiting membrane and bending behaviour. To restrict the use of excessive plate thickness of the shell, stiffeners have been used. Surface modeling is carried out for each of the parts of the bifurcation separately and then the assembly is obtained.
Meshing of the bifurcation should be carried out very carefully. There are two possible approaches-meshing the total assembly or meshing the parts and assembling the meshes. Both the approaches exhibit difficulties in ensuring compatibility along common intersection curves if automated meshing is employed. Hence this approach was abandoned and manual meshing has been carried out. This ensured very accurate location of the common nodes along intersecting boundaries and the node collapse procedure [1] is used to ensure compatibility. The closely spaced stiffeners have been modelled both as shell and eccentric beam.
Detailed analysis of stress distribution at the shell-shell, shell-ring beam and shell-sickle plate functions are carried out using ABAQUS software [2]. Preliminary results for the geometric nonlinear analysis [4] under external pressure loading are also reported.

REFERENCES

[1] *SDRC/I-DEAS Manuals and User's Guide*, SDRC Publications, (2000).
[2] *ABAQUS manuals and User's Guide*, Hibbitt, Karlsson & Sorensen, Inc. (2002).
[3] O. C. Zienkiewicz, R. L. Taylor, *The finite element method, 5th edition, volume 2*, Butterworth-Heinemann, London, (2000).
[4] A. Lazar, *Transient dynamic finite element analysis of plate and shell structures with finite rotation,* PhD Thesis, IIT Delhi, (1996).

COMPUTATIONAL MECHANICS (Abstracts)
WCCM VI in conjunction with APCOM'04, *Sept. 5-10, 2004, Beijing, China*

An Efficient Quadrilateral Shell Element for Analysis of Thick and Thin Plates/Shells

Guojian Shao *, Jiashou Zhuo, Jingbo Su

College of Civil Engineering, Hohai University, Nanjing, 210098, China
e-mail: gjshao@163.com

Abstract

The popular plate and shell elements at present are developed generally based on Reissner-Mindlin plate theory. Mindlin plate elements can avoided that the interpolation functions of field variations require C^1-continuity based on classical Kirchhoff thin plate theory, and require only C^0-continuity. So, it will be very convenience to build the interpolation functions of field v ariations. But it will result in shear locking to apply directly for thin plate and thin shell structures. In order to eliminate the phenomenon, some methods have been established thus as the reduced and selective integration method and the assumed strain method. However, the use of these methods, though successful in some cases, produces the further problem of rank deficiency of the stiffness matrix or the spurious zero energy modes. In this paper, an efficient and versatile 4-node quadrilateral shell element is introduced for shell application. The element is composed of the Mindlin plate bending element and the membrane element with drilling degrees of freedom, named GSQ24 shell element. The proposed GSQ24 shell element has six degrees of freedom per node.

In the bending element, based on the potential energy functional of the Mindlin plate theory and the shear locking mechanism, the hierarchical displacement interpolation functions in this paper are used for lateral displacement and the bilinear interpolation functions are used for rotation. Besides, bilinear interpolation functions are essential for shear strains. In the membrane element, the skew-symmetric part of the displacement gradient is defined as the real rotation and the modified Reissner functional is used. In order to match the flat bending element, the displacement and rotation interpolations of the membrane element should correspond with the lateral displacement and rotation interpolations of the bending element.

GSQ24 shell element can avoid the problem of ill conditioning matrix when elements are co-plane. Moreover, order matching of displacement interpolation functions between bending element and membrane element can be solved properly. Thus, modeling of complex joint of shell surface sections and compatibility with other elements with rotation degrees of freedom can be easily made. GSQ24 shell element can pass the patch tests for both the bending and the membrane, and has no spurious zero energy modes and no locking. Detailed numerical studies show the excellent performance of GSQ24 shell element for thick and thin plates and shells.

REFERENCES

[1] A. Ibrahimbegovic, *Quadrilateral finite elements for analysis of thick and thin plates*, Comp. Meth. App. Mech. Eng., 110, (1993), 195-209.

[2] T. J. R. Hughes, F. Brezzi, *On drilling degrees of freedom,* Comp. Meth. App. Mech. Eng., 72, (1989), 105-121.

[3] T. H. H. Pian, C. C. Wu, *A rational approach for choosing stress terms for hybrid finite element formulation,* Int. J. Num. Meth. Eng., 26, (1988), 2331-2343.

COMPUTATIONAL MECHANICS (Abstracts)
WCCM VI in conjunction with APCOM'04, *Sept. 5-10, 2004, Beijing, China*

Failure Analysis of RC Shell Structures Using Volume Control Method

Ha-Won Song*, Sang-Hyeok Nam, June-Hee Lee, Keun-Joo Byun

School of Civil & Environmental Engineering, Yonsei University, Seoul 120-749, Korea
e-mail: song@yonsei.ac.kr, salute@cmme.yonsei.ac.kr, nuke0308@cmme.yonsei.ac.kr, byun@yonsei.ac.kr

Abstract

By introducing a pressure node which has an increment of pressure as an additional degree of freedom of a shell finite element, the so-called volume control method was developed by utilizing the pressure node added into the finite shell element to overcome the drawbacks of conventional load control method and displacement control method. In this study, a so-called path-dependant volume control method which improves the volume control method for the analysis of path-dependant behaviors of RC shell structures subjected to cyclic loading is proposed. The path-dependant volume control method is a technique which calculates control -volume change of the RC structure depending on the cyclic loading conditions like loading, unloading, and reloading and overcomes the limitation of controlling the volume change of the RC structure according to the loading history. For the finite element failure analysis of RC shell structures, each structure is discretized with multi-layered shell elements and in-plane two dimensional constitutive equations for concrete and reinforcements are implemented for each layer of the shell elements. The results of the analysis for RC slab, RC tank, and RC box culvert subjected to cyclic load using the path-dependant volume control method are verified with experimental results (See Fig.1 and Fig.2).

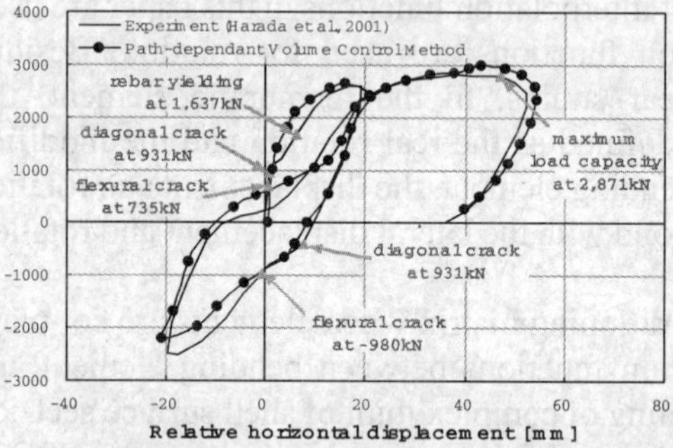

Fig. 1 Analysis of RC tank

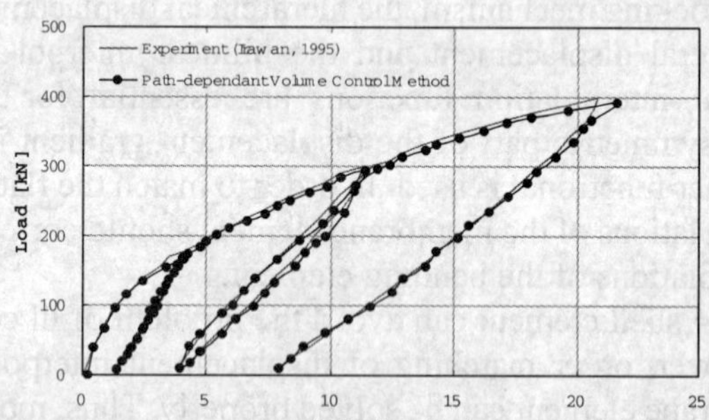

Fig. 2 Analysis of RC box culvert

REFERENCES

[1] H.-W. Song, S. H. Shim, K. J. Byun, K. Maekawa, *Failure analysis of RC shell structures using layered shell element with pressure node*, Journal of the Structural Engineering, 128(5), (2002), 655-664.

[2] H.-W. Song, S. H. Shim, K. J. Byun, *Finite element analysis of prestressed concrete containment vessel subjected to internal pressure*, Proceeding of the Third DIANA World Conference, Tokyo, Japan, (2002).

[3] K. Maekawa, H. Okamura, A. Pimanmas, *Nonlinear mechanics of reinforced concrete*, Spon Press, New York, USA, (2003).

COMPUTATIONAL MECHANICS (Abstracts)
WCCM VI in conjunction with APCOM'04, *Sept. 5-10, 2004, Beijing, China*

Exact Solutions for Vibration of Stepped Circular Cylindrical Shells

L. Zhang [1*], Y. Xiang [1, 2]

[1] *Centre for Construction Technology and Research, University of Western Sydney, Penrith South DC, NSW 1797, Australia*
[2] *School of Engineering and Industrial Design, University of Western Sydney, Penrith South DC, NSW 1797, Australia*
e-mail: 13001429@scholar.uws.edu.au, y.xiang@uws.edu.au

Abstract

This paper investigates the vibration behaviour of the cylindrical shells with variable thickness. The thickness variation of shells can be classified as change of thickness in axial direction or circumferential direction. The variation of the thickness of shells may be linear, parabolic or step-wise. Based on the Flügge thin shell theory, this paper is concerned with exact solutions for the vibration characteristics of circular cylindrical shells with step-wise thickness variations in the axial direction. A shell is first sub-divided into segments at the locations of the step variations and the state-space technique is employed to derive a homogenous differential equation system for each segment [1-2]. The shell boundary conditions and continuous conditions between segments are utilized to form the governing eigenvalue equation for the vibration of the stepped shell. A program is developed in this study to compute the eigenvalues and eigenvectors as governed by the differential equations of the Flügge shell theory. To ensure the correctness of the present results, comparisons are made with one paper available in the open literature based on the Donnell-Mushtari theory [3]. Exact frequency parameters are obtained and presented in tables and design charts for several stepped circular cylindrical shells. Shells with different edge support conditions and various combinations of step-wise thickness variations can be analysed by the proposed method. Furthermore, detailed consideration is given to the changes in natural frequencies for various modes of a shell as the step thickness ratio, locations of the steps, number of steps and shell length to radius ratio vary. The exact vibration results can serve as an important benchmark values for researchers to validate their numerical methods for such circular cylindrical shells.

REFERENCES

[1] Y. Xiang, Y. F. Ma, S. Kitipornchai, C. W. Lim, C. W. H. Lau, *Exact solutions for vibration of cylindrical shells with intermediate ring supports*, International Journal of Mechanical Sciences, 44, (2002), 1907-1924.

[2] Y. Xiang, K. M. Liew, S. Kitipornchai, *Exact buckling solutions for composite laminates: proper free edge conditions under in-plane loadings*, Acta Mechanica, 117(1-4), (1996), 115-128.

[3] J. Zhou, B. Yang, *Distributed transfer function method for analysis of cylindrical shells*, AIAA Journal, 33, (1995), 1698-1708.

COMPUTATIONAL MECHANICS (Abstracts)
WCCM VI in conjunction with APCOM'04, *Sept. 5-10, 2004, Beijing, China*

Bounds on Plastic Deformations for Structures in Plastic Shakedown

F. Giambanco, L. Palizzolo *, A. Caffarelli

Department of Structural Engineering & Geotechnics, University of Palermo, Viale delle Scienze, Palermo, 90128, Italy
e-mail: fgiamban@diseg.unipa.it, palizzol@diseg.unipa.it, alecaffa@msn.com

Abstract

In many cases of practical interest, elastic plastic structures are required to operate beyond their elastic limit when subjected to a combination of fixed and cyclic loads which can vary within a given load domain of convex hyperpolyedrical shape with vertices corresponding to a set of n independent basic loads, but with time history not specified. Under such conditions, the elastic shakedown theory provides useful tools in studying the behavior of the relevant structure; in particular, the so-called shakedown limit load multiplier provides an effective safety factor for the structure as well as the so-called bounding techniques provide suitably chosen limits on the plastic deformation related to the transient phase of the elastic shakedown [see, e.g., 1-5]. When the load multiplier is above the elastic shakedown limit, then the structure is exposed to a plastic shakedown behavior or to an incremental collapse condition. In both cases the steady-state response of the structure possesses the same periodicity features as the loads and can be determined by solving n linear complementarity problems (one for each basic load condition). Unfortunately, no information can be deduced in order to evaluate the plastic deformation occurring during the initial transient phase.
The present paper concerns the computation of bounds on suitable measures of the plastic deformations which characterize the transient phase response of elastic perfectly plastic structures subjected to a combination of fixed and cyclic loads in condition of plastic shakedown. A suitable bounding theorem is proposed and proved. It can be interpreted as a generalization of analogous theorems valid for elastic shakedown.

REFERENCES

[1] J. A. König, *On upper bounds to shakedown loads,* Z. Angew. Math., 59, (1979), 349-354.
[2] M. Capurso, L. Corradi, G. Maier, *Bounds on deformations and displacements in shakedown theory,* Matériaux et structures sous chargement cyclique, Ass. amicale des ingénieurs anciens eléves de l'E.N.P.C., Paris, (1979), pp. 231-244.
[3] C. Polizzotto, *A unified treatment of shakedown theory and related bounding techniques,* S. M. Archives, 7(1), (1982), 19-75.
[4] F. Giambanco, L. Palizzolo, *Optimal bounds on plastic deformations for bodies constituted by temperature-dependent elastic hardening material,* J. Appl. Mech., ASME, 64(3), (1997), 510-518.
[5] F. Giambanco, L. Palizzolo, *Computation of bounds on chosen measures of real plastic deformation for beams,* Comp. & Struct., 61(1), (1996), 171-182.

COMPUTATIONAL MECHANICS (Abstracts)
WCCM VI in conjunction with APCOM'04, *Sept. 5-10, 2004, Beijing, China*

An Algorithm of Hybrid Type Shakedown Analysis

Shun-ichi Kobayashi *, Masaki Nishikawa

Department of Civil and Earth Resources Eng., Kyoto University, Sakyo, Kyoto, 606-8501, Japan
e-mail: koba@basewall.kuciv.kyoto-u.ac.jp, nishikawa@basewall.kuciv.kyoto-u.ac.jp

Abstract

In spite of the developments of evolutional (time stepping) elasto-plastic analysis, shakedown analysis, recognized as a special and general case of limit analysis, is sometimes utilized to estimated safety factors for repeated loads because of its time and cost efficiency. From a mathematical point of view, shakedown analysis belongs to convex programming. As a consequence of this, both static shakedown analyses based on Melan's theorem and kinematic shakedown analysis based on Koiter's theorem, are dual to each other. These preferable features can be utilized for hybrid type shakedown analysis which deals with both static and kinematic variables simultaneously.

One of the authors developed a hybrid type rigid-plastic finite element method based on the interior point method, called Primal-dual Rigid-Plastic Finite Element Method (PDRPFEM). Firstly, he proposed Lagrangian for rigid plastic boundary value problems. Then, the duality is derived by infimum - supremum and supremum – infimum operations to the Lagrangian. According to his formulation, both static and kinematic variables are solved simultaneously under the constraints of a statically and plastically admissible stress field, a kinematically and plastically admissible velocity field and a complementarity condition. Numerical calculation is based on the interior point method for nonlinear optimization. During the iterative procedure, by introducing a barrier parameter which approximates the complementarity condition, a duality gap of objective functions of the upper and lower bound methods is monotonically converged to zero.

The abovementioned ideas are extended and developed to shakedown analysis in this paper. The authors propose an algorithm of hybrid type shakedown analysis based on the interior point method. Theoretically speaking, it should be noted that a dimension of a matrix used in the iterative procedure can be reduced to the size of PDRPFEM. This is a great advantage for numerical calculations in conjunction with the merits of the interior point method. Since the interior point method is recognized as the mighty numerical tool for large scaled problems in the field of optimization, the authors expect that the proposed algorithm is suitable for large scaled problems including three dimensional problems and fine finite element meshes.

REFERENCES

[1] S. Kobayashi, *Development of hybrid rigid plastic finite element method based on primal-dual interior point method*, Journal of Applied Mechanics, JSCE, 6, (2003), 95-106, (in Japanese).

[2] S. Kobayashi, *Limit and shakedown design in geotechnical engineering*, Doctoral Thesis, Kyoto University, Japan, (2004).

COMPUTATIONAL MECHANICS (Abstracts)
WCCM VI in conjunction with APCOM'04, *Sept. 5-10, 2004, Beijing, China*

Numerical Limit Analysis of Pressure Pipings with Local Wall-Thinnings

Yinghua Liu [*], Bingye Xu

Department of Engineering Mechanics, Tsinghua University, Beijing, 100084, China
e-mail: yhliu@mail.tsinghua.edu.cn, byxu@mail.tsinghua.edu.cn

Abstract

Limit analysis of structures is a practically useful branch of plasticity, which can determine the load-carrying capacities of structures and provide the theoretical foundations for structural designs and integrity assessments. In this paper, a numerical procedure is developed for plastic limit analysis of 3-D elastic-perfectly plastic bodies under complex loads. The method is based on the lower-bound limit theorem and von Mises yield criterion so that the lower-bound limit analysis can be conducted by solving a nonlinear mathematical programming problem. A SQP algorithm and a dimension reduction-based technique are used to solve the discretized finite element optimization formulation. A conception of active constraint set is introduced, so that the number of constraints can be reduced greatly. The basis vectors of reduced residual stress spaces are constructed by performing an equilibrium iteration procedure of elasto-plastic finite element analysis.

The numerical procedure is applied to carry out the plastic limit analysis of pressure pipings with local wall-thinnings under internal pressure, bending moment and axial force. Local wall-thinning caused by corrosion, erosion, mechanical damage or by grinding to eliminate surface cracks is a common volumetric defect on the surfaces of pipings. It may reduce the load-carrying capacities and even lead to severe failure accidents of pipings. 3-D finite element models of local wall-thinned pipings are established and used here for limit analysis. Based on a large number of calculated examples, the effects of different shapes and sizes of small area, axial, circumferential, and large area local wall-thinnings on plastic limit loads and failure modes of pressure pipings under the combined loads of internal pressure, bending moment and axial force are investigated and analyzed. The plastic limit load estimation formulae of local wall-thinned pressure pipings are derived. Some typical failure modes of local wall-thinned pipings are studied. By comparing the calculated results with the piping design criteria, an engineering safety assessment method for pressure pipings with local wall-thinnings is presented. In most cases, this assessment method needn't conduct the complicated strength calculation of piping systems by the finite element method and can be used easily and conveniently.

REFERENCES

[1] E. Stein, G. Zhang, J. A. König, *Shakedown with Nonlinear Hardening Including Structural Computation Using Finite Element Method*, Int. J. Plasticity, 8, (1992), 1-31.

[2] K. Schittkowski, *NLPQL, a Fortran subroutine solving constrained nonlinear programming problems*, Annals of Operations Research, 5, (1985), 485-500.

COMPUTATIONAL MECHANICS (Abstracts)
WCCM VI in conjunction with APCOM'04, *Sept. 5-10, 2004, Beijing, China*

Static Limit Analysis of Orthotropic Structures

Hongtao Zhang*, Yinghua Liu, Bingye Xu

Department of Engineering Mechanics, Tsinghua University, 100084, Beijing, China
e-mail: zhanght99@mails.tsinghua.edu.cn, {yhliu,byxu}@mail.tsinghua.edu.cn

Abstract

Limit analysis problem, coupled with finite element method and mathematical programming method, is a direct method to predict the load bearing capacity of solids and structures. Many work have been devoted to isotropic structures based on von Mises yield criterion. This paper turns to consider the orthotropic structures. Based on Hill's yield criterion [1], a numerical method is presented for the static limit analysis of 3-D orthotropic structures, where the temperature parameter method is used to construct the self-equilibrium stresses.
The application of this method is listed. First, the lower bound of the punch problem is studied and compared with the upper bound got by Capsoni et al [2], who obtained it by using kinemetic limit theory. The comparison illustrates that this method releases the severe constraints of Hill's yield criterion imposed on the kinematic method and avoids the limitation of the upper bound method. Next, the cylindrical-conical- cylindrical combined shell under internal pressure are solved, and compared with the upper bound. The lower bound limit loads of orthotropic pipeline under internal pressure, axial tension and bending moment are calculated, and the interactive curves are plotted. Numerical results show the applicability of this method.

REFERENCES

[1] R. Hill, *A theory of the yielding and plastic flow of anisotropic metals*. Proceedings of the Royal Society of London, Series A, 193, (1947), 281-297.
[2] A. Capsoni, L. Corradi, P. Vena, *Limit analysis of orthotropic structures based on Hill's yield condition*, International Journal of Solids and Structures, 38, (2001), 3945-3963.
[3] J. Pastor, S. Turgeman, J. P. Boehler, *Solution of anisotropic plasticity problems by using associated isotropic problems*, International Journal of Plasticity, 6(2), (1990), 143-68.

COMPUTATIONAL MECHANICS (Abstracts)
WCCM VI in conjunction with APCOM'04, *Sept. 5-10, 2004, Beijing, China*

3-D Finite Element Analysis of Egg-Shaped Digester and Reliability Analysis Based on Probability Density Evolution Method

Huaming Chen *, Jie Li, Jianbing Chen

Department of Building Engineering, School of Civil Engineering, Tongji University, 200092, Shanghai, China
e-mail: chenghuamin@sohu.com

Abstract

In the recent years, the construction of the s ewage treatment plant has increased day by day and the capacity of the construction of sewage treatment has increased too. So adopting reasonable, economical, reliable digester seems very important.
The egg-shaped digester is large-scale shell structure whose thickness of cross section always changes and the fluid-solid interface of the pre-stressed egg-shaped digester is curve. When the digester is vibrating, fluid will slosh following with vibration of the digester. Fluid Sloshing exerts an influence to the vibration of digester on the contrary. So it is comparatively complicated for calculation of internal force. There are two kinds of the simplified finite element method in documents, but they can not meet the demand of precision. So the more accurate method is using the three-dimensional finite element to analyze the egg-shaped digester in fact. This paper uses ANSYS finite element software to consider fluid-solid coupling vibration and calculates static response, modal analysis and time-history dynamic analysis for the egg-shaped digester.
In fact, not only the load but also the physical parameters of the structure are uncertain during using period of the structure. When analyzing service reliability of structure, it is necessary to consider the physical parameters of structure and random characteristics of load simultaneously. The stochastic structural analysis becomes the foundation of service reliability analysis of the structure. In the research in recent years, one probability density evolution method of the stochastic structural response develops. Through this method, the family of curve of probability density evolution of the structural response quantificationally is provided accurately Therefore according to appointed response limits of displacement, it can calculate and provide the reliability of stochastic structure at random loading directly. After introducing the probability density evolution method briefly, the paper gives the analysis method based on density evolution and c onsiders the dynamic reliability analysis of the physical parameters of the egg-shaped digester.

REFERENCES

[1] Jin Wang, Guangrui Dong, Dezhan He, *The technical application of the egg-shaped digester in China*, China Water and Wastewater, (9), (2001), 65-66, (in Chinese).
[2] Yonggang Wang, Yinguo Lv, Duihong Zhang, *Seismatic research of the liquid storage tanks*, Gas Deposition and Transportation, 18(6), (1999), 1-7, (in Chinese).
[3] Jianjun Wang, Qihan Li et al, *A review of numerical methods for fluid-structure Interaction with large surface sloshing*, Chinese Quarterly of Mechanics, 22(4), (2001), 447-454, (in Chinese).

COMPUTATIONAL MECHANICS (Abstracts)
WCCM VI in conjunction with APCOM'04, *Sept. 5-10, 2004, Beijing, China*

Dynamic Response and Reliability Analysis of Stochastic Structures

J. B. Chen*, J. Li

Department of Building Engineering, Tongji University, Siping Road 1239, Shanghai, 200092, China
e-mail: chenjb@mail.tongji.edu.cn, lijie@mail.tongji.edu.cn

Abstract

The probability density evolution method for dynamic response and reliability analysis of stochastic structures is proposed. As is known that in the traditional reliability assessment, the randomness of the parameters, such as the geometric size, the strength and so on, are taken into account. On the other hand, the deterministic response analysis techniques are usually used to obtain the loading effects. The inconsistence that the randomness is considered in the reliability assessment but neglected in the structural analysis, which may lead to error in the structural performance evaluation, may be bridged with a stochastic structural analysis. However, the existing stochastic structural analysis approaches, including the Monte Carlo simulation, the random perturbation technique and the orthogonal polynomials expansion, etc., are still incompetent for the purpose.

In contrast to the existing methods which are usually able to obtain only the up to second order statistical quantities, the newly proposed probability density evolution method outlined in the present paper can give the probability density functions of arbitrary response quantities of interest. In the method, based on the principle of preservation of probability, an uncoupled, one-dimensional and deterministic partial differential equation governing the evolutionary probability density function of the stochastic response is deduced and can be numerically solved. In addition, to assess the dynamic reliability, no additional assumption is needed, excepting that an absorbing boundary condition corresponding to the failure criterion is imposed on the probability density evolution equation. The dynamic reliability can be evaluated through solving the initial-boundary-value problem and integrating over the safe domain. With the proposed dynamic reliability assessment method, the joint probability density function of the structural response is not needed; neither the Poisson nor Markov assumption for the level-crossing process whose reasonability has been well supported. The numerical algorithm for the proposed method is studied, combining the time integration method and the finite difference method with TVD schemes. A frame with random physical parameters is taken as an example, where the dynamic response and the reliability are evaluated and compared with the Monte Carlo method. The investigations demonstrate that the proposed method is of high accuracy and efficiency.

REFERENCES

[1] G. I. Schueller ed., *A state-of-the-art report on computational stochastic mechanics*, Probabilistic Engineering Mechanics, 12(4), (1997), 198 -321.

[2] J. Li, S. T. Liao, *Response analysis of stochastic parameter structures under non-stationary random excitation*, Computational Mechanics, 27 (1), (2001), 61 -68.

[3] J. Li, J. B. Chen, *Probability density evolution analysis of nonlinear dynamic response of stochastic structures*, Acta Mechanica Sinica, 35(6), (2003), 716-722, (in Chinese).

[4] J. D. Anderson, Jr., *Computational Fluid Dynamics*, McGraw-Hill, New York, (1995).

COMPUTATIONAL MECHANICS (Abstracts)
WCCM VI in conjunction with APCOM'04, *Sept. 5-10, 2004, Beijing, China*

Application of Artificial Neural Network in Monte Carlo Simulation Analysis of Stochastic Structure

Taicong Chen*, Dajian Han, Cheng Su

Department of Civil Engineering, South China University of Technology, Wushang Road, Guangzhou, 510640, China
e-mail: cvchentc@scut.edu.cn, ardjhan@scut.edu.cn, cvchsu@scut.edu.cn

Abstract

In the analysis of stochastic structure, the Monte Carlo simulation method is well known to be capable to provide the most straightforward and accurate solutions to various structural problems. However, the time-consuming simulation of lots of random digital samples restricts it in a narrow use. Several methods, including the importance sampling technique and the Neumann expansion technique, have been developed to reduce the computing cost of the MC method, whereas the best improvement is observed to be only several times. In this paper, use is made of the intelligent tool, Artificial Neural Network, to replace the deterministic FEM solver in MC simulation. The idea is that FEM is only used to generate the input-output pairs needed for training ANN and that the trained ANN can map out all the MC samples instantly for the structural responses. Numerical results on the analysis of a bending plate have shown that the proposed MC-ANN method is able to improve the computing efficiency of the MC method by tens of times with little loss in accuracy.

REFERENCES

[1] A. Saltelli, et al, *Sensitivity analysis,* John Wiley & Sons Ltd., Chichester, England, (2000).
[2] M. Shinozuka, *Neumann expansion for stochastic finite element analysis,* J. Eng. Mech., ASCE, 114, (1988), 1335-1354.
[3] I. Flood, N. Kartam, *Neural networks in civil engineering: principles and understanding,* J. Comput. Civil Eng., ASCE, 8, (1994), 131-162.
[4] K. Bani-Hani, et al, *Experimental study of identification and control of structures using neural network, part 1: identification,* Earthquake Eng. Struct. Dyn., 28, (1999), 995-1018.
[5] J. C. Jan, et al, *Neural network forecast model in deep excavation,* J. Comput. Civil Eng., ASCE, 16, (2002), 59-65.
[6] M. Papadrakakis, et al, *Structural reliability analysis of elastic-plastic structures using neural networks and Monte Carlo simulation,* Comput. Methods Appl. Mech. Eng., 136, (1996) 145-163.
[7] J. E. Hurtado, *Analysis of one-dimensional stochastic finite elements using neural networks,* Probab. Eng. Mech., 17, (2002), 35-44.
[8] E. Vanmarcke, *Random Fields: Analysis and Synthesis,* MIT Press, Cambridge, (1983).
[9] W. Q. Zhu, et al, *Stochastic FEM based on local averages of random vector fields,* J. Eng. Mech., ASCE, 118, (1992), 496-511.
[10] M. T. Hagan, M. B. Menhaj, *Training feedforward networks with the Marquardt algorithm,* Trans. Neu. Netw., IEEE, 5, (1994), 989-993.
[11] T. C. Chen, et al, *Acceleration of Levenberg-Marquardt training of neural networks with variable decay rate,* Proc. Int. Joint Conf. Neu. Netw., IEEE, Portland, (2003), 1873-1878.
[12] F. Yamazaki, et al, *Neumann expansion for stochastic finite element analysis,* J. Eng. Mech., ASCE, 114, (1988), 1335-1354.

COMPUTATIONAL MECHANICS (Abstracts)
WCCM VI in conjunction with APCOM'04, *Sept. 5-10, 2004, Beijing, China*

Finite Element Solution of the Fokker Planck Equation Arising in Non-Linear Random Vibration Analysis

Jorge Crempien*

College of Civil Engineering, University of los Andes, San Carlos de Apoquindo 2200, Las Condes, Santiago, Chile
e-mail: jcrempie@uandes.cl

Abstract

The problem of linear structural systems subjected to earthquake induced base motion has usually been studied by using nonstationary random processes to model the excitation [1, 2]. In those cases the problem has been well studied by many researchers in the past [3]. However, when the system is a nonlinear structure subjected to non stationary random excitation, there are few studies. In this work a method to compute the statistics of the response assuming that the response can be modeled as a Markov process approach is presented. The Markov process assumption leads first to the Chapman Kolmogorov equation which in turn leads to a Fokker Plank equation involving the probability density function of the displacement and velocity response of the structure in any time.
Some authors have used moment functions to estimate a solution of this problem. In the present work Finite Element formulation is derived by writing first the F-P equation in its weak form, and then linear quadrilateral finite elements are implemented where the node variables is the combined probability for displacement and velocity. The results obtained for an elastic perfectly plastic stiffness type of structure depends on the initial natural period of the structure, its damping and also of the expected frequency content of the earthquake. This approach has also been used by other authors but for stationary or white random processes in other type of structures like duffing oscillator [4].

REFERENCES

[1] G. R. Saragoni, G. Hart, *Simulation of artificial earthquakes*, J. Earthquake Engineering and Structural Dynamics, 2(3), (1974), 249-267.
[2] A. Der Kiureghian, J. Crempien, *An evolutionary model for earthquake ground motion*, Structural Safety, 6, (1989), 235-246.
[3] T. K. Caughey, H. J. Stumpf, *Transient response of dynamic systems under random excitation*, J. Applied Mechanics, ASME, 28, (1961), 563-566.
[4] R. S. Langley, *A finite element method for the survival probability of nonlinear oscillations*, Computers and Structures, 27, (1987), 611-617.

COMPUTATIONAL MECHANICS (Abstracts)
WCCM VI in conjunction with APCOM'04, *Sept. 5-10, 2004, Beijing, China*

Multiscale Stochastic Simulation of Building Demolition

Petr Kabele *, Marie Kalousková

Faculty of Civil Engineering, Czech Technical University in Prague, Thákurova 7, Praha 6, 166 29, Czech Republic
e-mail: petr.kabele@fsv.cvut.cz, kalouskova@fsv.cvut.cz

Abstract

Since numerous precast concrete buildings are attaining their intended lifetime, many have to be deconstructed. To this end, use of controlled explosions appears as a suitable approach in many aspects. Computational simulations may provide a way for validation and conscious design of efficient and safe deconstruction procedures.
When analyzing a precast R/C structure during demolition, it is necessary to deal with phenomena such as material fracturing and yielding on the scale of individual structural members/joints (mesoscale). At the same time, on the scale of an entire building (macroscale), it is necessary to simulate finite dynamic motion of the structure and its fragments. A deterministic multiscale computational methodology for efficient solution of such a class of problems was proposed earlier [1]: upon identifying typical failure mechanisms of structural members and joints through detailed FE computations, a simplified mesoscale model of the failing sections was constructed. Resulting nonlinear bending and axial rigidity relations were implemented into beam elements, which were employed for a transient dynamic analysis of the whole structure collapsing in the course of a demolition.
In real structures, the input parameters of the mesoscale model (such as concrete and reinforcement strength) exhibit inherent scatter and uncertainties. Thus, it is desirable to consider them as random variables. In the proposed stochastic approach, these variables are randomly generated while considering their probability density functions and correlations. With these parameters, the rigidity relations are evaluated number of times using the simplified mesoscale model. In order to obtain statistically accurate results even with a relatively small number of evaluations, the Latin Hypercube Sampling [2] technique is employed.
In the next step, rigidity relations are assigned to the macroscale structural FE model in accordance with their statistical characteristics obtained through the stochastic analysis. To simulate a demolition, a transient dynamic analysis is performed with this model. The macroscale analysis is performed several times in order to check robustness and safety of a proposed demolition technique.

REFERENCES

[1] P. Kabele, T. Pokorný, *Computational analysis of precast concrete building deconstruction*, in H. A. Mang, F. G. Rammerstorfer, J. Eberhardsteiner eds. Proceedings of the Fifth World Congress on Computational Mechanics (WCCM V), July 7-12, 2002, Vienna, Austria, Vienna University of Technology, Austria, http://wccm.tuwien.ac.at, (2002).

[2] D. Novák, B. Teplý, Z. Keršner, *The role of Latin hypercube sampling method in reliability engineering*, in Proceedings of ICOSSAR-97: Structural Safety and Reliability, Kyoto, 1997, Balkema, Rotterdam, (1998), pp. 403-409.

COMPUTATIONAL MECHANICS (Abstracts)
WCCM VI in conjunction with APCOM'04, *Sept. 5-10, 2004, Beijing, China*

Semidefinite Programming for the Analysis of Structural Systems with Material and Geometric Uncertainties

Jiafan Zhang *, Zhichu Huang

School of Mechanical and Electrical Engineering, Wuhan University of Technology, Wuhan, 430070, China
e-mail: jfzhang01@21cn.com

Abstract

A semidefinite programming (SDP) approach is presented to estimate bounds on static responses of structural systems with uncertain material and geometric parameters that are described as interval numbers. Under the constraint of static governing equations of the structural system, an ellipsoid of minimal size in the sense that minimizes the sum of squares of semi-axis lengths is sought to contain the exact solution set of the displacement responses. The computation is formulated as a (convex) SDP problem based on the results from semidefinite relaxation techniques proposed in references [1-2], which can be solved very efficiently using interior-point algorithms. The approach presented here considers the modeling of the structural stiffness matrix containing uncertain structural parameters using linear fractional representation (LFR). We present a procedure to formulate the LFR of the structural stiffness matrix, combining with the finite-element method. It is shown by numerical examples that the approach can give reasonable estimates on the solution intervals, and tends to overestimate the bounds of some response quantities with the increasing number of the uncertain parameters. Using an interval-truncation procedure proposed in the reference [3], more accurate results can be given.

REFERENCES

[1] L. El. Ghaoui, H. Lebret, *Robust solution to uncertain semidefinite programs,* SIAM Journal of Optimization, 1, (1998), 33-52.
[2] L. El. Ghaoui, G. Calafiore, *Worst-case simulation of uncertain systems,* in Lecture Notes in Control and Information Sciences, Vol. 245, Springer-Verlag, London, (1999), pp. 134-146.
[3] S. S. Rao, L. Berke, *Analysis of uncertain structural systems using interval analysis,* AIAA Journal, 4, (1997), 727-735.

COMPUTATIONAL MECHANICS (Abstracts)
WCCM VI in conjunction with APCOM'04, *Sept. 5-10, 2004, Beijing, China*

A Quasi-Arch Length Method and Its Computer Implementation for NFEM

S. Cai [1*], X. Shao [2], P. Shen [2], J. Ye [3], Y. Zhang [2]

[1] *College of Resource and Environment Science, Hunan Normal University, Changsha, 410082, China*
[2] *College of Civil Engineering, Hunan University, Yuelushan, Hexi, Changsha, 410082, China*
e-mail: s.cai@hunnu.edu.cn, shaoxd@vip.163.com, shenying@public.cs.hn.cn, j.ye@leeds.ac.uk, zhangbridge@163.com

Abstract

Based on the previous nonlinear FE procedure, this work presented an improved constraint equation for solution of nonlinear FE equations in the load-displacement space of $\boldsymbol{R}^{n+1}$. The new algorithm, named quasi-arch length method, is linear and simpler for its constraint equation and a little bit less cost of iteration, but also able to trace the structural equilibrium paths with snap-through and snap-back. The nonlinear solution for structural equilibrium represented by FE mesh can be written as $\boldsymbol{R}(\boldsymbol{u})=\boldsymbol{P}_0+\lambda\boldsymbol{P}_t$, In which $\boldsymbol{u}$ denotes the structural displacement vector, $\boldsymbol{R}$ the structural resistance vector, $\boldsymbol{P}_0$ the initial structural loading, $\boldsymbol{P}_t$ the current structural load vector, λ the loading factor, n the total structural DOF. In collaboration with the new constraint equation $\boldsymbol{c}^T(\boldsymbol{u}-\boldsymbol{u}_t)+c_{n+1}(\lambda-\lambda_t)=\Delta l^t$, the iteration procedures for next equilibrium point $t+1$ of the nonlinear path are then proposed on the base of the previous point $t-1$ and the current point t as:

$$\begin{bmatrix}\boldsymbol{K}_T^m & -\boldsymbol{P}_t\\ \boldsymbol{c} & c_{n+1}\end{bmatrix}\begin{bmatrix}\Delta\boldsymbol{u}^{m+1}\\ \Delta\lambda^{m+1}\end{bmatrix}=-\begin{bmatrix}\boldsymbol{R}(\boldsymbol{u}^m)-\boldsymbol{P}_0-\lambda^m\boldsymbol{P}_t\\ \boldsymbol{c}(\boldsymbol{u}^m-\boldsymbol{u}_t)+c_{n+1}(\lambda^m-\lambda_t)-\Delta l^t\end{bmatrix},\quad \begin{cases}\boldsymbol{u}^{m+1}=\boldsymbol{u}^m+\Delta\boldsymbol{u}^{m+1}\\ \lambda^{m+1}=\lambda^m+\Delta\lambda^{m+1},\end{cases}\quad m=0,\ 1,\ 2,\cdots$$

where

$$\boldsymbol{c}=\boldsymbol{C}/\|\boldsymbol{C}\|,\quad c_{n+1}=C_{n+1}/\|\boldsymbol{C}\|,\quad \boldsymbol{C}_n=\frac{1}{u_0}(\boldsymbol{u}_t-\boldsymbol{u}_{t-1}),\quad C_{n+1}=\lambda_t-\lambda_{t-1},\quad \boldsymbol{C}=(\boldsymbol{C}_n^T\ \ C_{n+1})^T,$$

and $\boldsymbol{K}_T$ the structural tangent stiffness matrix, u_0 the structural maximum displacement, Δl^t prescribed incremental displacement, $\boldsymbol{u}_t$, $\boldsymbol{u}_{t-1}$ and λ_t, λ_{t-1} the previous and current displacement vector and the load factor. The computer implementation of the procedure has been successfully carried out for a truss element in tracing nonlinear paths with snap-through and snap-back. Numerical examples were presented and in good agreement with others results.

REFERENCES

[1] M. A. Crisfield, *Nonlinear Finite Element Analysis of Solids and Structures, vol. 2: Advanced Topics*, John Wiley & Sons Ltd., Chichester, England, (1997).

[2] J. L. Meek, H. S. Tan, *Geometrically nonlinear analysis of space frames by an incremental iterative technique.* Comput. Meths. Appl. Mech. Engrg, 47, (1984), 261-282.

[3] E. Riks, *An incremental approach to the solution of snapping and buckling problems.* International Journal of Solids and Structures, 15, (1979), 529-551.

[4] S. Cai, P. Shen, *On the methods solving nonlinear equations.* Journal of Hunan University, 27(3), (2000), 86-91.

COMPUTATIONAL MECHANICS (Abstracts)
WCCM VI in conjunction with APCOM'04, *Sept. 5-10, 2004, Beijing, China*

Stability and Bifurcations of Single-Impact Periodic Motions of an Impact-Forming Machinery

Yingxiang Chang *, Yandong Chu, Huaiwei Lu, Yanlong Zhang

School of Mathematics and Physics, Lanzhou Jiaotong University, Lanzhou, 730070, China
e-mail: changyx@mail.lzjtu.cn, cyd@mail.lzjtu.cn, luhw@mail.lzjtu.cn, zyl44@etang.com

Abstract

An impact-forming machinery shown in Fig.1 is considered. The components of the vibrating system collide with each other when the difference of their displacements equals the gap Δ. Such models play an important role in the studies of dynamics of mechanical systems with impacting components. Dynamics of the impact-forming system are studied with special attention to stability of period n single-impact motion, bifurcations and chaotic motion, etc. The Poincaré section associated with the state of the impact-forming system, just immediately after the impact, is chosen to establish Poincaré map of the impact-forming machinery. The period n single-impact motion is derived analytically, and then the disturbed map of single-impact periodic motion is established. Stability and local bifurcations of the periodic motion with one impact are analyzed by using the Poincaré map. Some representative time trajectory, projected portraits of Poincaré map and global bifurcation diagrams of the impact-forming machinery are plotted in the paper, see Fig.2. In designing impact tools it is of great interest to achieve the desired periodic impact velocities. In order to facilitate such design, global diagrams are useful. The global bifurcation diagrams for the impact velocity versus the forcing frequency enable the practicing engineer to select excitation frequency ranges in which stable responses with one impact can be expected to occur, and to predict the larger relative impact velocities of such responses.

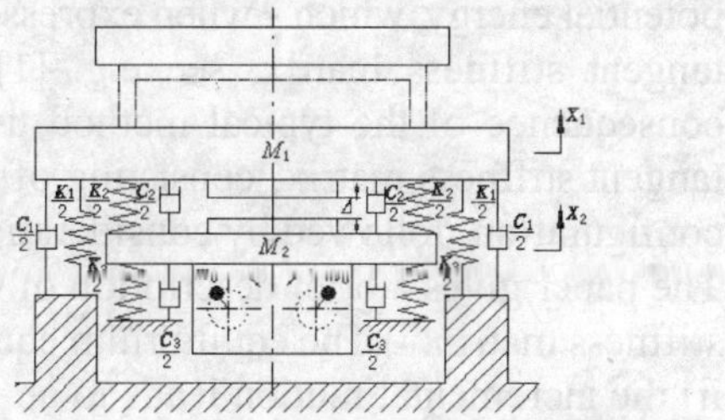

Fig.1 Schematic of the impact-forming machinery with double masses

The project is supported by National Natural Science Foundation (10172042) and Science Foundation of Gansu Province of China (ZS-031-A25-007-Z).

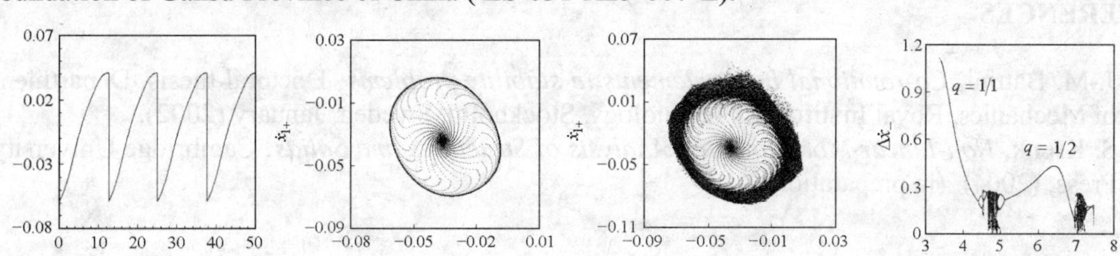

Fig.2 Representative time trajectory, projected portraits of Poincaré map and global bifurcation diagram: (a) the time trajectory of period two single-impact motion; (b) the quasi-periodic attractor associated with period two single-impact motion; (c) chaos; (d) global bifurcation diagram

COMPUTATIONAL MECHANICS (Abstracts)
WCCM VI in conjunction with APCOM'04, *Sept. 5-10, 2004, Beijing, China*

Symmetry of Co-Rotating Elements with Finite Rotations

S. Krenk *

Department of Mechanical Engineering, Technical University of Denmark, Building 101E, Studentertorvet, DK-2800 Kgs. Lyngby, Denmark
e-mail: sk@mek.dtu.dk

Abstract

The co-rotating element format is based on a procedure in which the element is described in a local element based frame of reference that co-rotates with the element. The purpose is to separate the effect of the potentially large motion of the element frame from the local element deformation within this frame, which is often described by a lower order theory. For beams and shells this lower order theory is typically a fully linear theory or a theory which includes the effect normal or in-plane forces. The advantage is to permit use of an element based on low order theory for solution of kinematically fully nonlinear problems.

It is a common finding that the co-rotation contribution to the stiffness matrix of e.g. beam and shell elements in three dimensions turn out to be non-symmetric, even when based on a theory with a potential energy, which - when expressed in terms of suitable parameters - should posses a symmetric tangent stiffness matrix, see e.g. [1] and references given there. The lack of symmetry is a consequence of the typical method used for the derivation of the co-rotation contribution to the tangent stiffness matrix, consisting of statement of the element equilibrium equations in the current configuration, followed by consideration of a small incremental motion.

The paper gives a brief description of the steps needed to obtain a consistent symmetric co-rotation stiffness matrix: i) The equilibrium equation of the element is expressed as the external virtual work, ii) the incremental motion leads to an increment in the external virtual work that includes a special contribution from the increment of the variational rotation, and iii) the stiffness matrix is complemented by contributions from bending and torsion deformations of the element. The method is illustrated by beam elements with two nodes, but follows general principles valid also for other co-rotating elements. Detailed background on finite rotations and their use in non-parametric formulation of beam and shell elements is given in [2].

REFERENCES

[1] J.-M. Battini, *Co-rotational beam elements in stability problems*, Doctoral thesis, Department of Mechanics, Royal Institute of Technology, Stockholm, Sweden, January (2002).

[2] S. Krenk, *Non-Linear Modelling and Analysis of Structures and Solids*, Cambridge University Press, (2004), (in preparation).

COMPUTATIONAL MECHANICS (Abstracts)
WCCM VI in conjunction with APCOM'04, *Sept. 5-10, 2004, Beijing, China*

Assessing the Ability of Structural Changes to Create Spectral Gaps

Ray Lawther *

School of Civil and Environmental Engineering, The University of New South Wales, Sydney 2052, Australia
e-mail: r.lawther@unsw.edu.au

Abstract

Structures vibrate with discrete natural frequencies, dividing the frequency spectrum into frequency-free ranges, or spectral gaps. We may require that a structure has a specified spectral gap, and if this range contains frequencies then the structure must be changed. The extent of the changes depends on the number of natural frequencies that are contained in the required spectral gap. If the required spectral gap contains n frequencies then changes of at least rank n are needed to remove all. This paper considers the problem 'if a required spectral gap contains n natural frequencies, can changes to the structure stiffness at given freedoms $f_1, f_2 \ldots, f_n$ remove them?'
A simple criterion is developed for answering this question, and if the answer is 'yes', methods for calculating successful changes are presented. The criterion for assessing the possibility is positive definiteness of a particular matrix. If positive definite, this matrix defines a matrix function $\mathbf{M}(\mathbf{X})$, giving structural changes which are successful in moving all unwanted frequencies if all eigenvalues of $\mathbf{X}$ are in [0, 1]. Possibilities are particularly abundant if the successful $\mathbf{M}(\mathbf{X})$ includes $\mathbf{0}$.

REFERENCES

[1] R. Lawther, *Enlarging spectral gaps of a structure using spring-mass oscillators unpublished*, available at www.civeng.unsw.edu.au/staff/ray.lawther/ (2003).
[2] B. N. Parlett, *The symmetric eigenvalue problem*, Prentice Hall, (1980).
[3] J. H. Wilkinson, *The algebraic eigenvalue problem*, Oxford UP, Oxford, (1965).
[4] Lord Rayleigh, *Theory of sound*, Dover, New York, (1945).

COMPUTATIONAL MECHANICS (Abstracts)
WCCM VI in conjunction with APCOM'04, *Sept. 5-10, 2004, Beijing, China*

Finite Element Analysis of Safety Factor for Wet-Sprayed Fiber Reinforced Concrete Lining Structure for Railway Tunnel within Hard Rock

Junying Lian [1*], Taiquan Zhou [1], Yuan Hua [1], Renhe Wang [2]

[1] *Department of Civil Engineering, Southern Yangtze University, Wuxi, 214063, China*
[2] *Department of Civil Engineering, Anhui University of Science and Technology, Huainan, 232001, China*

e-mail: zhoutaiquan@sytu.edu.cn, rhwang@aust.edu.cn

Abstract

The anchor-sprayed is an important tunnel lining structure, reflecting the New Austrian Tunnel Method main idea of the utmost utilization of the rock self-bearing capacity. The wet-sprayed fiber reinforced concrete lining structure has higher strength, more flexible deformation capacity, easier construction and can be used as both temporary and permanent lining structure. The evaluation of the limit bearing capacity of the anchor-sprayed lining structure is an important issue to be considered in the lining structure design and construction. There remains no evaluation specification on the fiber reinforced concrete lining structure according to the current tunnel design and construction codes. The definition of the safety factor of such tunnel lining structure is used to evaluate the limit bearing capacity of the lining structure. A nonlinear finite element analysis is performed on the calculation of safety factor for the wet-sprayed fiber reinforced concrete lining structure, whose definition is the ratio between the limit bearing capacity to the required bearing capacity of the lining structure. The finite element model includes the rock, the anchor bolt, and the sprayed concrete. Based on the incremental nonlinear finite element analysis, the incremental load-displacement method is used to obtain the nonlinear deformation process and the limit bearing capacity state of the lining structure. The nonlinear material constitutive relationship for the rock, the anchor, the rock and the interaction between the tunnel and the rock are considered in detail in the analysis. As a case for study, the nonlinear analysis for the safety factor evaluation is performed on an existing rail tunnel lining structure using the wet-sprayed fiber reinforced concrete. The limit bearing capacity and the corresponding safety factor for the lining structure is studied using the nonlinear analysis. Also the limit failure mode of the anchor-sprayed concrete lining structure is obtained. The numerical analysis result for the tunnel lining structure safety factor shows that the current tunnel lining structure is safe enough to be in use.

REFERENCES

[1] G. M. Smith, *Probability and Statistics in Civil Engineering*, Collins Press, (1986).
[2] O. C. Zienkiewicz, A. H. C. Chan, M. Pastor, B. A. Shrefler, *Computational Geomechanics*, John Wiley & Sons Ltd., Chichester, England, (1999).
[3] K. J. Bathe, *Finite Element Procedures in Engineering Analysis*, Prentice-Hall, Englewood Cliffs, (1982).

COMPUTATIONAL MECHANICS (Abstracts)
WCCM VI in conjunction with APCOM'04, *Sept. 5-10, 2004, Beijing, China*

Spatial Geometric Nonlinear Analysis Method Based on Tension-Compression and Torsion-Warping Stability Interpolation Function

Jian Liu [1, 2*], Xiangyun Huang [2], Ming Xu [1], Kaixi Li [1], Fulin Zhou [2], Minhui Zhou [3]

[1] *Chongqing University, Chongqing 400045, China*
[2] *Guangzhou University, Guangzhou 510405, China*
[3] *The Architectural Design and Research Institute of Guangdong Prov., Guangzhou, 510010, China*

Abstract

Conception and computational formula on torsion-warping stability function is brought forward for the first time. Based on the theory of continuum mechanics, geometric nonlinear analysis FEM model of multistorey and high-rise lightweight spatial steel frame was established considering shearing deformation effect according to stability interpolation function. Making use of updated Lagrangian formulation, starting from increment virtual work equilibrium equation, the explicit expresses of beam-column elements' stiffness matrix for spatial steel construction framing was derived. Meanwhile, the model on beam-column theory of geometric nonlinear analysis of multistorey and high-rise spatial steel frame was also discussed considering shearing deformation effect according to stability interpolation function. Conception and computational formula on torsion-warping stability function is brought forward for the first time, based on tension-bending model. For the first time the torsion-warping tangent stiffness matrix of bi-axisymmetric open thin-wall and lightweight beam-column element is derived, based on torsion-warping stability interpolation function, and at the same time spatial geometric tangent stiffness matrixes for beam-column element of lightweight steel structure based on tension-compression and torsion-warping stability interpolation function is presented for the first time.

COMPUTATIONAL MECHANICS (Abstracts)
WCCM VI in conjunction with APCOM'04, *Sept. 5-10, 2004, Beijing, China*

The Stability Behavior of the Large Span Deep Arch-Beam Structures

Y. F. Luo [1*], Z. Y. Shen [1], Q. Jing [1], Q. Z. Zhu [2]

[1] *College of Civil Engineering, Tongji University, 1239 Siping Road, Shanghai, 200092, China*
[2] *Post Department of Shanghai, 458 Wuding Road, Shanghai, 200041, China*
e-mail: yfluo93@mail.tongji.edu.cn

Abstract

The linear and nonlinear stability behavior of a single plane deep arch structure, a sub-arch-beam structure composed of parallel plane arches and perpendicularly linked beams and a practice engineering deep arch-beam structure used in a long span structure with 202m×50m plane size is analyzed and compared in the paper.
Firstly, the stability behavior of the single deep steel arch used in the practice engineering structure is analyzed. The valuable ideas about the arch-in-plane stability of the perfect and the imperfect single deep arches are obtained. Then, the stability behavior of a sub-arch-beam structure, which composed of the parallel plane deep arches and the beams perpendicularly linked to the arches, is analyzed. One sub-structure unit (two arches and the beams between the arches) and more sub-structure units (more than two arches and the beams between every two arches) are analyzed respectively for comparison. Two sub-structure systems, in which cross braces are positioned or not positioned at two ends of the sun-structures, are considered in stability analysis of the more sub-structure units. The change of the stability behavior of the arch and the sub-arch-beam structures is observed and explored from the numerical analyses. Finally, the stability behavior of the perfect and the imperfect integral engineering structure of the steel arch-beam system is presented. The numerical results of the linear stability, the geometric nonlinear stability of the perfect and the imperfect arch-beam structures, the elasto-plastic stability of the structures are listed and compared.
The numerical results show that the effects of the geometric nonlinearity and the geometric imperfections on the arch in-plane stability behavior of the deep arch structure are not so obvious, but the effects of the plasticity on the stability behavior of the structure are obvious. The stability behavior of the integral arch-beam structure is different from the stability behavior of the single deep arch or the sub-arch-beam structure, which is a part of the integral arch-beam structure.
Compared the numerical results of the different arch-beam structures, it can be concluded: It is not safe if justify the stability of a integral arch-beam structure by means of the stability of the single arch or the sub-arch-beam structure which belongs to the integral arch-beam structure. And the cross braces should be positioned at least between one couple columns and the corresponding roof area for increasing the out-arch-plane stability of the arch-beam structure.

COMPUTATIONAL MECHANICS (Abstracts)
WCCM VI in conjunction with APCOM'04, *Sept. 5-10, 2004, Beijing, China*

Dynamic Buckling of Cylindrical Shells under Torsion Stress Waves

Yuan Ma*, Xinsheng Xu

State Key Laboratory of Structure Analysis of Industrial Equipment and Department of Engineering Mechanics, Dalian University of Technology, Dalian,116024, China
e-mail: mayuan945@163.com , xsxu@dlut.edu.cn

Abstract

The dynamic buckling of cylindrical shells has been studied since the 1960s. R. J. Mao [1] analysed the elastic-plastic buckling of cylindrical shells under torsion with a deep thick-shell model under various boundary conditions. D. Karagiozova [2] discussed some characteristic featuers of the dynamic in elastic buckling behaviour of cylindrical shells subjected to axial impact loads. But it is difficult to solve the question of dynamic buckling of cylindrical shells subjected to torsion impact loads, because there are intersectant items.

This paper discusses the dynamic buckling of cylindrical shells in propagating and reflecting of torsion stress waves. The investigation is based on the theory of stress wave propagation and bifurcation conditions. The Hamiltonian system is introduced to the dynamic buckling of structures. The problem can be described mathematically in the symplectic space. The solutions of Hamiltonian dual equations, which are shown by canonical variables, are obtained and the problem is come down to eigensolutions. A new computational method is presented for the torsion buckling of cylindrical shells. As numerical example, phenomena of axisymmetric and non- axisymmetric dynamic buckling are revealed under torsion impact. Critical buckling loads and modes are given. It is showed that critical buckling loads of low stage is not always less than high stage. In propagating of torsion stree waves, cylindrical shell presents first the critical buckling of optimum stage, which agrees with experimental results.

REFERENCES

[1] R. J. Mao, G. Lu, *A study of elastic-plastic buckling of cylindrical shells under torsion*, Thin-Walled Structures, 40(12), (2002), 1051-1071.

[2] D. Karagiozova, N. Jones, *On dynamic buckling phenomena in axially loaded elastic-plastic cylindrical shells*, International Journal of Non-linear Mechanics, 37(7), (2002), 1223-1238.

[3] C. Pozrikidis, *Buckling and collapse of open and closed cylindrical shells*, Journal of Engineering Mathematics, 42(2), (2002), 157-180.

COMPUTATIONAL MECHANICS (Abstracts)
WCCM VI in conjunction with APCOM'04, *Sept. 5-10, 2004, Beijing, China*

Geometric Nonlinear Model of Thin-walled Spatial Steel Frames

G. J. Nie*, Z. Zhong

Key Laboratory of Solid Mechanics of MOE, Tongji University, Shanghai, 200092, China
e-mail: ngj8269@sina.com

Abstract

Thin-walled spatial steel frames are widely used in buildings. In spite of having relatively stronger stiffness than concrete, the steel structures are predominantly affected by instability due to the buckling of structural members composed of very slender plates. In order to use thin-walled structures effectively, an understanding of all the deformations and stresses caused by flexure and torsion is required [1~2]. While the thin-walled beam generates the torsion deformation, it is usually accompanied by the warping deformation. If the warping deformation is restrained, the normal stress and shearing stress caused by warping will generate in the beam. At the same time the additional warping shearing strain will affect the warping deformation again. So it is necessary to introduce the warping deformation as the independent degree of freedom during the finite element analysis of thin-walled spatial steel frames and it is essential to consider the second order effect of axial force, shearing force, biaxial bending moment and bimoment in the geometric nonlinear model of thin-walled element [3].

Based on the space beam model of six degrees of freedom at each end, the new thin-walled element analytical model has seven degrees of freedom in every node because the restraint torsion of the element is considered. The pattern of the torsion deformation uses the cubic polynomial and the pattern of the warping deformation uses a warping function instead of the slope of the line of torsion deflection due to considering the effect of the secondary shear stress. The patterns of other deformations adopt the same function as the space beam model. According to the principle of virtual work, the elastic stiffness matrix and geometrical stiffness matrix of the element can be obtained. The elastic stiffness matrix of this new model is 14×14. The effect of axial force, shearing force, biaxial bending moment and bimoment of the element is involved in the geometrical stiffness matrix and simultaneously the second order effect of axial force is related to the ratio of the slenderness of the member.

Using the new model derived in this paper, the corresponding program is worked out and simultaneously the behavior of the geometric nonlinear of thin-walled spatial steel frames is discussed. The numerical examples indicate the effectiveness and precision of this geometrical nonlinear model. It can be used in the geometric nonlinear analysis of arbitrary thin-walled spatial steel structures.

REFERENCES

[1] H.G. Kwak, D. Y. Kim, H. W. Lee, *Effect of warping in geometric nonlinear analysis of spatial beams, Journal of constructional steel research*, 57, (2001), 729-751.

[2] H. Shakourzadeh, Y. Q. Guo, J. L. Batoz, *A torsion bending element for thin-walled open and closed cross sections*, Comput. & Struct., 55, (1995), 1045-1054.

[3] G. J. Nie, *Research on the nonlinear analytical model and analytical methods of spatial steel member structures*, Shanghai: Tongji University, (2002), (in Chinese).

COMPUTATIONAL MECHANICS (Abstracts)
WCCM VI in conjunction with APCOM'04, *Sept. 5-10, 2004, Beijing, China*

Stability of Elastic and Viscoelastic Systems under a Stochastic Parametric Excitation

Vadim D. Potapov *

Department of Structural Mechanics, Moscow University of Means Communication, Obraztsov Str., 15, Moscow, 127994, Russia
e-mail: potapov.vd@relcom.ru

Abstract

Problems of the stability of elastic and viscoelastic systems under action of random loads, first of all of columns, subjected to the longitudinal force which is a stochastic stationary process, were considered by many authors. A sufficiently thorough survey of these works is contained in monograph [1]. The greatest number of results were obtained for that case, when the stationary process is supposed as a Gaussian white noise. If parametric forces are stationary wide-band processes then the solution of the stability problem becomes significantly more complicated. In such a case mainly sufficient conditions of the almost sure stability were obtained. It should be underlined that the estimation of stability boundaries, which are obtained with help of these criterions usually rather rough. If mentioned processes are arbitrary enough then for the solution of the stability problem more correct results, from the point of view of stability boundaries, can be obtained with help of a method of the statistical simulation of stochastic functions in a combination with computational methods for the solution of the problem.

The present work is devoted to the investigation of the stability of elastic and viscoelastic systems, excited by random stationary parametric loads in the form of colored noises. Further the simulation routine is based on the Runge-Kutta method of the fourth order. For such a case a scheme of the simulation of the white noise is very important. Concrete versions of this simulation will be considered on examples below. The principal purpose of the present work is the investigation of the stability of system with respect to statistical moments. Since the closed system of equations for the moments of the displacements $y_j(t)$ in the case of colored noise could not be obtained, the method of statistical data processing is applied. The estimation of moments $\left\langle y_j^p \right\rangle$ for the instant t_n can be obtained as a result of statistical average of values $\left\langle y_j^p \right\rangle$, derived from the solution of equations, describing the behavior of the considered system, for the enough large number of realizations. Using the procedure, suggested in the work [1], the estimation of the top Liapunov exponent can be obtained. The fulfilled calculations allow to estimate the influence of different characteristics (of random stationary loads, of viscous properties of the material) on top Liapunov exponents and consequently on the stability with respect to statistical moments of the different order.

REFERENCES

[1] V. D. Potapov, *Stability of stochastic elastic and viscoelastic systems,* Wiley, Chichester, England, (1999).

COMPUTATIONAL MECHANICS (Abstracts)
WCCM VI in conjunction with APCOM'04, *Sept. 5-10, 2004, Beijing, China*

On Computational Dynamic Buckling Criteria for Shells under Transient Loads

Juan C. Virella, Luis A. Godoy *

Department of Civil Engineering, University of Puerto Rico, Mayaguez PR 00681-9041, Puerto Rico
e-mail: jvirella@ce.uprm.edu, lgodoy@uprm.edu

Abstract

This paper reports on the dynamic buckling of thin -walled steel storage tanks under a horizon tal excitation at the base. The structure is modeled using finite elements and a geometrical and material nonlinear analysis to compute the transient response in the time domain [1]. The liquid stored in the tank is modeled taking into account the impulsiv e action. The excitation is introduced in terms of inertial forces with a time variation equal to the acceleration induced by the earthquake. The load factor is the horizontal peak ground acceleration of an earthquake.
The dynamic buckling criteria initial ly employed is due to Budiansky and Roth [2], according to which dynamic buckling occurs whenever a small increase in the load parameter leads to a large increase in the transient displacements. The paper shows that dynamic buckling problems may occur in two ways: when there is a sudden change in the displacements and the deflected shape of the shell during the oscillation (i.e. the deflected mode) changes due to dynamic buckling, then a "bifurcation behavior" may be said to occur [3]. The original mode of small vibration changes to a different mode under large amplitude vibrations. But there is a second way, in which the transient displacements increase progressively and there is no change in the mode of vibration. However, there is a significant change in the slope of the curve of load parameter versus first maximum amplitude of oscillation. This indicates a change in the stiffness at some load level, but with the same mode of vibration. Such process may be called "limit point behavior" in the sense that instability preserves the shape of deflections.

REFERENCES

[1] H. D. Hibbit, B. I. Karlsson, P. Sorensen, *ABAQUS user manual,* version 6.3, Pawtucket, RI, USA, (2002).

[2] B. Budiansky, S. Roth, *Axisymmetric dynamic buckling of clamped shallow spherical shells*, NASA Collected Papers on Stability of Shell Structures, TN-1510, (1962), 597-606.

[3] W. Auli, F. G. Rammerstorfer, *On the dynamic instability of shell structures-criteria and algorithms,* in T. J. R. Hughes, E. Hinton eds. Finite Element Methods for Plate and She ll Structures, 2, Pineridge Press, Swansea, (1986), 58-82.

COMPUTATIONAL MECHANICS (Abstracts)
WCCM VI in conjunction with APCOM'04, *Sept. 5-10, 2004, Beijing, China*

Interval method for structural reliability analysis

Denggang Wang*

School of Civil Engineering and Mechanics, Shanghai Jiaotong University, Shanghai, 200240, China
e-mail: wangdenggang@263.net

Abstract

Probabilistic method, fuzzy method, and interval method are the three major approaches to model imprecision or uncertainty present in engineering problem. The interval method is more suitable than probabilistic and fuzzy approach when there is no sufficient experimental data to obtain priori probability density function or subjective function of structural parameters. By representing the uncertain structural parameters as interval numbers and based on the theories of interval mathematics, an interval method for structural reliability analysis is put forward. Overestimation is a major drawback in conventional interval computations. In order to overcome this drawback, the structural functional function was calculated by using of two optimization problems to obtain its' lower and upper bounds respectively. Moreover, an intellective algorithm named as real-code genetic algorithm was introduced to locate the global optima of the optimization problems. Examples show that the presented method is feasible and efficient.

REFERENCES

[1] R. G. Sexmith, *Probability-based safety analysis—value and drawbacks*, Structural Safety, 21, (1999), 303-310.

[2] H. Q. Zhang, *Grey system identification in machinery reliability*, Chinese Journal of Mechanical Engineering, 28, (1992), 102-108, (in Chinese).

[3] F. C. Xiao, X. P. Zhang, Study for grey reliability analysis and its applications. Structure & Environment Engineering, (2001), 38-42, (in Chinese).

[4] Y. Ben-Haim, *A non-probabilistic concept of reliability*, Structural Safety, 14, (1994), 227- 245.

[5] Y. Ben-Haim, *A non-probabilistic measure of reliability of linear systems based on expansion of convex models*, Structural Safety, 17, (1995), 91-109.

[6] I. Elishakoff, *Discussion on: a non-probabilistic concept of reliability, Structural Safety*, 17, (1995), 195-199.

[7] S. X. Guo, Z. Z. Lu, Y. S. Feng, *A non-probabilistic model of structural reliability based on interval analysis*, Chinese Journal of Computational Mechanics,18, (2001), 56-60, (in Chinese).

[8] G. Alefeld, D. Claudio, *The basic properties of interval arithmetic, its software realizations and some applications*, Computers and Structures, 67, (1998), 3-8.

[9] Q. Y. Li, Z. Z. Mo, L. Q. Qi, *Numerical solution of nonlinear equations*, Science Press, Beijing, China, (1999), (in Chinese).

[10] D. G. Wang, J. Li. *A reliable approach to compute the static response of uncertain structural system*, Chinese Journal of Computational Mechanics, 20, (2003), 662-669. (in Chinese).

[11] Z. Michalewicz. *Genetic algorithm+data structures=evolution programs*, Springer-Verlag, Berlin Heidelberg, (1992, 1994, 1996).

COMPUTATIONAL MECHANICS (Abstracts)
WCCM VI in conjunction with APCOM'04, *Sept. 5-10, 2004, Beijing, China*

Explicit Numerical Analyses for Nonlinear Vibration and Stability of Rotors under Unbalanced Magnetic Excitations

Yuefang Wang[1*], Lihua Huang[2]

[1] *Department of Engineering Mechanics, Dalian University of Technology, Dalian 116024, China*
[2] *School of Hydraulic and Civil Engineering, Dalian University of Technology, Dalian 116024, China*
e-mail: yfwang@dlut.edu.cn, llihuah@hotmail.com

Abstract

Electric machines operate in non-uniform magnetic fields where an excitation called unbalanced magnetic pull (UMP) is generated in gaps between rotors and stators, causing severe vibration of the rotors in some occasions [e.g. 1-4]. The nonlinear vibration of the rotors having large eccentricities in non-uniform gap magnetic fields has not been fully attacked. In this paper, a prototyped, Jeffcott rotor is studied for its nonlinear free and forced vibration responses. The magnetic potential in the gap is analytically expressed, and the UMP is modeled through the energy conversion from the magnetic potential to mechanical energy. Following that, the equations of motion are developed considering the UMP and mass unbalance excitations. For weakly nonlinear free vibrations, the system is simplified and compared to a Duffing oscillator. Also, the steady-state motion caused by mass unbalance is obtained in closed form. Using Routh-Hurwitz's criterion, the sufficient condition of stability of the forced vibration is derived and the possibility of jump phenomenon is pointed out.

When the initial eccentricity or the magnetic flux is large, the vibration becomes strongly nonlinear. In this case, the equations of motion are rewritten as first-order state equations. The displacements and velocities are solved through an eighth-order Runge-Kutta method with adaptive step-sizes. Using this approach, we locate the rotor's cross-section center and plot its complicated trajectories for given initial eccentricity and magnetic flux density. By means of the Fast Fourier Transformation (FFT) analysis, we numerically determine the higher-order harmonic terms regarding the excitation frequency. Also, the FFT results are used to determine the bifurcation at certain values of parameters, e.g. the gap width and flux density. In the near future, the stability region for parametric vibration response regarding damping will be included both analytically and numerically.

REFERENCES

[1] J. Fruchtenight, H. Jordan, O. Seinsch, *Running Instability of Cage Induction Motors Caused by Harmonic Fields due to Eccentricity*. Archives of Electrotech, 65, (1982), 271-292.

[2] R. Belmans, A. Vandenput, W. Geysen, *Influence of Unbalanced Magnet Pull on the Radial Stability of Flexible-Shaft Induction Machines*, Proc. of the IEE, Part B, 134, (1987), 101-109.

[3] A. C. Smith, D, G. Dorrel, *Calculation and Measurement of Unbalanced Magnetic Pull in Cage Induction Motors with Eccentric Rotor*, IEE Proc. Electrical Power Applications, 143, (1996), 193-201.

[4] D. Guo, F. Chu, D. Chen, *The Unbalanced Magnetic Pull and Its Effects on vibration in a Three-Phase Generator with Eccentric Rotor*, J. Sound and Vibration, 254(2), (2002), 297-312.

COMPUTATIONAL MECHANICS (Abstracts)
WCCM VI in conjunction with APCOM'04, *Sept. 5-10, 2004, Beijing, China*

Buckling of Rectangular Mindlin Plates on Non-Homogeneous Elastic Foundations

Y. Xiang *, N. Palaparthy

School of Engineering and Industrial Design, University of Western Sydney, Penrith South DC NSW 1797, Australia
e-mail: y.xiang@uws.edu.au

Abstract

This paper is concerned with the buckling behaviour of rectangular Mindlin plates resting on non-homogenous elastic foundations that consist of multi-segment Winkler-type elastic foundations. A rectangular plate is assumed to be simply supported on two parallel edges and the two remaining edges may have any combination of free, simply supported or clamped conditions. The plate is first divided into subdomains along the interfaces of the multi-segment foundations. For each segment, the governing differential equations for buckling of a Mindlin plate on a Winkler-type elastic foundation are solved by employing the Levy solution approach associated with the state space technique [1-2]. The domain decomposition method is used to cater for the continuity and equilibrium conditions at the interfaces of the subdomains. An eigenvalue equation for buckling of a rectangular Mindlin plate on a non-homogenous elastic foundation is derived through an appropriate application of the plate boundary and interface conditions.
Rectangular plates subjected to uni- and bi-axial in-plane loads are considered. First-known exact solutions for buckling of rectangular Mindlin plates on a non-homogenous elastic foundation are obtained. The buckling of square Mindlin plates partially resting on an elastic foundation is studied in details. The influence of the foundation stiffness parameter, the foundation length ratio and the plate thickness ratio on the buckling factors of square Mindlin plates is discussed. The effect of the non-homogenous elastic foundations on the buckling mode shapes of the plates is also examined. The exact buckling solutions presented in this paper may be used as benchmarks for researchers to check their numerical methods for such a plate buckling problem.

REFERENCES

[1] Y. Xiang, K. M. Liew, S. Kitipornchai, *Exact buckling solutions for composite laminates: proper free edge conditions under in-plane loadings,* Acta Mechanica, 117, (1996), 115-128.
[2] Y. Xiang, *Vibration of rectangular Mindlin plates resting on non-homogenous elastic foundations,* Int. J. Mech. Sci., 45, (2003), 1229-1244.

COMPUTATIONAL MECHANICS (Abstracts)
WCCM VI in conjunction with APCOM'04, *Sept. 5-10, 2004, Beijing, China*

Elasto-Plastic Stability Analysis for Concrete-Filled-Steel-Tube Arch Bridges

Quansheng Yan*, Jinyong Xu

Department of Civil Engineering, South China University of Technology, Wushan, Guangzhou, 510640, China
e-mail: cvqshyan@scut.edu.cn , celxu@163.com

Abstract

In the last decade the concrete-filled-steel-tube arch bridges have been widespread used in the bridge practice in China because of the benefits of combining the two construction materials, low cost and short construction time as the void steel tube served as the false work. Due to the large compression in concrete-filled-steel tube arch ribs under external loads, the inelastic stability behavior of CFST arch bridge, which is particularly important in limit state design, is not yet thoroughly understood. With the fiber section model, an approach for the second-order inelastic stability analysis of the CFST arch bridge is presented. The arch ribs are modeled as 3-D beam elements. Both the geometric and material nonlinearities are taken into account. For the steel tube a perfectly elastic-plastic material model is implemented for in the formulation. And segmental polynomial expressions for the stress-strain curve of the confined concrete in the steel tube are employed. The elastic-plastic stiffness matrix is calculated employing the method of Newton-Cotes' integration with the direct stresses and strains at the centers of the segments across section. Thus the development of plasticity both alongside the longitude of the element and the section is traced. With the proposed approach, a program has been developed for the nonlinear large displacement analysis of CFST arch bridges.
In order to verify the theoretical approach, two test models of CFST arch rib are computed. The results show that the predicted limit loads and displacements agree well with the measured values of modeling test. And the behaviors of 12 axial and eccentrically loaded columns are also analyzed. The numerical results of the load-deflection curve and ultimate loads agree well with the measured results. Finally the ultimate load analysis for a practical CFST arch bridge is carried out.

REFERENCES

[1] A. S. Nazmy, *Stability and load-carrying capacity of three dimensional long-span steel arch bridges*, Computers and Structures, 65(6), (1997), 857-868.
[2] J. F. Hajjar, P. H. Schiller, A. Molodan, *A distributed plasticity model for concrete-filled steel tube beam-columns with interlayer slip*, Engineering Structures, 20(4-6), (1998), 398-412.
[3] L. H. Han, *Concrete-filled-steel-tube structure*, Science Press, Beijing, (2000), (in Chinese).
[4] A. M. E. Spacone, S. El-Tawil, *Nonlinear analysis of steel-concrete composite structures: state of the art*, Journal of Structural Engineering, 130(2), (2004), 159-168.

COMPUTATIONAL MECHANICS (Abstracts)
WCCM VI in conjunction with APCOM'04, *Sept. 5-10, 2004, Beijing, China*

Equivalent Material Parameter Study on Steel Member Strengthened by Carbon Fiber

Gang Yang [*1], Aifeng Zhang [1], Baolin Xiong [2], Lili Jiang [2]

[1] *Dalian Maritime University, No.1, Linghai Road, Dalian, 116026, China*
[2] *Shenyang Architectural and Civil Engineering University, Shenyang, 110168, China*
e-mail: yanggang@newmail.dlmu.edu.cn, af418@21cn.com

Abstract

The carbon fiber as a kind of high -strength composite has been widely used in consolidating concrete construction and increasing bearing capacity of steel member. However, owing to complication of structure glued with carbon fiber, composite is generally considered as shell or solid element and member wrapped as solid element when analyzing spatial construction and performing FEM calculation in theoretical study. Furthermore, calculation work for more complicated structure is enormous and it is sometimes impossible to make the calculation. The work will be greatly reduced if whole structure is considered as beam elements.
From this the expressions of cross-sectional area and moment of inertia are deduced again and the formulation of calculating elastic modulus of I, C and L-type member is given by the equivalent ideas. For example, E_{Eq} of I-type steel is

$$E_{Eq} = \frac{4(b+2t+t_2-\frac{1}{2}t_1)A_{11}+(2bt_2+ht_1-2t_2t_1)E_s}{2(b+2t)(t_2+2t)+t_1(h-2t_2-2t)}$$

Where A_{11} is principal rigidity of composite, E_s is elastic modulus of steel, t is thickness of composite. Others are cross-sectional geometrical parameters of rolled steel.
Here the percent increase in buckling load is used to discuss the method presented. It is independent of end conditions and length of member. Generally, wrapped position is selected by considering effect to buckling, for example, only glue carbon fiber to the surface of flanges for I-type member.
For verifying the theory, take two kinds of I-type steel for example. Three kinds of end conditions (both ends pinned, one fixed and one free, both ends fixed) are considered. The examples show that the results of this paper are very good agreement with 3D FEM analysis. All errors are in 1%. It indicates that the developed method is an efficient choice in calculating structural deformation and buckling analysis of member.

REFERENCES

[1] Luc R. Taerwe, Stijn Matthys, *FRP for Concrete Construction*, Concrete International, October, (1999), 34-36.
[2] Qilin Zhao, *Analysis on the 'Reverse-Arch Prestressed Technique' and the Improved Bearing Capacity for Steel Structure Strengthened by Carbon-Fiber*, Steel Construction, 17(3), (2002), 51-54, (in Chinese).
[3] Dong Wang, *FEM of concrete column reinforced with FRP against seism*, J. Shenyang Architecture and Civil Engineering Univ., 17(4), (2001), 259-261,264, (in Chinese).

COMPUTATIONAL MECHANICS (Abstracts)
WCCM VI in conjunction with APCOM'04, *Sept. 5-10, 2004, Beijing, China*

Analysis on Buckling and Post-Buckling Performances of Cold-Formed Thin-Wall Rectangular Columns

Zhijun Yuan *, Guoqing Gui, Jiqun Hu

School of Architecture Engineering of Nanchang University, Nanchang, 330029, China
email: sandy_ncu@hotmail.com, tmgc@ncu.edu.cn

Abstract

The cold-formed thin-walled structures are applied more widely in the actual engineering project, while it is absent of researches on the buckling and post -buckling performances of cold-formed thin-walled rectangular columns. Based on the present situat ion the following issues are discussed in this research:

1) Based on the Marguerre theory of large-deflection plate with initial deformation, the Galerkin analysis method is used for the theoretical development on buckling and post -buckling performances of these columns considering the mutual action of the group of plates and a non -linear equilibrium is established to analyze this member.

2) An experiment research is presented in this paper and the comparison is drawn between the theoretical analysis and ex periment research.

3) Some conclusion about the performance of the buckling and post -buckling performances of cold-formed thin-walled rectangular columns are given in this paper such as the relation between the ratio of the length to width and the mutual a ction of the group of plates and so on.

REFERENCES

[1] Shaorong Yan, Guoqing Gui, Zulong Sun, *Post-buckling analysis for channel sections under nonuiform compression in their webs*, Research on Steel Structural Engineering, Vol. 2, (in Chinese).

[2] R. Sziland, *Theory and analysis of plates – classical and numerical methods*. Prentice-Hall, Inc, (1974).

[3] Yaoqian Yang, *Flat sheet theory,* Railway publishing company, (1980), (in Chinese).

[4] Zhijun Yuan, *Analysis and experimental research on buckling and post-buckling performances of cold-formed thin-wall rectangular columns*, Master Thesis, Nanchang University, Jiangxi, China, (2001).

COMPUTATIONAL MECHANICS (Abstracts)
WCCM VI in conjunction with APCOM'04, *Sept. 5-10, 2004, Beijing, China*

Model Updating Incorporating Damping Matrix

Vikas Arora *, T. K. Kundra, S. P. Singh

Department of Mechanical Engineering, Indian Institute of Technology Delhi, Hauz Kaus, New Delhi, 110016, India
e-mail: vikingarora@yahoo.com, mer02007@ccsun50.iitd.ernet.in

Abstract

Model updating techniques are used to update a finite element model of the structure so that an updated model predicts more accurately the dynamics of the structure. A number of model updating methods have been proposed in recent years and details of them can be found in text [1]. Almost all these methods neglect damping to prevent complexity. So these methods can be used up to the point of predicting the natural frequencies and real modes. But these cannot use for complex Frequency Response functions (FRFs) and complex modes. All the real structures have some form of damping. This paper presents a FRF-based finite element model updating procedure in which the updating variables also include damping parameters. The procedure assumes structural damping mechanism.

$[m]\ddot{x} + [k]x + i[d]x = \{f\}$

where d represents structural damping

$d = \alpha[m] + \beta[k]$

α and β are taken as damping updating parameters. The FRFs sensitivities are calculated using updating parameters, which includes damping parameters. The updating parameters are defined to the elemental level so that parameters are adjusted locally and facilitate the physical interpretation of the adjusted model. The comparison of the proposed method is done with the two popular updating methods namely Inverse eigensensitivity method and Response function method on the basics of computer simulated experimental data of a free-free beam is used. The test cases for complete, incomplete and noisy experimental data are considered. The updated models are compared on the basis of some error indices constructed to quantify error in mode shapes and response functions.

REFERENCES

[1] J. E. Mottershead, M. I. Friswell, *Finite element model updating in structural dynamics*, Kluwer Academic Publishers, Dordrecht, (1995).

[2] J. D. Collins, G. C. Hart, T. K. Hasselman, B. Kennedy, *Statistical identification of structures,* AIAA Journal, 12(2), (1974), 185-190.

[3] R. M. Lin, D. J. Ewins, *Model updating using FRF data*, Proceedings, 15th International modal analysis seminar, K.U. Leuven, Belgium, (1990), 141-163.

COMPUTATIONAL MECHANICS (Abstracts)
WCCM VI in conjunction with APCOM'04, *Sept. 5-10, 2004, Beijing, China*

Semi-Analytical Analysis of Free Vibration of Super Tall Building Mega Frame Structure Considering Subgrade Deformation

Yaoqing Gong*, Xiangdong Xie

Department of Civil Engineering, Jiaozuo Institute of Technology, Jiaozuo, 454000, China
e-mail: gongyq@263.net

Ab*stract*

The superstructure and its foundation of a super tall building space-mega-frame structure were, in the paper, simplified equivalently and continuously to a three-dimensional model, which is a combination of stiffened-thin-walled tubes on semi-infinite elastic body. The foundation of the building was taken as a part of the superstructure that in fact is the extension of the superstructure to the underground. The superstructure, since a space-mega-frame structure is composed of two grades of frames, the first one is the mega-frame, the main body of load bearing; and the second one is the supplementary frame within the mega-frame, which is also play a role of structure. The geometric dimension (cross sectional area and inertial moment etc.) of the mega-frame is very large, but that of the supplementary is very small, both of which don't belong to the same order of magnitude. In order to carry out the effective simplified analysis to these kinds of structure with three dimensional model, it is possible to transform the structures equivalently and continuously into the combination of several closed stiffened-thin-walled tubes with different stiffness and different mass by means of equivalent principle of rigidity and mass.[1] Thusly, the members of the supplementary frame are transformed to a combination of thin-walled tubes and the members of the mega-frame are transformed to the stiffened ribs of the tubes. The tubers concert with their ribs. When establishing the motion equations of free vibration for the above mentioned stiffened-thin-walled tube combination on the semi-infinite elastic subgrade, unknown functions of the longitudinal nodal lines and the centroidal lines of the sections were taken as generalized coordinate(infinite degree of freedom), and semi-discretizing along periphery with interpolation functions between nodal lines was made. So the motion field of the stiffened-thin-walled tubes could be expressed by the generalized coordinates, which is a function set. The free vibration equations then were obtained by Hamilton's principle. And the equations could be solved by ordinary differential equation solver, a kind of computational software named COLSYS. [2] A computational example was also presented in the paper and some valuable conclusions are obtained through analyzing the reasonable results .

REFERENCES

[1] Yaoqing Gong, *Tall Building Structures on Elastic Subgrade and Research of Semi-Analytical Method*, Beijing: Tsinghua University, 1999, (in Chinese).

[2] Si Yuan, *Introduction of a common compute program for boundary problems of ordinary differential equation* — COLSYS, Computational Mechanics and Application, 2, (1990), 104-105, (in Chinese).

COMPUTATIONAL MECHANICS (Abstracts)
WCCM VI in conjunction with APCOM'04, *Sept. 5-10, 2004, Beijing, China*

Free Vibration Analysis of Rotating Cantilever Flexible Plates Using the *p*-Version of The Finite Element Method

S. M. Hamza-cherif*, A. Houmat [1]

Department of Mechanical Engineering, University Abou Bekr Belkaid, Tlemcen, 13000, Algeria
e-mail: smhamzacherif@caramail.com

Abstract

Vibration analysis of rotating cantilever plates is an important subject of study in mechanical engineering. There are many examples in astronautic and aeronautic which can be modelized as rotating cantilever plates, such as solar panels of satellite, turbo -engine blades and helicopter blades. In this study a *p*-version of the finite element method is applied to free vibration analysis of rotating cantilever plates in conjunction with the new modeling dynamic method using the arc - length stretch deformation. The flexible and the rigid body DOF are supposed uncoupled, the linear equations of motion are derived including coupled terms between the in -plane and the transverse bending vibration, caused by the gyroscopic effect. The hybrid displacements may be expressed as the combination of the in-plane and out-of-plane shape functions formulated in terms of linear and cubic polynomials functions used generally in FEM plus a variable number of trigonometric shapes functions representing the internals DOF for the plate element. Trigonometric enriched stiffness, mass, gyroscopic and centrifugal stiffness matrices are derived using symbolic computation.
The convergence properties of the rotating plate Fourier *p*-element proposed and the results are in good agreement with the work of other investigators. The influences of rotating speed and coupling effect on the natural frequencies are investigated. It is shown that by this element the order of the resulting matrices in the FEM is considerably reduced leading to a significant decrease in computational effort.

REFERENCES

[1] M. A. Dokainish, S. Rawtani, *Vibration analysis of rotating cantilever plates*, Int. J. Num. Meth. Eng., 3, (1971), 233-248.

[2] V. Ramamurti, R. Kielb, *Natural frequencies of twisted rotating plates*, J. Sound Vib., 97(3), (1984), 429-449.

[3] H. H. Yoo, C. Pierre, *Modal characteristic of a rotating rectangular cantilever Plate*, J. Sound Vib., 259(1), (2003), 81–96.

[4] A. Houmat, *A sector Fourier p-element applied to free vibration analysis of sectorial plates*, J. Sound Vib., 243(2), (2001), 269-282.

[5] S. M. Hamza-cherif, A. Houmat, *Natural frequency of rotating flexible beams by using hierarchical finite element method*, in J. B. Castillero, M. Ostoja-Starzewsky, R. R. Ramos eds. *American Academy of Mechanics*, Proc. 8th Pan Amer. Cong. of App. Mech., Havana, (2004), pp. 101-104.

[6] N. S. Bardell, *The application of symbolic computing to hierarchical finite element method*, Int. J. Num. Meth. Eng., 28, (1989), 1181-204.

COMPUTATIONAL MECHANICS (Abstracts)
WCCM VI in conjunction with APCOM'04, *Sept. 5-10, 2004, Beijing, China*

Bending-Moment-Eigenvector Expansion of Free End Displacement of Cantilever Rigid Frame

Guanghao Liu [1*], Qiang Li [2], Meiyun Xue [1]

[1] *School of Civil Engineering, Hebei University of Technology, Tianjin, 300132, China*
[2] *Transportation College, Southeast University, Nanjing, 210096, China*
e-mail: guanghaoliu@eyou.com, liqiang@xpms.com.cn, xuemeiyun@scse.hebut.edu.cn

Abstract

This paper is Bending-Moment-eigenvector expansion of end-displacement of cantilever rigid frame of which be made of several poles of stiffness be constant. Those are from two parts. First are properties of Bending-Moment-eigenvector and of bar-end-displacement bending moment diagram[2]. And second is basic physical relationship between internal force and end displacement of straight cantilever beam that is out from relationship of end-displacement and internal force of beam [1, 3]. The shown of free end displacement of frame is in terms of Bending-Moment-eigenvector of each pole. In a pole it needs only one Bending-Moment-eigenvector of which at characteristic-point of pole. The parameters of expansion are line-stiffness of pole and location of pole standing to free end of frame. And shear force has been expunction from equation. The characteristic-point of pole is determined by the position projection of the free-end on the pole. The characteristic-point position on range [1/3, 1/2] of pole, if projection of free-end of frame is on direction prolong of pole. The characteristic-point position on [0, 1/3] and/or [1/2, 1] respectively, if the projection is on opposite and it is divided by region [−1/3, −2/3]. There is no characteristic-point and no Bending-Moment-eigenvector for need of this expansion, if the projection is in (−1/3, −2/3) of pole. There is a straight beam of having a static equilibrium of several concentrated force. In which force equals to ratio of bending-moment-eigenvector to stiffness of pole. When each force is at midpoint of each segment of the beam the calculation of end bending moment is dual with calculation of expansion.

REFERENCES

[1] Guanghao Liu, *Study on end-displacement relationship with internal force of beam and analysis bending moment*, Proc. 7th Nat. Conf. Str. Eng., Vol. 1, Shijiazhuang, 1998, Eng. Mech., Beijing, (1998), pp. 536-539, (in Chinese).
[2] Guanghao Liu, *Characteristic bending moment method for each displacement of beam*, J. Hebei Univ. Tech., 29(2), (2000), 95-97, (in Chinese).
[3] Guanghao Liu, *Research of generalized diagram multiplication method and the properties of some internal force diagram*, J. Hebei Univ. Tech., 25(2), (1996), 115-120, (in Chinese).

COMPUTATIONAL MECHANICS (Abstracts)
WCCM VI in conjunction with APCOM'04, *Sept. 5-10, 2004, Beijing, China*

A Numeric Approach Combining Finite Element with Transfer Matrix for Vibration of Magnetostrictive Rod

Xinchun Shang [1*], Mingyu Jin [2], Xiang Han [1]

[1] *Dept. of mathematics and mechanics, Beijing Univ. of Science and Technology, Beijing, 100083 China*
[2] *North China Institute of Computing Technology, Beijing 100083, China*
e-mail: shangxc@public3.bta.net.cn

Abstract

A numerical approach combined the finite element method with the transfer matrix method is applied for solving the vibration problem on a magnetostrictive rod. The engineering background of the problem is from the Terfenol-D (a kind of magnetostrictive material containing a rare earth compound) linear actuator. The work behavior of the actuator can be modeled as the vibration of a rod. The vibration is excited by the harmonic current running through a solenoid around the rod, as magnetostrictive material expands or contracts in response to variation of magnetic field. The material constitutive relation of the rod is assumed as cubic nonlinear. One end of the rod is restricted by prestress linear spring and another end is fully fixed. The total energy of kinetic, strain and magnetic energy stored in the rod is expressed as a functional of displacement variable. By discretization and linear interpolation for displacement in the space dimension and using Harmilton variational principle, the governing finite element equations is formulated as a system of linear second-order ordinary differential equations for the nodal values of the displacement. The global mass matrix is constant, while the global stiffness matrix and the nodal force vector are dependent on the time variable. Introducing nodal velocities the governing equations are then transformed into a system of linear first-order differential equations for the nodal displacements and velocities. The numeric solutions of the nodal displacements and velocities are obtained by the transfer matrix method associated with matrix exponent. As a numeric example, the forcing vibration of Terfenol-D rod excited by harmonic current is analyzed. The numerical solution shows that the response of displacement is periodic and the frequency of response is roughly as twice as that of the exciting current. This result is also in accordance with the experiment one. The numeric computation confirms that the present method is simple and efficient.

REFERENCES

[1] H. J. Busch-Vishniac, *Electromechanical sensors and actuators,* Mechanical engineering series, Springer-Verlag, New York (1999).

[2] M. B. Moffet, A. Clark, M. Wun-Fogle, *Characterization of Terfenol-D for magnetostrictive transducers.* J. Acoust. Soc. Amer., 89, (1991), 1448-1455.

[3] Y. Yamamoto, H. Eda, T. Mori, A. Rathore, *Smart vibration sensor using giant magnetostrictive materials,* JMSE Int. J. Series C, 40, (1997), 260-266.

[4] J. N. Reddy, J. I. Barbosa, *On vibration suppression of magnetostrictive beam.* Smart Mater. Struct., 9, (2000),49-58.

[5] J. R. Partt, S. S. Oueini, A. H. Nayfeh, *Terfenol-D nonlinear vibration absorber,* J. Intelligent Mater. Systems Struct., 10, (1999) 29-35.

[6] Y. Wan, D. Fang, K. Hwang, *Nonlinear constitutive relations for magnetostrictive materials,* Int. J. Non-linear Mech., 38, (2003), 1053-1065.

[7] X. Shang, J. J. Grannell, in M. D. Gilchrist eds. *Modern practice in stress and vibration analysis*, Proc. 3rd Int. Conf., Dublin, A. A. Balkema, (1997), pp. 463-468.

COMPUTATIONAL MECHANICS (Abstracts)
WCCM VI in conjunction with APCOM'04, *Sept. 5-10, 2004, Beijing, China*

Free Vibration Behaviour of a Range of High Aspect Ratio Aircraft Wings Using the Dynamic Stiffness Method

Huijuan Su *, J. Ranjan Banerjee

School of Engineering and Mathematical Sciences, City University, London, Northampton Square, London, EC1V 0HB, U.K.
e-mail: h.su@city.ac.uk, j.r.banerjee@city.ac.uk

Abstract

The primary aim of this paper is to in vestigate the free vibration behaviour of a range of high aspect ratio aircraft wings using the dynamic stiffness method. In order to achieve this, the aircraft wings are idealised as beam elements for which dynamic stiffness matrices are available in literature. The method of dynamic stiffness matrix is well established. This procedure gives exact results because the shape function used is derived from the governing differential equation of motion of the structural element and hence it is exact [1]. Thus the accuracy of results in the dynamic stiffness method is independent of the number of elements used in the analysis. For instance, one single structural element can be used to compute any number of natural frequencies without compromising numerical accura cy. This is, of course, impossible in the finite element and other approximate methods. Generally the dynamic stiffness method is applied in conjunction with the Wittrick-Williams algorithm [2, 3] to yield natural frequencies and mode shapes of a vibrating structure. This paper focuses on the aeronautical application of the dynamic stiffness method by investigating the free vibration problems of aircraft wings. In particular, six different aircraft wings with varying degree of complexities in their construc tion are considered in the analysis. Out of these six wings, two are for sailplanes, two are lightweight trainer aircraft and the other two are for medium to high range transport airliners. The results of the analysis are compared and contrasted. The first six natural frequencies and mode shapes are shown for each of the six aircraft wings. The results show interesting distribution of bending and torsional modes in the free vibratory behaviour of the aircraft wings. The investigation provides an insight int o the free vibration characteristics of a wide range of aircraft wings and reveals many interesting features. This study is particularly important from an aeroelastic point of view because the free vibration analysis is generally a fundamental pre-requisite before carrying out a flutter analysis when nomal mode method is used.

REFERENCES

[1] J. R. Banerjee, *Dynamic stiffness formulation for structural elements: a general approach*, Computers and Structures, 63, (1997), 101 -103.
[2] F. W. Williams, and W. H. Wittrick, *A general algorithm for computing natural frequencies of elastic structures*, Quart. J. Mech. Appl. Math., 24, (1971), 263 -284.
[3] F. W. Williams, and W. H. Wittrick, *Exact buckling and frequency calculations surveyed*, Journal Structural Engineering, ASCE , 109, (1983), 169-187.

COMPUTATIONAL MECHANICS (Abstracts)
WCCM VI in conjunction with APCOM'04, *Sept. 5-10, 2004, Beijing, China*

Time Collocation Methods for Structural Dynamic Problems

Chen Tang [1, 2*], Haiqing Yan [1], Zhanqing Chen [3], Hao Zhang [1], Ming Liu [1], Guimin Zhang [1]

[1] *Department of Applied Physics, University of Tianjin, Tianjin, 300072, China*
[2] *State Key lab. of Nonlinear Mechanics, Institute of Mechanics, Chinese Academy of Sciences, Beijing, 100080, China*
[3] *Department of Mechanical Engineering and Automation, Harbin Institute of Technology, Harbin, 150086, China*
e-mail: tangen@eyou.com, yanhg@eyou.com

Abstract

It is usually difficult to evaluate the response analytically for time-dependent problems, so a lot of research has been performed in the area of the numerical solution of s tructural dynamic problems. The numerical method treated in this paper is the so -called "collocation method". In the last decade, various numerical schemes have been constructed based on collocation method. The spectral collocation methods were constructed for partial differential equations, e.g. [1]. The spatial collocation method were applied to the shallow water equations (SWE) [2], where the SWE are discretized in time with the semi-implicit leapfrog method, and in space with the cubic spline collocation method on a skipped latitude–longitude grid. The time collocation methods (TCM) were presented for heat equations, e.g. in [3]. Hermite collocation methods were applied efficiently in one space dimension to steady and unsteady di fferential problems with steep solutions [4]. The Legendre collocation method was used to solve numerically Cahn –Hilliard (C–H) equation with Neumann boundary conditions [5].

Although collocation methods were applied to structural dynamic problems, e.g. [6], formulae for the polynomial coefficients seem to be complex, and schemes were only applied to the systems whose prescribed load is independent on time. In this paper, we derive new formulae for the polynomial coefficients, which are more practical and succinct than those previously given. The new scheme can be applied to systems whose the prescribed load is dependent on time. The work is an extension of earlier work [7]. We test the present method on two examples, and compare with exact solution and well studied and widely used four-order Runge-Kutta method. In all cases, the present method gives better results than the conventional Runge -Kutta method. In addition, an important advantage of the method is the simplicity in software programming.

REFERENCES

[1] O. V. Vasilyev, S. Paolucci, J. Comput. Phys., 125, (1996), 498 -512.
[2] A. T. Layton, J. Comput. Phys. 179, (2002) 578-592.
[3] J. Kouatchou, Numer. Methods Partial Di fferential Equations, 17, (2001), 54-63.
[4] A. W. Lang, D. M. Sloan, J. Comput. Appl. Math. 140, (2002), 499–520.
[5] X. D. Ye, J. Comput. Appl. Math. 150, (2003) 87-108.
[6] T. C. Fung, Comput. Methods Appl. Mech. Engrg., 190, (2000), 1651-1662.
[7] J. Kouatchou, J. Comput. Appl. Math., 150, (2003), 129–141.

COMPUTATIONAL MECHANICS (Abstracts)
WCCM VI in conjunction with APCOM'04, *Sept. 5-10, 2004, Beijing, China*

Theoretical Analysis for the 3-D Seismic Response of a Base-Isolated High-Rise Building

D. M. Wei [1*], J. X. Kang [2]

[1] *Department of Civil Engineering, South China University of Technology, Guangzhou 510640, China*
[2] *Department of Civil Engineering, Taiyuan University of Technology, Taiyuan, 030024, China*
e-mail: weidemin@21cn.com

Abstract

With the development of technology and the rising of the living standards of the people, more and more questions for the earthquake resistant behavior and reliability of high-rise buildings emerged. A new design theory which is very different from the traditional ones, that is, the control theory for structure vibration, has been developed. The control methods can be divided into active control, semi-active control, passive control and mixed control. In these control techniques, the passive control techniques are widely used in practice because they do not need any external energy and are easily constructed. Base-isolation is the most practical and economical measure of passive-control techniques. Many buildings and bridges with base-isolators had been built and a lot of theoretical researches in this field had been done. However, there are still some urgent problems need to be solved, for example, the practical arrangement and the reliability of base-isolation techniques for high-rise buildings, the evaluation of anti-seismic efficiency, and nonlinear analyses of seismic response of isolated buildings etc.

In this paper, the 3-D seismic response of the highest frame-shear wall building in China with rubber-isolators and its seismic response in the non-isolate conditions are investigated by using the finite element method. Through the comparison of the computational results, the difference between the maximum interstorey shears, the lateral displacements and the acceleration responses of the isolated building and its corresponding non-isolated pattern are given. And through computations and analyses, some useful results and conclusions are obtained.

REFERENCES

[1] D. M. Zeng, J. Y. Su et al, *Development and application of seismic base isolation technology of buildings*, Earthquake Resistant Engineering, 3, (1996), 37-41.
[2] H. N. Li, *Study on vibration damping properties of pendulum-structures system*, Engineering Mechanics, 13(3), (1996), 123-129.
[3] J. X. Tang, *Stipulations on design of seismic isolation and energy dissipation for building*, Earthquake Resistant Engineering, 3, (1999), 13-17.
[4] B. L. Hou, *New developments of earthquake isolation and applications of earthquake control techniques in Japan*, Industrial Construction, 11, (2000), 74-78.
[5] X. D. Liu, J. Y. Su, Choice and Measurement of Ground motion Record used for Time-History analysis, Building Structures, 22(5), (1992), 8-12.
[6] Y. Y. Wang, *A review of seismic response spectra time history analysis and energy method*, Journal of Building Structures, 21(1), (2000), 21-28.
[7] D. M. Wei, L. L. Dai, The seismic response of a base-isolated building of the masonry and concrete composite structure, Journal of South China University of Technology 29(10): 73-76.
[8] J. X. Kang, The 3-Dimension seismic response of a base-isolated high-rise structure, Taiyuan University of Technology, Taiyuan, China, (2002).

COMPUTATIONAL MECHANICS (Abstracts)
WCCM VI in conjunction with APCOM'04, *Sept. 5-10, 2004, Beijing, China*

High Order Iterative Algorithms and Lagrange Multipliers for Unilateral Contact Problems

W. Aggoune*, H. Zahrouni, M. Potier-Ferry

Laboratoire de Physique et Mécanique des Matériaux, University of Metz, Ile du Saulcy, Metz, 57045 Cedex 01, France
e-mail: {aggoune, zahr, potier-ferry}@lpmm.univ-metz.fr

Abstract

This work is a contribution to the development of efficient numerical methods for unilateral contact problems. We propose new algorithms based upon the Asymptotic numerical Method (ANM) which combines a perturbation technique with the finite element method, to solve this kind of problems. The first step of the method consists in expanding the unknowns of a non linear problem in the form of series. This allows to transform the initial non linear problem into a recursive sequence of linear problems which have the same tangent operator. Then the linear problems are solved using the finite element method. Since the contact law is not analytic and not adapted to asymptotic expansions, we replace it by a hyperbolic relation between forces and displacements. The contact reactions are considered as external forces and determined in the form of series. So as to discretize the unilateral contact problem, we have chosen a mixed formulation involving Lagrange multipliers. To describe the whole solution curve, we use a path following technique which is based upon rational fractions called Padé approximants.

The path following technique is controlled by a parameter which defines the length of each step. In order to improve the reliability of the ANM algorithms, we propose to correct the solution at high order when this becomes necessary. We propose new high order prediction-correction algorithms for contact problems. They allow to compute large computation steps with corrections when the residual is no more satisfying. A high order predictor is combined with a high order corrector which is built up by associating series expansions, a homotopy technique and Padé approximants. Two strategies have been used to define the prediction-correction algorithms. The first algorithm can be seen as a high order algorithm of Newton type and involves the treatment of two matrices per step : one for the prediction and one for the correction. In the second, the prediction curve is computed by using the matrix decomposed during the previous correction phase. This allows us to decompose only one matrix per step.

Numerical tests on 2D frictionless contact problems are presented to assess the efficiency of the proposed algorithms : contrarily to other algorithms, the number of steps does not vary with the mesh and the number of tangent matrices is much lower. The reliability and the efficiency of the classical ANM are improved as well.

REFERENCES

[1] W. Aggoune, H. Zahrouni, M. Potier-Ferry, *High order prediction-correction algorithms for unilateral contact problems*, J. Comp. Appl. Math., (in press).

[2] H. Lahmam, J. M. Cadou, H. Zahrouni, N. Damil, M. Potier-Ferry, *High order predictor-corrector algorithms*, Int. J. Num. Meth. Eng., 55, (2002), 685-704.

COMPUTATIONAL MECHANICS (Abstracts)
WCCM VI in conjunction with APCOM'04, *Sept. 5-10, 2004, Beijing, China*

A Velocity and Level-Sets Based Approach of Dynamic Contact Problems with Applications

Hachmi Ben Dhia*, Chokri Zammali

Labo. Mss-Mat, Ecole Centrale Paris, Châtenay-Malabry 92295, France
e-mail: bendhia@mssmat.ecp.fr, zammali@mssmat.ecp.fr

Abstract

Impact problems are nonlinear, irregular and multiscale. They have to be approximated numerically with a particular caution. Hughes et al. have designed numerical schemes since the late seventies and increasing efforts have been since furnished by the computational contact community to address this numerical issue. Important advances have this way been done. But to our best knowledge and due to the aforementioned intrinsic difficulties of the impact problems (strong nonlinearities, space and time irregularities and multiscale character), none of the available numerical schemes is yet sufficiently robust and efficient to address for instance dynamic contact problems such as those involved by vibrating structures under impact. This work is a contribution to this wide and complex theme.

Following a pioneering idea of J.J. Moreau [1], the unilateral contact laws are written in terms of positions, velocities and percussions. The obtained *Signorini-Moreau* contact model is then converted to multi-valued equalities by using two unknown *Level-Sets* type fields standing for a location of the positions of contact surfaces with respect to each other and for the *Sign* of the normal velocity jump field on the interface. An original continuous weak-strong mixed formulation of a dynamic 3-D contact problem is then carried out in a straightforward manner. The associated discrete nonlinear systems are derived by using three discretization methods, namely a θ-time discretization scheme, a *Galerkin* method and a collocation one. The numerical algorithm used to solve the nonlinear and irregular discrete problems is described and numerical tests exemplifying our methodology are given. It is shown in particular that our numerical solutions do not exhibit pathologies such as spurious oscillations of the numerical mechanical fields (velocities and contact loads). These developments are extended in a second step to take into account friction phenomena.: a global dynamic frictional contact approach using an additional *Level-Set* field (whose 1-isovalues characterize the sticking contact regions) is presented and discretized. Significant dynamic frictional contact tests are given. Moreover, some promising applications of the multiscale *Arlequin* method [2] to the multiscale impact problems are shown.

REFERENCES

[1] J-J. Moreau, *In nonsmooth mechanics and applications*, CISM courses and lectures, 1-82. Springer-Verlag, Wien/NY, (1988).

[2] H. Ben Dhia, *Multiscale mechanical problems: the Arlequin method*, CRAS 236, série Iib, 1998. 899-904.

COMPUTATIONAL MECHANICS (Abstracts)
WCCM VI in conjunction with APCOM'04, *Sept. 5-10, 2004, Beijing, China*

A Robust Contact Model for Non-Smooth Rigid Bodies

Y. T. Feng *, K. Han, D. R. J. Owen

Civil & Computational Engineering Centre, University of Wales Swansea, SA2 8PP, UK
e-mail: {y.feng, k.han, d.r.j.owen}@swansea.ac.uk

Abstract

Modelling of contact between non-smooth objects, such as polygons/polyhedra, is a challenging topic in both finite element based computational contact mechanics and discrete element simulations [1-2]. In the context of penalty based contact solution approaches, the fundamental problem associated with the conventional node/segment(facet) contact models for the so-called corner/corner contact situations, as illustrated in Figure 1, lies in two related aspects: 1) the gap function is not well defined; and 2) the normal contact direction may also be ambiguous. Consequently the contact forces (including both directions & magnitudes) are difficult to determine, and often not evolved in a smooth way but present a discontinuous jump when only a small relative movement occurs between the two objects in contact. This numerical defect may introduce a certain amount of artificial energy into the computation, which, when accumulated and propagated, could cause severe numerical errors or result in a total simulation failure. The present work presents a novel contact model for rigid convex polygons based on the following assumptions: 1) a potential energy (W) is associated with the contact which is a monotonic increasing function of the overlap area (A); and 2) the normal force ($\mathbf{F}_n$) acts along the direction which reduces the contact energy with the maximum rate:

$$W = W(A); \quad \mathbf{F}_n = -\nabla W(A)$$

It can be proved [2] that the above assumptions uniquely define the normal, tangential contact directions and reference contact point. The model is simple and elegant, and can easily handle complex contact scenarios such as those shown in the Figure. More significantly, it provides a generic contact model for all types of objects. Numerical examples are presented to demonstrate the robustness of the model developed. A particular application to super-quadrics is given in [3].

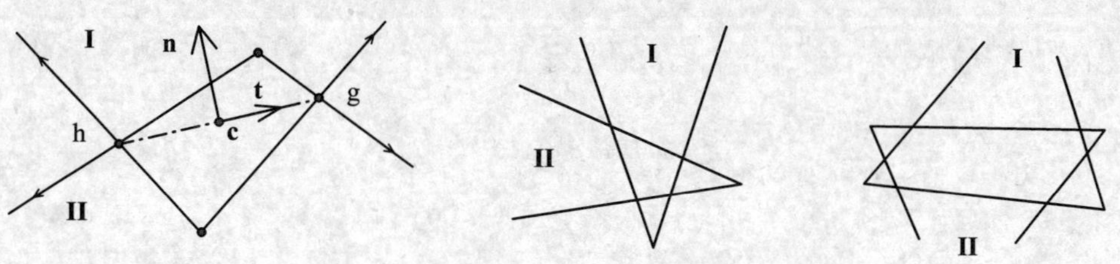

Fig. 1 Three corner/corner contact situations

REFERENCES

[1] P. Wriggers, *Computational Contact Mechanics*, John Wiley & Sons Ltd., Chichester, (2002).

[2] Y. T. Feng, D. R. J. Owen, *A 2D polygon/polygon contact model: algorithmic aspects*, Engineering Computations, (in press).

[3] K. Han, Y. T. Feng, D. R. J. Owen, *A robust contact model for rigid super-quadrics*, WCCM VI, Beijing, China, Sept. 5-10, (2004).

COMPUTATIONAL MECHANICS (Abstracts)
WCCM VI in conjunction with APCOM'04, *Sept. 5-10, 2004, Beijing, China*

Energy Dissipation by Friction in Dynamic Multibody Contact Problems

Z. Q. Fcng *, B. Magnain, J. M. Cros, P. Joli

Energy Dissipation by Friction in Dynamic Multibody Contact Problems
e-mail: feng@iup.univ-evry.fr

Abstract

The bi-potential method has been successfully applied for the modeling of frictional contact problems in static cases. This paper presents the application of this method for dynamic analysis of impact problems with multiple deformable bodies. Instead of second order algorithms, a first order algorithm is applied for the numerical integration of the time-discretized equation of motion. The numerical results show clearly the physical energy dissipation introduced by frictional effects between the solids in contact.

COMPUTATIONAL MECHANICS (Abstracts)
WCCM VI in conjunction with APCOM'04, *Sept. 5-10, 2004, Beijing, China*

The Mortar Method applied to 2D Large Deformation Finite Element Contact Problems

Kathrin A. Fischer *, Peter Wriggers

Institute of Mechanics and Computational Mechanics, University of Hanover, Appelstrasse 9a, 30167 Hannover, Germany
e-mail: fischer@ibnm.uni-hannover.de, wriggers@ibnm.uni-hannover.de

Abstract

The mortar method is a special discretization technique. Its application to contact mechanics is motivated by the limitations of well known elements, for example the node-to-segment element. Coupling of this contact element with higher order shape functions is inconsistent, and there is a loss of accuracy in the transmission of contact stresses.
In contrast, the mortar method consists of a variational projection of the displacements from one contacting body to the next, where unstructured grids are connected using a Lagrange multiplier formulation. Therefore, consistent coupling with higher order continuum elements is possible, which is of high interest regarding engineering applications.
The mortar method originates from the domain decomposition technique [1], where the physical domain is separated into regions which can be modeled by different discretizations. Adopted to contact mechanics this special technique enforces the constraints in the discretized system for non-matching contact partners.
Large deformation contact comprehends curved surfaces and large relative motion with finite deformations especially in the contact area. This heavily nonlinear boundary value problem is a great challenge regarding the numerical treatment.
In this context, mortar elements including a special Lagrange multiplier formulation, see [2], with linear and quadratic shape functions is presented.
Additionally, a penalty formulation in the style of the mortar method is shown.
The different formulations are compared with respect to convergence and accuracy.

REFERENCES

[1] C. Bernardi, Y. Maday, A. T. Patera, *A new nonconforming approach to domain decomposition: the mortar element*, in H. Brezis & J.-L. Lions eds. Collège de France Seminar XI, Pitman, (1994), pp. 13-51.

[2] B. I. Wohlmuth, *A Mortar finite element method using dual spaces for the Lagrange multipliers*, SIAM Journal of Numerical Analysis, 38(3), (2000),: 989-1012.

COMPUTATIONAL MECHANICS (Abstracts)
WCCM VI in conjunction with APCOM'04, *Sept. 5-10, 2004, Beijing, China*

A Robust Contact Model for Rigid Super-Quadrics

K. Han*, Y. T. Feng, D. R. J. Owen

Civil & Computational Engineering Centre, University of Wales Swansea, SA2 8PP, UK
e-mail: {k.han, y.feng, d.r.j.owen}@swansea.ac.uk

Abstract

Disks and spheres are the most commonly used geometric objects in discrete element simulations, mainly due to their geometric simplicity and highly computational efficiency. However, a drawback of disk/spherical elements is that they completely lack angularity, i.e. cannot provide any resistance to rolling motion. It has been increasingly recognised that the angularity of arbitrary shaped objects generally plays a fundamental role in the dynamic behaviour of many practical problems. Consequently a number of non-circular/spherical discrete elements have been used among which super-quadrics [1] are the most general. A 3D super-quadric is defined by the following implicit function:

$$f(x,y,z)=\left(\frac{x}{a}\right)^m+\left(\frac{y}{b}\right)^m+\left(\frac{z}{c}\right)^m-1=0$$

where parameters *a, b, c* and *m* determine the shape of the super-quadric. Note that disk/sphere and ellipse/ellipsoid are all special cases of super-quadrics. Despite its mathematic elegancy and smooth boundary, super-quadrics are not widely used because of a high computational cost associated with its contact detection. An overlap check between two super-quadrics may involve numerically obtaining the intersections of two nonlinear functions, which is a very expensive operation and may fail in certain circumstances. In order to overcome these difficulties, Williams & Pentland [2] proposed a more efficient discretised version but with a degraded accuracy. The present work follows a similar idea. The main steps of the proposed algorithm include: 1) an adaptive sampling procedure is employed to represent a super-quadric by a polygon; 2) based on the overlap zone of the bounding boxes of the two super-quadrics, the two polygons are clipped to exclude those edges lying outside the overlap zone; 3) an effective polygon intersection algorithm is adopted to find the intersections and overlap area of the two clipped polygons, if they exist; and 4) a robust polygon/polygon contact model [3] is then applied to determine the contact forces between the two objects. The algorithm can be applied to any super-quadrics, and is efficient and robust. The accuracy is controlled by the number of points used in the polygon to represent the original super-quadric. Several examples are provided to demonstrate the performance of the model developed.

REFERENCES

[1] *A. Barr, Superquadrics and angle-preserving transformations*, IEEE Comput. Graph. App., 1, (1981), 1-20.

[2] J. R. Williams, A. Pentland, *Superquadrics and model dynamics for discrete elements in interactive design*, Engineering Computations, 9, (1992), 115-128.

[3] Y. T. Feng, K. Han, D. R. J. Owen, *A robust contact model for non-smooth rigid bodies*, WCCM VI, Beijing, China, Sept. 5-10, (2004).

COMPUTATIONAL MECHANICS (Abstracts)
WCCM VI in conjunction with APCOM'04, *Sept. 5-10, 2004, Beijing, China*

Wear Simulation Based on Node-to-Segment Element

Feng Liu [1*], P. Wriggers [2]

[1] *Institute of Engineering Mechanics, Guangdong University of Technology, Guangzhou, 510643, China*
[2] *Institut für Baumechanik und Numerische Mechanik, University of Hannover, 30167, Germany*
e-mail: liutong96@hotmail.com, wriggers@ibnm.uni-hannover.de

Abstract

Wear phenomenon concerned with two sliding contactors is simulated by the nonlinear finite element method. Holm-Archard wear law is used to build model for mild oxidation and delaminating wear. A modified wear law is implemented to calculate wear profile, which is affected strongly by the initial surface topology. The node-to-segment contact element is applied to obtain contact stresses and sliding distance between two contact bodies. A simple geometry update process is implemented to test if more accurate variables about the system could be obtained. Two illustrative examples of indentation and pin-on-disk test show the ability of the proposed method to simulate wear phenomenon.

REFERENCES

[1] A. Cantizano, A. Carnicero, G. Zavarise, *Numerical simulation of wear-mechanism maps,* Computational Materials Science, 25, (2002), 54-60.
[2] Niclas Stroemberg, *A Newton method for three-dimensional fretting problems,* International Journal of Solids and Structures, 36, (1999), 2075-2090.
[3] Dae-Cheol Ko, Dong-Hwan Kim, Byung-Min Kim, *Finite element analysis for the wear of Ti-N coated punch in the piercing process,* Wear, 252, (2002), 859-869.
[4] P. Wriggers, *Finite element algorithms for contact problems,* CIMNE, Barcelona, 2, (1995).
[5] O. Zienkiewicz, R. Taylor, *The finite element method,* 1, McGraw-Hill, (1989).
[6] J. F. Archard, *Contact rubbing flat surfaces,* Journal of Applied Physics, 24(8), (1953), 981-988.
[7] E. Rabinowicz, *Friction and wear of materials, second edition*, Wiley, New York, (1995).
[8] J. Hallquist, *Nike2d: an implicit finite deformation, finite element code for analyzing the static and dynamic response of two-dimensional solids,* Report UCRL-52678, UC-Lawrence Livermore National Laboratory, (1979).
[9] A. Curnier, *A theory of friction,* Int. J. Solids Struct. 20, (1984), 637-647.
[10] P. Wiggers, C. Miehe, *Recent advances in the simulation of thermomechanical contact processes,* in D. R. J. Owen, E. O. Hinton eds. *Proceedings of COMPLAS III,* Swansea, Pineridge Press, (1992).
[11] T. Laursen, J. Simo, *A continuum-based finite element formulation for the implicit solution of multibody large deformation frictional contact problems,* Int. J. Numer. Meth. Eng., 36, (1993), 3451-3485.
[12] P. Wriggers, *Nichtlineare finite-element-methoden,* Springer, (2001), (in German).
[13] P. Wriggers, *Computational contact mechanics,* Springer, (2002).

COMPUTATIONAL MECHANICS (Abstracts)
WCCM VI in conjunction with APCOM'04, *Sept. 5-10, 2004, Beijing, China*

Friction Contact Analysis of Elastomer Seals in Fuel Cell Structures

Ping Zhou [*], Chengwei Wu

State Key Laboratory of Structural Analysis for Industrial Equipment, Department of Engineering Mechanics, Dalian University of Technology, Dalian, 116024 China
e-mail: vigorouszp@sina.com, cwwu@dlut.edu.cn

Abstract

Fuel cells convert chemical energy of hydrogen and oxygen directly into electricity with high efficiency and low emissions. This makes fuel cells prime candiates for powering the next generation of electric vehicles and for a varity of stationary applicat ions including distributed residential power generation. The micro-scale fuel cells have aslo gained the attention of cellular phone and laptop manufactuers. Fuel cells are called the most propectable green energy in the new century.
Elastomers are widely used in the seals of fuel cells. This paper, using the Moony-Revlin constiutive model and finite element method, analyzes the relationship between the sealing tight force, sealed pressure and the big deformation of the elastomers, considering the contact s urface friction and temperature effect. Relationships between the critical seal pressure, normal force, surface friction coefficient, and compression ratio are given. Under assumption of non -apparent slippage, the maximum theoretical seal pressure reaches 31 of atmosphere pressure, which is high enough to meet the requirements in fuel cell seals. In order for design engineers to easily use the analysis results, most of the theoretical results are fitted by experiential equations. It is found that (a) the critical seal pressue is only dependent on the compression ratio, but independent of the elastomer across section size; (b) the normal force is proportional to the radius of the elastomer. The studied temperature range is between –20°C and 85°C, which meets the seal requirements of the proton-exchange-membrane fuel cell (PEMFC), one of the most successful fuel cells for powering electric vehicles. A strong non-linear temperature effect on the seal behavior is also found. In fuel cell seals such a temperatur e effect should be considered.

REFERENCES

[1] G. Cacciola, V. Antonucci, S. Freni, *Technology up date and new strategies on fuel cells*, J. Power Sources, 100, (2001), 67-79.

[2] A. Riga, *Failure analysis and quality control of automotive elastomeric seals by thermal analysis techniques*, Thermochimica Acta, 357-358, (2000), 217-223.

[3] K. J. Bathe, E. Ramm, E. L. Wilson, *Finite element formulations for large dynamic analysis*, J. Mech. Eng., 9, (1975), 353-386.

[4] Y. C. Gao, T. J. Gao, *Large deformation contact of a rubber notch with a rigid wedge*, Int. J. Solids & Struct., 37, (2000), 4319-4334.

COMPUTATIONAL MECHANICS (Abstracts)
WCCM VI in conjunction with APCOM'04, *Sept. 5-10, 2004, Beijing, China*

Sliding Mode Based Fuzzy Control of Magnetorheological Damper

T. S. Aravindhan *, Kshitij Gupta

Department of Mechanical Engineering, IIT Delhi, Hauz Khaz, New Delhi – 110016, India
E-mail: mer01017@ccsun50.iitd.ernet.in, aravindhan_ts@yahoo.com

Abstract

Magnetorheological (MR) dampers, which are semi -active devices that use MR fluids to produce controllable forces is used in the simulation study. A novel sliding mode based fuzzy l ogic controller for magnetorheological damper for semi -active suspension system is proposed. A quarter -car active suspension system is controlled to reduce the vertical acceleration and displacement to a level of, skyhook reference model. A two degree -of-freedom quarter car model is considered. A mathematical model of MR fluid damper is adopted. Compared to the passive system, the acceleration of the sprung mass is significantly reduced for the system with a controlled MR damper. A sliding mode controller with a reference model is used to control the vibrations. A look -up table for the rule base of the fuzzy logic controller has been adopted.
The Fuzzy logic controller inputs are the sliding surface, and its velocity. Based on this a fuzzy logic controller is developed and the magnetorheological damper is controlled. The membership functions are fine tuned, by trial and error method to get the suspension displacement and acceleration in par with the reference model adopted. The vehicle vibrations are reduced considerably with a sliding mode based fuzzy logic controller to enhance the riding comfort, while riding uncertain terrains.

REFERENCES

[1] A. J. Healey, et al, *An analytical and experimental study of automobile dynamics with random roadway inputs*, ASME Journal of Dynamic Systems, Measurement and Control, 99 (4), (1977), 284-293.
[2] Andrew J Barr, Jeffery L. Ray, *Control of an active suspension using fuzzy logic*, Fuzzy Systems, Proceedings of the Fifth IEEE International Conference on fuzzy systems, 1, (1996), pp. 42- 48.
[3] C. Y. Lai, W. H. Liao, *Vibration control of a suspension system via a magnetorheological fluid damper*, Journal of Vibration and Control, 8 , (2002), 527-547.

COMPUTATIONAL MECHANICS (Abstracts)
WCCM VI in conjunction with APCOM'04, *Sept. 5-10, 2004, Beijing, China*

Finite Element Based Optimal Design of Smart Lightweight Structures for Vibration and Noise Control

Ulrich Gabbert *, Heinz Köppe, Jean Lefevre, Tamara Nestorovic

Institut für Mechanik, Otto-von-Guericke Universität Magdeburg, Universitätsplatz 2, 39106 Magdeburg, Germany
e-mail: ulrich.gabbert@mb.uni-magdeburg.de

Abstract

Smart lightweight structures have to fulfill specific design criteria during operation even if undesired disturbances are exciting the structure [1]. In such dynamic applications distributed piezoelectric sensor/actuator patches provide an excellent basis to actively cancel out specific undesired vibration modes. In the paper our finite element based simulation tool for smart structures is briefly presented first. This tool contains several multi-physics finite elements including a new layered shell type finite element, which consists of any passive and active layers, where in the piezoelectric layers the fully coupled electromechanical constitutive equations are used [2]. In acoustic applications the structure is coupled with the acoustic fluid and, consequently, the pressure in the fluid can also be taken into account. Then our approach to calculate the best positions of actuators/sensors at a given structure is presented. A mathematical optimization technique has been developed, where the positions of the piezoelectric actuators and sensors represent the design variables. In our experiences the minimization of the transfer function between structural points defines a quite useful objective function. For the controller design the Matlab/Simulink tool is used, which is connected with our finite element software via a bidirectional data interface. The quality of the optimization approach is mainly depending on the accuracy of the sensitivity analysis [3]. Different methods to calculate these values are compared. Since the controller is included in the design step, the calculation of the partial derivations of the control law is also part of the sensitivity analysis. In the paper the design concept is applied to a plate structure attached with thin piezoelectric patch modules. As an interesting result it will be shown how significantly the control parameters influence the optimal actuator positions. The results of the numerical simulations are also compared with measurements. Finally, some recent engineering applications of our finite element based design concept for smart lightweight structures are briefly discussed.

REFERENCES

[1] H. Janocha, *Adaptronics and smart structures — basics, materials, design, and applications*, Springer Berlin, (1999).

[2] U. Gabbert, H. Köppe, F. Seeger, H. Berger, *Modeling of smart composite shell structures*, J. of Theoretical and Applied Mechanics, 40(3), (2002), 575-593.

[3] F. Seeger, U. Gabbert, *Optimal placement of distributed actuators for a controlled smart elastic plate*, PAMM, 2(1), (2003), 262-263.

COMPUTATIONAL MECHANICS (Abstracts)
WCCM VI in conjunction with APCOM'04, *Sept. 5-10, 2004, Beijing, China*

Effect of Support Conditions on the Behaviour of the laminated Glass Plates

Mehmet Zulfu Asik *

Department of Engineering Sciences, Middle East Technical University, Ankara, 06531, Turkey
e-mail: azulfu@metu.edu.tr

Abstract

Laminated glass elements of modern structures consist of two or more thin glass plates and an interlayer PVB (polyvinyl butyral) show very complex behaviour. They can easily undergo large displacements even under their own weight. Their behavior under large displacements are non linear, and the derivation of the governing nonlinear differential equations is based on the large deflection theory in order to explain their true behavior.

Vallabhan [1] in his paper introduced a solution technique to study the behavior of a single glass plate under large deflection. Later Vallabhan et al. [2] extend his studies on laminated glass units by developing a mathematical model. They compared the results of the model with the results of experiments conducted at the Glass Research and Testing Laboratory at Texas Tech University to verify their mathematical model. Asik [3] in his paper presented the nonlinear behaviour of laminated glass plates in detail. Vallabhan et al. [4] and Asik et al. [5] made contributions by their research on the true behavior of glass units.

In the present study, a mathematical model is developed to study the laminated glass plates having different type of the boundary conditions. Glass units considered in this study have two thin glass plates and an interlayer PVB bonding them together. Mathematical model is based on the large deflection theory. Five nonlinear partial differential equations obtained through variational principle for lateral and axial displacements. The coupled governing differential equations are discretized by using finite difference method and solved numerically by employing iterative procedure.

REFERENCES

[1] C. V. G. Vallabhan, *Iterative analysis of rectangular glass plate,* J. of Struct. Engrg., ASCE, 109(2), (1983), 489-502.

[2] C. V. G. Vallabhan, Y. C. Das, M. Magdhi, M. Z. Aşık, J.B. Bailey, *Analysis of laminated glass units,* J. of Struct. Engrg., ASCE,, 119(5), (1993), 1572-1585.

[3] Asik MZ, *Laminated glass plates: revealing of nonlinear behaviour,* J. of Computers & Structures, 81(28-29), (2003), 2659-2671.

[4] C. V. G. Vallabhan, M. Z. Aşık, K. Kandil, *Analysis of structurally glazed systems,* J. of Computers & Structures, 65(2), (1997), 231-239.

[5] M. Z. Aşık, C. V. G. Vallabhan, *On the convergence of nonlinear plate equations*, J. of Computers & Structures, 65(2), (1997), 225-229.

COMPUTATIONAL MECHANICS (Abstracts)
WCCM VI in conjunction with APCOM'04, *Sept. 5-10, 2004, Beijing, China*

Dynamic Loading Capacity of Stiffened Delaminated Plates

Ruixiang Bai [1*], Haoran Chen [1], Man Wang [1, 2]

[1] *State Key Laboratory of Structural Analysis for Industrial Equipment, Dalian University of Technology, Dalian, 116024, China*
[2] *Department of Mechanical Engineering, Dalian Institute of Light Industry, Dalian, 116034, China*
e-mail: bairx@dlut.edu.cn, haoranch@sina.com, wm6085@sina.com

Abstract

In designing of composite laminated structures, one must take into consideration such as imperfections, delamination, that are often preexisting or generated by external low velocity impact by foreign objects during the manufacturing process and/or in the service life. The initial-delamination and the progressive failure under dynamic load can significantly alter the dynamic properties and response of composite structures. Therefore, stiffener is one of the extremely effective aspects in increasing the load capacity of laminated composite structures and can restrain the delamination in progressive process [1-3]. This paper is to study on dynamic response and loading capacity of the stiffened penetration delaminated composite plates. A formula of element stiffness and mass matrices for the composite laminated plate and beam are deduced by using the first-order shear deformation theory. A damping model is constituted on the basis of Rayleigh damping model using Adams' strain energy method (MSE). A delaminated constrained model and a virtual interface element are also developed for avoiding the overlap and penetration phenomenon between the upper and lower sub-laminates at the delaminated region. The failure analysis method for the stiffened delaminated plates under dynamic load is established by Tsai's failure criterion and corresponding reduced stiffness role. By some numerical examples, the effects of frequency of dynamic load, delamination depth, stiffener location and reduction of structure stiffness during the failure process upon dynamic behavior of the delaminated composite laminates are discussed. For composite structures analyzing and predicting the dynamic properties and load capacity of stiffened laminated plates is significant engineering subject need to be solved urgently. Whereas, most of the previous studies have been focused on the prediction of the dynamic characteristics for the perfect and delaminated composite laminates without stiffeners. Therefore, the investigation of this paper is meaningful for the composite structures designers.

REFERENCES

[1] R. D. Adams, D. G. Bacon, *Effect fiber-orientation and laminate geometry on properties of CFRP*, J. Comp. Mater., 7 (1973), 402-428.

[2] P. M. Mujumdar, S. Suryanarayan, *Flexural Vibrations of Beams with Delaminations*, J. SOUND VIBRATION, 125(3) , (1988), 441-461.

[3] D. A. Saravanos, D. A Hopkins, *Effects of Delaminations on the Damped Dynamic Characteristics of Composite Laminates: Analysis and Experiments*, J. Sound Vibration, 195(5), (1996), 977-993.

COMPUTATIONAL MECHANICS (Abstracts)
WCCM VI in conjunction with APCOM'04, *Sept. 5-10, 2004, Beijing, China*

On Composite Laminate Model with Hierarchical Warping Functions

Zhongmin Deng [1*], Chuanyue Huang [2], Youshi Hong [3]

[1] *School of Space Technology, Beijing University of Aeronautic and Astronautics (BUAA), Beijing, 10083, China*
[2] *Chinese Helicopter Research and Design Institute, Jingdezhen, 333001, China*
[3] *LNM, Institute of Mechanics, Chinese Academy of Sciences, Beijing, 100080, China*
e-mail: dengzm9742@263.net, chuanyueh@163.com, hongys@imech.ac.cn

Abstract

The theory with hierarchical warping functions had been used to analyze composite thin-walled structure, laminated beam and had good results. In the present paper, a series of hierarchical warping functions are developed to analyze the cylindrical bending problems of composite lamina. These warping functions which refine through-the-thickness variation of displacements can be obtained by the stages: In the first stage, a basic through-the-thickness displacement field is assumed while corresponding composition force and moment of stress can be equal to exterior load; then, by using constitutive relation, equilibrium equation and displacement kinematics relation, the distribution displacement function is corrected successively while corresponding stress are self-equilibrium. The in-plane displacement thickness distribution is composed of basic and corrective functions while anisotropic, material discontinues, and transverse shear and normal strain are considered. Then the hierarchical finite element method is used to form a numerical algorithm. The distribution of the displacements, in-plane stresses, transverse shear stresses and transverse normal stress for composite laminate are analyzed with the present model. The results show that this model has more precise mechanical response than the first deformation transverse theory, and the corrective order affects the accuracy of result.

REFERENCES

[1] H. Matsunaga, *Assessment of a global higher-order deformation theory for laminated composite and sandwich plates*, Composite Structures, 56, (2002), 279-291.

[2] A. K. Noor, W. S. Burton, *Assessment of shear deformation theories for multilayered composite plates*, Appl. Mech. Rev, 42, (1989), 1-12.

[3] Z. M. Deng, D. C. Zhu, *High-order warping theory for twist analysis of composite thin-Walled beams*, Acta Aeronautic et Astronautical Sinica, 20, (1999), 458-461, (in Chinese).

[4] C. Y. Huang, *A model of transverse distribution displacement functions for composite laminate*. The dissertation for Ph.D. of Beijing University of Aeronautic and Astronautic, Beijing, China, (2002).

[5] D. C. Zhu, *Development of hierarchical finite element methods at BIAA*, in Yagawa G. eds. *Proc. of Int. Conf. on Computation Mechanics*, Tokyo, 1986, Springer-Verlag, (1986), Vol. 1, pp. 123-128.

[6] N. J. Pagano, *Influence of shear coupling in cylindrical bending of anisotropic laminates*, J Compos. Mat., 4, (1970), 330-343.

COMPUTATIONAL MECHANICS (Abstracts)
WCCM VI in conjunction with APCOM'04, Sept. 5-10, 2004, Beijing, China

Design of Laminated Cylinders Using Exact Elasticity Solution

Pavel Tabakov*, Evan Summers

Centre for Advanced Materials, Design and Manufacturing Research, Durban Institute of Technology, Durban, South Africa
e-mail: pashat@dit.ac.za, evan@ananzi.co.za

Abstract

An approach for the optimal design of thick laminated cylindrical pressure vessels is given. The maximum burst pressure is computed using an exact elasticity solution and subject to the Tsai-Wu failure criterion. The design method is based on an accurate three-dimensional stress analysis. Exact elasticity solutions are obtained using the stress function approach [1] where the radial, circumferential and shear stresses are determined taking the closed ends of the cylindrical shell into account. The derived system of the governing algebraic equations allows us easily and accurately to analyse a multilayered pressure vessel with an arbitrary number of layers and any thickness. The approach used is elegant and straightforward compared to other three-dimensional solutions found in the literature. The solution does not contain any approximations like series or other numerical techniques and, thus, makes the optimizing procedure much simpler and very fast.
Because of the anisotropy in composites and the presence of curvature in cylindrical pressure vessels, the design optimization of such structures presents considerable challenge and difficulty. Moreover, in the case of five and more layers it is hardly possible to find the optimum fibre orientation and their best combination through the thickness when using any known classical method. In the present study the genetic algorithm is used [2]. It is implemented and optimized to serve this particular problem. This approach differs from those known in the literature and seems to be much more effective for some engineering problems [3]. Both continuous and discrete optimising variables are used. The biggest problem in such cases, the premature "convergence", was overcome by using a parabolic penalty function introduced by the authors.
Due to the wide use of the cylindrical composite pressure vessels in various engineering applications, including aviation and space, the accurate analysis and design of such structures is imperative.

REFERENCES

[1] S. G. Lekhnitskii, *Theory of elasticity of an anisotropic elastic body*, Transl. by P. Fern, Holden-Day, Inc., San Francisco, (1963).
[2] J. H. Holland, *Adaptation in natural and artificial systems*, The University of Michigan Press, Ann Arbor, MI, USA, (1976).
[3] P. Y. Tabakov, M. Walker, *An improved genetic algorithm in parametric optimization of engineering structures*, Proceedings of the International Conference on Artificial Intelligence, Durban, South Africa (1999).

COMPUTATIONAL MECHANICS (Abstracts)
WCCM VI in conjunction with APCOM'04, *Sept. 5-10, 2004, Beijing, China*

The Frequencies of Two-Dimensional Laminated Beam by Several Laminated Composite Models

Jinmei Tian*, Yufeng Xing

The Solid Mechanics Research Center, Beijing University of Aeronautics & Astronautics, Beijing, 100083, China
e-mail: xingyf@buaa.edu.cn

Abstract

The beam is assumed to be in a state of plane stress. Exact solutions to the inherent characteristics of orthotropic laminated beams with simply-supported ends are obtained by means of theory of elasticity. Among the high-order theories of laminated composite beam, some consider the shear warping of the cross section, some take the thickness effect of the beam into account. But only some of essential and natural boundary conditions are satisfied by all the approximate methods. In this paper, the exact frequencies of two-dimensional orthotropic laminated beam are compared with the results of several high-order theories, and according to the comparison the high-order theories are evaluated.

Because the length of the beam is larger than the height, it can be conclude that the effects of the boundary conditions of two supported ends are less than that of the bottom and top surfaces. MIXED theory satisfies the stress boundary conditions on the upper and bottom surfaces, so the solutions got by this theory agree well with the exact results obtained here.

And based on the exact frequencies and mode functions of the two-dimensional beam, the indirect mode superposition method, in which the impactor and the target are considered as two independent bodies, is adopted to give the analytical solutions of the problem of a particle impact against simply-supported two-dimensional beam at middle position. The characteristics of the longitudinal and shear stress waves traveling in the beam are investigated with respect to different span-to-depth ratios. And there are two conclusions that with the decrease of the length-to-height ratio of the beam, the proportion of the mode number along the thickness direction of the beam among the totally needed mode number becomes larger, and the effect of the transverse wave on the impact response becomes more apparent than that of longitudinal wave with the increase of the length-to-height ratio.

REFERENCES

[1] Y. F. Xing, D. C. Zhu, Y. S Qiao, *The inherent vibration characteristics of composite laminated beams and elastic beams,* Acta Mechanica Sinica, 30, (1998), 628-634.

[2] T. Kant, S. R. Marur, G. S. Rao, *Analytical solution to the dynamic analysis of laminated beams using higher-order refined theory,* Comp. Struct. 40, (1998), 1-9.

[3] M. Kameswara Rao, Y. M. Desai, M. R. Chitnis, *Free vibration of laminated beams using mixed theory*, Composite Structure, (2001), 149-160.

[4] C. T. Sun, W. J. Liou, *Investigation of laminated composite plates under impact dynamic loading using a three-dimensional hybrid stress finite element method*, Comput. and Struct., 33, (1998), 879-884.

COMPUTATIONAL MECHANICS (Abstracts)
WCCM VI in conjunction with APCOM'04, *Sept. 5-10, 2004, Beijing, China*

Characterization of Mechanical Properties on Ductile Iron Modified by Laser with Space Array

Fahai Ba [1*], Gang Yu [2], Guoqiang Yan [1], Cuihua Gan [2]

[1] *Shanghai Research Institute of Materials, Shanghai, 200437, China*
[2] *State key Laboratory of Nonlinear Mechanics, Institute of Mechanics, Chinese Academy of Sciences, Beijing, 100080, China*
e-mail: bafahai@hotmail.com, bafahai@163.com

Abstract

Pulsed laser beams with two-dimensional array (3×3 and 7×7) distribution by binary optical elements were used to modify surface processing for ductile iron. The mechanical properties of modified surface were analyzed and discussed in detail. The treated area usually shows a semilunar shape; the usual test method used is that microhardness is measured along the line in which the case depth of modified layer is deepest. But the experiment and analysis shows that the microhardness distribution is involved in the measured sites. Based on above, the statistical method of Microhardness test is proposed to describe the characters of microhardness distribution of modified area, in which the grids are divided into with a space between columns 30μm×30μm or 30μm×50μm in the transect of modified area, and the size of the grid is only determined by indentation size. The Microhardness is tested in the intersections of grids along the laser beam till the Microhardness approximating 300HV (the matrix). As a result, the contour map drawn out by suffer soft shows the practical Microhardness distribution by means of the Microhardness distributive data obtained, let the length of grids as y-axis along the depth in modified surface and the width as x-axis, respectively. The area within Microhardness beyond 400HV is called modified area, and the case depth will be determined by following steps. The special line in parallel with modified surface (X-axis) may be drawn out, in which Microhardness of more than 50% intersections beyond 400HV, then the distance from X-axis to which is so called case depth. The analysis shows the statistical method for microhardness test was apt to determine case depth of modified layer and describe the characters of microhardness distribution in comparison with that of conventional method. On the other hand, the hardest area often lies under some depth from surface according to the experiment above. However, the routine test method for wearing resistance is only suitable for testing that of homogeneous modification layer or surface layer, so the routine test cannot stand for the practical wearing resistance. The wearing experimental shows that there lies a limit of 60-120 h for ductile iron treated by laser beam in the case of ring-block wearing. Based on the experiment, the effectors on wearing test were analyzed and discussed, and finally a suitable method for wearing test was put forward.

REFERENCES

[1] Fahai Ba, Cuihua Gan, Gang Yu, *Microstructure evolution of laser pulse modified ductile iron surfaces*, Third APLS, Osaka, Japan, (2002).
[2] Chunlin Gao, Gang Yu, *Design of binary phase gratings with specific intensity distribution of output*, Chinese Journal of Laser, A28(4), (2001), 365 -368, (in Chinese).
[3] Fahai Ba, Cuihua Gan, Gang Yu, *Cow-eye microstructure evolution during laser pulse processing ductile iron*, SPIE 2002, Beijing, (2002), pp. 679 -68.

COMPUTATIONAL MECHANICS (Abstracts)
WCCM VI in conjunction with APCOM'04, *Sept. 5-10, 2004, Beijing, China*

A Numerical Study on Flow Induced Localized Corrosion of a Plate

Y. F. Chen [1, 2*], L. S. Niu [1]

[1] *Deparment of engineering mechanics, Tsinghua University, Beijing 100084, China*
[2] *School of materials science and engineering, University of Science and Technology Beijing, Beijing 100083, China*
e-mail: niulsh@mail.tsinghua.edu.cn

Abstract

Flow induced localized corrosion (FILC) is one of the corrosion types found in petroleum pipelines which is severe to the safety of the pipelines. In this paper, some achievements in the area of the flow induced localized corrosion and corrosion model were reviewed and commented on. The stress corrosion problem of a plate was carefully studied based on the finite element method. A phenomenological and quasi-static corrosion model was established here to avoid more concerns of the corrosion mechanism. A method of vital and dead element was introduced to suit the movable boundary conditions through a redevelopment of the FEM code MARC. The relationships between the fluid parameters and the intrinsic mechanism of the corrosion were analyzed. The rule of element death was also discussed.

COMPUTATIONAL MECHANICS (Abstracts)
WCCM VI in conjunction with APCOM'04, *Sept. 5-10, 2004, Beijing, China*

Solutions of Materially Nonlinear Analysis Obtained by Strain Combinations Control

Jamile S. Fuina [1*], Roque L. Pitangueira [1], Raul R. Silva [2]

[1] *Department of Structural Engineering, Federal University of Minas Gerais, Av. do Contorno, 842, Belo Horizonte, 30110-060, Brazil*

[2] *Department of Civil Engineering, Pontifical Catholic University of Rio de Ja neiro, Rua Marquês de São Vicente, 225, Rio de Janeiro, 22453-900, Brazil*

e-mail: jamile@dees.ufmg.br, roque@dees.ufmg.br, raul@civ.puc-rio.br

Abstract

This paper discusses the determination of the equilibrium paths in the nonlinear finite element structural analysis. A brief study about the incremental-iterative procedures, that use load and displacement combinations as the analysis controlling parameters, is presented.

The limitations of the classical methods in materially nonlinear problems are discussed. A method, that includes the mechanics of the material deterioration process, using strain combinations in subdomains of the problem, is proposed. The combinations are strain measures such as mean, invariants, eigenvalues, among others. The subdomains are characterised as groups of integration points, once in the Finite Element Method, normally, the strains are obtained in these points. Such groups can be chosen as the whole finite element mesh, one or more finite elements or just one integration point. The proposed method permits the change of the control subdomain (one integration point or one finite element) during the analysis process. This change is based on the search for region that sampled the biggest increase on the standing control combination in the last incremental step.

The computational implementation details of the proposed model into a object oriented finite element program is discussed.

Some numerical simulations of materially nonlinear problems are presented. The analysis of the obtained results permits to discuss the adequacy of the classical and proposed methods in the solution of the problems.

COMPUTATIONAL MECHANICS (Abstracts)
WCCM VI in conjunction with APCOM'04, *Sept. 5-10, 2004, Beijing, China*

Finite Deformation Analysis Using Natural Strain (Relation of Principal Axis of Stress and Principal Axis of Elastic Deformation after Pre-Deformation)

Y. Kato *, Y. Moriguchi

Department of Mechanical Engineering, College of Science & Technology Nihon University, 1-8-14 Kanda-Surugadi Chiyoda-Ku, Tokyo, 101-8308, Japan
e-mail: kato@mech.cst.nihon-u.ac.jp

Abstract

The Natural Strain is obtained by integrating infinitesimal strain increment on an identical line element over the whole process of the deformation path. Consequently, the shear strain becomes pure angular strain, which is obtained by removing the rigid body rotation from the rotating angle of a line element. Since the expression of the Natural Strain is different from the strain expression of ordinary rate type, this theory has merits that can satisfy the additive law of strain on an identical line element. In this paper, the anisotropy that occurs by a large pre-deformation is chosen as the subject of this study. Especially, two types of pre-deformations, i.e. uni-axial tension and large shear deformation, are treated as the first step of this study. Using the test piece of pure copper, which is given these pre-deformations, the proportional loading test by tension and torsion is carried out in the elastic region. The direction of the principal axis of the elastic deformation is obtained from the extensional strain components of the tri-axial strain gage in consideration of the Natural Strain theory, and the relation between the direction of the principal axis of stress and the direction of the principal axis of elastic deformation is confirmed experimentally.

On the other hand, the numerical model of anisotropy caused by pre-deformation is proposed in this paper. And, we estimate the relation of both principal axes from calculated results and compare with the experimental results.

REFERENCES

[1] A. E. Green, W. Zerna, *Theoretical elasticity*, Oxford at the Clarendon Press, (1968).
[2] E. H. Lee, R. L. Mallett, T. B. Wertheimer, *Stress analysis for anisotropic hardening in finite-deformation plasticity,* Trans. ASME, J. Appl. Mech., 50, (1982), 554-560.
[3] S. Stören, J. R. Rice, *Localized necking in thin sheets,* J. Mech. Phys. Solids, 23, (1975), 421-441.
[4] M. Goto, *An inquiry into the spin of a deforming body,* Trans. JSME, A, 52(476), (1986), 1134-1141.
[5] E. H. Lee, *Elastic-plastic deformation at finite strain,* J. A. Mechanica, 36, (1969), 1-6.
[6] Y. Kato, T. Nishimura, *The stress analysis using the rate type formulation of natural strain,* ASME PVP&ICPVT, 326, (1996), 159-166.

COMPUTATIONAL MECHANICS (Abstracts)
WCCM VI in conjunction with APCOM'04, *Sept. 5-10, 2004, Beijing, China*

Modelling Equations and Numerical Methods of Elastic Plastic Anisotropic Materials

Abduvaly A.Khaldjigitov [1*], Abduqaim Kadirov [1], Abduwaili Abudureheman [2]

[1] *Department of Computer Science, National University of Uzbekistan, Vuzgorodok NUUs, Tashkent, 174, Uzbekistan*
[2] *Department of mathematics, Xinjiang Cultural University, St.Nyanchan-42, Urumchi, 830052, China*
e-mail: khaldabd@yandex.ru

Abstract

Mathematical modelling of nonlinear deformation process of elastic plastic composite materials is one of the fundamental problems of solid mechanics. Usually composite material on the basis of any averaging method may be considered as a homogeneous anisotropic material. We shall remind that fibrous and layered composite materials at averaging replace with transversely isotropic or orthotropic materials [3].
The given work is devoted to development of the adequate modelling equations and effective numerical methods for solution of elastic plastic problems for transversely isotropic materials. The strain space and stress space plasticity theories for transversely isotropy materials are offered. On the basis of comparison of experimental and theoretical curves are shown validity of plasticity theories for description transversely isotropic materials. Experimental curves received on tension (compression) of the samples cut out of fibrous composites under various angles to the direction of fibres are compared with theoretical ones.
The stain space theory has a number of theoretical and numerical advantages as against stress space theory, unequivocally allowing describing softening materials. The constitutive relations for strain space plasticity theory are based on the Il'jushin's plasticity postulate [1] and allow expressing an increment of stress tensor through strain tensor and its increment. Such relationship is very convenient for the formulation plasticity boundary problems, and allows to study softening materials [2]. Strain space theory very conveniently for the numerical solution of elastic plastic problems on FEM as there is no necessity for calculation of stress tensor at each increment of external loading. In connection with stain space theory it is possible to modify initial stress method and to classify other methods from the point of view of convergence. Test numerical examples about equilibrium of 2-D and 3-D plasticity problems for transversely isotropy rectangle and a parallelepiped under various regional conditions are solved.

REFERENCES

[1] A. Il'jushin, *About plasticity postulate*, PMM, 25(3), (1961), 503.
[2] O. C. Zienkiewicz, Valliappan, I. P. King, *Elastoplastic solutions of engineering problems-initial stress finite element approach*, Int. J. Num. Meth. Engng, 1, (1969), 75-100.
[3] A. A. Khaldjigitov, U. Adambaev, M. Babadjanov, *Plasticity theories for transversely isotropic materials*, Seventh Int. Conf. Comput. Plasticity, CIMNE, Barcelona, April 7-11, (2003), p.292.

COMPUTATIONAL MECHANICS (Abstracts)
WCCM VI in conjunction with APCOM'04, *Sept. 5-10, 2004, Beijing, China*

Numerical Simulation of 3D Powder Compaction Processes Using Cap Plasticity Model

Amir R. Khoei *, Shaia Azizi

Department of Civil Engineering, Sharif University of Technology, 11365-9313, Tehran, Iran
e-mail: arkhoei@sharif.edu

Abstract

In this paper, recent activities to develop a simulation code for 3D finite element modeling of powder forming processes are presented. The compaction forming of metal powder is a process involving large deformations, large strain, non-linear material behavior and friction. Consequently, the numerical analysis of such a highly non-linear process is a formidable computational problem. For the successful modeling of powder compaction process several requirements have to be met [1-3]. Firstly, as the compaction process involves a very large reduction in volume, the formulation adopted must be capable of representing this physical process. Secondly, since powder is a frictional and compressible material, the densification is dependent on the compaction stage and the material model needs to reflect the frictional and yielding characteristics of the powder. Thirdly, the involvement of two different materials, which have contact and relative movement in relation to each other, must be considered.

In reference [1-3], the 2D FE model was developed for describing the isothermal deformation behavior of metal powder. The purpose of the present study is to extend the simulation of powder forming processes to 3D finite element analysis. As the forming process in powder compaction problems is a large deformation process, an elasto-plastic analysis often utilizes a large strain model. A large displacement formulation, based on a total and updated Lagrangian formulation is utilized. In order to describe the constitutive model of the highly non-linear behavior of powder materials, a cone-cap plasticity model is applied based on a hardening rule to define the dependence of the yield surface on the degree of plastic straining. This model reflects the yielding, frictional and densification characteristics of powder along with strain and geometrical hardening, which occur during the compaction process. Finally, the powder behavior during the compaction of a 3D multi-level component is numerically simulated. The predictive compaction forces at different displacements are computed and compared with the available experimental details reported in the literature. The simulation of the deformation is shown as well as the distribution of relative density and stress contours at different time stages. It is shown that the proposed 3D model using cap plasticity theory is capable of simulating the metal powder during compaction.

REFERENCES

[1] A. R. Khoei, R. W. Lewis, *Adaptive finite element remeshing in a large deformation analysis of metal powder forming*, Int. J. Num. Meth. Eng., 45, (1999), 801-820.

[2] R. W. Lewis, A. R. Khoei, *A plasticity model for metal Powder forming processes*, International Journal of Plasticity, 12, (2001), 1659-1692.

[3] A. R. Khoei, A. Bakhshiani, M. Mofid, *An endochronic plasticity model for finite strain deformation of powder forming processes*, Finite Elem. Analy. Des., (2003), (in press).

COMPUTATIONAL MECHANICS (Abstracts)
WCCM VI in conjunction with APCOM'04, *Sept. 5-10, 2004, Beijing, China*

Finite Element Analysis on Moisture Absorption in Plastic Electronic Packages

Quanyong Li *, Yubing Gong, Daoguo Yang

Center of Micro-electric Package and Assembly Technology, Guilin Univ. of Electronic Technology, Guilin 541004, China
e-mail: hmlqy@gliet.edu.cn, gybjustdoit@sina.com, cepat@gliet.edu.cn

Abstract

The phenomenon of "popcorn" cracking of plastic electronic packages during reflow soldering is widely attributed to the effect of moisture absorbed by the plastic molding compound resin prior to the solder reflow process. By comparing the characteristic of humidity problem and heat problem, the equivalent relations between the two types of problem is brought out in this paper. Aiming to a specified type of electronic packages, the three-dimensional moisture diffusion modeling and coupled humidity-mechanical modeling are established using ANSYS software. The complete distribution of moisture in the specific electronic packages during the industry process is obtained. And the corresponding humidity-mechanical stress in the packages is also obtained.
From the computing result, it can be seen that after 84 hours pre-absorbing moisture process, the moisture in the packages is saturated. After the reflow process, the humidity in the surface is dramatic decreased. But the difference of humidity in the interface remains. From the stress analysis, it can be founded that the margin of the molding compound is the area which has the maximum humidity-mechanic stress.
It must be noted that the analysis method used in this paper is also available for more general plastic electronic packages and the results induced here are helpful to understand the influence of moisture absorption on the reliability of plastic electronic packages.

REFERENCES

[1] Sheng Liu, Yuhai Mei, *Behavior of delaminated plastic IC packages subjected to encapsulation cooling, moisture absorption and wave soldering*, IEEE Electronic Components and Technology Conference, (1995).
[2] A. A. O. Tay, T. Y. Lin, *Influence of temperature, humidity and defect location on delamination in plastic IC packages*, IEEE InterSociety Conference on Thermal Phenomena, (1998).
[3] Shijian Luo, C. P. Wong. *Influence of moisture absorption in underfill materials*, IEEE Electronic Components and Technology Conference, (2002).
[4] E. Jesse Galloway, Barry M. Miles, *The prediction of moisture absorption and desorption in PBGA packages*, IEEE Electronic Components and Technology Conference, (1997).
[5] Ee Hua Wong, Yong Chua Teo, Thiam Beng Lim, *Moisture diffusion and vapour pressure modeling of IC Packaging*, IEEE Electronic Components and Technology Conference, (1998).

COMPUTATIONAL MECHANICS (Abstracts)
WCCM VI in conjunction with APCOM'04, *Sept. 5-10, 2004, Beijing, China*

Two-Phase Model for Fluid-Saturated Porous Media

Ying Liu, Lingtian Gao*

Department of Mechanics and Engineering Science, Peking University, Beijing, 100871, China
e-mail: y_liu@mail.china.com

Abstract

Three-dimensional two-phase models, for analyzing the problems of stress wave propagation in fluid-saturated porous media, are proposed by introducing the interface-coupling concept of the fluid-solid interaction mechanics. The difference for two-phase models proposed here to the usually used theories for wave propagation in fluid-saturated porous media is that the investigated object is the true structure of the fluid-saturated porous media, instead of the substitutive models. Based on those models, and adopting direct coupling technique, a three-dimensional fluid-solid mixture explicit dynamic finite element scheme is developed. Using Galerkin weighted residual method and considering coupling effects on the fluid-solid interface, solution domain is divided into solid elements, fluid elements and fluid-solid coupling elements. The three-dimensional stress wave propagation in the fluid-saturated porous media due to impact loading is numerically simulated. The present results show that the wave speeds obtained by our models accord well with the speeds given by Biot's theory, but our models can further reflect the influence of the true structure of the porous media on the propagation characteristic of the stress waves. Moreover, the effects of the porosity and the shapes of the pores on the propagation characteristic of the main waves are discussed in detail.

REFERENCES

[1] M. A. Biot, *Generalized theory of acoustic propagation in porous dissipative media*, J. Acoust. Soc. Am., 54, (1962), 1254-1264.

[2] R. M. Bowen, *Compressible porous media models by use of the theory of mixtures*, Int. J. Eng. Sci., 20, (1982), 697-735.

[3] Y. Liu, *Stress wave propagation in fluid-saturated porous media under impact loading*, PhD Thesis, Dalian University of Technology, China, (2002), (in Chinese).

[4] B. A. Schrefler, *Mechanics and thermodynamics of saturated/unsaturated porous materials and quantitative solutions*, Appl. Mech. Rev., 55, (2002), 351-388.

[5] A. Turgut, T. Yamamoto, *Measurement of acoustic wave velocities and attenuation in marine sediments*, J. Acoust. Soc. Am., 87, (1990), 2376-2383.

[6] T. Yamamoto, A. Turgut, *Acoustic wave propagation through porous media with arbitrary pore size distributions*, J. Acoust. Soc. Am., 83, (1988), 1744-1751.

COMPUTATIONAL MECHANICS (Abstracts)
WCCM VI in conjunction with APCOM'04, *Sept. 5-10, 2004, Beijing, China*

Numerical Analysis for the Application of Metallic Foams for the Protection of Structural Elements

Marc A. Quint *

Structural Design, Charlottenstrasse 94, D-72764 Reutlingen, Germany
e-mail: marc.quint@xperteez.de

Abstract

The use of metallic foams for energy dissipation in crashworthiness analysis on the vehicle side is under development for quite some time and has gained wide recognition. Aluminum foams are currently used in various vehicle structures for the energy absorption during crash events.
This paper considers metallic foams preferably with metal cover sheets for the use as shields to protect structural elements such as columns or walls from impacting vehicles or dynamic forces in general by energy dissipation and damping.
The newly developed ferrous foams give a better perspective over the aluminum foams because of their elasticity under low loading and larger energy absorption for high impacting forces.
The general behavior of columns and massive walls in standard constructions is examined under low, medium and large dynamic loads, simulating the impact of vehicles at different speeds.
The investigation uses transient dynamic analyses with simplified elastic -plastic material properties and impulse loading substituting the impacting bodies to characterize the response of the structural elements. The results obtained from FE analyses are compared to loads defined in building codes for vehicle impact.
Future application of the results may lie in strengthening existing structures as well as the assessment of e.g. forklift use in existing buildings by reducing the effective stresses, strains and resulting forces in the exposed structural elements.

REFERENCES

[1] U. Mohr, W. Bleck, P. F. Scholz, *Metallic foams produced by slipcasting*, Advanced Engineering Materials, Wiley-VCH, 4, (2002), 28-33.

[2] U. Mohr, W. Bleck, *Processing and first characterization of ferrous foams*, Metal Matrix Composites and Metallic Foams, EUROMAT, in T.W. Clyne, F. Simancik, Wiley-VCH, 5, (2000), 28-33.

COMPUTATIONAL MECHANICS (Abstracts)
WCCM VI in conjunction with APCOM'04, *Sept. 5-10, 2004, Beijing, China*

A Numerical Scheme Combining Conventional FEM and Gradient-Dependent Inelastic Models

Xinpu Shen[1*], Guoxiao Shen[1], Lixin Chen[2]

[1] *College of Architectural Engineering, Shenyang University of Technology, 110023, Shenyang. China*

[2] *Engng Center for Coal-beded Methane, Northeastern Bureau of Coal Geology Shenyang, 110035, China*

e-mail: xinpushen@vip.sina.com, guoxiao_shen@yahoo.com, clx-holley@163.com

Abstract

Gradient-enhanced plasticity and damage model is one of the major measures in dealing with failure analysis of concrete structures, and is widely used by various researchers [1]. There are two major disadvantages of the existing models: (1) special FEM formulation has to be formulated for this purpose; and (2) there are additional boundary conditions caused by the Laplacian terms.

In this study, the Taylor series expansion and Moving Least Square method [2] were used in the derivation of the numerical scheme for the calculation of the Laplacian of a damage field. Equations were presented on the basis of the Taylor series expansion for both the 2-dimensional and 3-dimensional cases. Numerical results suggests that the algorithmic approach is valid and very efficient compared to other existing methods.

By adopting the approach proposed in this study, it is possible to extend existing conventional Finite Element Method programs to include gradient -dependent constitutive models to properly regularise the solutions in a failure analysis of solids. A numerical example was calculated for the purpose of model validation with a double notched specimen. This preliminary study suggests that the method is robust.

The above formulation is not limited to the calculation of the Laplacian of a damage field, but it is also applicable to other kind of inelastic variables, such as equivalent plastic strain, hardening internal variables etc, which might appear in a gradient -enhanced inelastic constitutive model.

REFERENCES

[1] E.C. Aifantis, *On the role of gradient in the localization of deformation and fracture*. Int. J. Engng. Sci. 30, (1992), 1279-1299.

[2] Villon, P., 1991, *Contribution to the Optimization*, Doctoral Thesis, University of Technology of Compiegne, France, (in French).

COMPUTATIONAL MECHANICS (Abstracts)
WCCM VI in conjunction with APCOM'04, *Sept. 5-10, 2004, Beijing, China*

Transient Temperature and Stress Analysis for Thermo-Viscoelastic SRM Grains

Xinqi Xu *, Aiping Tian

Naval Aeronautical Engineering Academy, Yantai, 264001, China

Abstract

The transient temperature distribution in a solid rocket motor grain due to the effect of environment temperature is calculated with finite element method. Based on the viscoelastic integral constitutive relation, the thermal stress and strain distribution is calculated. The variation curve of stresses and strains with time for dangers node is presented. It provides scientific basis for structural integrity analysis.

COMPUTATIONAL MECHANICS (Abstracts)
WCCM VI in conjunction with APCOM'04, *Sept. 5-10, 2004, Beijing, China*

Numerical Simulation of Elastic Behavior and Failure Processes in Heterogeneous Material

Xiaoping Zheng *, Zhenhan Yao

Department of Engineering Mechanics, Tsinghua University, Beijing, 100084 China
e-mail: zhengxp@mail.tsinghua.edu.cn, demyzh@tsinghua.edu.cn

Abstract

A general numerical approach is developed to model the elastic behaviour and failure processes of heterogeneous materials. Although a two-parameter Weibull statistical distribution is used to simulate initial heterogeneity of materials, the model can be modified to use multi-parameter Weibull distributions or other suitable statistical distributions (see, for example, Andrew and Price, 1979). Since the techniques developed in spring network method and the volume average concept (RVE) are employed, the propose method simulates crack initiation and propagation without significant computational complexity. Moreover, by adopting a displacement-type finite element method, the described method is well suited for easy implementation in standard finite element codes.
A selection of applications of the present method to plane stress problems is presented. The numerical results indicate that the method is very suitable to simulate strain softening, pre-peak and post-peak nonlinear phenomena occurring in the failure process of heterogeneous materials. The results also indicate that the more homogeneous the stiffness or strength of the materials at the microscale level, better effective stiffness and strength occur at the macroscale level. Furthermore, even though the same statistical distribution is taken to describe stiffness or strength in heterogeneous materials, because of the different space configurations of properties different failure strengths and paths occur in the materials.
The present approach, after suitable modification, provides a method to investigate effective properties of various practical heterogeneous materials, optimize and design ingredients and space configurations in composite materials, etc. It also allows examination of nonlinear phenomena in the structures or materials by combing simple constitutive relations and heterogeneous model in contrast to the more conventional approach by combining complicated constitutive model and homogeneous model. In this respect, the presented approach readily describes realistic materials.

REFERENCES

[1] Z. Hashin, *Analysis of composite material - a survey*, J. Appl. Mech., 50, (1983), 481-505.
[2] M. Ostoja-Starzewski, *Random field models of heterogeneous materials*, Int. J. Solids Structures, 35, (1998), 2429-2455.
[3] K. Terada, M. Hori, T. Kyoya, N. Kikuchi, *Simulation of the multi-scale convergence in computational homogenization approaches*, Int. J. Solids Structures, 37, (2000), 2285-2311.
[4] Y. Q. Chen, Z. H. Yao, X. P. Zheng, *Theoretical statistical solution and numerical simulation to heterogeneous brittle materials*, ACTA Mechanica Solida Sinica, 19(3), (2003), 276-284.

COMPUTATIONAL MECHANICS (Abstracts)
WCCM VI in conjunction with APCOM'04, *Sept. 5-10, 2004, Beijing, China*

On the Use of Embedded Discontinuities in the Framework of a Discrete Crack Approach

J. Alfaiate [1*], L. J. Sluys [2]

[1] *Department of Civil Engineering and ICIST, Instituto Superior Técnico, Av. Rovisco Pais, Lisboa, 1049-001, Portugal*

[2] *Department of Civil Engineering And Geosciences, Delft University of Technology, P.O. Box 5048 GA Delft, Delft, 2600, The Netherlands*

e-mail: alfaiate@civil.ist.utl.pt, L.J.Sluys@citg.tudelft.nl

Abstract

In this paper a strong embedded discontinuity formulation is adopted to study fracture in quasi-brittle materials. The technique of embedding discontinuities in finite el ements is discussed in the framework of a discrete crack approach: the formation of a new embedded discontinuity is not derived from the continuum properties as done in [1]; instead, it is assumed that microcracking localizes into a discontinuity surface with initial zero thickness. Thus, the introduced kinematics is similar to the one adopted using interface elements to model the discontinuities and this is why this formulation is called a discrete strong embedded discontinuity approach [2]. In this numerical model, some particular aspects are distinct from previous embedded formulations, such as: 1) a non-homogeneous displacement jump field is adopted within each parent element; 2) the jumps are obtained at additional nodes which are located at the embedde d discontinuity surface; 3) these additional degrees of freedom can be considered either as global, in which case continuity of the jumps across the element boundaries is automatically enforced, or as local to the parent element, in which case the bandwidt h of the stiffness matrix is kept constant during calculations but the jumps are discontinuous across element boundaries; 4) the approximated displacement field is discontinuous and 5) within each parent element, the additional displacements induced by the jumps correspond to rigid body motions; in consequence, these displacements are not *smeared* over the entire or part of the element and the work done by the stresses on the enhanced strains is null . A comparative study of different strong discontinuity formulations is presented, namely the discrete-interface approach, as well as previous embedded formulations and one parti tion of unity model [3].

REFERENCES

[1] J. Oliver, A. E. Huespe, M. D. G. Pulido, E. Chaves, *From continuum mechanics to fracture mechanics: the strong discontinuity approach*, Engineering Fracture Mechanics, 69, (2002), 113-136.

[2] J. Alfaiate, L. J. Sluys, *A discrete strong embedded discontinuity approach*, in Victor C. Li, Christopher K.Y. Leung, Kasper Willam, Sarah Billington eds. *Fracture Mechanics of Concrete and Concrete Structures* (*Framcos 5*), Proc. 5th Int. Conf., Vail Colorado, USA, (2004).

[3] G. N. Wells, L. J. Sluys, *A new method for modelling cohesive cracks using finite elements*, Int. J. Num. Meth. Eng., 50(2), (2001), 2667-2682.

COMPUTATIONAL MECHANICS (Abstracts)
WCCM VI in conjunction with APCOM'04, *Sept. 5-10, 2004, Beijing, China*

LEFM and Cohesive-Crack Approaches to Safety Evaluation of Concrete Gravity Dams

Gabriella Bolzon [*]

Department of Structural Engineering, Politecnico di Milano, piazza Leonardo da Vinci 32, Milano, I 20133, Italy
e-mail: gabriella.bolzon@polimi.it

Abstract

Linear Elastic Fracture Mechanics (LEFM) is often used to analyze crack propagation and safety margin of large structures such as concrete gravity dams; see, e.g., [1 -2]. LEFM is an undoubtedly attractive approach when results have to be transferred from prototype to real dimensions, as in the case of centrifuge modeling [2]. In this case, in fact, t he nominal strength of self-similar structures can be related to one dimensional parameter only, the material characteristics being captured by fracture energy; see, e.g., the comprehensive treatise [3] and references therein. On the other hand, cohesive-crack models, eventually resting on simplified hypothesis leading to direct methods as outlined in [4], can be used for safety assessment in this context and more naturally account for peculiar features such as friction in mixed mode fracture.
For increasing structural size and brittleness, LEFM and cohesive -crack approaches are expected to return equivalent results; nevertheless, the latter method might be more reliable and computationally competitive when significant weight contribution overcomes the desta bilizing effects of water pressure in crack propagation, as shown in [5] for pure opening mode.
In the presence of significant compressive stresses, failure is mostly induced by Mode II fracture (sliding mode). Comparative merits of the aforementioned appr oaches are discussed in this contribution with specific reference to this situation.

REFERENCES

[1] G. A. Plizzari, F. Waggoner, V. E. Saouma, *Centrifuge modeling and analysis of concrete gravity dams*, ASCE J. Struct. Eng., 121, (1995), 1471-1479.
[2] G. A. Plizzari, *LEFM applications to concrete gravity dams*, ASCE J. Eng. Mech., 123, (1997), 808-815.
[3] Z. P. Bažant, *Scaling of structural strength*, Hermes Penton Science, London, UK, (2002).
[4] G. Bolzon, G. Cocchetti, *Direct assessment of structural resistance against pressurized fracture*, Int. J. Num. Anal. Meth. Geomech., 27, (2003), 353-378.
[5] G. Bolzon, *Size effects in concrete gravity dams: a comparative study*, Eng. Fract. Mech., 71, (2004), 1891-1906.

COMPUTATIONAL MECHANICS (Abstracts)
WCCM VI in conjunction with APCOM'04, *Sept. 5-10, 2004, Beijing, China*

Electroelastic Fracture Analysis of Piezoelectric Materials with a Central Crack

Zongjie Cao [1, 2*], Zhenbang Kuang [1]

[1] *Department of Engineering Mechanics, Shanghai Jiaotong University, Shanghai, 200240, China*
[2] *Department of Aeronautic Mechanical Engineering, Aeronautic University of Air Force, Changchun, 130022, China*
e-mail: zjcao96@sina.com, Zbkuang@mail.sjtu.edu.cn

Abstract

Because piezoelectric materials in mechanic behavior are brittle, flaws, voids and microcracks inevitably exist in the manufacturing process of materials and operations. So the strength of the piezoelectric materials is weakened by the presence of defects. Many efforts have been done to study the fracture behaviors of piezoelectric materials in order to understand mechanic and electric failure phenomena and to improve the device's design and performance. It is necessary to be computed the high stresses or stress intensity factors for the defects in order to estimate the reliability of components in systems. Most of the analytical solutions are often restricted to some special loading and boundary conditions, such as uniform loading and infinite boundary conditions while in engineering the problems are always finite size subjected to general mechanical electric loading. So efficient computational approaches are developed to suitable for general piezoelectric problems. The special singular finite element has been considered to be a good alternative for treating piezoelectric problems. In this paper, the local solutions of displacement fields and potential fields in cracks of piezoelectric materials[1] are first used to construct a new electric potential modelling and a new displacement modelling of crack problems of piezoelectric materials [2-3]. With the minimum potential energy principle and variational principle, piezoelectric finite element formulations with a central crack are proposed. A new singular finite element is constituted then a new method for calculating stress intensity factors of crack problems of piezoelectric materials is obtained. Numerical example of PZT-4 piezoelectric ceramics with a crack embedded in a piezoelectric materials subjected to uniform loading is considered to illustrate the validity of the present method. Results obtained by present method are good agreement with that of Shindo's solutions [4]. With the different selections of the structural dimensions, it illustrates that the results calculat ed with the present method are stable, reliable and highly accurate, high efficiency and easy use in practical program operation.

REFERENCES

[1] Z. B. Kuang, F. S. Ma, *The crack tip fields*, Xi'an Jiaotong University Press, (2002), (in Chinese).
[2] Z. C. Wang, L. S. Zhang, C. T. Liu, *A new formulated of a quasi-compatible finite element and its application in fracture mechanics*, Engineering Fracture Mechanics, 37(6), (1990), 1195-1201.
[3] Z. J. Cao, B. C. Wen, Z. C. Wang, *A new method for calculation of three dimensional crack stress intensity factors*, Chinese Quarterly of mechanics, 22(4), (2001), 401-407, (in Chinese).
[4] Y. Shindo, K. Watanabe, F. Narita, *Electroelastic analysis of a piezoelectric ceramic strip with a central crack*, International Journal of Engineering Science, 38, (2000), 1-19.

COMPUTATIONAL MECHANICS (Abstracts)
WCCM VI in conjunction with APCOM'04, *Sept. 5-10, 2004, Beijing, China*

A Comparative Study between Discrete and Continuum Models to Simulate Concrete Fracture

E. W. V. Chaves *, R. C. Yu, G. Y. Ruiz

E.T.S. Caminos Canales y Puertos ..., University of Castilla-La Mancha, Avd. Camilo José Cela, s/n 13071 Ciudad Real, Spain
e-mail: Eduardo.Vieira@uclm.es, rena@uclm.es, Gonzalo.Ruiz@uclm.es

Abstract

We present a comparative study between models that follow the two main trends to study fracture of quasi-brittle materials like concrete. On the one hand we focus on a discrete model that represents the fracture process by cohesive elements that are inserted in the original mesh only when the opening condition is met [1, 3]. Such elements also implement contact and friction algorithms. Besides, mesh size is selected so as to represent the inter-locking effect between the crack surfaces. Everything considered, this approach leads to an explicit and multiscale modeling of fracture. On the other hand we analyze a continuum model based on the Strong Discontinuity Approach (SDA) [4-5]. It localizes damage in bands that are narrowed to the limit of having null width and thus simulate the fracture surface. The initiation and propagation processes are modeled by a bifurcation analysis that searches for the surfaces where damage can be localized at every step. Finally, we use both approaches to do 3D simulations of fracture tests in concrete that allow an evaluation of their relative performance.

REFERENCES

[1] M. Ortiz, A. Pandolfi, *A class of cohesive elements for the simulation of three-dimensional crack propagation*, Int. J. Numer. Meth. Eng, 44, (1999), 1267-1282.
[2] J. Simo, J. Oliver, *Modelling strong discontinuities by means of strain softening constitutive equations*, in H. Mang et al eds. Proc. EURO-C 1994 Computer Modeling of concrete structures, Swansea, Pineridge Press, (1994), pp. 363-372.
[3] J. Oliver, A. E. Huespe, M. D. G. Pulido, E. Samaniego, *On the strong discontinuity approach in finite deformation settings*, Int. J. Numer. Meth. Engng, 56, (2003), 1051-1082.

COMPUTATIONAL MECHANICS (Abstracts)
WCCM VI in conjunction with APCOM'04, *Sept. 5-10, 2004, Beijing, China*

One-Dimensional Nonlinear Model for Prediction of Crack Spacing in Concrete Pavements

G. Chen *

Department of Structural Engineering, Building Research Institute, 1 Tachihara, Tsukuba, Ibaraki 305-0802, Japan
e-mail: gongfa@kenken.go.jp

Abstract

Shrinkage cracking is commonly observed in concrete flat structures such as highway pavements, slabs, and bridge decks. Crack spacing due to shrinkage has received considerable attention for many years [1-2]. However, some aspects concerning the mechanism of crack spacing still remain un-clear. Though it is well known that the interval of the cracks generally falls with a range, no satisfactory explanation has been put forward as to why the minimum spacing exists.

In a previous work, Chen and Baker [3-4] proposed a one-dimensional linear model and investigated the minimum and maximum crack spacings in concrete pavements from the energy viewpoint and explores mechanisms that control spacing. Unfortunately, the linear model fails in some examples. Following the previous work, this paper presents a one-dimensional non-linear model to predict crack spacing in reinforced concrete slabs due to shrinkage. The pavement is represented by a series of sub-structures that consist of a cohesive crack and an elastic bar restrained by distributed elastic springs. The concrete softening constitutive law is assumed as exponential instead of a bi-linear relation as in linear model. First, an analytical model, consisting of two cohesive cracks, is employed to investigate the variation of the cracking pattern with the length of the elastic bar (i.e., the spacing between the two possible cracks). Then, the force acting on the cohesive cracks and the energy profiles are investigated. The limitation of the previous linear model is investigated by comparing the results of both the linear and non-linear models.

The numerical examples demonstrate that the proposed non-linear model overcomes the disadvantage of the linear model and is able to predict the crack spacings in concrete pavements.

REFERENCES

[1] B. F. McCullough, *Criteria for the design, construction and maintenance of continuously reinforced concrete pavement*, Australian Road Research, 13(2), (1983), 79-99.

[2] W. Shen, D. J. Kirkner, *Distributed thermal cracking of AC pavement with frictional constraint*, Journal of Engineering Mechanics, 125(5), (1999), 554-560.

[3] G. Chen, G. Baker, *A simple model for predication of crack spacing in concrete pavements*, in H. P. Lee, K. Kumar eds. *Recent advances in computational science & engineering*, Imperial College Press, Singapore, (2002), pp. 561-564.

[4] G. Chen, G. Baker, *An analytical model for predication of crack spacing due to shrinkage in concrete pavements*, Journal of Structural Engineering, (2003), (accepted).

COMPUTATIONAL MECHANICS (Abstracts)
WCCM VI in conjunction with APCOM'04, *Sept. 5-10, 2004, Beijing, China*

Meshless Analysis of Crack Propagation in Multiphase Micromorphic Solid

Youping Chen [1], Azim Eskandarian [1*], Morton S. Oskard [2], James D. Lee [3]

[1] *Center for Intelligent System Research, The George Washington University, Ashburn, VA 20147, USA*
[2] *Turner-Fairbank Highway Research Center, Federal Highway Administration, U. S. Department of Transportation, McLean, Virginia 22101, USA*
[3] *Department of Mechanical and Aerospace Engineering, The George Washington University, Washington, DC, 20052, USA*
e-mail: ypchen2@gwu.edu, eska@gwu.edu, Mort.Oskard@fhwa.dot.gov, jdlee@gwu.edu

Abstract

The material body considered in this work belongs to a class of multiphase micromorphic solids. In Eringen's micromorphic continuum theory [1], a material body is envisioned as a collection of a large number of finite-size and deformable particles. Each particle has nine degrees of freedom representing stretches and rotations, in addition to the three classical degrees of freedom accounting for the motion of its centroid, and may be considered as a polyatomic molecule, a primitive unit cell of a crystalline solid [2,3], or a chopped fiber in a composite, etc. General micromorphic theory contains specialized micromorphic theory for covalent and ionic crystals [4], micropolar theory, and classical continuum theory as its subcases. In this work, attention is focused on multiphase material such as polycrystalline solid consisting of grains and grain boundaries. Due to the existence of material inhomogeneity, material forces and Eshelby stresses [5] are obtained, that indicate the location of crack initiation. Based on meshless method [6], the problem of mixed mode crack propagation in multiphase micromorphic solid is investigated. Numerical results, showing the path of crack propagation and the ultimate strength of the materials, are presented and discussed.

REFERENCES

[1] A. C. Eringen, E. S. Suhubi, *Nonlinear theory of simple micro-elastic solids – I*, Int. J. Engng Sci., 2, (1964), 189-203.

[2] Y. Chen, J. D. Lee, *Connecting molecular dynamics to micromorphic theory, part I: instantaneous mechanical variables*, Physica A, 322, (2003), 359-376.

[3] Y. Chen, J. D. Lee, *Connecting molecular dynamics to micromorphic theory, part II: balance laws*, Physica A, 322, (2003), 377-392.

[4] Y. Chen, J. D. Lee, *Determining material constants in micromorphic theory through phonon dispersion relations*, Int. J. Engng. Sci., 41, (2003), 871-886.

[5] G. A. Maugin, *Material forces: concepts and applications*, Appl. Mech. Rev., 48 , (1995), 213-245.

[6] Y. Chen, J. D. Lee, A. Eskandarian, *Dynamic meshless method applied to nonlocal cracked problems*, Theoretical and Applied Fracture Mechanics, 38, (2002), 293-300.

COMPUTATIONAL MECHANICS (Abstracts)
WCCM VI in conjunction with APCOM'04, *Sept. 5-10, 2004, Beijing, China*

The Application of the Linear Matching Methods to Structural Problems with Cracks

M. Salahuddin Habibullah *, A. R. S. Ponter

Department of Engineering, University of Leicester, University Road, Leicester, LE1 7RH, U.K.
e-mail: asp@le.ac.uk

Abstract

The life assessment of structures containing cracks requires an understanding of the relationship between the behaviour at the crack tip as well as the overall behaviour of the structure. The paper discusses direct methods for the evaluation of the range of load or temperature variation based upon two alternative criteria; that the crack tip field satisfies a fracture criterion and the overall behaviour does not exceed a ratchet boundary.

The procedures are based upon the Linear Matching Method, previously applied to problems where the elastic stress field does not contain a singularity and a shakedown limit exists [1,2]. When a crack is present the shakedown limit is trivially non-existent and the significant limit is always the ratchet limit where a reverse plasticity condition occurs at the crack tip [3]. By applying a two stage process, we show that the ratchet limit may be evaluated by first evaluating the variation of a changing residual stress field at the crack tip. The application of a further constant load to the body takes the loading to the ratchet limit. To evaluate loads corresponding to a prescribed crack tip field condition the accumulated strain over a load cycle is evaluated at particular points near to the crack tip and relating to the standard HRR crack tip fields. The Linear Matching Method may then be used to converge to load and temperature range values that correspond to a prescribed crack tip condition. The methods are applied to the problem of a circumferentially cracked tube subjected to internal pressure and through-thickness temperature variation, producing interaction diagrams corresponding to a ratchet limit and to a prescribed crack tip field fracture criterion.

REFERENCES

[1] H. F. Chen, A. R. S. Ponter, *A method for the evaluation of a ratchet limit and the amplitude of plastic strain for bodies subjected to cyclic loading*, European Journal of Mechanics A/Solids, 20, (2001), 555-572.

[2] A. R. S. Ponter, H. F. Chen, *A minimum theorem for cyclic loading in excess of shakedown, with applications to the evaluation of a ratchet limit*, European Journal of Mechanics A/Solids, 20, (2001), 539-554.

[3] M. Salahuddin Habibullah, A. R. S. Ponter, *Application of the linear matching method to cracked bodies*, Proceedings of the American Society of Mechanical Engineering, PVP2003, 441, (2003), 185-196.

COMPUTATIONAL MECHANICS (Abstracts)
WCCM VI in conjunction with APCOM'04, *Sept. 5-10, 2004, Beijing, China*

Analysis of Carbon and Polypropylene Hybrid Fiber Reinforced Cement-Based Composites Using Continuum Damage Mechanics

Yuan Hua [1*], Junying Lian [1], Taiquan Zhou [1], Renhe Wang [2]

[1] *Department of Civil Engineering, Southern Yangtze University, Wuxi, 214063, China*
[2] *Department of Civil Engineering, Anhui University of Science and Technology, Huainan, 232001, China*

e-mail: zhoutaiquan@sytu.edu.cn, rhwang@aust.edu.cn

Abstract

Cement-based materials in general and concrete in particular, require reinforcements because of their low tensile strength and strain capacities. The improvement in the strength and energy absorption capacity of the material, due the addition of randomly oriented and distributed fibers to a cementitious matrix, is usually accompanied by a strong nonlinear behavior of the composite. Therefore, appropriate nonlinear theories are needed to make realistic predictions for structural applications of such material. Based on the method of nonlinear finite element analysis, and guided by test observations, damage mechanics has been proven to be a very rational approach when dealing with the nonlinear behavior of carbon and polypropylene hybrid fiber reinforced cement-based composites, especially prior to the phenomenon of localization. The response of four-point bending beams has been investigated for four volume fractions of carbon fiber (0.4, 0.6, 0.8 and 1.0%) and four volume fractions of polypropylene fiber (0.10, 0.15, 0.20 and 0.25%) and four length and diameter ratio (l/d=550, 600, 650, 700) of carbon fiber and three length and diameter ratio (l/d=750, 850, 1000) of polypropylene fiber. It was found that this approach was very suitable for the prediction of the test behavior for carbon and polypropylene hybrid fiber reinforced cement-based composites.

REFERENCES

[1] A. Mazars, G. Pijaudier-Cabot, *Continuum Damage Mechanics: Application to Concrete*, Journal of Engineering Mechanics, ASCE, 115, (1989), 345-365.

[2] Y. Hua, H. B. Oi, Z. Q. Jiang, S. Z. Huang, S. B. Zhang, *Study on the bending fatigue damage of the carbon and the polypropylene hybrid fiber reinforced concrete*, Key Engineering Materials, 183, (2000), .571-576.

[3] J. Neter, W. Wasserman, M. H. Kutner, *Applied Linear Statistical Models*, Irwin Publications, Boston, (1990).

[4] S. T. Dai, J. F. Labus, *Damage and Failure Analysis of Brittle Materials by Acoustic Emission*, Journal of Materials Civil Engineering, 9, (1997), 200-205.

COMPUTATIONAL MECHANICS (Abstracts)
WCCM VI in conjunction with APCOM'04, *Sept. 5-10, 2004, Beijing, China*

Prediction of Anisotropy Affected Fatigue Crack Growth Paths on Pure Aluminum Sheets

Yudy Surya Irawan *, Yoshihiko Hagiwara, Shin-ichi Ohya

Department of Mechanical System Engineering, Musashi Institute of Technology, 1-28-1 Tamazutsumi, Setagaya-ku, Tokyo, 158-8557, Japan
e-mail: g0372002@sc.musashi-tech.ac.jp, yhagisan@sc.musashi-tech.ac.jp, sin008@sc.musashi-tech.ac.jp

Abstract

It is common that the anisotropy of rolling plate does not affect the fatigue crack propagation paths. Thus, when CCT-specimens of rolled plate with any rolling direction are applied by tensile fatigue load, the cracks propagate to mode I direction. However, in case of CCT specimens of pure aluminum sheets A1100P with rolling direction 30 degrees towards loading direction, the fatigue crack does not grow to mode I direction but to rolling direction as a mixed mode I-II crack [1]. In order to understand the behavior of this anisotropy affected fatigue crack and improve the previous estimation technique [2, 3], basic experimental data was obtained from fatigue test using no-holed CCT-specimens with 30 degrees of rolling direction. The stress intensity factors and energy release rates were calculated using two-dimensional finite element analysis. Based on the basic experimental results, the anisotropy affected cracks were simulated as follow: 1) an initial mode I crack grows in the direction with maximum energy release rate and the crack starts to gradually change its direction to rolling direction in the certain value of mode I stress intensity factor; 2) On rolling direction, the crack continues to propagate to rolling direction in the certain value of total stress intensity factor and the ratio of mode II and mode I energy release rate, G_{II}/G_I is higher than 2.0; and 3) The crack change the direction from rolling direction to mode I direction, when maximum mode II stress intensity factor is reached. This purposed method can predict crack propagation paths in no-hole specimens and 6-type holed specimens with 0, 30, 45 and 60 degrees of rolling direction. The simulation results were confirmed well to be coincided with experimental results.

REFERENCES

[1] Y. Hagiwara, D. Sohret, *Effect of anisotropy and sheet thickness on the fatigue crack propagation of aluminum plate specimens*, Proceedings of the 37th Japan Congress on Material Research, JSMS, Kyoto, (1994), pp. 46-51.

[2] Y. Hagiwara, D. Sohret, Estimation of the fatigue crack propagation path of anisotropic aluminum plate specimen, Transactions of Japan Society of Mechanical Engineers, 60, (1994), 915-920, (In Japanese).

[3] Y. Hagiwara, D. Sohret, *Quantitative estimation on the fatigue crack propagation of anisotropy aluminum plate specimen*, Acta Mechanica Solida Sinica, 8, (1995), 340-345.

COMPUTATIONAL MECHANICS (Abstracts)
WCCM VI in conjunction with APCOM'04, *Sept. 5-10, 2004, Beijing, China*

Effect of Bugling Deformation on Fracture Analysis of Aircraft Panel Containing Damage and Automation Calculation

Jinsan Ju*, Xiugen Jiang, Jinzhao Zhuang

Department of Civil Engineering, China Agricultural University, Qinghuadong Road, Beijing, 100083, China
e-mail: jujinsan@tsinghua.org.cn, jiangxiugen@tsinghua.org.cn

Abstract

Due to the complication of stress distribution around the crack tips, it is hard to obtain the fracture parameters in a 3-dimension model. Some kinds of methods to simplify the model or calculation procedure are carried out. Basically for the crack in the thin shell structure such as aircraft panel, the parameters can be obtained finally from 2-dimentional local models by these methods because it is relatively simple. However these methods ignored the effect of 3-D deformation on the fracture parameters. This paper investigates the effects of 3-dimentional bugling deformations on the stress intensity factor of aircraft panel. It describes the difference of fracture parameters among 2-dimention, 3-dimention models with bugling deformation and without bucking deformation of aircraft panel through comparing the calculation results; the effect of size of the local model on SIF caused by bulging deformation is also considered. It was found the effect of 3-D bugling deformation on fracture parameters is significant and cannot be ignored.

Software packages ABAQUS and ANSYS are both used to solve this problem using different meshed models. The results from these two soft wares are compared with each other and that of theoretical ones. The results show the calculations in this paper are good. In addition, a high efficient program system is made to transfer model information including node, element, load and boundary conditions from ANSYS to ABAQUS automatically for the analysis in ABAQUS. This system can also solve this problem in ANSYS automatically from model creating, adding load and boundary condition, solving to post processing.

REFERENCES

[1] T. L. Anderson, *Fracture mechanics fundamentals and applications*, Florida, USA, CRC Press, (1995), pp. 60-64.

[2] S. N. Atluri, *Structural integrity and durability*, Georgia, USA, Tech Science Press, (1997), 328-390.

[3] H. Kawai, L. Wang, D. S. Pipkins, K. O'Sullivan, P. E. O'Donoghue, S, N. Atluri, *Automation of global-intermediate-local analysis based on feature-based modeling techniques and the finite element alternating method*, Computer Modeling and Simulation in Engineering, 4(1), (1999), 37-41.

[4] L. Wang, W. T. Chow, H. Kawai, S. N. Atluri, *Prediction of widespread fatigue damage thresholds in aging aircraft*, Journal of American Institute of Aeronautics and Astronautics, 36, (1998), 457-464.

[5] L. Wang, W. T. Chow, H. Kawai, S. N. Atluri, *Residual strength of aging aircraft with multiple site damage/multiple element damage*, Journal of American Institute of Aeronautics and Astronautics, 36, (1998), 840-847.

COMPUTATIONAL MECHANICS (Abstracts)
WCCM VI in conjunction with APCOM'04, *Sept. 5-10, 2004, Beijing, China*

Stress Intensity Factors for Deep Circumferential External Surface Cracks in Hollow Cylinders

W. K. Lao *, K. P. Kou

Department of Civil and Environmental Engineering, Faculty of Science and Technology, University of Macau, Macao, China
e-mail: wklao@126.com, kpkou@umac.mo

Abstract

For the case of circumferential external surface cracks in hollow cylinders, Mettu, Raju and Forman [1] have presented finite element solution to the stress intensity factors at three aspect ratios ($a/c = 0.6$, 0.8 and 1.0) and three crack depth to thickness ratios ($a/t = 0.2$, 0.5 and 0.8). However, stress intensity factors for low aspect ratios ($a/c = 0.2$ and 0.4) were obtained by extrapolation because of distortion of the transformed mesh. As a result, accurate assessment of fracture strengths and life prediction of fatigue cracks cannot be achieved at low aspect ratios and high crack depth to thickness ratios ($a/t >$ 0.80).

In this paper, stress intensity factors for a new circumferential external surface crack configuration are determined, especially for deep cracks ($a/t > 0.8$), by finite element method. Three-dimensional linear elastic finite element models were established and analysed with ABAQUS [2]. The collapsed, 20-noded, singular solid elements were used around the crack front to simulate the stress singularity near the crack tip. The J contour integral method was adopted to determine the stress intensity factors along the crack front. To verify the validity of the stress intensity factors at the deepest point, the limited displacement extrapolation technique [3] was also employed. Stress intensity factors for surface cracks subjected to tension or bending were available for various geometric parameters such as internal radius to thickness ratio (R/t), crack aspect ratio (a/c) and crack depth to thickness ratio (a/t). Moreover, distribution of stress intensity factors along the crack front and variations of stress intensity factors with geometric parameters were investigated. Finally, an equation for stress intensity factors was developed for deep cracks, and the equation is within ±3.09% of the finite element results. It can be concluded that the stress intensity factors provide new and valuable data, especially for deep cracks, for assessment of fracture strengths and life prediction of fatigue cracks for the crack configuration considered.

REFERENCES

[1] S. R. Mettu, I. S. Raju, R. G. Forman, *Stress intensity factors for part-through surface cracks in hollow cylinders*, JSC report 25685/LESC report 30124, NASA Lyndon B. Johnson Space Center/Lockheed Engineering and Sciences Co. Joint Publication, (1992).

[2] HKS Inc., *ABAQUS/Standard user's manual*, Version 6.3-1, Pawtucket, RI, USA, (2002).

[3] I. L. Lim, I. W. Johnston and S. K. Choi, *Comparison between various displacement-based stress intensity factor computation techniques*, Int. J. Fracture, 58, (1992), 193-201.

COMPUTATIONAL MECHANICS (Abstracts)
WCCM VI in conjunction with APCOM'04, *Sept. 5-10, 2004, Beijing, China*

Crack Detection with Strain Gages and the Body Force Method

H. Noguchi [1*], T. Harada [2]

[1] *Department of Mechanical Engineering Science, Faculty of Engineering, Kyushu University, 6-10-1 Hakozaki, Higashi-ku, Fukuoka 812-8581, Japan*
[2] *Department of Mechanical Engineering, Kurume National College of Technology, 1-1-1 Komorino, Kurume 830-8555, Japan*
e-mail: nogu@mech.kyushu-u.ac.jp, harada@kurume-nct.ac.jp

Abstract

Previously, we proposed a method [1] for estimating stress fields of a notched region in a real structure using stress values by strain gages near the notch and the general stress functions. This method was successful for estimation of stress fields and crack detection [2] in 2-dimensional structures. However, it cannot be available for 3-dimensional structures. In this study, the idea called "Pseudo-3-dimensional problem" is proposed in order to apply the above method for 3-dimensional structures. Based on this idea, a method is proposed for estimating the stress fields using a few surface strain data, obtained from strain gages and the body force method (BFM). A notch in a 3-dimensional structural surface is considered to be an observed domain. The observed domain is treated as a pseudo- 3-dimensional plate without thickness, not a part of a 3-dimensional body. In other words, the body force is distributed in the observed domain to add a 3-dimensional influence to a plane stress problem. In the present study, the results by elastic finite-element method (FEM) analysis are used instead of the strain gage outputs on actual structures. It is possible to obtain the maximum stress value within 10% error by self-assessment of the result with the present method. Moreover we apply this method to crack detection in notched structures. A characteristic stress field due to the crack can be expected in the observed domain. Crack detection is then tried by showing the characteristic stress field. Detection of an ideal crack with FEM, which is a through crack or a quarter-circular crack on the hidden side, is studied. In both a through crack and a quarter-circular hidden crack, the characteristic stress field can be shown; the characteristic stress field means that two stress concentrations exist along the notch. Finally, the application limit of the present method is shown.

REFERENCES

[1] T. Harada, H. Noguchi, *Estimation with a few strain gages of stress fields near a notch root of actual structures under in-plane loading,* J. Tes. Eva., 27(2), (1999), 122-130.

[2] T. Harada, H. Noguchi, *Fatigue crack detection with estimation of stress fields near a notch root of thin plate structures in in-plane loading,* J. Tes. Eva., 31(2), (2003), 154-161.

COMPUTATIONAL MECHANICS (Abstracts)
WCCM VI in conjunction with APCOM'04, *Sept. 5-10, 2004, Beijing, China*

A Continuum Shape Sensitivity Method for Fracture Analysis of Functionally Graded Materials

B. N. Rao [*], S. Rahman

Center for Computer-Aided Design, The University of Iowa, Iowa City, IA 52242-1000, USA
e-mail: bnrao@yahoo.com, rahman@engineering.uiowa.edu

Abstract

In recent years, special type of composite materials known as functionally graded materials (FGMs) [1] have been introduced and applied to the development of structural components that are exposed to non-uniform service requirements. Although the absence of sharp interfaces in FGMs largely reduces material property mismatch, cracks do occur when FGMs are exposed to external loading and environments. The overall mechanical and thermal-mechanical responses and reliability of FGMs are largely determined by the fractures induced by such cracks. Hence, fracture mechanics methodologies are required to quantify appropriate crack-driving force so that FGMs with improved toughness can be developed to meet current and future industrial needs.

This paper presents a new continuum shape sensitivity method [2] for calculating mixed-mode stress-intensity factors for a stationary crack in two-dimensional, linear-elastic, isotropic FGMs with arbitrary geometry. The method involves the material derivative concept taken from continuum mechanics, the mutual potential energy release rate, and direct differentiation. Since the governing variational equation is differentiated prior to discretization, resulting sensitivity equations are independent of approximate numerical techniques, such as the finite element method, boundary element method, mesh-free method, or others. The discrete form of the mutual potential energy release rate is simple and easy to calculate, as it only requires multiplication of displacement vectors and stiffness sensitivity matrices. By judiciously selecting the velocity field, the method only requires displacement response in a subdomain close to the crack tip, thus making the method computationally efficient. Finite-element based numerical examples, which comprise mode-I and mixed-mode deformations and/or single or multiple interacting cracks, are presented to evaluate the accuracy of the fracture parameters calculated by the proposed method. Comparisons have been made between stress-intensity factors predicted by the proposed method and available reference solutions in the literature, generated either analytically or numerically using various other fracture integrals or analyses. Excellent agreement is obtained between the results of the proposed method and previously obtained solutions.

REFERENCES

[1] S. Suresh, A. Mortensen, *Fundamentals of Functionally Graded Materials*, IOM Communications Ltd., London, (1998).

[2] B. N. Rao, S. Rahman, *Continuum Shape Sensitivity Analysis of Mode-I Fracture in Functionally Graded Materials*, Proceedings of 2003 ASME Pressure Vessels and Piping Conference, Cleveland, OH, July 20-24, (2003).

COMPUTATIONAL MECHANICS (Abstracts)
WCCM VI in conjunction with APCOM'04, *Sept. 5-10, 2004, Beijing, China*

The Analysis of Three-dimensional Stress Singularities by Using a Quasi-spherical Coordinate System

Haitao Wang [1*], K. Y. Sze [2]

[1] *Department of Civil Engineering, South China University of Technology, Wushan Road, Guangzhou, 510641, China*
[2] *Department of Mechanical Engineering, University of Hong Kong, Pokfulam Road, Hong Kong SAR, China*
e-mail: htwang@graduate.hku.hk, kysze@hku.hk

Abstract

Cracks, notches and multi-material joints are often found in modern engineering structures and materials. Thorough understanding of the singular stress fields around them is needed for the reliability prediction of these structures and materials during their service life. Many studies about singular asymptotic stress fields in linear elastic materials have been carried out in recent two decades. Most of the researches in this area have focused on two-dimensional geometry. The present concern is three-dimensional singularities of stress in three-dimensional conical configurations. According to Nakamura and Park [1], in the case of through thickness cracks in thin plates, 3% of thickness is dominated by three-dimensional stress singularities, and sometimes they are even graver than −1/2.

Several numerical procedures have been proposed for the determination of three-dimensional stress singularities ([2-5]). But they are all developed in conventional spherical coordinates. To avoid the inherent singularity of the spherical coordinates at their poles, a quasi-spherical coordinate system is developed. In this coordinate system, a finite element procedure is proposed to determining the eigensolutions at three-dimensional vertices in which the field variables are proportional to the $(\lambda+1)$-th power of the distance from the vertices. The resulting global equation is a second order characteristic matrix equation. Two demonstrating problems are investigated. It can be seen that the formulation yield satisfactory results.

REFERENCES

[1] T. Nakamura, D. M. Parks, *Three-dimensional elastic stress field near the crack front of a thin plate*, J. Appl. Mech., 55, (1988), 805-813.

[2] Z. P. Bazant, *Three-dimensional harmonic functions near termination or intersection of gradient singular lines: a general numerical method,* Inter. J. Engng., 12, (1974), 221-243.

[3] Z. P. Bazant, L. F. Estenssoro, *Surface singularity and crack propagation*, Inter. J. Solids & Structures, 15, (1979), 405-426.

[4] J. P. Benthem, *The quarter-infinite crack in a half-Space: alternative and additional solutions*, Inter. J. Solids & Structures, 16, (1980), 119-130.

[5] F. Ghahremani, C. F. Shih, *Corner singularities of three-dimensional planar interface cracks*, J. Appl. Mech., 59, (1992), 61-68.

COMPUTATIONAL MECHANICS (Abstracts)
WCCM VI in conjunction with APCOM'04, *Sept. 5-10, 2004, Beijing, China*

Crack-Cutting Method for Solving the Bending Problems of Bars with Cracks

Xiaochun Wang*

College of Environment and Civil Engineering, Chengdu University of Technology, Chengdu 610059, China
e-mail: wxiaoch@sina.com

Abstract

It is well known that bending problems of prismatic bars are one of the important problems in engineering. There have been an exact solution for bending problems of bars, but the corresponding results of cracked bar's bending problems are seldom published. There has not yet been a convenient and general method to deal with the bending problem of bars with non-circular cross section with cracks. This paper is devoted to CCM (crack-cutting method) for solving the bending problems of non-circular cross section bars with cracks. The so-called CCM is a method that by use of the results given for a cylinder with cracks, and further letting the cracks expand to a definite limit, a polygonal cross section can be obtained by cutting the original circular cross section, and then the corresponding results are derived. In 1987 and 1988 (Tang & Wang, [1] and Wang & Tang, [2]), CCM was successfully used to solve the rectangular prism's torsion problem with or without cracks. In the present paper discussing bending problems, our objective is to derive the integral equations and expressions of shear stresses τ_{xz} and τ_{yz} for an arch-shaped bar and a rectangular bar with cracks, respectively.

Based on the general solution for a cylinder with multi-cracks [3] and the idea of limit cutting, the singular integral expressions of shear stresses of bars with arch-shaped and rectangular cross section with cracks are derived. A method of numerical implementation is also presented for the singular integral expressions. By means of these expressions, it is convenient to do numerical calculation of bending shear stresses of bars with cracks. Besides, owing to the generality of the proposed method, it does not result in any new difficulty in solving the bending problem, no matter whether the bar's cross section has cracks or not. And, it is easy to see that CCM can be used to deal with bending problems of bars with an arbitrary circular inscribed polygonal cross section with cracks.

REFERENCES

[1] Renji Tang, Xiaochun Wang, *Crack-cutting method to solve torsion problem for a cracked cylinder with multi-angular cross section*, Proc. Int. Conf. on Fracture and Fracture Mechanics, Shanghai, China, (1987), pp. 130-133.

[2] Xiaochun Wang, Renji Tang, *On the torsion of a cylinder with several cracks*, Applied Mathematics and Mechanics, 9, (1988), 745-754.

[3] Xiaochun Wang, *Bending integral expressions of a cylinder with cracks*, Applied Mathematics and Mechanics, 15, (1994), 163-165.

[4] N. I. Muskhelishvili, *Some basic problems of the mathematical theory of elasticity*, the second English edition, Leyden, Noordhoff International Publishing, The Netherlands, (1977).

COMPUTATIONAL MECHANICS (Abstracts)
WCCM VI in conjunction with APCOM'04, *Sept. 5-10, 2004, Beijing, China*

Advantage of the Fractal Finite Element Method for Two-Dimensional Crack Problems

Jianfan Xie [1*], S. L. Fok [2], A. Y. T. Leung [2]

[1] *School of MMMEM, University of Nottingham, University Park, Nottingham, NG7 2RD, UK*
[2] *The Manchester School of Engineering, University of Manchester, Oxford Road, Manchester M13 9PL, UK*
e-mail: jianfan.xie@nottingham.ac.uk, alex.fok@man.ac.uk, Andrew.leung@man.ac.uk

Abstract

The Fractal Finite Element Method (FFEM) for calculating 2-D stress intensity factors is modified by making the similarity ratio in the construction of the fractal mesh a variable. Compared with other numerical methods, FFEM has several advantages. Firstly, only conventional finite elements are required for modelling the singularity at the crack tip. No new singular elements have to be derived. Secondly, the implementation and coding for this method can be achieved easily by modifying a standard finite element program, as only inversion and transformation of small matrices are needed. Finally, by employing a suitable analytical solution as a global interpolation function, SIF can be obtained directly and precisely, thus avoiding the need of special integration schemes.

REFERENCES

[1] A. Y. T. Leung, K. L. Su, *Fractal two level finite element method for 2-D crack problems*, Proc. 2nd Asian-Pacific Conf. Computational Mech., Brookfield, Rotterdam, (1993), pp. 675-679.

[2] M. L. Williams, *Stress singularities resulting from various boundary conditions in angular corners of plates in extensions*, J. Appl. Mech., 24, (1957), 109-114.

[3] L. Ying, *Infinite element methods*, Peking University Press, (Beijing) and Vieweg Pu blishing, (1995).

[4] A.Y.T. Leung, K. L. Su, *Fractal two-level finite element method for two-dimensional cracks*, Microcomputers in Civil Engrg, 11, (1996), 249-257.

[5] H. Tada, P. C. Paris, G. R. Irwin, *The stress analysis of cracks handbook, Del Research Corporation*, Pennsylvania, (1973).

COMPUTATIONAL MECHANICS (Abstracts)
WCCM VI in conjunction with APCOM'04, *Sept. 5-10, 2004, Beijing, China*

Computing Bounds to Mixed Mode *J*-integral in Elastic Fracture Mechanics

Z. C. Xuan [1*], J. Peraire [2], B. C. Khoo [3]

[1] *Programme of HPCES, Singapore-MIT Alliance E4-04-10, 4 Engineering Drive 3, 11757, Singapore*

[2] *Department of Aeronautics and Astronautics, Massachusetts Institute of Technology, Cambridge, MA 02139, USA*

[3] *Department of Mechanical Engineering, National University of Singapore, 10 Kent Ridge Crescent, 119260, Singapore*

e-mail: smaxzc@nus.edu.sg

Abstract

We compute the bounds to the mixed mode *J*-integral in two-dimensional linear elasticity with a novel and efficient a-posteriori bounds method. The *J*-integral, which is typically expressed as a contour integral, is recast as a surface integral which yields a quadratic continuous functional of the displacement. By expanding the quadratic output about an approximate finite element solution, the output is expressed as a known computable quantity plus linear and quadratic functionals of the solution error. The quadratic component is bounded by the energy norm of the error scaled by a continuity constant, which is determined explicitly. The linear component is expressed as an inner product of the errors in the displacement and in a computed adjoint solution, and bounded using standard a-posteriori error estimation techniques. The *J*-integral in mixed mode will be separated into two forms that are related to the stress intensity factors of open mode and shear mode. The method is illustrated with two fracture problems in plane strain elasticity.

REFERENCES

[1] Z. C. Xuan, K. H. Lee, A. T. Patera, J. Peraire, *Computing upper and lower bounds for the J-integral in two-dimensional linear elasticity*, Singapore-MIT Alliance Symposium, (2004).

[2] J. Peraire, A. T. Patera, *Bounds for linear-functional outputs of coercive partial differential equations: Local indicators and adaptive refinement*, in P. Ladeveze, J. T. Oden eds. *Proc. of the Workshop on New Advances in Adaptive Computational Methods in Mechanics*, Elsevier, (1998).

[3] M. Paraschivoiu, J. Peraire, A. T. Patera, *A posteriori finite element bounds for linear-functional outputs of elliptic partial differential equations*, Computer Methods in Applied Mechanics and Engineering, 150, (1997), 289-312.

[4] M. Ainsworth, J. T. Oden, *A posteriori error estimation in finite element analysis*, Computer Methods in Applied Mechanics and Engineering, 142 , (1997), 1-88.

COMPUTATIONAL MECHANICS (Abstracts)
WCCM VI in conjunction with APCOM'04, *Sept. 5-10, 2004, Beijing, China*

Modeling of Mixed Mode Debonding along FRP-Concrete Bond Interface

J. Yin [1*], Z. S. Wu [2], T. Andou [1], K. Kato [1]

[1] *Advanced Simulation Technology of Mechanics (ASTOM), RIKEN 2-1, Hirosawwa, Wako, Saitama 351-0198, Japan*

[2] *Dept. of Urban and Civil Engineering, Ibaraki University, 4-12-1 Nakanarusawa-cho, Hitachi, 316-8511, Japan*

e-mail: yin@astom.co.jp, zswu@mx.ibaraki.ac.jp, andou@astom.co.jp, kato@astom.co.jp

Abstract

Interfacial debonding is one of the most significant fracture behaviors in FRP-strengthened concrete structures. Although the macroscopic failure mode is almost prescribed along the FRP-concrete bond interface, the mixed mode fracture of interfacial concrete adjacent to bond interface is frequently observed in the fields and labs experiments. The loading process of the interfacial concrete is rather complicated, which could vary constantly with the propagating fracture. It cannot be simply treated as mode I fracture in conventional smeared crack models, even at onset of cracking. Therefore, a more applicable fracture model is required to deal with such a mixed mode debonding on both fracture initiation and subsequent evolution.

A fracture criterion based on the tractions acted on the potential failure plane, proposed by Weihe [1], is referred and modified in this paper. The initial failure constitutive relationship is defined on a generic plane of degradation (POD) as

$$F_0 = -(\sigma_n - f_{t0} - a_0)^2 + \sigma_t^2 / (\tan^2 \phi) + a_0^2 ,$$

An inelastic displacement $\mathbf{u}^{ck} = \left[u_n^{ck}, u_t^{ck} \right]^{\mathrm{T}}$ across the POD is formulated, analogous to classical flow theory of plasticity. In Weihe's formulation, the stresses are derived from the elastic strains by subtracting inelastic part $\mathbf{u}^{ck}/h$ from the total strain, where h is crack band width. In this paper, a significant modification is to treat the inelastic displacement $\mathbf{u}^{ck}$ as the real discontinuity on POD and the stresses are computed by the elastic displacements $\mathbf{u}^{e}$ by removing $\mathbf{u}^{ck}$ on POD from the total displacements. In addition, the rotation of POD orientation can also be permitted once the stress state breaks the current evolved failure constitutive relationship. A FE formulation for 2-dimensional CST element is given specifically. A backward Euler integration scheme is utilized to have the stresses follow the evolutional degraded constitutive relationship.

A numerical example of single lap shear test FRP-strengthened concrete prism is provided [2], which concerns only the debonding failure of interfacial concrete. The simulation results of POF evolution, and bonding load capacity and shear stress distribution within adhesive layer are given.

REFERENCES

[1] S. Weihe, *Modelle der fiktiven Rißbildung zur Berechnung der Initiierung und Ausbreitung von Rissen: Ein Ansatz zur Klassifizierung*, Ph. D. Thesis, Universität Stuttgart, Germany, (1995), (in German).

[2] Z. S. Wu, H. Yuan, H. D. Niu, *Stress transfer and fracture propagation in different kinds of adhesive joints*, Journal of Engineering Mechanics, ASCE, 128(5), (2002), 562-573.

COMPUTATIONAL MECHANICS (Abstracts)
WCCM VI in conjunction with APCOM'04, *Sept. 5-10, 2004, Beijing, China*

A New Boundary Element Method for Half-plane Elastic Bodies Containing Cracks

Zh. X. Zhang [1*], Y. B. Wang [2], J. Su [1], Sh. H. Yang [3]

[1] *Department of Mechanics, Jiangsu University, Zhenjiang, 212013, China*
[2] *Engineering College, Ocean University of China, Qingdao, 266003, China*
[3] *Library of Lanzhou University, 730000, China*
e-mail: aame@sohu.com, wangyb@ouc.edu.cn, sujun@ujs.edu.cn

Abstract

This paper studies the crack problems of a half-plane body, which material is considered to be isotropic and elastic. Taking the dislocation densities on the crack surfaces as unknown quantities, which is denoted by $w(t)$ and further transformed to $H(t)$ that is a combination of the traction and dislocation density, a new boundary integral equation (BIE) is obtained for solving crack problems of a half-plane body. Compared with the conventional boundary integral equations, the new equation containing $1/r$ singularity can solve crack problems which have arbitrary shapes, where is the distance measured from the singular boundary points, and no hypersingular terms appear. This is helpful for analyzing crack problems because of the lower-singularity and the higher precision in calculation. To get the numerical solution of the new BIE, this article presents singular interpolation functions of order $r^{1/2}$ for the discretization of the crack near the crack tips, such that stress singularity can be modeled appropriately, a new formulation is also given for stress intensity factors (both K_{I} and K_{II}) in terms of $H(t)$ exactly. Some typical examples of cracks located in half-plane bodies have been calculated here, which shows the effectiveness of the new boundary element method for half-plane crack problems.

REFERENCES

[1] F. Delale, F. Erdogan, *On the mechanical modeling of the interfacial region in bonded half-planes,* Journal of Applied Mechanics, 55, (1988), 317-324.
[2] N. I. Muskhelishvili, *Singular integral equations,* Noordhoff, Groningen, The Netherlands (1953).
[3] F. Erdogan, K. Arin, *A half plane and a strip with an arbitrarily located crack,* International Journal Fracture, 11(2), (1975), 191-204.
[4] K. T. Chau, Y. B. Wang, *A new boundary integral formulation for plane elastic bodies containing cracks and holes,* International Journal Solids Structures, 36, (1999), 2041-2074.
[5] Y. B. Wang, K. T. Chau, *A new boundary element for plane elastic problems involving cracks and holes,* International Journal of Fracture, 87, (1997), 1-20.

COMPUTATIONAL MECHANICS (Abstracts)
WCCM VI in conjunction with APCOM'04, *Sept. 5-10, 2004, Beijing, China*

Prediction of High Cycle Crack Initiation Fatigue Life by Damage Mechanics-Finite Element- Additional load Method

Hassan Ayman, Deyu Cui *, Xing Zhang

Research Institute of Solid Mechanics, Beihang University, Box 508, Beijing, 100083, China
e-mail: aymanh68@hotmail.com

Abstract

The problems of prediction for high cycle fatigue crack initiation lives of structural members are encountered in engineering. Formerly, this kind of problems was solved on the basis of a great deal of experimental work and mathematical statistics. Those methods are not only very expensive but also time consuming. The computational damage mechanics approach looks like acceptable for engineering application. However, there are still three difficulties in this approach as follows:
1) The increments of loading cycles are taken as computational step length, but the fatigue life of structure member is unknown in advance. 2) By variable stiffness method it must be reassemble the stiffness matrix for each step of computation to consider the deterioration of materials under repeated loading. 3) The scatter of fatigue test data can not be described quantitatively.
We improve the conventional damage mechanics–finite element methods as following:
1) The damage increments at critical element is considered as the step length instead of the increments of load cycles, where the value of damage parameter, D, lies between 0 and 1. 2) To consider the effect of stress-damage coupling, a finite element-additional loads method is developed to predict fatigue life. In this method, the stiffness matrix keeps the same as that of a sound body. The damage-induced additional loads are introduced to avoid the repeated assembly of global stiffness matrix once the material is deteriorated during cyclic loading. 3) The initial damage is introduced to consider the scatter of test data quantitatively by computational mechanics.
To verify the availability of the damage mechanics-finite element-additional loads method presented, the theoretical results of fatigue life for axisymmetric specimen are computed and compared with the experimental ones. A good agreement is achieved from the comparison.

REFERENCES

[1] J. Lemaitre, *A course on damage mechanics*, Springer-Verlag, New York, (1992).
[2] J. Lemaitre, *Micro-mechanics of crack initiation*, Int. J. Fract., 42, (1990), 1-6.

COMPUTATIONAL MECHANICS (Abstracts)
WCCM VI in conjunction with APCOM'04, *Sept. 5-10, 2004, Beijing, China*

Damage Mechanics-Finite Element Analysis for Median Cycle Fatigue Crack Initiation Life Prediction

Hassan Ayman, Xing Zhang*, Deyu Cui

Research Institute of Solid Mechanics, Beihang University, Box 508, Beijing, 100083, China
e-mail: aymanh68@hotmail.com

Abstract

Fatiguc is onc of the principal modes of failure about structural members under repeated loading. For many cases, the maximum equivalent stress produced by the initial loading may be greater than the initial yielding stress and the range of variation of equivalent stress caused by the subsequent loading is still within the subsequent elastic range. This kind of fatigue failure is called median cycle fatigue (MCF). For the life prediction of MCF, the first important problem is to determine the maximum equivalent stress within elasto-plastic zone. This problem is solved on the basis of deformation theory, finite element analysis and additional loading-iteration method. With the application of additional load method, the analysis of displacement and stress fields for elasto-plastic range can be carried out as that for an elastic body. The modified maximum equivalent stress can be solved by iteration method. At the subsequent cyclic loading of MCF, because of the strain hardening of material, the structure member can still work within linearly elastic range with the modification in stress ratio. For the prediction of fatigue life, it is simple to find the modified equivalent maximum and minimum loads from the ordinary ones by the method of proportion. Then, the prediction of fatigue can be performed by the linearly elastic damage mechanics approach applied in of HCF [1] After above modification about loading, the damage evolution equation are developed into four cases and the initial damage of finite element caused by plastic deformation is considered as a power function of plastic work. The above method is applied successfully at sheet specimen and the numerical results are acceptable in engineering application.

REFERENCES

[1] X. Zhang, *Applied fatigue damage mechanics of metallic structural members*, National Defense Press, China, (1998).

[2] Hassan Ayman, Deyu Cui, Xing Zhang, *Prediction of high cycle fatigue crack initiation life by damage mechanics-finite element-additional load method*, (to be published).

COMPUTATIONAL MECHANICS (Abstracts)
WCCM VI in conjunction with APCOM'04, *Sept. 5-10, 2004, Beijing, China*

Model for the Analysis of the Seismic Behaviour of Reinforced Concrete Highway Bridges

Alex H. Barbat [1*], Sergio Oller [1], Consuelo Gómez [2]

[1] *Department of Structural Mechanics, Technical University of Catalonia, Jordi Girona 1-3, Modulo C1, Barcelona, 08034, Spain*
[2] *UAM, Distrito Federal, México*
e-mail: alex.barbat@upc.es, sergio.oller@upc.es, cgomez@correo.azc.uam.mx

Abstract

The evaluation of damage caused by seismic action to existing bridges has received an important attention in recent years, because it is the first step towards reducing of casualties and economic losses. In damage detection and evaluation, the application of simple and reliable models has been prioritized, because they are necessary in the multi-analyses required by Monte Carlo simulations. A simplified vulnerability evaluation model characterizing the expected damage due to seismic action in RC highway bridges with simple pier bents is therefore proposed in this paper. The model is based on the hypothesis of the flexible pier-rigid deck behaviour of the structure subjected to transversal seismic loads. The piers are modeled as continuous elements with distributed mass and infinite axial stiffness, while the girders are perfectly stiff elements concentrating the mass at the top of the piers. The bearings of the girders are modeled as equivalent short elements that work to shear. The soil-structure interaction effect is considered by means of linear springs that represent the rotational stiffness of the soil. Accordingly the model has as many degrees of freedom as the transversal displacements at the top of the piers.
The study of the damage produced by the earthquake load is centered on the piers of the bridge, while the dynamic study of the deck can be performed after the structural analysis of the piers in an uncoupled way. The simplified model was developed based on the correlation between the dynamic responses of the proposed model and of the real structure in its first few vibration frequencies. It was also calibrated by using the results of linear and non-linear finite element analyses of the Warth Bridge, Austria.
The damage caused by the seismic action is defined by applying an isotropic damage model based on continuum damage mechanics. Using this model, the inertia of the damaged cross section at the base of each bridge pier is obtained and the evaluation of the damage under seismic actions is based on the modification of the cross sectional inertia of the bridge piers. The maximum damage of the piers is the principal aim of the proposed structural evaluation methodology. A local damage index is used for this purpose, which describes the state of the material at each point of the structure and is based on a constitutive damage law. The proposed model permits a simple, reliable and efficient structural analysis. Examples of application of the proposed model to analyze the seismic behavior of the Warth Bridge demonstrate its reliability and efficiency.

COMPUTATIONAL MECHANICS (Abstracts)
WCCM VI in conjunction with APCOM'04, *Sept. 5-10, 2004, Beijing, China*

Finite Element Based Micromechanical Model of Ductile Intergranular Cracks in Columnar Two-Phase Systems

Helmut J. Böhm [1*], Dominik Duschlbauer [1], Thomas Drabek [1], Christian M. Chimani [2]

[1] *Christian Doppler Laboratory for Functionally Oriented Materials Design, Institute of Lightweight Design and Structural Biomechanics, Vienna University of Technology, Gußhausstr. 27-29, A-1040 Vienna, Austria*

[2] *VOEST–Alpine Industrieanlagenbau, Turmstr. 44, A–4031 Linz, Austria*

e-mail: hjb@ilsb.tuwien.ac.at

Abstract

A Finite Element based micromechanical approach is employed for studying the initiation and propagation of intergranular cracks in columnar austenite–ferrite two-phase systems. Such cracks, which can play an important role in the hot ductility behavior of certain steel grades [1], can occur when thin bands of ferrite form at the boundaries of austenitic grains as the second ductility minimum is traversed during cooling down. Because the initial yield stress of ferrite is less than half that of austenite, plastic strains tend to concentrate markedly in the ferritic regions. This can lead to ductile failure in the ferrite even at moderate overall strains, with the resulting cracks following the grain boundaries and showing dimpled faces.

This behavior is investigated by micromechanical models in which the microstructure is idealized as austenitic regions of regular or perturbed hexagonal shape that are surrounded by thin layers of ferrite that have a volume fraction of a few percent. The nucleation and growth of ductile cracks is described by an element elimination model that is triggered by a ductile damage indicator [2-3], the algorithm being implemented into the Finite Element code ABAQUS/Standard (HKS Inc., Pawtucket, RI). The marked mesh dependence that is typically found in local continuum models of this type is strongly reduced by employing a nonlocal averaging scheme.

The micromechanical analyses show that even though a ductile failure mechanism is active locally in the ferrite bands the overall response of the material under uniaxial tensile displays a very limited ductility because cracks can progress easily in the highly strained ferrite.

REFERENCES

[1] C. M. Chimani, K. Mörwald, *Micromechanical investigation of the hot ductility behavior of steel*, ISIJ Int., 39, (1999), 1194-1197.

[2] S. R. Gunawardena, S. Jansson, F. A. Leckie, *Transverse ductility of metal matrix composites*, in K. Haritos, G. Newaz, S. Mall eds. *Failure Mechanisms in High Temperature Composite Materials*, ASME, New York (1991).

[3] F. D. Fischer, O. Kolednik, G. X. Shan, F. G. Rammerstorfer, *A note on calibration of ductile failure damage indicators*, Int. J. Fract., 73, (1995), 345-357.

COMPUTATIONAL MECHANICS (Abstracts)
WCCM VI in conjunction with APCOM'04, *Sept. 5-10, 2004, Beijing, China*

Novel Model for Material Failure of Massive Structures

D. Brancherie [1,2*], A. Ibrahimbegovic [2], P. Villon [1], A. Rassineux [1], P. Breitkopf [1]

[1] *Département Génie des Systèmes Mécaniques, Université de Technologie de Compiègne, Centre de Recherche Royallieu, Compiègne, 60 205, France*

[2] *Secteur Génie Civil, LMT-Cachan, Ecole Normale Supérieure de Cachan, 61 avenue du Président Wilson, Cachan, 94 235, France*

e-mail: delphine.brancherie@ lmt.ens-cachan.fr, ai@lmt.ens-cachan.fr

Abstract

In order to achieve a better evaluation of the limit load of a complex structural system, it is needed to reach a reliable description [1] of the softening branch of the material behavior. Mesh dependency of the computed response when using classical continuum mechanics is, by now, well known and has led to the development of a number of remedies known as 'localization limiters '.

In this work, we aim at describing the behavior of massive structures until rupture by using localization limiters [2, 3] developed recently by following 'strong discontinuity approach' [1]. The vast majority of works consider only problems where dissi pation in the structure is only attributed to strain localization zones. Such an approach is sufficient only for thin structures but reveals to be insufficient for massive structures where diffuse dissipative mechanisms cannot be neglected. The main novelty of this work is to present a model capable of taking into account both the contribution of diffuse dissipative mechanisms (cf. development of micro-cracks) accompanied by strain hardening as well as localized damage modes with strain softening effects. T his is achieved by introducing displacement discontinuities coupled with a continuum damage mechanics model. The combination of the two types of dissipative mechanisms is taken into account by building adequate continuum model to describe diffuse dissipati on and an adequate localized or discrete model to deal with the dissipation taking place in localized zones. The continuum damage model is here considered as isotropic and constructed in the same way as plasticity models [2]. The discrete model associated to the displacement discontinuity is constructed as a multi-surface model. Such a construction involves anisotropy (e.g. unilateral crack opening). In this way, the approach can be seen as a new way to construct a predictive and robust anisotropic damage m odel. Several numerical examples with non-structured and adaptive meshing will be presented to illustrate a very satisfying performance of this model.

REFERENCES

[1] J. C. Simo, J. Oliver, F. Armero, *An analysis of strong discontinuity induced by strain softening solutions in rate-independent solids*, J. Comp. Mech., 12, (1993), 277-296.

[2] A. Ibrahimbegovic, D. Brancherie, *Combined hardening and softening constitutive model of plasticity: precursor to shear slip line failure*, Computational Mechanics, 33, (2003), 88-100.

[3] D. Brancherie, A. Ibrahimbegovic, *Fracture of massive structures: a novel approach to constructing an anisotropic damage combining continuum-type hardening and discrete-type softening models*, (submitted).

COMPUTATIONAL MECHANICS (Abstracts)
WCCM VI in conjunction with APCOM'04, *Sept. 5-10, 2004, Beijing, China*

A RC Beam Model Derived from a full Triaxial Coupled Damaging-Plastic Quasi-Isotropic Model for Concrete

M. Cuomo *, A. Nicolosi

Department of Civil and Environmental Engineering, University of Catania, viale A. Doria 6, Catania, 95125, Italy
e-mail: mcuomo@dica.unict.it, anicolo@dica.unict.it

Abstract

While for traditional concrete beams successful simplified constitutive models have been developed for analyzing their structural performances, the introduction of advanced mixtures, such as those employed in High Strength Concrete, and the need of accurate damage control for sensitive applications, require more accurate modeling of the non-linear, irreversible, behaviour of concrete.
A full triaxial coupled damaging-plastic model has been recently contributed by the authors [1], that accounts for three distinct dissipation modes related to fracture evolution, plastic crushing and volumetric hardening, coupled with damaging mechanisms. The model has shown to be quite successful in predicting the evolution of damage in complex stress paths.
The purpose of the present study is to develop a particularization of the model for a one-dimensional beam element, so that it can be used in the analysis of RC beam members, also in the presence of secondary reinforcements and/or confinement.
The case of slender beams with prevailing bending deformations is addressed, within the framework of the Euler-Kirchhoff beam theory. The formulation includes static and dynamic analyses, in a 3D setting, for a general beam cross section. The effect of confinement is accounted for. The model has been implemented in a numerical algorithm that allows evaluating the stress and strain distribution within the cross section as a function of the internal forces and of the load history, as well as the damage evolution along the beam.
Furthermore, it is possible to evaluate interaction domains for combined loading conditions, as a function of damage, together with the relevant ductility of the cross section and the peak values of the internal forces. It is stressed that the damage model used for concrete predicts the whole stress-strain curves also in the softening region, provided the limit linear state of the material, i.e., the stress level at which damage starts, is known.

REFERENCES

[1] M. Cuomo, L. Contrafatto, A. Nicolosi, *Elastic-plastic damaging model for concrete under multiaxial stress states,* XVI Congresso AIMETA di Meccanica Teorica ed Applicata, Ferrara, (2003).

COMPUTATIONAL MECHANICS (Abstracts)
WCCM VI in conjunction with APCOM'04, *Sept. 5-10, 2004, Beijing, China*

Numerical Simulation for Perforation Damage of Concrete Slab Impacted by Conventional Weapon Header

Guoqiang Deng*, Jun Dong, Xiumin Yang

Canbao Building Design Institute, No. 24 Taiping Road, Haidian District, Beijing, 100850, China
e-mail: hnjia@sina.com.cn

Abstract

Impacted by conventional weapon header, concrete target slab will produce various damage shapes, which closely depend on the slab thickness, concrete strength, weapon header velocity and so on. According to test results, all these damage shapes can be classified as penetration, impact spallation, penetration perforation and impact perforation (Fig.1). Generally, the damage process includes three parts that is front surface crating, middle penetrating and back surface spalling. The three parts are induced by different damage mechanism, which is very difficult for theoretical analysis. Experiment research is expensive with limited application range. Nowadays, numerical simulation method turns into a strong tool in solving damage effect of concrete slab under impact and explosive loading.

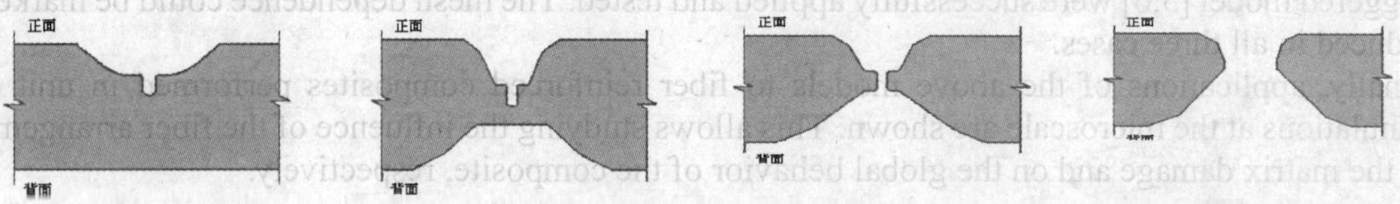

(a) impact crater (b) impact spall (c) penetration perforation (d) impact perforation

Fig.1 Damage characteristic shapes of concrete slab under projectile impacting

Concrete is constituted by solid grain, cement and air, and its mechanics behavior is complex which distinguishes from isotropy metal, so right material constitutive is the key in simulating the damage process of concrete slab under conventional weapon header impacting. That can describe the action of concrete under high pressure and high strain ratio. Concrete damage constitutive model (CDM), equation of status (EOS) and spallation rule are selected, which can well describe these damage phenomena, including micro cracks expand and interaction, spallation etc under impact loading. Based on concrete material damage model, impact perforation damage progress of reinforced concrete impacted by conventional weapon header is systemically simulated by employed finite element program Ls-dyna3d. The phenomena of catering, penetrating and spalling are clearly displayed (Fig.2), and physical mechanisms of concrete impact damage are revealed. The numerical simulated results agree well with that of experiment.

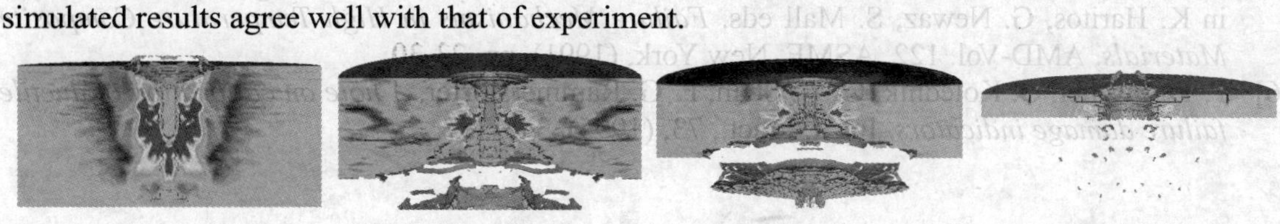

(a) impact crater (b) impact spall (c) penetration perforation (d) impact perforation

Fig.2 Results of numerical simulation

COMPUTATIONAL MECHANICS (Abstracts)
WCCM VI in conjunction with APCOM'04, *Sept. 5-10, 2004, Beijing, China*

Implementation of Nonlocal Damage Models in Commercial FE-Codes

T. Drabek*, H. J. Böhm

Christian Doppler Laboratory for Functionally Oriented Materials Design & Institute of Lightweight Design and Structural Biomechanics, Vienna University of Technology, Gußhausstraße 25-27, Vienna, A-1040, Austria
e-mail: drabek@ilsb.tuwien.ac.at, hjb@ilsb.tuwien.ac.at

Abstract

The simulation of the initiation and propagation of cracks in ductile materials requires appropriate models for describing the nucleation, growth and coalescence of voids in a smeared out manner. A number of models of this type can be found in the literature, but they typically have the drawback of showing a marked mesh dependence of the solution. This can be resolved by regularization techniques, e.g. nonlocal smoothing procedures [1] as used in the present contribution.
The feasibility of and strategy for implementing nonlocal ductile damage models into the commercial finite element code ABAQUS/Standard are shown. Nonlocal versions of the Gurson-Tvergaard-Needleman-model [2-3], Rousselier-model [4] and of a ductile damage indicator triggered model [5,6] were successfully applied and tested. The mesh dependence could be markedly reduced in all three cases.
Finally, applications of the above models to fiber reinforced composites performed in unit cell simulations at the microscale are shown. This allows studying the influence of the fiber arrangement to the matrix damage and on the global behavior of the composite, respectively.

REFERENCES

[1] V. Tvergaard, A. Needleman, *Nonlocal effects in ductile fracture by cavitation between larger voids,* in D. R. J. Owen, E. Oñate, E. Hinton eds. *Computational Plasticity: Fundamentals and Applications*, Pineridge Press, Swansea, UK, (1995), pp. 963-973.
[2] A. L. Gurson, *Continuum theory of ductile rupture by void nucleation and growth: part I – yield criteria and flow rules for porous ductile media,* J. Engng. Mater. Technol., 99, (1977), 2-15.
[3] A. Needleman, V. Tvergaard, *An analysis of ductile rupture modes at a crack tip,* J. Mech. Phys. Sol., 35, (1987), 151-183.
[4] G. Rousselier, *Ductile fracture models and their potential in local approach of fracture,* Nucl. Engng. Design, 105, (1987), 97-111.
[5] S. R. Gunawardena, S. Jansson, F. A. Leckie, *Transverse ductility of metal matrix composites,* in K. Haritos, G. Newaz, S. Mall eds. *Failure Mechanisms in High Temperature Composite Materials*, AMD-Vol. 122, ASME, New York, (1991), pp. 23-30.
[6] F. D. Fischer, O. Kolednik, G. X. Shan, F. G. Rammerstorfer, *A note on calibration of ductile failure damage indicators,* Int. J. Fract., 73, (1995), 345-357.

COMPUTATIONAL MECHANICS (Abstracts)
WCCM VI in conjunction with APCOM'04, *Sept. 5-10, 2004, Beijing, China*

Gradient Hardening and Damage in Crystal (Visco-) Plasticity

Magnus Ekh [1*], Thomas Svedberg [1], Kenneth Runesson [2]

[1] *Dept. of Applied Mechanics, Chalmers University of Technology, Gothenburg, Se 41296, Sweden*
[2] *Dept. of Structural Mechanics, Chalmers University of Technology, Gothenburg, Se 41296, Sweden*
e-mail: mekh@am.chalmers.se, keru@sm.chalmers.se

Abstract

In this paper a formulation of a model concept that, within the framework of continuum thermodynamics and finite strains, describes crystal (visco-) plasticity, crystal damage and non-local hardening of the gradient type. The model will in future be applied in simulations of duplex stainless steel (DSS) in order to predict how e.g. grain size, grain geometry and volume fraction of the two phases in DSS affect its fatigue properties. The crystal damage is based on the concept of fictitious (undamaged) configuration [1], and it is assumed to be driven by inelastic slip in each slip system. Furthermore, the non-local hardening, e.g. [2], in each slip system gives a contribution on each slip system which is added to the well established local hardening. The paper is concluded by a numerical study of how model parameters (such as grain size, damage rate) will influence the stress strain response of a simplified model of a polycrystal.

REFERENCES

[1] M. Ekh, R. Lillbacka, K. Runesson, *A model framework for anisotropic damage coupled to crystal (visco-) plasticity*, Int. J. Plasticity, (in press).

[2] T. Svedberg, K. Runesson, *A thermodynamically consistent theory of gradient-regularized plasticity coupled to damage*, Int. J. Plasticity, 13(6-7), 669-696, (1997).

COMPUTATIONAL MECHANICS (Abstracts)
WCCM VI in conjunction with APCOM'04, *Sept. 5-10, 2004, Beijing, China*

Derivation of Finite Strain Damage Model for Simulation of Low-Cycle Fatigue of Steel Structures

J. Korelc*, P. Skuber, D. Beg

Faculty of Civil and Geodetic Engineering, University of Ljubljana, p.p. 3422, Ljubljana, Slovenia
e-mail: {jkorelc, pskuber, dbeg}@fgg.uni-lj.si

Abstract

Low-cycle fatigue of constructural steel is closely related to the accumulation of plastic strain under cyclic loading and eventually leads to the formation of cracks or even failure of a structure. Low-cycle fatigue is very important in seismic design where for economical reasons localized damage is allowed and structure is subjected to cyclic loading. Usually very rough models are used to access damage accumulation e.g. accumulated plastic strain or accumulated plastic work. In the paper a new approach to the assessment of low-cycle fatigue based on damage mechanics is presented. A Pedersen-Tvegaard material model for tool materials, that combines kinematic and isotropic hardening with damage evolution is extended for finite strains. Owing to specific properties of constructural steel the material model had to be additionally modified. A three-dimensional finite element for large strain is developed based on the derived material model. Shear and volumetric locking is treated by Taylor expansion of shape functions. Formulation leads to the system of 28 nonlinear algebraic equations for each material (integration) point that have to be solved and consistently linearized. Symbolic manipulations, more precisely, automatic generator of numerical codes *AceGen*, has been used to that end. The generator is a *Mathematica* package able to resolve the mayor problem of symbolic generation of nonlinear finite elements arrays (e.g. stiffness matrix, etc.), which is an exponential growth of the size of derived expressions. An approach, implemented in *AceGen*, avoids this problem by combining: symbolic and algebraic capabilities of *Mathematica*, automatic differentiation technique, simultaneous optimization of expressions and theorem proving . The generator translates final symbolic formulas in a compiled language and incorporates the code into a nonlinear finite element analysis environment.
The 3D numerical analysis of low-cycle fatigue of the structural details for which test results are available is presented. The numerical analysis show good agreement with the experiments in the case of small plastic strains as well as in the case of local buckling at the locations of plastic hinges. The presented numerical model permits a life prediction of the structural details for arbitrary loading histories.

REFERENCES

[1] T. O. Pedersen, *Cyclic plasticity and low cycle fatigue in tool materials*, PhD Thesis, Technical University of Denmark, DCAMM Report S82, (1998).
[2] J. Korelc, *Multi-language and multi-environment generation of nonlinear finite element codes*, Engineering with Computers, 18, (2002), 312-327.

COMPUTATIONAL MECHANICS (Abstracts)
WCCM VI in conjunction with APCOM'04, *Sept. 5-10, 2004, Beijing, China*

Damage of a Reinforced Concrete Wall under Earthquake: a Robust Approach

S. Lignon *, L. Jezequel

Laboratoire de Tribologie et Dynamique des Systèmes, Ecole Centrale de Lyon, 36 avenue Guy de Collongues, 69134 Ecully Cedex, France
e-mail: sylvain.lignon@ec-lyon.fr, louis.jezequel@ec-lyon.fr

Abstract

Damage analysis of structures during earthquake is very complex, because of the diversity of structures and excitations. The study consists of developing a robust analysis of the damage, in order to allow the definition of a criterion determining if a damaged structure can be repaired or must be bulldozed.

The analysed structure is a heavily reinforced concrete shear wall, which is supposed to be representative of nuclear engineering structures. Previous experimental results have proved that only the first mode of such a structure is excited during an earthquake, that's why a simplified model has been developed based on this consideration [2]. This simplified model consists in a one degree of freedom mechanical system with modal characteristics changes related to damage increase. Moreover, for such a reinforced concrete structure, damage is assumed to be related to the maximum displacement at the top of the wall.

The robust approach is introduced in this model by assuming firstly that the excitation is uncertain, but centered on a nominal accelerogram depending on the considered site. Secondly, the stiffness of the structure itself is assumed to be uncertain: this uncertainty can be a consequence either of an erection defect, or of previous damage. Thirdly we combine the two previous kind of uncertainty. In all cases, we look for the temporal envelope of the response which allows an estimation of the maximum imaginable damage.

The major risk of such a robust analysis is an overestimation of the response, which can lead to an oversizing in the developmental stage or an abusive destruction of an existing structure. The goal is then to define the possible variations of the parameters the more accurately as possible. As these parameters are supposed to be contained in convex hull, we speak of convex analysis [1].

REFERENCES

[1] A. Baratta, I. Elishakoff, G. Zuccaro, M. Shinozuka, *A generalization of the Drenick-Shinozuka model for bounds on the seismic response of a single-degree-of-freedom system*, Earthquake Engng. Struct. Dyn., 27, (1998), 423-437.

[2] M. Brun, J. M. Reynouard, L. Jezequel, *A simple shear wall model taking into account stiffness degradation*, Engineering Structures, 25, (2003), 1-9.

COMPUTATIONAL MECHANICS (Abstracts)
WCCM VI in conjunction with APCOM'04, *Sept. 5-10, 2004, Beijing, China*

Crack Propagation Analysis in Elastic Structures Using a Continuum Damage Mechanics

Heraldo Costa Mattos [1], Stella Maris Pires Domingues [1*], Fernando Alves Rochinhar [2]

[1] *Universidade Federal Fluminense, Departamento de Engenharia Mecânica, Laboratório de Mecânica Teórica e Aplicada, Rua Passo da Pátria, 156, Niterói, 24210-240, Brasil*
[2] *Universidade Federal do Rio de Janeiro, Departamento de Engenharia Mecânica, EE-COPPE/ UFRJ, CP 68503, Rio de Janeiro, 21945-970, Brasil*
e-mail: stella@mec.uff.br, heraldo@mec.uff.br

Abstract

This paper discuss the possibility of structural failure prediction through a special kind of gradient-enhanced damage theory in which the material is considered to possess a substructure or microstructure. In these theories the microscopic movements are accounted for by the damage variable and a reformulation of the kinematics (to include the possible "micro-motions") and of some basic governing principles of the classical Continuum Mechanics is necessary. The theory allows an adequate description of the strain-softening and localization behaviors due to the material degradation, Costa-Mattos (1992) and (1995), Frémond (1996), Domingues (1996) and (1999), for example. First, a continuum damage model for quasi-brittle materials is presented. Then, a simple numerical technique, based on the finite element method, is proposed to approximate the solution of the resulting nonlinear mathematical problems. Within this framework, a numerically predicted macro-crack is the set of points in the structure where the damage variable has reached its critical value. One of the main features of such model is that it is valid for any kind of geometry and external loading. This procedure is applied to the analysis of macro-crack initiation and propagation in several structures and the results are compared with predictions of fracture Mechanics.

REFERENCES

[1] H. Costa-Mattos, R. Sampaio, *Analysis of the fracture of brittle elastic materials using a continuum damage model*, Structural Engineering and Mechanics, 3(5), (1995), 411-428.

[2] H. Costa Mattos, M. Fremond, E. N. Mamiya, *A simple model of the mechanical behavior of ceramic-like material*, Int. J. for Solids and Structures, Vol. 20, N° 24, pp. 3185-3200, 1992.

[3] S. M. P. Domingues, *Análise de materiais elásticos frágeis através de um modelo de dano contínuo,* Doctor Thesis, Pontifícia Univ. Católica do Rio de Janeiro - PUC-Rio, (1996), (in Portuguese).

[4] S. M. P. Domingues, H. Costa-Mattos, F. A. Rochinha, *Nonlinear elastic brittle problems: operator split methods in numerical solution*, Journal of Theorical and Applied Mechanics, Poland, 37(4), (1999), 847-861.

[5] M. Frémond, B. Nedjar, *Damage, gradient of damage and principle of virtual power*, Int. J. Solids Structures, 33(8), (1996), 1083-1103.

COMPUTATIONAL MECHANICS (Abstracts)
WCCM VI in conjunction with APCOM'04, *Sept. 5-10, 2004, Beijing, China*

Effects of Interphase Properties on the Mechanical Behavior of Trinary Polymer-Based Composites

Esmaeili K. Naghi *, Wei Lu, Yoshihiro Tomita

Graduate School of Science and Technology, Kobe University, 1-1, Rokkodai-cho, Nada, Kobe, 657-8501, Japan
e-mail: esmaeili@solid.mech.kobe-u.ac.jp, riku@solid.mech.kobe-u.ac.jp, tomita@mech.kobe-u.ac.jp

Abstract

A third phase, engineered elastic interph ase, is introduced between the particle and the matrix in the glass particle blended polymer to improve the mechanical behavior of the primary composite. The constitutive equations employed for matrix and particle are molecular chain network model for glassy polymer and the linear elasticity models for glass and interphase, respectively. A parametric study was performed to clarify the effects of the interphase elastic properties and the thickness on the localized plastic deformation and the maximum mean str ess in the matrix, normal stress along the particle surface, the volumetric strain of the interphase and the macroscopic response of the composite, under the macroscopic strain ratios from $\Gamma_1/\Gamma_2 = 0.0$ to $\Gamma_1/\Gamma_2 = 1.0$. Different types of shear bands are observed, depending on the macroscopic strain ratio and the mechanical and geometrical properties of the interphase. The shear band patterns that appear in the different stages of deformation are responsible for the macroscopic yielding of the unit cell and mean stress in the matrix. On the basis of the results of simulations, it is suggested that it is better for the stiffness of the interphase to be lower than that of the matrix, while Poisson's ratio of the interphase may be between that of the filler and the matrix. A rather thick interphase layer contributes to the reduction of the maximum mean stress in the matrix and maximum normal stress at the particle surface, but is also accompanied by a drop in the macroscopic yielding stress. The results obtained confirmed that a suitable combination of the interphase properties leads to an improvement in the mechanical characteristics of the blended polymer.

REFERENCES

[1] L. Gao, E. Mader, *Characterization of interphase nanoscale property variations in glass fibre reinforced polypropylene and epoxy resin composites*, Composites, 33, (2001), 559-576.
[2] S. D. Gardner, S. U. Pittman, R. M. Hackett, *Residual thermal stresses in filamentary polymer-matrix composites containing an elastomeric interphase*, J. Compos. Mater., 27, (1993), 830-859.
[3] Y. Tomita, S. Tanaka, *Prediction of deformation behavior of glassy polymers based on molecular chain network model*, Int. J. Solids Struct., 32, (1995), 3423-3434.
[4] J. Llorca, M. Elices, Y. Termonia, *Elastic properties of sphere-reinforced composites with a meshophase*, Acta Mater., 48, (2000), 4589-4597.

COMPUTATIONAL MECHANICS (Abstracts)
WCCM VI in conjunction with APCOM'04, *Sept. 5-10, 2004, Beijing, China*

A New Plastic Damage Model with Plastic Hardening for Concrete

Xinpu Shen[1], Lu Yang[1,2*], Fusheng Zhu[2]

[1] *College of Architectural Engineering, Shenyang University of Technology, 110023, Shenyang. China*
[2] *College of Mineral Resources and Civil Engineering, Northeastern University, 110006, Shenyang, China*
e-mail: xinpushen@vip.sina.com, yanglu515@163.com

Abstract

The aim of this investigation is to build a constitutive model that can simulate the history-dependent plastic damage behaviour of massive structural concrete under confinement. The model adopted here is constructed on the basis of the plasticity-based damage model reported by [1]. Although this model was proposed for the simulation of plastic damage phenomena of metals, it is still an attractive model for the purpose of this study because of the following advantages over the other available damage models: firstly, in this model, the damage evolution is closely connected to the increase of plastic strain, but it is also influenced explicitly by the elastic strain; secondly, damage evolution is coupled with increase of plastic strain; finally, this model is relatively easy to modify to make it suitable for the simulation of plastic damage phenomena of concrete, which is quite different from metals. Generalized Drucker-Prager criterion for plastic loading together with its plastic potential for non-associated plastic flow rule is referred here.

The characteristics of this study are: (1) A new plasticity-based damage model was proposed, and Drucker-Prager -type plasticity was adopted in the formulation of the constitutive equations. (2) Numerical validations were performed for the proposed model with a driver subroutine developed for validation of constitutive models in this study. Numerically predicted behaviour with the proposed model seems reasonably accurate for the uniaxial tension and uniaxial compression compared with the experimental data of references. (3) Numerical tests of multi-axial compression under various hydrostatic confinement stresses were carried out. Results indicated the validity of the proposed model for simulating plastic damage phenomena under various loading paths.

REFERENCES

[1] K. Saanouni, C. Forster, F.B. Hatira, *On the anelastic flow with damage*. Int. J. Dama. Mech., 3, (1994), 140-169.

COMPUTATIONAL MECHANICS (Abstracts)
WCCM VI in conjunction with APCOM'04, *Sept. 5-10, 2004, Beijing, China*

Parameter Identification of a Plastic Damage Model for Concrete

Xinpu Shen[1], Guoxiao Shen[1*], Lixin Chen[2]

[1] *College of Architectural Engineering, Shenyang University of Technology, 110023, Shenyang. China*
[2] *Engng Center for Coal-beded Methane, Northeastern Bureau of Coal Geology Shenyang, 110035, China*
e-mail: xinpushen@vip.sina.com, guoxiao_shen@yahoo.com, clx-holley@163.com

Abstract

Constitutive parameter identification at local level has been investigated by quite a few researchers [1] in the recent years. Studies on parameter identification of a plasticity-based damage constitutive model for concrete are presented in this paper. Differences between the stress-strain curves experimentally obtained and the ones numerically obtained at local level were used as the objective functional of a regularized least square method [2]. For solving the inverse problem, the solution of the direct problem at local level obtained by a driver subroutine developed by [3] was embedded in the iterative solution procedure for the inverse problem proposed in this paper. Sensitivity matrix was calculated numerically by finite difference method.
Numerical examples are given for fitting of a set of experimental results of stress-strain curves. These experimental curves were obtained by tri-axial compressions of standard cubic specimen under various values of hydrostatic pressures. Three parameters, among which two parameters are for describing scalar damage evolution and one parameter for triaxiality-dependent plastic hardening, were chosen as unknowns to be identified. Parameter identification under various loading cases which include uniaxial extension and tri-axial compression with different confining stresses were performed. Results indicate that the numerical scheme is stable and has reasonable accuracy.

REFERENCES

[1] T. Furukawa, T. Sugata, S. Yoshimura, M. Hoffman, *An automated system for simulation and parameter identification of inelastic constitutive models*. Comput. Methods Appl. Mech. Engrg, 191, (2002), 2235-2260.
[2] X. Shen, Z. Cen, B. Xu, *Numerical scheme for solving elasto-plastic identification problem of material properties in geotechnical engineering*. Chinese J Geotech Engng, 17, (1995), 66-71.
[3] X. Shen, L. Yang, F. Zhu, *A new plastic damage model with plastic hardening for concrete.* (submitted to WCCM-VI, Beijing, 2004).

COMPUTATIONAL MECHANICS (Abstracts)
WCCM VI in conjunction with APCOM'04, *Sept. 5-10, 2004, Beijing, China*

Numerical Simulation of Thermo-Damage-Stress Fields in Steel Structural Members under Thermal Loading

Masakazu Takagaki [1*], Yutaka Toi [1], Tai Asayama [2]

[1] *Institute of Industrial Science, University of Tokyo, 6-1 Komaba-4chome Meguro-ku, Tokyo, 153-8505, Japan*

[2] *Japan Nuclear Development Institute, Oarai Engineering Center, 4002 Narita, Oarai, Ibaraki, 311-1393, Japan*

e-mail: mtak@iis.u-tokyo.ac.jp, toi@ iis.u-tokyo.ac.jp and asayama@oec.jnc.go.jp

Abstract

Steel structural members are widely used in various loading environments. For example, they are exposed to high temperature environment in the heat treatment; therefore it is possible that damage is occurred in the structural member due to the heat loading. And in the case of high temperature equipments, the stiffness of them is deteriorated by thermal fatigue with the operation of equipments. Though these deteriorations can be evaluated experimentally in the case of simple members, it is evidently difficult to evaluate damage of actual structural members. Therefore, we have thought that the development of analysis methods for estimating damage is indispensable.

In the present study, the finite element method using the concept of continuum damage mechanics have been developed and the damage estimates of several problems have been conducted. Stiffness reduction due to damage is expressed by applying a scalar damage variable. This analysis method contains some modules to solve heat conduction, liquid metal embrittlement and electro-magnetic field in order to conduct the analyses of hot-dip galvanization, induction hardening and thermal fatigue. As for heat conduction analysis, it is necessary to consider the effect of damage because a distribution of the heat conductivity is influenced by the microdefects. Therefore, the thermal conductivity considering the effect of damage with a damage variable is used to calculate temperature distribution in the structures containing damages. Then we have confirmed the validity of the present analysis method based on the comparison of each result of experiments and analyses.

REFERENCES

[1] J. Lemaitre, J. L. Chaboche, *Mechanics of solid materials*, Cambridge University Press, (1990).

[2] J. Skrzypek, A. Ganczarski, *Modeling of damage effect on heat transfer in time-dependent non-homogeneous solids*, J. of Thermal Stresses, 21(3), (1998), 205-231.

[3] T. Inoue, S. Nagaki, T. Kishino, M. Monkawa, *Description of transformation kinetics, heat conduction and elastic-plastic stress in the course of quenching and tempering of some steels*, Ingenieur-Archiv, 50, (1981), 315-327.

[4] Y. Toi, K. Kobashi, T. Iezawa, *Finite element analysis of thermal elasto-plastic behaviours of bridge girders in hot-dip galvanization*, 53, 6, (1994), 1307-1316.

[5] Y. Toi, J.-M. Lee, *Thermal elasto-viscoplastic damage behavior of structural members in hot-dip galvanization*, Int. J. of Damage Mechanics, 11(2), (2002), 171-185.

COMPUTATIONAL MECHANICS (Abstracts)
WCCM VI in conjunction with APCOM'04, *Sept. 5-10, 2004, Beijing, China*

A Computational Model for Simulation of Fractured Materials Using Continuum-Based Discrete Element Method (CDEM)

Manhong Zhao [1*], Shihai Li [1], Yuannian Wang [1], J. G. Wang [2]

[1] *Institute of Mechanics, Chinese Academy of Science, Beijing 10080, China*
[2] *Tropical Marine Science Institute, National University of Singapore, 119260 Singapore*
e-mail: mh_zhao@21cn.com, shli@imech.ac.cn, deepen_eye@yahoo.com.cn, tmswjg@nus.edu.sg

Abstract

Based on the continuum-based discrete element method (CDEM), this paper presents a computational model to simulate fractured materials. The relationship of macroscopic failure of material and internal rupture strain strength is exploited. First, the CDEM treats the 8-node solid isoparametric element in FEM as a fundamental system composed of 8 particles and interconnecting springs. A complete process from continuum to discontinuum under loading can be simulated by CDEM. Second, a spring changes its properties if this spring is intersected by a fracture: 1, the shear and tension forces are equal to zero on fracture plane when the spring in normal direction of the fracture plane is in tension. 2, When the spring is in compression, the normal force in the fracture plane is the projection of spring force in normal direction, and the shear force in the fracture plane is the one in tangential direction. When the normal and shear forces satisfy the Mohr-Coulomb failure criterion, the spring fails and the spring stiffness should be revalued. Third, the stress to strain curve for fractured samples are tested under tension and compression. It is found that the material performs linear elasticity when fracture density is small and external load is low. However, nonlinearity occurs with the larger load. Furthermore, the 'hardening' and 'softening' property of fractured material is simulated under displacement loading test.

REFERENCES

[1] C. H. Dowding, T. B. Belytschko, H. J. Yen, *Dynamic computational analysis of opening in joint rock J of Geotechnical Engineering*, ASCE, 109(12), (1983), 1551-1556.
[2] L. Jing, J. A. *Hudson Numerical methods in rock mechanics, Int. J of Rock Mechanics and Mining Sciences*, 39(4), 2002, 409-427.

COMPUTATIONAL MECHANICS (Abstracts)
WCCM VI in conjunction with APCOM'04, *Sept. 5-10, 2004, Beijing, China*

Estimation of Shear Band Propagation Velocity in Metal Specimens and Failure Phenomenon under Impact Loading

Adam Glema[1*], Tomasz Lodygowski[1], Piotr Perzyna[2]

[1] *Inst. of Struct. Eng., Poznan University of Technology, ul. Piotrowo 5, 60-965 Poznan, Poland*
[2] *Inst. Fundamental Tech. Research, ul. Swietokrzyska 21, 00-049 Warszawa, Poland*
e-mail: Adam.Glema@put.poznan.pl, Tomasz.Lodygowski@put.poznan.pl, pperzyna@ippt.gov.pl

Abstract

The propagation of dynamic shear bands in inelastic solids generated by impact-loaded adiabatic processes is studied. The aim of this study is to estimate the velocity of shear band propagation in the specimen loaded by high impact [1]. The attention is focused on evolution of material properties (temperature and porosity dependence) and its influence on the velocity of plastic zones propagation. The phenomenon is associated with the foregoing failure mode — the shear band overtakes the material segregation and crack formation.
The motivation of computations comes from the experimental observations presented in Refs. [2, 3]. A photograph of an arrested shear band in the specimen represents the plastic strain localization, the width of shear band zone and the temperature evolution.
The elastic-viscoplastic material model with the strain rate dependence is applied within the thermodynamic framework. The internal state variables consisting of equivalent plastic deformation (thermomechanical couplings) and volume fraction porosity (microdamage mechanisms) are employed to represent the hardening/softening effects of dissipation and dispersion. The identification procedure for the material functions and parameters involved in the constitutive model is considered for a set of metal specimens tested in the experiments. First, we used the results from the twisting test of the thin tubular steel specimen [4]. Next, the numerical investigation concerns the plate tension tests [2] and pre-notched plate investigated in Ref. [3].
We take here advantage of our previous numerical investigation presented in Ref. [4], focusing now on the estimation of the shear band propagation velocity. The validation of material parameters is controlled by the comparison with experiments the shear band width, temperature rise and deformation and/or material point velocity until the final failure.

REFERENCES

[1] A. S. Bonnet-Lebouvier, A. Molinari, P. Lipiński, *Analysis of the dynamic propagation of adiabatic shear bands*, Int. J. Solids and Structures, 39(16), (2002), 4249-4269.
[2] A. Marchand, J. Duffy, *An experimental study of the formation process of adiabatic shear bands in a structural steel*, J. Mech. Phys. Solids, 36, (1988), 251-283.
[3] P. R. Guduru, A. J. Rosakis, G. Ravichandran, *Dynamic shear bands: an investigation using high speed optical and infrared diagnostic*, Mechanics of Materials, 33, (2001), 371-402.
[4] A. Glema, W. Kakol, T. Lodygowski, *Numerical modelling of adiabatic shear band formation in a twisting test*, Engineering Transactions, 45(3-4), (1997), 419-431.

COMPUTATIONAL MECHANICS (Abstracts)
WCCM VI in conjunction with APCOM'04, *Sept. 5-10, 2004, Beijing, China*

A Simplified Vertebrae Disk Model for Computations of Human Lumbar Spine Segment

Adam Glema [1], Tomasz Lodygowski [1*], Witold Kakol [1], Marcin Wierszycki [1], M. B. Ogurkowska [2]

[1] *Institute of Structural Engineering, Poznan University of Technology, Piotrowo 5, 60 -965 Poznan, Poland*

[2] *Department of Biomechanics, University of School of Physical Education, K rolowej Jadwigi 27/35, Poznan, Poland*

e-mail: {Adam.Glema, Tomasz.Lodygowski, Witold.Kakol, Marcin.Wierszycki}@put.poznan.pl, ogurkmal@man.poznan.pl

Abstract

The FEA modeling of a human lumbar spine s egment (L4-L5) is considered. The segment consists of two vertebral bodies, facet joints, disc, posterior elements and ligaments. The inter -vertebral disc and ligaments play the essential role in the motion of spine kinematics and in contribution to genera l stiffness of spine segment. The aim of the paper is to replace the nonlinear response of the disc by a connector type element which properties are obtained by computer simulations of the isolated segment. The simplified model can be used further in simulation of surgery, analyses of spinal equilibrium and stability. The geometry data of vertebrae is acquired from the Computer Tomography, the CAD model is generating using different techniques and then imported into FE pre-processor and meshed [1]. The mech anical properties of tissues, loading are recognized for the finite element method computations. The results concerning fields of deformation, strains, stresses and contact interactions are obtained and allow building a tangent stiffness matrix of the analyzed spine segment. How to overcome the difficulties in every step were discussed by the authors in detail [2]. The simplified model was validated for four loading schemes: axial compression, two bending in vertical plains, and torsion. The axial displacem ent and disc bulge for compression, and rotation for bending and an axial rotation for torsion were compared with the experimental data. Taking into consideration the complexity of the single segment the equivalent stiffness can be further simplify to mode l the whole spine. The detailed description of the approach will be presented during the conference.

REFERENCES

[1] M. Rychlik, M. Morzynski, M. Nowak, W. Stankiewicz, T. Łodygowski, M. B. Ogurkowska, *Acquisition and transformation for measurements data of biomedical objects to CAD systems*, CMM-2003 Conference on Computer Methods in Mechanics, 9th Int. Conference on Numerical Methods in Continuum Mechanics, Zilina, Slovak Republic, Sept., (2003).

[2] W. Kakol, T. Lodygowski, M. B. Ogurkowska, M. Wierszycki, *Are we able to support medical diagnosis or rehabilitation of human vertebra by numerical simulation?* CMM-2003 Conference on Computer Methods in Mechanics, Gliwice, Poland, June 3 -6, (2003).

COMPUTATIONAL MECHANICS (Abstracts)
WCCM VI in conjunction with APCOM'04, *Sept. 5-10, 2004, Beijing, China*

Interaction between a Circular Inclusion and a Nearby Moving Crack by BIEM

Jun Lei [1*], Yuesheng Wang [1], Dietmar Gross [2]

[1] *Institute of Engineering Mechanics, Beijing Jiaotong University, Beijing 100044, China*
[2] *Institute of Mechanics, TU Darmstadt, Hochschulstrasse 1, D-64289 Darmstadt, Germany*
e-mail: drleij@tom.com, yswang@center.njtu.edu.cn

Abstract

In this paper, the interaction between a circular and a nearby moving crack embedded in an elastic medium is studied by boundary integral equations method (BIEM).
To deal this problem, multi-region technique and two kinds of time-domain boundary integral equations (BIEs) are introduced. The system is divided into two parts along the interface between the inclusion and the matrix medium. Each part is linear, elastic, homogeneous and isotropic. For the crack surfaces, a non-hypersingular traction boundary integral equation is applied; for the interface, a traditional displacement boundary integral equation is used. In the numerical solution procedure, square root shape functions are adopted as to describe the proper asymptotic behavior in the vicinity of the crack-tips. The integration over time can be analytically computed via linear or constant temporal interpolation functions. The crack growth is modeled by adding new elements of constant length to the moving crack tip, which is controlled by the fracture criterion based on the maximum circumferential stress. In each time step, it is evaluated to determine the direction and the speed of the crack advance.
The numerical results of the crack growth path, speed, dynamic stress intensity factors (DSIFs) and dynamic interface tractions for various material combinations and geometries are presented. The effect of inclusion on the moving crack is discussed.

REFERENCES

[1] L. B. Freund, *Dynamic Fracture Mechanics*, Cambridge University Press, Cambridge, (1990).
[2] Th. Seelig, D. Gross, *Analysis of dynamic crack propagation using a time-domain boundary integral equation method*, Int. J. Solids Structures, 34, (1997), 2087-2103.
[3] J. Lei, Y. S. Wang, D. Gross, *Dynamic interaction between a sub-interface crack and the interface in a bi-material: time-domain BEM analysis*, Archive of Applied Mechanics, 73, (2003), 225-240.
[4] J. Lei, Y. S. Wang, D. Gross, *Time-domain BEM analysis of a rapidly growing crack in a bi-Material*, Int. J. Fract., (in press).

COMPUTATIONAL MECHANICS (Abstracts)
WCCM VI in conjunction with APCOM'04, *Sept. 5-10, 2004, Beijing, China*

A Pseudo-Incident Wave Technique for Treating Interacting Inhomogeneities in Advanced Composites

Xiaodong Wang *

Department of Mechanical Engineering, University of Alberta, Edmonton, AB, T6G 2G8, Canada
e-mail: xiaodong.wang@ualberta.ca

Abstract

Composite materials are receiving more and more attention from the research and engineering communities for their applications in the design of advanced structures. One of the fundamental issues regarding the effective use of these composites is the determination of their local or micro-scale property for design and failure avoidance. The static mechanical behaviour of fibre reinforced composites has been extensively investigated using either the continuum mechanics model or the discrete micromechanical approach [1]. In contrast, the dynamic behaviour of composite materials has received much less attention [2], especially when the coupling between mechanical field and electric field exists, as in the case of piezoelectric fibre-reinforced composites. It is important to note, however, that most fibre reinforced composites are sensitive to low velocity impact, which may occur accidentally during the service. In addition, composites are currently being considered for use in situations involving dynamic loading, such as active control of structural vibration, active control of structural acoustic transmission and reflection.

It is, therefore, the primary objective of the current article to provide a comprehensive micromechanical model for treating the plane elastodynamic interaction between arbitrarily located and oriented inhomogeneities (inclusions and/or microdefects) in advanced composite materials. The formulations governing the steady state interaction problem are greatly simplified by the use of fundamental solutions of individual inhomogeneities in conjunction with self-consistent Pseudo-incident wave method. According to this method, the original problem is decomposed into fundamental solutions involving simpler single inhomogeneity subproblems. These subproblems are summed up in a consistent manner to provide a system of linear algebraic equations. The resulting linear equations are solved to provide the dynamic field in the composites. The method is applied to both traditional fibre-reinforced composites and piezoelectric fibre-reinforced composites. Numerical examples are provided to show the effect of the location and size of cracks, inhomogeneities, piezoelectric fibres, and the loading frequency upon the dynamic local stress and electric fields.

REFERENCES

[1] M. Kachanov, *Elastic solids with many cracks: a simple method of analysis,* Int. J. Solids Structures, 23, (1987), 23-44.

[2] X. D. Wang, S. A. Meguid, *Diffraction of SH-wave by interacting matrix crack and an inhomogeneity,* J. Appl. Mech., 64, (1997), 568-575.

COMPUTATIONAL MECHANICS (Abstracts)
WCCM VI in conjunction with APCOM'04, *Sept. 5-10, 2004, Beijing, China*

Elasto-plastic Dynamic Analysis for Concrete Filled Steel Tubular Arch Bridges

Feng Xiong [1*], Haowu Liu [2]

[1] *Department of Civil Engineering, Sichuan University, Chengdu, 610065, China*
[2] *Department of Hydraulic Engineering, Sichuan University, Chengdu, 610065, China*
e-mail: fengxiong99@hotmail.com, haowuliu@hotmail.com

Abstract

Recently concrete filled steel tubular (CFST) arch bridges develop rapidly. There have been over one hundred bridges constructed just in China. A number of long span arch bridges are located in earthquake zones. To investigate the seismic behavior a shaking-table test of CFST arch bridge was carried out. The acceleration, displacement and strain responses are obtained. CFST arch bridges are generally flexible and have a little stiffness. Under earthquake loading, their nonlinear properties are very obvious even under weak earthquakes. So it is very necessary for CFST arch bridge design to apply nonlinear time-historical analysis.

To analyze the earthquake response of the model bridge, a three dimensional elastic-plastic model is developed in this paper, which considers concrete filled steel tube as a composite material. Firstly the yield equation of concrete filled steel tube material is formulated. It is a special curve surface including compress, bending and twist relation. Followed the elastic-plastic constitutive relation is founded based on elastic-plastic theory and elastic-plastic stiffness matrix is obtained. In analysis, the arch bridge is divided into three dimensional beam elements. The fiber model is employed for steel tube and core concrete in the section. When the structure is in elastic state, the elastic element stiffness matrix is combined from steel and concrete fiber by plane-section assumption to consider the composite effect of two materials. Finally the dynamic time-historical analysis of the model bridge is completed by using the model and the case analysis is studied. The calculation results have a good agreement with finds from shaking-table test. It shows that the model is appropriate to CFST arch bridges.

REFERENCES

[1] Graham H. Powell, Paul Fu-song Chen, *3D Beam-column Element with Generalized Plastic Hinges*, J. of Engineering Mechanics ASCE, 112(7), (1986).

[2] Hajjar J. F., *Concrete-Filled Steel Tube Columns under Earthquake Loads,* Progress in Structural Engineering and Materials, Vol. 2, No. 1, (2000), pp. 72-82.

[3] Ichinohe Y., Matsutani T. and Nakajima M., *Elasto-Plastic behavior of Concrete Filled Steel Circular Columns*, Proc. 3rd International Conf. On Steel-Concrete Composite Construction, Fukuoka, Japan (1991), pp.131-136.

COMPUTATIONAL MECHANICS (Abstracts)
WCCM VI in conjunction with APCOM'04, *Sept. 5-10, 2004, Beijing, China*

Numerical Simulation on Failure of Internally Pressured Cylinder Shell under Laser Irradiation

Yixia Yan *, Gang Chen, Sizhong Li, Yuze Chen

Institute of Southwest Structural Mechanics, Mianyang, 621900, China
e-mail: pu-yf@163.com, caep_chengang@sina.com, lisizhong@263.com

Abstract

The engine shell of the guide missile subjects to high-density internally pressure at normal active state. If it bears high-density laser irradiation, the temperature of the laser spot and adjacent materials will raise localized thermal soft. The shell localized large deformation results in entirely bursting during thermal and mechanical acting. It is important academic and engineering worth to research failure of internally pressured cylinder shell under laser irradiation. Many investigators have contributed to theoretical analyses, numerical simulation and experimental measurement on it [1-2]. Most of the numerical simulation studies can't describe the whole phenomena, including thermal soften, large deformation and bursting of cylinder shell. To research the failure phenomenon, the thermal-mechanical coupling FEM is used in the paper. Heat conduction analysis coupled with strain rate is applied and factors such as the physical properties of material changing with the temperature are considered. The effective plastic strain failure criterion is applied to modeling the failure of the shell. The FEM results describe the whole phenomena, including thermal soften, large deformation and bursting of cylinder shell. And the time of shell failure with different internally pressure are obtained. Figure 1 and Figure 2 is the cracked shell obtained from the experiment [2] and numerical simulation result of the paper, respectively. The figures show the burst feature of the numerical result tally with those of the experiment.

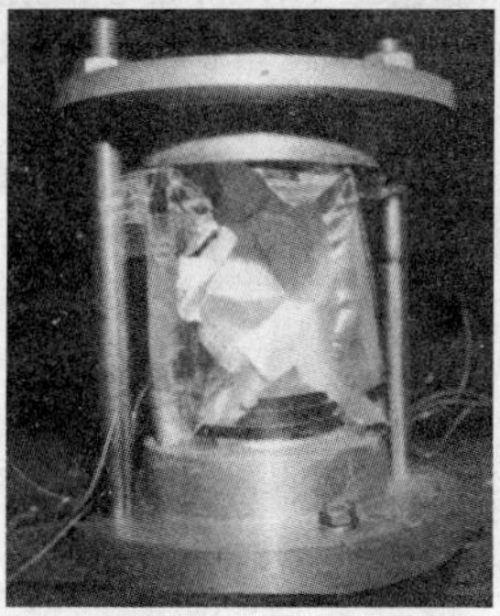

Fig. 1 The experiment result

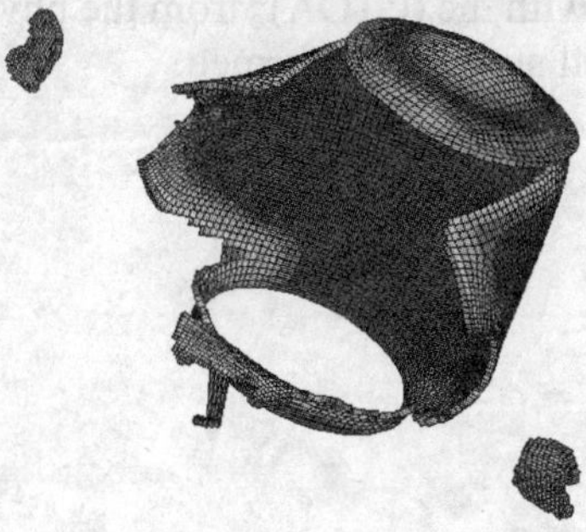

Fig. 2 The numerical result

REFERENCES

[1] Zheng Ji, Yonggang Liu, *A thermal failure numerical model of internal pressured cylindrical shell under laser irradiating*, High Power Laser and Particles Beams, 8(1), (1996), 99-104, (in Chinese).

[2] S. Li, Y. Chen, G. Zhang, Z. Li, *The failure of shells subjected to internal pressure and irradiated by CW CO_2 laser beam*, 25th AIAA Plasmadynamics and Lasers Conference, June 20-23, Colorado Springs, CO, USA, (1994).

COMPUTATIONAL MECHANICS (Abstracts)
WCCM VI in conjunction with APCOM'04, *Sept. 5-10, 2004, Beijing, China*

The Finite Element Simulation on Dynamic Crack in Curved Shells

Z. Zhuang [1*], X. B. Yang [1], J. Mamtimin [2], X. C. You [1]

[1] *Department of Engineering Mechanics, Tsinghua University, Beijing, 100084, China*
[2] *Mechanical Engineering School, Xinjing University, Wulumuqi, 830008, China*
e-mail: zhuangz@mail.tsinghua.edu.cn

Abstract

In cases of important shell structures where the load case or the environment is complicated, and where regular crack inspection fails to be implemented, as a result of which, the cracking process cannot be controlled effectively. A study of crack propagation and arrest obtains its significance in seeking out the mechanism of avoiding crack propagation, so as to set up a second defense line against disaster.
In the arrest study of fracture, previous work has focused on finding what the toughness values of materials should be to make shells have enough arrest capability. On the other hand, a numerical method for predicting the crack propagation route has been developed using moving finite element (FE) for dynamic crack in in-plane elements and FE with enhancement function for static crack in the shells, respectively. The shortcomings of these works are that they could not simulate dynamic crack propagating in the curved shells.
Against the background, the paper, relying on FE computation, tries to find the crack moving route and solve the arrest prediction on the shells. An enhancement function has been deduced in 4 nodes quadrilateral in-plane and shell elements, respectively, with reduced integration. An explicit program has been developed to simulate crack propagation in the curved shells and crack moving zigzag in in-plane elements. The process of the crack propagation with non-uniform speed is simulated. Geometry measurement of crack tip opening angle (CTOA) is obtained by using FE simulation. Compared with the $(CTOA)_c$ from the physical test, the geometric criterion to determine the fracture of steel shell structure is formed.

COMPUTATIONAL MECHANICS (Abstracts)
WCCM VI in conjunction with APCOM'04, *Sept. 5-10, 2004, Beijing, China*

Simulation of Texture Evolution in Polycrystalline Materials by Using Multi-Scale Crystalline Homogenization FEM

Yiping Chen [1*], Eiji Nakamachi [2], Shinji Iihoshi [2]

[1] *Department of Mechanics, Huazhong University of Science and Technology, Wuhan 430074, China*
[2] *Department of Mechanical Engineering, Osaka Institute of Technology, 5-16-1Omiya, Asahiku, 535-8585, Osaka, Japan*
e-mail: ypchen88@mail.hust.edu.cn, nakamati@med.oit.ac.jp

Abstract

A homogenization procedure is introduced to predict macro-continuum plastic properties by considering a micro-structural inhomogeneity characterized by using the crystalline morphological data. A "semi-implicit" elastic/crystalline viscoplastic finite element (FE) method based on a "crystallographic homogenization" theory is newly formulated for the multi-scale analysis. In the homogenization procedure, the asymptotic series expansion is introduced to define the displacement in the micro-continuum. This homogenization FE analysis is aimed at predicting the plastic deformation induced texture evolution of polycrystalline materials, the constituent micro-structure of which is represented by an assembly of single crystal grain. The rate dependent crystal plasticity model is used for the constitutive equation in the micro-structure. Its displacement is decomposed into two parts: The homogenized deformation defined in the macro-continuum and the perturbed one in the micro-continuum. This multi-scale analysis code makes it possible to carry out an alternative transition from a representative micro-structure to the macro-continuum. This homogenization procedure satisfies both the compatibility and the equilibrium in the micro-structure. In our formulation, a recapitulation of the crystalline constitutive equations for the single crystal grains is first presented and then followed by the derivation of governing equations for macro- and micro-continua, respectively. This FE code is applied to predict the texture evolution and the performance of the code is demonstrated by numerical examples of texture evolution of FCC sheet metals.

REFERENCES

[1] J. Guedes, N. Kikuchi, *Preprocessing and postprocessing for materials based on the homogenization method with adaptive finite element methods*, Computer Methods in Applied Mechanics and Engineering, 83, (1990), 145-198.

[2] C. Miehe, J. Schotte, M. Lambrecht, *Homogenization of inelastic solid materials at finite strains based on incremental minimization principle, application to the texture analysis of polycrystals*, Journal of the Mechanics and Physics of Solids, 50, (2002), 2123-2167.

[3] C. Miehe, J. Schroder, J. Schotte, *Computational homogenization analysis in finite plasticity, simulation of texture development in polycrystalline materials*, Comput. Meth. Appl. Mech. Eng., 171, (1999), 387-418.

[4] K. Terada, K. Yuge, N. Kikuchi, *Elasto-plastic analysis of composite materials by using the homogenization method (I): Formulation*, Journal of Japan Society for Mechanical Engineering (A), 61(590), (1995), 91-97.

COMPUTATIONAL MECHANICS (Abstracts)
WCCM VI in conjunction with APCOM'04, *Sept. 5-10, 2004, Beijing, China*

Field Representation of Atomic N-Body Dynamics

Youping Chen *, James Lee

School of Engineering and Applied Science, The George Washington University, Washington DC 20052, USA

Email: ypchen2@gwu.edu, jdlee@gwu.edu

Abstract

This paper aims to formulate a field theory that exactly represents classical N-body dynamics system and is able to work as an alternative to, but is computationally more efficient than, molecular dynamics simulation in studying statistical and finite temperature properties of materials. In this paper, multi-length and multi-time scale material behavior is addressed from the viewpoint of crystal dynamics. Atomistic definition and the corresponding field representations of fundamental physical quantities, such as stress, heat flux, internal energy, temperature, polarization, etc. are obtained and discussed. Time evolutions of conserved physical quantities under external fields are derived in terms of atomic variables, and are expressed in terms of field quantities. This then results in the mathematical representation of conservation laws of the new field theory. Nonlinear nonlocal, linear nonlocal, and linear local constitutive relations are derived from atomistic definitions of physical quantities. Strain measures are analytically obtained. They are homogeneous lattice strain and inhomogeneous internal atomic strains. All material parameters involved are obtained through the formulation. The correspondence between the multiscale field theory and the atomic-level MD simulations are analytically achieved. The potential applications of the field theory are discussed.

REFERENCES

[1] Y. Chen, J. D. Lee, A. Eskandarian, *Examining physical foundation of continuum theories from viewpoint of phonon dispersion relations*, International Journal of Engineering Science, 41, (2002), 61-83.

[2] Y. Chen, J. D. Lee, *Connecting molecular dynamics to micromorphic theory, part I: instantaneous mechanical variables*, Physica A, 322, (2003), 359-376.

[3] Y. Chen, J. D. Lee, *Connecting molecular dynamics to micromorphic theory, part II: balance laws*, Physica A, 322, (2003), 377-392.

[4] Y. Chen, J. D. Lee, A. Eskandarian, *Atomistic viewpoint of the applicability of microcontinuum theories*, International Journal of Solids and Structures, 41, (2004), 2085-2097.

[5] Y. Chen, J. D. Lee, A. Eskandarian, *Atomistic counterpart of micromorphic theory*, Acta Mechanica, 161, (2003), 81-102.

COMPUTATIONAL MECHANICS (Abstracts)
WCCM VI in conjunction with APCOM'04, *Sept. 5-10, 2004, Beijing, China*

Damage Percolation Simulation in Aluminum Alloy Sheet Forming

Zengtao Chen[1*], Michael J. Worswick[1], Stephen F. Corbin[1], A. Keith Pilkey[2], David J. Lloyd[3]

[1] *Department of Mechanical Engineering, University of Waterloo, 200 University Avenue West, Waterloo, N2L 3G1, Canada*

[2] *Department of Mechanical Engineering, Queen's University, 99UniversityAvenue,Ki ngston, K7L 3N6, Canada*

[3] *Kingston Research and Development Centre, Alcan International Limited, 945 Princess Street, Kingston, K7L 5L9, Canada*

e-mail: {ztchen, worswick}@lagavulin.uwaterloo.ca, scorbin@mecheng1.uwaterloo.ca, pilkey@me.queensu.ca, David.Lloyd@alcan.com

Abstract

A so-called damage percolation model [1] has been proposed by the authors in which ductile damage development within microstructures at the individual particle level is accommodated and vi sualized. This model has been recently further extended by means of a "loose coupling" with a damage -based finite element model so that ductile fracture in real sheet forming operations can be handled [2]. The coupled approach utilizes a Gurson -based yield surface [3] to account for the global softening effects of void damage, while the local damage development and void coalescence events are modeled using the damage percolation code.

In the current study, stretch flange forming, a typical sheet forming pr ocess of automotive aluminum AA5182 and AA5754 is modeled using the coupled approach. Deformation gradients of sheet forming operations are taken into account through the Gurson -based finite element calculation. A mesh area of interest (AOI) is determined based on the damage distribution determined by the Gurson-based finite element calculation. Microstructural features are characterized through a matrix erosion tessellation technique [4] and particle size and spatial distribution is mapped onto the AOI through the damage percolation code. Formability of stretch flanges is predicted through the onset of fracture within AOI using the coupled approach. Void areal fraction within the AOI is quantified using the coupled modeling and compared to the metallographi c results.

REFERENCES

[1] M. J. Worswick, Z. T. Chen, A. K. Pilkey, D. Lloyd, S. Court, *Damage characterization and damage percolation modeling in aluminum alloy sheet*, Acta Mater., 49, (2001), 2791-2803.

[2] Z. T. Chen, M. J. Worswick, N. Cinotti, A. K. Pilkey, D. Lloyd, *A linked FEM-damage percolation model of aluminum alloy sheet forming*, Int. J. Plasticity, 19, (2003), 2099-2120.

[3] A. L. Gurson, *Plastic flow and fracture behaviour of ductile materials incorporating void nucleation, growth and interaction*, PhD. Thesis, Brown University, Providence, R. I., (1975).

[4] A. K. Pilkey, *Effect of second-phase particle clustering on aluminum-silicon alloy sheet formability*, PhD. Thesis, Carleton University, Ottawa, Canada, (1997).

COMPUTATIONAL MECHANICS (Abstracts)
WCCM VI in conjunction with APCOM'04, *Sept. 5-10, 2004, Beijing, China*

Modelling of Contact-friction Interactions in Entangled Fibrous Materials

D. Durville *

LMSSMAT, Ecole Centrale Paris / CNRS UMR 8579, Grande Voie des Vignes, 92290 Chatenay-Malabry, France
e-mail: durville@mss.ecp.fr

Abstract

Contact-friction interactions between entangled fibres play a crucial part in structures or materials like ropes, textile fabrics, or glass wool. In order to simulate the mechanical behaviour of such structures, a special attention has to be put on the modelling of these contact-friction interactions. The proposed finite element model, considering each fibre as a 3D beam, is capable to handle large collections of fibers undergoing large displacements. The method is based on the construction of intermediate geometries in all regions where contact is likely to occur. Contact elements, constituted by pairs of material particles, are automatically generated from these geometries, providing a symmetrical treatment with respect to interacting beams.

Simulation results for samples of both textile fabrics and glass wool, will be given to show the efficiency of the model.

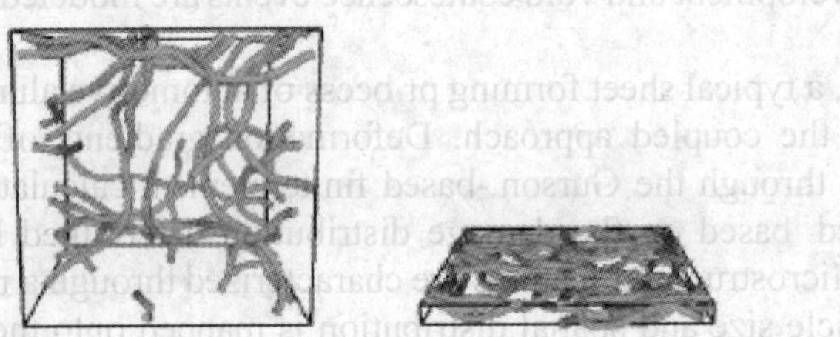

Fig. 1 Compression of a random glass wool sample

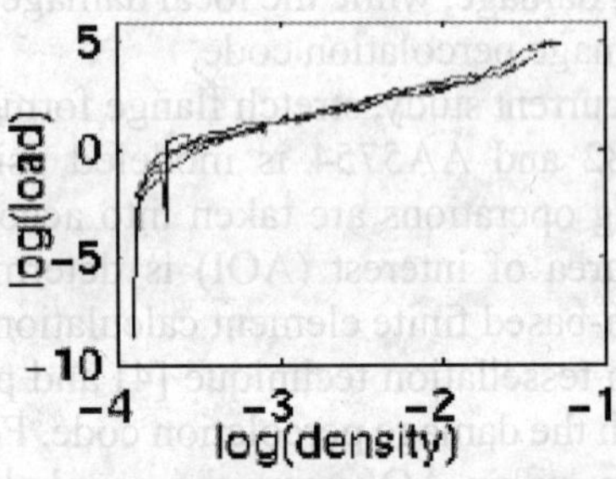

Fig. 2 Comparison of loading curves for four random glass wool samples

REFERENCES

[1] D. Durville, *Numerical simulation of the mechanical behaviour of textile structures at a mesoscopic scale*, in T. E. Onate, B. Kröplin eds. *Textile composites and inflatablestructures*, Barcelona, CIMNE, Barcelona, (2003), pp. 310-315.

[2] D. Durville, *Finite element modelling of a unit textile composite cell*, Fifth International Conference on Textile Composites, Leuven, Belgium, (2000).

[3] G. Zavarise, P. Wriggers, *Contact with friction between beams in 3-D space*, International Journal for Numerical Methods in Engineering, 49, (2000), 977-1006.

COMPUTATIONAL MECHANICS (Abstracts)
WCCM VI in conjunction with APCOM'04, *Sept. 5-10, 2004, Beijing, China*

Bauschinger Effects Analysis on Metals with High and Low Grain Boundaries under Tension and Compression

Hongbing Fang [1], Mark F. Horstemeyer [1, 2*], Xianwu Ling [1]

[1] *Center for Advanced Vehicular Systems, Mississippi State University, MS State, MS 39762, USA*
[2] *Department of Mechanical Engineering, Mississippi State University, MS State, MS 39762, USA*
e-mail: hfang@cavs.msstate.edu, mfhorst@me.msstate.edu, xling@cavs.msstate.edu

Abstract

Research shows that Bauschinger effects exist in nickels of single crystals, low angle grain boundary (GB), and high angle GB. Under simple shear deformation, Bauschinger effects were found to be the largest for the case of high angle GB and the lowest for the single crystal. In this study, atomistic simulations are performed on three types of materials, single crystal, low angle GB, and high angle GB nickels under tension and compression using the modified embedded atom method (MEAM). For each of the materials, we first apply a tensile load followed by a reverse, compressive load after yield in tension, all at a constant strain rate of 10^9 s^{-1}. The stress-strain relationships for the three types of materials are then compared with each other to examine the Bauschinger Effects. In order to examine the load sequence effect, simulations are also performed using a compressive load followed by a reverse, tensile load after yield, also at a constant strain rate of 10^9 s^{-1}. The Bauschinger effects are compared on the same material between the two load sequences for all the three types of materials. The samples of single crystal, low angle GB, and high angle GB have approximately nine thousand atoms, and are periodic *Y* and *Z* directions. The simulations are performed with the parallel program WARP running on SGI Origin 2000, which has a total of 64 195 MHz R10000 processors and 32 GB RAM. A single simulation of 200 ps loading takes approximately five hours with eight processors.

REFERENCES

[1] H. Fang, M. F. Horstemeyer, M. I. Baskes, K. Solanki, *Atomistic simulations of Bauschinger effects of metals with high angle and low angle grain boundaries*, Computer Methods in Applied Mechanics and Engineering, (in press).

[2] M. F. Horstemeyer, *Damage influence on Bauschinger effect of a CAST A356 aluminum alloy*, Scripta Materialia, 39, (1998), 1491-1495.

[3] M. F. Horstemeyer, M. I. Baskes, S. J. Plimpton, *Computational nanoscale plasticity simulations using embedded atom potentials*, Theoretical and Applied Fracture Mechanics, 37, (2001), 49-98.

COMPUTATIONAL MECHANICS (Abstracts)
WCCM VI in conjunction with APCOM'04, *Sept. 5-10, 2004, Beijing, China*

Atomistic Simulation of Crack Propagation and Its Size-Dependent Behavior

Yafang Guo [*], Yuesheng Wang

Institute of Engineering Mechanics, Northern Jiao Tong University, Beijing, 100044, China
e-mail: guoyafang@btamail.net.cn, yswang@center.njtu.edu.cn

Abstract

The atomistic simulation is an effective method to study the atomic-scale processes of fracture, and the boundary conditions of atomic region are usually given by the continuum elastic field of crack according to continuum mechanics [1-2]. However, even the biggest simulation system that we can obtain, which contains about thousands millions of atoms [2], it is still too small comparing with the size of real materials. Thus the singularity of stress field around crack tip and the size-dependent behavior of atomistic simulation should be considered.

In this paper the displacement and stress distribution at crack tip in bcc-iron are analyzed by using both discrete atomistic simulation method and linear elastic continuum mechanics [3]. The results indicate that the singularity of stress field at crack tip no longer exists when discrete atomistic simulation method is applied. The radius of the nonlinear region at the crack tip region is about 100Å, and the truncated distance of continuum elastic field is about 100Å away from the crack tip. The further investigations of the energy and atomic structure indicate that the elastic field of continuum mechanics cannot affect the atomic processes at crack tip effectively through the common region when the size of atomistic region is less than 300Å. We can assume the range of the system size that can bridge continuum with discrete region is at the scale of about 300Å.

Based on these, the behavior of crack propagation at different strain rate and temperature in bcc-iron is studied by use of the combined atomistic-continuum method. The results show that at low temperature, the cleavage fracture occurs and the twin formation is cooperative process in brittle fracture. At high temperature, the twinning becomes weakened and vanishes at the brittle-to-ductile transition temperature (BDTT) accompanied dislocation nucleation, which is perpendicular to the crack surface. The BDTT we have obtained is about 250K, which is consistent with the experimental result [4].

REFERENCES

[1] B. DeCelis, A. S. Argon, S. Yip, *Molecular dynamics simulation of crack tip processes in alpha-iron and copper*, J. Appl. Phys., 54, (1983), 4864-4878.

[2] F. F. Abraham, D. Schneider, B. Land, D. Lifka, J. Skovira, J. Gerner, M. Rosenkrantz, *Instability dynamics in three-dimensional fracture: an atomistic simulation*, J Mech Phys Solids., 45, (1997), 1461-1471.

[3] G. C. Sih, H. Liebowitz ed. *Methematical Fundamentals, in Fracture: An Advanced Treatise*, Vol. 2, Academic, New York, USA, (1968).

[4] F. A. McClintock, A. S. Argon, *Mechanical Behavior of Materials*, charpter 17, Reading, Addison-Wesley, Massachusetts, USA, (1966).

COMPUTATIONAL MECHANICS (Abstracts)
WCCM VI in conjunction with APCOM'04, *Sept. 5-10, 2004, Beijing, China*

Multi-Scale Modeling and Simulation for Computational Alloys Design and a Quasi-Particle Dynamics Approach

S. Hao [1, 2*], B. Moran [1], G. B. Olson [2]

[1] *Department of Mechanical Engineering, Northwestern University, Evanston, IL 60208, USA*
[2] *Department of Materials Science and Engineering,, Northwestern University, Evanston, IL 60208, USA*
e-mail: suhao@northwestern.edu, b-moron@northwestern.edu, olson@elmo.tech.nwu.edu

Abstract

In the computational design of high toughness steels as hierarchically structured materials, a multiscale, multiphysics methodology is developed for a "ductile fracture simulator". The key-issue is to establish the relationships between up-atomic mechanisms and sub-atomic physics. At the nanometer scale, the method unites continuum mechanics with quantum physics, using first-principles calculations to predict the force-distance and momentum-angular rotation laws for the adhesions on interface and within bulk materials, which has been applied in the cell models for the inclusions at two length scales. A "Particle Dynamics"(PD) approach has been proposed that integrates molecular dynamics with quantum mechanics computation by representing an atomic system as a particle system through lumping fixed number of atoms into a super-atom, which we call a "quasi-particle", while preserving the essential properties of the atomic system via a proposed "equivalent stiffness rule". This rule requires that the particle system has the same crystal structure and stiffness as the original atomic system but with a larger inter-particle spacing that is determined from the scale of interest, see Fig. 1. The original physics is preserved through transforming the inter-atomic potential into an inter-particle potential by ensuring the same elastic constants for both systems. At macroscopic scale, a hierarchical constitutive model has been developed by combining conventional plasticity, microstructural damage, strain gradient effects and transformation induced plasticity. For steels, a Toughness-Strength-Adhesion diagram is obtained. The proposed concept of creative alloys design can be stated as the exploration of the effective connections among the five key-components: elements selection, process design, micro/nanostructure optimization, desirable properties and industrial performance by virtue of innovations and inventions.

REFERENCES

[1] G. B. Olson, *Computational design of hierarchically structured materials*, Science, 277(5330), (1997), pp. 1237-1242.

[2] A. J. Freeman, *Materials by design and the exciting role of quantum computation/simulation*, Journal of Computational and Applied Mathematics, 149(1), (2002), pp. 27-56.

[3] S. Hao, W. Liu, B. Moran, G. Olson, *Multi-scale constitutive model and computational framework for the design of ultra-high strength, high toughness steels*, Computer Methods for Applied Mechanics and Engineering, 193(17-20), (2004), 1865-1908.

[4] S. Hao, B. Moran, W. Liu, G. Olson, *A hierarchical multi-physics model for design of high toughness steels*, J. Computer-Aided Materials Design, (in press).

COMPUTATIONAL MECHANICS (Abstracts)
WCCM VI in conjunction with APCOM'04, *Sept. 5-10, 2004, Beijing, China*

Multi-Scale Modeling for Corroded Reinforced Concrete and Soil Foundation

Tetsuya Ishida, Ken-Ichiro Nakarai, Kukrit Toongoenthong, Koichi Maekawa*

Department of Civil Engineering, The University of Tokyo, Hongo, Bunkyo-ku, Tokyo 113, Japan
e-mail: {ishi, nakarai, kukurit1, maekawa}@concrete.t.u-tokyo.ac.jp

Abstract

Multi-scale modeling of structural concrete performance is presented as a systematic knowledge base of coupled cementitious composites, geo-technical materials and structural mechanics in a unified manner. The multi-scale modeling of micro-pores of cementitious composite like concrete and cemented soil takes up a primary scheme to simulate aging and varying quality of materials and structures [1-2]. This framework is thought to be extensible to soil foundation with larger scales of voids among soil grain particles. Here, we can treat ion dissolution and transport in between cementitious solids and soil foundation for environmental assessment. This multi-scale approach is also integrated with structural mechanics of crack nonlinearity [2-3] and functionality of the damaged reinforced concrete can be assessed with cracking caused by volumetric expansion of corroded steel (see *Fig. 1*). This study shows possibility of an enlarged scheme of multi-scale-physical and multi-layered mechanical modeling of non-organic cementitious composites and soil foundation, and presents some engineering application of performance and environmental assessment. This is a trial to couple material oriented knowledge with structural mechanics.

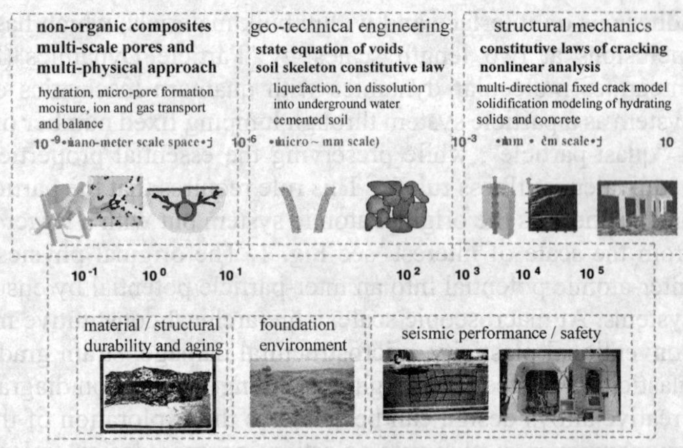

Fig.1 Multi-scale micro-pore and interlinked meso-scale voids.

REFERENCES

[1] K. Maekawa, R. P. Chaube, T. Kishi, *Moeling of concrete performance,* Spon Press, London, (1999).

[2] K. Maekawa, T. Ishida, T. Kishi, *Multi-scale modeling of concrete performance – integrated material and structural mechanics* -, Advanced Concrete Technology, (http://www.j-act.org), Vol.2, JCI, (2003).

[3] K. Maekawa, A. Pimanmas, H. Okamura, *Nonlinear mechanics of reinforced concrete*, Spon Press, London, (2003).

COMPUTATIONAL MECHANICS (Abstracts)
WCCM VI in conjunction with APCOM'04, *Sept. 5-10, 2004, Beijing, China*

Atomic-*S*cale Finite Element Method in Multiscale Computation

B. Liu [1*], Y. Huang [1], H. Jiang [1], S. Qu [1], K. C. Hwang [2]

[1] *Department of Mechanical and Industrial Engineering, Univ. of Illinois at Urbana- Champaign, 1206 West Green Street, Urbana, IL 61801, USA*

[2] *Department of Engineering Mechanics, Tsinghua University, Beijing 100084, China*

e-mail: {binliu2, huang9, hqjiang, squ}@uiuc.edu, huangkz@mail.tsinghua.edu.cn

Abstract

The multiscale simulation is important to the development of nano-technology and to the study of materials and systems across multiple length scales. In order to develop an efficient and accurate multiscale computation method within a unified theoretical framework, we propose an order-*N* atomic-scale finite element method (AFEM). It is as accurate as molecular mechanics simulations, but is much faster than the widely used order-N^2 conjugate gradient method. The combination of AFEM and continuum finite element method provides a seamless multiscale computation method suitable for large scale problems.

REFERENCES

[1] E. B. Tadmor, M. Ortiz, and R. Phillips, *Quasicontinuum analysis of defects in solids*, Philosophical Magazine A-Physics of Condensed Matter Structure Defects and Mechanical Properties, 73(6), (1996), 1529-1563.

[2] R. E. Rudd and J. Q. Broughton, *Concurrent coupling of length scales in solid state systems*, Physica Status Solidi B-Basic Research, 217(1), (2000), 251-291.

[3] G. J. Wagner and W. K. Liu, *Coupling of atomistic and continuum simulations using bridging scale decomposition*, Journal of Computational Pysics, 190, (2003), 249-274.

[4] W. H. Press B. P. Flannery, S. A. Teukolsky, and W. T. Vetterling, *Numerical recipes*, Cambridge Univ. Press, New York, (1986).

COMPUTATIONAL MECHANICS (Abstracts)
WCCM VI in conjunction with APCOM'04, *Sept. 5-10, 2004, Beijing, China*

Finite Element Analysis of Hyperelastic Materials with MSC.Marc

Changhong Liu*, Sanjay Choudhry

MSC.Software Corporation, 500 Arguello Street, Suite 200, Redwood City, CA 94063, USA
E-mail: changhong.liu@mscsoftware.com, sanjay.choudhry@mscsoftware.com

Abstract

Hyperelastic materials are characterized by properly defined strain energy functions, formulated based on either the strain invariants or the principal stretches of deformation gradient. Typical examples of hyperelastic materials include foam and rubber materials. A framework has been carefully established in the general-purpose finite element program MSC.Marc so that a generalized hyperelastic material model can be easily implemented by simply defining the derivatives of the energy function with respect to the strain invariants or the principal stretches. The effects of viscoelastic damage and thermal expansion have been taken into account in the framework. This paper is to address the issues of implementation of the framework.

For compressible foam materials, four types of generalized strain energy functions are supported within the framework: 1) Invariants-based model, 2) Principal-stretches-based model, 3) Invariants-based model with volumetric and deviatoric split, 4) Principal-stretches-based model with volumetric and deviatoric split.

MSC.Marc uses conventional displacement elements for compressible foam models. No volumetric constraint is considered.

Because rubber-like materials are nearly incompressible, it is numerically more efficient to split the energy function into a volumetric part and a deviatoric part. A three-field mixed formulation, which treats hydrostatic pressure and the dilatation as independent variables, has been used in MSC.Marc to overcome the numerical difficulties coming from the volumetric constraints [1-2]. For rubber, two types of generalized strain energy functions are supported in the framework: 1) Invariants-based model with volumetric and deviatoric split, 2) Principal-stretches-based model with volumetric and deviatoric split.

Examples of using this framework through a user subroutine and Applications will be presented.

REFERENCES

[1] C. H. Liu, G. Hofstetter, H. A. Mang, *3D finite element analysis of rubber-like materials at finite strains*, Engineering Computations, 11, (1992), 111-138.

[2] J. C. Simo, R. L. Taylor, *Quasi-incompressible finite elasticity in principal stretches, continuum basis and numerical algorithms*, Computer Method in Applied Mechanics and Engineering, 85, (1991), 272-310.

COMPUTATIONAL MECHANICS (Abstracts)
WCCM VI in conjunction with APCOM'04, *Sept. 5-10, 2004, Beijing, China*

Computational Homogenization of Microheterogeneous Materials Including Delamination at Finite Strains

Stefan Löhnert *, Peter Wriggers

Institute of Mechanics and Computational Mechanics, University of Hanover, Appelstr. 9A, 30167 Hannover, Germany
e-mail: loehnert@ibnm.uni-hannover.de, wriggers@ibnm.uni-hannover.de

Abstract

During the last years, the development and application of new composite materials gained more and more importance. For engineering applications it is necessary to get effective material properties of such materials. In this paper we present aspects of computational homogenization procedures of microheterogeneous materials which can show decohesion in a cohesive zone around the inclusions. Applications to this are e.g. polymer coatings stiffened with sand or the development of other modern microheterogeneous materials like aluminum mixed with boron particles. Due to the delamination process which can often be observed in a cohesive zone between matrix and particle material, we get finite deformations and finite strains within the representative volume element (RVE). The geometrical and material nonlinearities as well as the immense numerical effort cause the main difficulties. The homogenization procedure leads to an effective stress strain curve for the RVE. It is necessary to do statistical tests in order to get a representative result. Here we set a special focus on the adaptive numerical model, the statistical testing procedure and the different boundary conditions applied on the RVE. In detail we look at pure displacement, pure traction and natural boundary conditions. A comparison of the results with these boundary conditions is shown.

REFERENCES

[1] J. F. W. Bishop, R. Hill, *A theory of the plastic distortion of a polycrystalline aggregate under combined stresses*, Phil. Mag., 42, (1951), 414-427.
[2] J. F. W. Bishop, R. Hill, *A theoretical derivation of the plastic properties of a polycrystalline face-centered metal.* Phil. Mag., 42, (1951), 1298-1307.
[3] R. Hill, *The elastic behavior of a crystalline aggregate,* Proc. Phys. Soc. (Lond.), A65, (1952), 349-354.
[4] A. Needleman, et al, *Matrix reinforcement, and interfacial failure,* in S. Suresh, A. Mortensen, A. Needleman eds. *Fundamentals of metal matrix composites*, Butterworth-Heineman Publishers, (1993).
[5] T. Zohdi, P. Wriggers, *Computational micro-macro material testing,* Archives of Computational Methods in Engineering, 8(2), (2001), 131-228.

COMPUTATIONAL MECHANICS (Abstracts)
WCCM VI in conjunction with APCOM'04, *Sept. 5-10, 2004, Beijing, China*

3D Finite Volume Scheme for Czochralski Crystal Growth

Jing Lu, Zibing Zhang, Qisheng Chen *

Institute of Mechanics, Chinese Academy of Sciences, 15 Beisihuanxi Road, Beijing, 100080, China
e-mail: qschen@imech.ac.cn

Abstract

Czochralski (Cz) technique, which is used for growing single crystals, has dominated the production of single crystals for electronic applications. The Cz growth process involves multiple phases, moving interface and three-dimensional behavior. Much has been done to study these phenomena by means of numerical methods as well as experimental observations. A three-dimensional curvilinear finite volume based algorithm has been developed to model the Cz process. A body-fitted transformation based approach is adopted in conjunction with a multizone adaptive grid generation (MAGG) technique to accurately handle the three-dimensional problems of phase-change in irregular geometries with free and moving surfaces. The multizone adaptive model is used to perform a three-dimensional simulation of the Cz growth of silicon single crystals.

Since the phase change interface are irregular in shape and they move in response to the solution, accurate treatment of these interfaces is important from numerical accuracy point of view. The multizone adaptive grid generation (MAGG) is the appropriate scheme for this purpose. Another challenge encountered is the moving and periodic boundary conditions, which is essential to the numerical solution of the governing equations. Special treatments are implemented to impose the periodic boundary condition in a particular direction and to determine the internal boundary position and shape varying with the combination of ambient physicochemical transport process and interfacial dynamics. As indicated above that the applications and processes characterized by multi-phase, moving interfaces and irregular shape render the associated physical phenomena three-dimensional and unsteady. Therefore a generalized 3D model rather than a 2D simulation, in which the governing equations are solved in a general non-orthogonal coordinate system, is constructed to describe and capture the features of the growth process.

All this has been implemented and validated by using it to model the low pressure Cz growth of silicon. Accuracy of this scheme is demonstrated by agreement of simulation data with available experimental data. Using the quasi-steady state approximation, it is shown that the flow and temperature fields in the melt under certain operating conditions become asymmetric and unsteady even in the absence of extrinsic sources of asymmetry. Asymmetry in the flow and temperature fields, caused by high shear initiated phenomena, affects the interface shape in the azimuthal direction thus results in the thermal stress distribution in the vicinity, which has serious implications from crystal quality point of view.

REFERENCES

[1] V. Prasad, H. Zhang, A. P. Anselmo. *Transport phenomena in Czochralski crystal growth*, in J. P. Harnett, Thomas F. Irvine Jr. eds. *Advances in heat transfer*, vol. 30, (1997).

[2] H. Zhang, V. Prasad, M. K. Moallemi, *A numerical algorithm using multizone adaptive grid generation for multiphase transport process with moving and free boundaries*, Num. Heat Transfer, Part B, 29, (1996), 339-421.

COMPUTATIONAL MECHANICS (Abstracts)
WCCM VI in conjunction with APCOM'04, *Sept. 5-10, 2004, Beijing, China*

A Variationally Consistent Multi-scale Modeling of Stressed Grain Growth

S. Mehraeen*, J. S. Chen

Department of Civil & Environmental Engineering, University of California, Los Angeles, Los Angeles, CA 90095, USA
E-mail: shafigh@seas.ucla.edu, jschen@seas.ucla.edu

Abstract

Understanding how the macroscopic external load affects the meso-scale grain growth process and how the grain structure influences the macro-scale property of the continuum represents an interesting multi-scale problem. The material velocity and grain boundary velocity are the two primary multi-scale variables involved in grain growth process; each plays different but complementary role in the evolution of grain growth. The material velocity represents the degree of grain deformation in response to the macroscopic external load. Due to the heterogeneity of the grain structure, the material velocity exhibits both coarse and fine scale characteristics. Consequently, the strain rate, stress, and strain energy density in each grain present multi-scale properties. The multi-scale behavior of the grain boundary velocity, on the other hand, is dictated by the multi-scale driving force motivated by the multi-scale strain energy density jump across the grain boundaries.
This paper presents a multi-scale mathematical and computational framework for analyzing the global and local behavior of stressed grain growth. This is achieved by introducing an asymptotic expansion of field variables into the variational equation based on the principle of virtual power. The expanded variational equation gives rise to multi-scale Euler equations describing evolution processes at different scales, the scale coupling relation, as well as the homogenized material properties. The multi-scale characteristics of grain boundary migration triggered by the multi-scale property of strain energy density jump across the grain boundaries are also discussed. A stressed grain growth example is analyzed to demonstrate multi-scale behavior of grain deformation and grain structure evolution under mechanical loading.

REFERENCES

[1] J. S. Chen, V. Kotta, H. Lu, D. Wang, D. Moldovan, D. Wolf, *A variational formulation and a double-grid method for meso-scale modeling of stressed grain growth in polycrystalline materials,* Computer Methods in Applied Mechanics and Engineering, 193, (2004), 1277-1303.

[2] J. S. Chen, and S. Mehraeen, *Variationally consistent multiscale modeling and homogenization of stressed grain growth,* Computer Methods in Applied mechanics and Engineering, (in press).

COMPUTATIONAL MECHANICS (Abstracts)
WCCM VI in conjunction with APCOM'04, *Sept. 5-10, 2004, Beijing, China*

Modeling of Coupled Self-diffusion and Mechanics in Polycrystalline Solids

Hashem M. Mourad, Krishna Garikipati *

Department of Mechanical Engineering, University of Michigan, 2350 Hayward St., Ann Arbor, Michigan 48109-2125, USA
e-mail: hmourad@umich.edu, krishna@umich.edu

Abstract

We present recent advances in the modeling of coupled self-diffusion and mechanics in polycrystalline solids. This work is based upon a coupled, atomically-based continuum formulation [1] that was applied to problems such as diffusional creep, hillock formation and electromigration. The formulation was extended to consider void formation [2]. In this paper we discuss models for the evolution of grain boundaries, voids and free surfaces. The treatment of interfaces and free surfaces is based on the Level Set Method. It has been enhanced with velocity projection techniques, re-initializing algorithms and stabilization methods. The latter allow the application of the finite element method to the level set equation. Since the finite element method is already the method of choice for diffusion and mechanics, this affords a particularly straightforward implementation of the coupled problem. A notable feature is that the fully coupled physics of self-diffusion and mechanics is accounted for in the presence of evolving interfaces. The nonlinearities inherent in the problem due to coupling will be highlighted. Additionally, the staggered algorithms used to solve the coupled equations and finite element methods to accurately and robustly treat interfaces will be discussed.

REFERENCES

[1] K. Garikipati, L. Bassman, M. Deal, *A lattice-based micromechanical continuum formulation for stress-driven mass transport in polycrystalline solids*, J. Mech. Phys. Solids, 49, (2001), 1209-1237.

[2] K. Garikipati, H. Mourad, *The Gibbs free energy density at grain boundaries of stressed polycrystals*, (submitted).

COMPUTATIONAL MECHANICS (Abstracts)
WCCM VI in conjunction with APCOM'04, *Sept. 5-10, 2004, Beijing, China*

Homogenization of Real-World Balanced Woven Composites with Imperfect Microstructure

Michal Šejnoha *, Jan Zeman

Department of Structural Mechanics, Faculty of Civil Engineering, Czech Technical University in Prague, Thákurova 7, Prague 6, 166 28, Czech Republic
e-mail: sejnom@fsv.cvut.cz, zemanj@cml.fsv.cvut.cz

Abstract

The remarkable material properties offered by composite materials such as high strength, light weight, corrosive resistance and affordability, etc., have resulted in their use in diverse high-performance engineering applications. Among the most prominent material systems complying with the aforementioned requirements remain polymer matrix systems reinforced either by aligned fibers, whiskers or fabrics. The popularity of latter materials, in particular, is under continuous rise due to advantageous strength/weight ratio, easiness of manipulation and low production costs.

Although all the above analyses rely on the idealized geometry of woven fabric composites, it is clear that the collective properties of these materials are to a great extent influenced by imperfections developed during the fabrication process. In this context the waviness, misalignment and/or non-uniform cross-sectional aspect ratio of tows in the longitudinal direction play the main role in assessing the overall behavior of such material systems. In this regard, the formulation of a reliable and accurate numerical model is of paramount importance.

A simple procedure was proposed in [1] to determine a Periodic Unit Cell (PUC) for balanced plain weave fabric composites with reinforcement imperfections. Geometrical parameters of the PUC are determined so that the resulting PUC resembles the real geometry of a composite, obtained from digitized micrographs of plain weave cross-sections, as close as possible. In particular, the quality of a unit cell is quantified by the least-square difference between proper statistical descriptors related to the original and an idealized unit cell, respectively. Once the desired geometrical parameters are determined, the finite element model of a woven composite is formulated and used to predict the overall response of the composite using homogenization methods.

The present contribution extends the results obtained in [1] in several aspects. In particular, more complex unit cell is formulated to incorporate a relative shift of individual layers in the model. This generalization on the theoretical level is complemented on the experimental level with acquisition and analysis of high-resolution images of real microstructures.

REFERENCES

[1] J. Zeman, M. Šejnoha, *Homogenization of plain weave composites with imperfect microstructure: Part I - Theoretical formulation*, International Journal of Solids and Structures, (2003), (accepted for publication).

COMPUTATIONAL MECHANICS (Abstracts)
WCCM VI in conjunction with APCOM'04, *Sept. 5-10, 2004, Beijing, China*

Towards a Robust Dynamic Analysis of Reinforced Concrete Structures

K. V. Spiliopoulos*, G. C. Lykidis

Department of Civil Engineering, National Technical University of Athens, Zografou Campus, Athens, 153 73, Greece
e-mail: kvspilio@central.ntua.gr, glykid@mail.ntua.gr

Abstract

The nonlinear analysis of concrete structures with three dimensional solid finite elements has been a subject of research for many decades and various models have appeared in the literature. Despite the fact that the triaxial behaviour of concrete has been thoroughly investigated at the material level, the applications to structures are mainly limited to monotonic loads. This seems to be due to the fact that the finite element modeling of reinforced concrete under cyclic loads usually exhibits extensive instabilities which lead to early divergence of the solution during the analysis. Even less numerical examples exist in the literature for the dynamic analysis of such structures.

In the present paper, an existing 3D finite element constitutive model for structural concrete [1] which is based on the smeared crack approach is extended for use in dynamic analysis of simple reinforced concrete structures. This particular model has proved to yield a close fit to experimental values for static monotonic cases. The crack formation in this model starts when the triaxial stress state reaches an ultimate strength surface.

In the present approach, the crack strategy for cyclic loading, developed in Ref. [2], has been altered to provide better stability and accuracy. The Menegotto – Pinto uniaxial steel model is used for the reinforcing bars. An additional improvement is the use of an embedded formulation of the steel bars [3] so as the exact arrangement of the reinforcement may be taken into account. The numerical scheme uses the Newmark's method in conjunction with the Newton – Raphson iterative technique. Examples of structures under ground motion, using the proposed computational method, are compared against experimental data. The numerical results show good agreement with the experimentally observed response and the proposed numerical strategy appears to have computational efficiency and stability.

REFERENCES

[1] M. D. Kotsovos, M. N. Pavlovic, *Structural concrete*, *finite element analysis for limit state design*, London: Thomas Telford, (1995).

[2] M. D. Kotsovos, K. V. Spiliopoulos, *Modelling of crack closure for finite element analysis of structural concrete*, Computers and Structures, 69, (1998) 383-398.

[3] Alau E. Elwi, Terry M. Hrudey, *Finite element model for curved embedded reinforcement*, ASCE, Journal of Engineering Mechanics, 115, (1989), 740-754.

COMPUTATIONAL MECHANICS (Abstracts)
WCCM VI in conjunction with APCOM'04, *Sept. 5-10, 2004, Beijing, China*

Influence of Interfaces in Multi-Scale Modeling of Woven Composites

M. Wierer *, M. Šejnoha, J. Zeman

Department of Structural Mechanics, Faculty of Civil Engineering, CTU Prague, Thakurova 7, Prague, 16629, Czech Republic
e-mail: martin.wierer@fsv.cvut.cz, sejnom@fsv.cvut.cz, zemanj@cml.fsv.cvut.cz

Abstract

As suggested by the title the general goal of this contribution is to address an effect of imperfect bonding on the response of woven composites usually supplied in the form of complex multilayered structures. The present geometrical arrangement promotes the analysis to be carried out on three levels (micro - level of fiber tows, meso - single ply, macro - laminated tube). However, since the emphasis is on the modeling of interfacial behavior between individual tows within a single ply we limit our attention to two lowest scales only (micro and meso-scale).
It is well known that the essential part in the modeling of composites is the formulation of representative volume element (RVE). While the periodic nature of a fiber-tows arrangement reduces the basic geometrical model on meso-scale to a certain periodic unit cell, the distribution of fibers within individual tows (micro-scale) assumes random character. Analyses of material systems with disordered microstructures are still a subject of ongoing research. Some promising directions are discussed in paper [3]. In the present contribution, this difficulty is handled by employing a computational model based on the extended Hashin-Shtrikman variational principles to provide estimates of the local fields on micro-scale. Numerical analysis on micro-scale is then followed by mesoscopic analysis of a certain periodic unit cell formed by individual fiber-tows. Standard homogenization technique based on periodic fields is implemented within the framework of the finite element method [1] and extended to account for the imperfect bonding along the fiber-tow interfaces. The constitutive modeling on both scales is completed by including the time dependent behavior of individual phases using the concept of eigen-stresses. In particular, the generalized Leonov model is implemented to govern the nonlinear viscoelastic response of the matrix phase. In addition, the tensile brittle failure of the polymeric matrix at the fiber matrix interface is assessed with the help of plastic-damage model discussed in [2].

REFERENCES

[1] J. C. Michel, H. Moulinec, P. Suquet, *Effective properties of composite materials with periodic microstructure: a computational approach*, Computer Methods in Applied Mechanics and Engineering, 172, (1999), 109-143.

[2] K. de Proft, *Combined experimental-computational study to discrete fracture of brittle materials*, Ph.D. Thesis, Vrije Universiteit Brussel, Brussel, Belgium, (2003).

[3] M. Šejnoha, J. Zeman, *Overall viscoelastic response of random fibrous composites with statistically quasi uniform distribution of reinforcements*, Computer Methods in Applied Mechanics and Engineering, 191, (2002), 5027-5044.

COMPUTATIONAL MECHANICS (Abstracts)
WCCM VI in conjunction with APCOM'04, *Sept. 5-10, 2004, Beijing, China*

A Class of Strain-Localization Finite Elements and Its Application to Dynamic Shear Banding

Qiang Yang *, Alejandro Mota, Michael Ortiz

Division of Engineering and Applied Science, California Institute of Technology, 1200 E. California Blvd., MC 105-50, Pasadena, CA 91125, USA
e-mail: yang@aero.caltech.edu, mota@aero.caltech.edu, ortiz@aero.caltech.edu

Abstract

We present a class of three dimensional finite-deformation strain-localization elements which enables the accurate and efficient simulation of strain-localization processes such as dynamic shear band formation and propagation. The element treats the strain localization process as occurring at the sub-grid scale and thus the localized deformations appear as displacement discontinuities on the scale of mesh. The elements take the form of a double surface and deform in accordance with an arbitrary constitutive law which, particularly, is the same as those governing the bulk behavior. Moreover, they allow for the development of displacement and velocity jumps across volume element boundaries. The thickness of the localized zone is set by an additional field variable which is either treated as an empirical constant or determined by recourse to a variational thermodynamic principle. The localization elements are inserted, and become active, only when localized deformations become energetically favorable. The strain-localization elements are endowed with full finite kinetics and are compatible with a conventional finite element discretization of the bulk material. The implementation presented in this work is three dimensional and allows for finite deformations. The versatility and predictive ability of the method are demonstrated through the simulation of the dynamic impact of a pre-notched C300 steel sample by a steel projectile [1]. The shear band velocity and thickness profile along the propagation path are well captured by the numerical simulations.

REFERENCES

[1] P. R. Guduru, A. J. Rosakis, G. Ravichandran, *Dynamic shear bands: an investigation using high speed optical and infrared diagnostics*, Mechanics of Materials, 33, (2001), 371-402.

COMPUTATIONAL MECHANICS (Abstracts)
WCCM VI in conjunction with APCOM'04, *Sept. 5-10, 2004, Beijing, China*

The Improved Numerical Method of Riemann-Liouville Fractional Derivative for FE formulation

Wei Zhang [1*], Nobuyuki Shimizu [2]

[1] *Institute of Applied Mechanics, Jinan University, Guangzhou, Guangdong 510630, China*
[2] *Department of Mechanical Engineering, Iwaki Meisei University, Iwaki, Fukushima, Japan*
e-mail: weezhang@public.guangzhou.gd.cn

Abstract

A numerical algorithm for the Riemann-Liouville fractional derivative is presented in this paper. There are several definitions of fractional derivatives [1]. Although these definitions are different in forms, they are the same in essence and can be transferred for each other under some conditions. As Riemann- Liouville definition of fractional derivative has its own characteristics and has now found its wide applications in many field [1-2], instead of using finite Grünwald definition of fractional derivative to approximate the Riemann-Liouville's, this work has developed a numerical algorithm directly from Riemann-Liouville's definition by taking advantages of the features of the integrand kernel, assuming the approximating function for the integrand and making use of Newmark-type numerical methods. The algorithm is used to analyze the transient dynamic response for a viscoelastic oscillator and to formulate the FE analysis procedures. The key formulation is summarized as follows:

The definition of Riemann- Liouville fractional derivative is taken as:

$$D^q f(t) = \frac{\mathrm{d}^m}{\mathrm{d}t^m}\int_0^t \frac{(t-\tau)^{m-q-1}}{\Gamma(m-q)} f(\tau)d\tau \qquad m-1 \le q \le m$$

The main idea of the numerical scheme for it is:

$$\begin{cases} D^{\nu} f(t_n) = \dfrac{1}{\Gamma(1-\nu)}\left[\dfrac{f(0)}{t_n^{\nu}} + \displaystyle\int_0^{t_{n-1}} \dfrac{Df(\tau)}{(t_n-\tau)^{\nu}}\mathrm{d}\tau + \int_{t_{n-1}}^{t_n} \dfrac{Df(\tau)}{(t_n-t)^{\nu}}\mathrm{d}\tau\right], \quad 0<\nu<1 \\ Df(\tau) = Df(t_{n-1}) + \eta(\tau)\left[Df(t_n) - Df(t_{n-1})\right] \\ Df(t_n) = Df(t_{n-1}) + (1-\alpha)\Delta t D^2 f(t_{n-1}) + \alpha\Delta t D^2 f(t_n) \\ D^2 f(t_n) = \dfrac{1}{\beta\Delta t^2}[Df(t_n) - Df(t_{n-1})] - \dfrac{1}{\beta\Delta t}Df(t_{n-1}) - (\dfrac{1}{2\beta} - 1)D^2 f(t_{n-1}) \end{cases}$$

where without the loss of generality q is taken as ν, $\eta(\tau)$ denotes the known and selected function for required accuracy and α, β, the parameters of Newmark-type integration method.

The sample results show that the algorithm possesses the advantages of fast convergence, higher accuracy and higher stability and easy for application and further modification.

REFERENCES

[1] N. Shimizu, W. Zhang, *Fractional Calculus Approach to Dynamic Problem of Viscoelastic Materials*, JSME International Journal, Series C, 42(4), (1999), 825-837.

[2] Yu. A. Rossikhin, M. V. Shitikova, *A new method for solving dynamic problems of fractional derivative viscoelasticity*, International Journal of Engineering Science, 39, (2001), 149-176.

COMPUTATIONAL MECHANICS (Abstracts)
WCCM VI in conjunction with APCOM'04, *Sept. 5-10, 2004, Beijing, China*

Mesolevel Simulation of the Elastic Modulus of Concrete

Jianjun Zheng [*], Lu Jiang, Xinzhu Zhou

School of Civil Engineering and Architecture, Zhejiang University of Technology, Hangzhou 310014, China
e-mail: jjzheng4335@sina.com

Abstract

The elastic modulus of concrete is a parameter of most practical significance for the design and assessment of concrete structures. At a mesoscopic level, concrete is usually considered to be a two-phase material for the prediction of its mechanical properties, being composed of aggregate particles dispersed in cement paste. More research, both experimental and analytical, has however shown that there exists an interfacial transition zone (ITZ) between the aggregates and cement paste of concrete with normal mixtures. Due to this ITZ, arguably concrete should be modelled as a three-phase material, consisting of aggregate particles, ITZ and cement paste. With this model, the influence of the largest aggregate diameter, the thickness of ITZ and the mechanical properties of ITZ on the elastic modulus of concrete can be evaluated.
The intention of this paper is to present a numerical method which can predict the elastic modulus of concrete with reasonable accuracy. Based on the mesostructural simulation method, the ITZ area fraction is evaluated by applying Monte Carlo algorithm. A parametric study is then conducted to investigate the effect of the largest aggregate diameter, ITZ thickness and aggregate gradation on the ITZ area fraction. Since the ITZ area fraction is usually much smaller than aggregate and cement paste ones, a special mesh generating technique is adopted to divide each phase as accuracy as possible. The finite element method is used to calculate the elastic modulus of concrete. Finally, the numerical results are compared with the experimental data, demonstrating a good agreement between them. A parametric study is conducted to investigate the effect of the largest aggregate diameter, ITZ thickness, aggregate gradation and the mechanical properties of ITZ on the elastic modulus of concrete.

COMPUTATIONAL MECHANICS (Abstracts)
WCCM VI in conjunction with APCOM'04, *Sept. 5-10, 2004, Beijing, China*

Micromechanical Modeling of Damage-Coupled Viscoelastic Behavior of Particulate Asphalt Materials

Qingli Dai*, Martin H. Sadd

Department of Mechanical Engineering and Applied Mechanics, University of Rhode Island, Kingston, RI 02881, USA
E-mail: daiq@egr.uri.edu, sadd@egr.uri.edu

Abstract

This study presents a micromechanical modeling scheme for the damage-coupled viscoelastic behavior of asphalt materials by using finite element methods. A unified approach is proposed using the Schapery nonlinear viscoelastic model for the brittle (rate-independent) failure and rate-dependent damage behavior. A FE incremental algorithm with a recursive relationship for three-dimensional (3D) viscoelastic damage behavior was developed. This algorithm was later used in a 3D user-defined material model for the asphalt mastic to predict global viscoelastic damage behavior of asphalt materials.

The microstructural model of particulate asphalt materials first incorporates an equivalent lattice network structure whereby intergranular load transfer is simulated through an effective asphalt mastic zone. And then combines the ABAQUS user-defined material subroutine with continuum elements for the effective asphalt mastic and rigid body defined with rigid elements for each aggregate. Cyclic loading responses of linear and damage-coupled viscoelastic materials were compared. The effect of loading rate on the material viscoelastic damage behavior was also investigated by using numerical simulations of model asphalt samples.

REFERENCES

[1] J. C. Simo, *On a fully three-dimensional finite-strain viscoelastic damage model: formulation and computational aspects,* Comput. Methods Appl. Mech. Eng., 60, (1987), 153-173.

[2] S. M. Park, Y. R. Kim, R. A. Schapery, *A viscoelastic continuum damage model and its application to uniaxial behavior of asphalt concrete,* Mech. of Materials, 24, (1996), 241-255.

[3] M. A. Zotch, S. E. Groves, *A three-dimensional finite element formulation for thermoviscoelastic orthotropic media,* Int. J. Numer. Methods Engng, 40, (1997), 2267-2288.

[4] R. M. Haj-Ali, A. H. Muliana, *A micromechanical constitutive framework for the nonlinear viscoelastic behavior of pultruded composite materials,* Int. J. Solids Struct., 40, (2003), 1037-1057.

COMPUTATIONAL MECHANICS (Abstracts)
WCCM VI in conjunction with APCOM'04, *Sept. 5-10, 2004, Beijing, China*

Many-Body Toner Simulation Using Parallel Hierarchical Tree Code

R. Hoffmann [1*], G. Wachutka [2]

[1] *Océ Printing Systems, Poing, 85581, Germany*
[2] *Institute for Physics of Electrotechnology, Munich University of Technology, Munich, 80333, Germany*
e-mail: rainer.hoffmann@ops.de, wachutka@tep.ei.tum.de

Abstract

Many industrial printing processes use electric fields to control the transport of charged particles across small air gaps. In our work we analyzed in detail the transfer of charged, densely distributed toner particles from one roller to a neighboring one by appropriately adjusted electric fiel d forces. To this end we performed microscopic many-body simulations using the discrete element method[1], where all relevant forces like Coulomb interaction, particle -roller adhesion and external electric field forces are included in the underlying physic al model.
Since all the particles are charged, the computational cost for the calculation of the inter -particle Coulomb forces rises as $O(N^2)$. As the resulting computation times become excessive and, hence, would severely restrict the number of particles c onsidered in a simulation run, the direct summation method was abandoned in favour of the hierarchical Barnes -Hut tree code algorithm [2]. This algorithm exploits the fact that the individual contributions of distant particles are small, so that these particles can be grouped to pseudo-particles. Thus the computational cost can be reduced to an $O(N \log N)$ scaling.
The original algorithm originating from astronomy, where collisions are virtually non -existent, had to be adjusted to a situation, where particle contacts occur frequently. To this end, an extra limitation had to be introduced, so that non -physical particle overlaps could be avoided.
For a further reduction of the computation time, the code has been parallelized so that the execution could be distributed over a cluster of computer units. Since the only non -parallelized part of the code is the tree-building phase, the achievable speed-up for a small cluster is up to 85 % of the ideal value.
The results obtained from the many-body simulations (amount of particles transferred, size of transfer area) conform well to the observed experimental findings. Since field-induced toner particle transfer is the central mechanism of any printing process, its detailed visualization and understanding on the basis of "virtual experiments" on the computer has proven to be very useful for the further improvement of existing printers (e.g. speed, print quality) and for the optimized design of new printer generations, such as color printers.

REFERENCES

[1] J. E. Barnes, P. Hut, *A hierarchical $O(n \log n)$ force-calculation algorithm*. Nature, 324(4), (1986), 446-449.

[2] P. Cundall, O. Strack, *A discrete numerical model for granular assemblies*, Geotechnique, 29(1), (1979), 47-65.

COMPUTATIONAL MECHANICS (Abstracts)
WCCM VI in conjunction with APCOM'04, *Sept. 5-10, 2004, Beijing, China*

Numerical Analysis of Smart Structures Using Fuzzy Finite Element Method

C. Chee*, S. Valliappan

School of Civil & Environmental Engineering, University of New South Wales, Sydney, 2052, Australia
e-mail: ck.chee@unsw.edu.au, s.val@unsw.edu.au

Abstract

Smart materials have been investigated extensively in the past decades as potential sensors and actuators for shape, vibration and noise control because of interaction property between mechanical and electrical systems. A wide range of application of smart materials has been found in aerospace and civil structures to actively control vibration and improve performance.

As a powerful tool for numerical analysis, finite element method is widely used in numerical analysis and design of the structural systems integrated with smart materials. In the finite element analysis of 'smart structures' adopted for structural vibration control, most of the existing investigations assume that the input parameters as deterministic variables and hence the responses of the system are also deterministic. However, it is well known that there are many uncertainties associated with the 'smart structures' such as the piezoelectric, mechanical and physical properties. Besides these, there are also uncertainties associated with the properties of the materials used for structural components, foundations, as well as the parameters associated with earthquake loading. Some of these uncertainties are vague and imprecise and hence the use of fuzzy set theory, initiated by Zadeh in 1965 [1], is most appropriate.

The key issues involved in fuzzy set theory are 1) Modelling of fuzzy input variables; and 2) Computation of fuzzy responses. Modelling of fuzzy input variables can be done by constructing a membership function for each variable with imprecise information. The next step is computing the fuzzy response quantity which will be based on the 'Extension Principle'. The number of fuzzy variables in the analysis of smart structural systems is generally very high and hence a full set of finite element analyses involving all the fuzzy parameters will be computationally expensive and time consuming. It is necessary to distinguish between the parameters which significantly affect the dynamic response of structures integrated with 'smart structures' and those which are not significant. Therefore, some form of optimization has to be adopted to minimize the number of finite element analyses.

A fuzzy finite element method is presented in this paper for analyzing analysis and design of 'smart structures' which can be used to control the structural vibration due to earthquakes. This unified approach combines the finite element technique and the concepts of fuzzy sets as well as optimization. Fuzzy variables are modelled by the triangular membership function and modified π membership function. The vertex method is employed for the solutions of fuzzy equations. The sensitivity index is developed for the optimization process to enhance the computation efficiency of the proposed method. The proposed fuzzy finite element method is validated by numerical examples.

REFERENCES

[1] L. A. Zadeh, *Fuzzy sets, Information and Control*, 8, (1965), 338-353.

COMPUTATIONAL MECHANICS (Abstracts)
WCCM VI in conjunction with APCOM'04, *Sept. 5-10, 2004, Beijing, China*

An Electro-Mechanical Coupled Mesh-Free Method for Fracture Analysis of Cracks in Piezoelectric Ceramic Materials

Xianghua Guo, Daining Fang [*]

Department of Engineering Mechanics, Tsinghua University, Beijing, 100084, China
e-mail: fangdn@mail.tsinghua.edu.cn

Abstract

Based on the moving least square interpolation, an electro-mechanical coupled element free Galerkin method is developed to analyze crack propagation in piezoelectric ceramic structures subjected to electro-mechanical coupled loading. The EFG method establishes discrete equations in term of the Gauss integration based on background, so it is accurate and stable. In this paper, the displacement and electric potential are employed as nodal variables and the general displacement boundary conditions are introduced into discrete equations by the penalty function method, so that the stiff matrix is sparse, symmetrical and positive definite, which can considerably reduce computational effort. In order to satisfy the discontinuity of both mechanical and electric displacements in nonconvex bodies, such as cracks and holes, the discontinuous diffraction functions [1] are introduced into weight functions. An infinite plate with a central circular defect is studied to verify the validity of our electro-mechanical coupled method. The agreement between the predicted stress and electric field distribution with those obtained from existing numerical simulation [2] and exact theoretical solutions [3]. In order to investigate both stress and electric field distributions at a crack tip in a piezoelectric material, a single edge-notched specimen has been modeled. In the discrete equations, some basis functions with singular items are adopted to catch the singularity of stress and electric field. In addition, the nodal refinement is carried out around the crack tip. The numerical results show that both the ratio of the electrical to the mechanical loads and the direction of electric field have important effects on the angle at which the hoop stress reaches its maximum value, which makes the crack has the tendency to propagate straightward or deviate away from the crack line. In some practical performances, the piezoelectric materials are often deposited on the substrates. Some cracks will be found in the interface of piezoelectric material and a substrate. In this paper, Lagrange multiplier is imposed at the variational level to ensure satisfaction of interface conditions. The influence of geometry and material parameters on the crack propagation trajectory also is studied. In another example involving the study of multiplayer actuators, the proposed model is found to be accurate in the simulation of stress and electric field concentrations due to the abrupt end of an internal electrode.

REFERENCES

[1] T. Belytschko, M. Fleming, *Smoothing, enrichment and contact in the element-free Galerkin method*, 71, (1999), 173-195.

[2] D. Fang, H. Qi, Z. Yao, *Numerical analysis of crack propagation in piezoelectric ceramics*, Fat. Fract. Engrg. Mater. Struct., 21, (1998), 1371-1380.

[3] H. Sosa, Y. Pak, *Three dimensional eigenfunction analysis of a crack in a piezoelectric material*, Int. J. Solids Struct., 26, (1990), 1-15.

COMPUTATIONAL MECHANICS (Abstracts)
WCCM VI in conjunction with APCOM'04, *Sept. 5-10, 2004, Beijing, China*

Topology Optimization for the Design of Piezoelectric Smart Structures: Static and Time-Harmonic Applications

Martin Kögl [1*], Emílio C. N. Silva [2]

[1] *Structural Analysis and Occupant Simulation (RBP/BS), DaimlerChrysler AG, HPC E222, 70546 Stuttgart, Germany*

[2] *Department of Mechatronics and Mechanical Systems Engineering, University of São Paulo, Av. Prof. Mello Moraes, 2231, CEP 05508-900 São Paulo, SP, Brazil*

e-mail: m_koegl@yahoo.com, ecnsilva@usp.br

Abstract

The performance of multi-layer piezoelectric smart structures [1] can be improved by optimising the integrated sensors and actuators. In the present paper, a novel approach for the design of piezoelectric transducers is described that uses topology optimisation to obtain an optimal distribution of piezoelectric material inside a given design domain. The topology optimisation procedure employed here uses the material distribution method [2] in conjunction with a new piezoelectric material model described in [3], which contains two sets of design variables: the pseudo-density ρ_1 and the piezoelectric polarisation ρ_2. Optimisation of the density ρ_1 for a given objective function yields simultaneously the optimal shape, size, position, and number of the piezoelectric transducers, while the optimal polarisation is determined by the design variable ρ_2.

In the *actuator* design process, an optimal distribution of piezoelectric material is sought which maximises the output displacement of the structure at a given point and in a given direction under the action of an electric charge applied to the actuators. In contrast, the *sensor* design approach determines an optimal distribution of piezoelectric material inside the design domain so as to maximise the output voltage of the sensors for a set of applied external loads. Numerical examples are given to demonstrate the application of the method to the optimisation of piezoelectric sensors/actuators in thin multi-layer smart structures (static and time-harmonic problems). The structural response is calculated with the Finite Element Method using newly developed locking-free piezoelectric plate and shell elements [4, 5]. The examples demonstrate the potential of the proposed method. Results for time-harmonic analyses and for sensor applications are given for the first time.

REFERENCES

[1] R. L. Clark, W. L. Saunders, G. P. Gibbs, *Adaptive structures*, John Wiley & Sons, New York (1998).

[2] M. P. Bendsøe, O. Sigmund, *Topology optimization, theory, methods and applications*, Springer-Verlag, Berlin, (2003).

[3] M. Kögl, M. L. Bucalem, E. C. N. Silva, *A new material model for the topology optimization of smart structures*, Proc. XXIV CILAMCE, Ouro Preto, Brazil, (2003).

[4] M. Kögl, M. L. Bucalem, *Piezoelectric MITC plate elements*, Proc. WCCM V, Vienna, Austria (2002).

[5] M. Kögl, M. L. Bucalem: *Locking-free piezoelectric shell elements*, in K.-J. Bathe ed. *Computational fluid and solid mechanics 2003*, Elsevier, Amsterdam, (2003), pp. 392-395.

COMPUTATIONAL MECHANICS (Abstracts)
WCCM VI in conjunction with APCOM'04, *Sept. 5-10, 2004, Beijing, China*

Nonlinear Shape Control Simulation of Piezolaminated Plates and Shells

S. Lentzen*, R. Schmidt

Institute of General Mechanics, RWTH Aachen University, Templergraben 64, Aachen, 52062 Germany
e-mail: lentzen@iam.rwth-aachen.de, schmidt@iam.rwth-aachen.de

Abstract

A geometrically nonlinear 5-parameter composite shell element with integrated piezoelectric layers is presented. The kinematics of the shell is described by the Reissner-Mindlin first order shear deformation theory and the strain-displacement relations are valid for small strains and moderate rotations [1]. To avoid the locking phenomenon the assumed natural strain approach proposed by Park and Stanley [2] is used. The finite element implementation is based on the total Lagrangian formulation which necessitates the introduction of new defined electric field quantities. Both the electric field as well as the electric displacement vector has to refer to the initial undeformed configuration and have to be energetically conjugated at the same time.
The developed finite element is validated on two examples which are discussed by Lee et al. The referred authors used a geometrically linear 6-parameter degenerated shell. The mentioned examples are further extended to deal with problems within the geometrically nonlinear regime. In literature the problem, which is dealt with in these examples, is referred to as shape control, which is in static cases a mere straightening of deformed structures. It is noticed that when the piezoelectric layers are applied as sensors, the predicted sensor voltages are remarkably different from the ones predicted by linear theory. Due to the stiffer behavior of the geometrically nonlinear structure the differences are of quantitative nature. Qualitative differences occur due the induced membrane stresses. The latter fact could become important in matters like optimal sensor placement of smart structures.

REFERENCES

[1] R. Schmidt, J. N. Reddy, *A refined small strain and moderate rotation theory of elastic anisotropic shells*, J. Appl. Mech., 55, (1988), 611-617.

[2] K. C. Park, G. M. Stanley, *A curved C^0 shell element based on assumed natural coordinate strains*, ASME, J. Appl. Mech., 53, (1986), 278-290.

[3] S. Lee, N. S. Goo, H. C. Park, K. J. Yoon, C. Cho, *A nine-node assumed strain shell element for analysis of a coupled electro-mechanical system*, Smart Mater. Struct., 12, (2003), 355-362.

COMPUTATIONAL MECHANICS (Abstracts)
WCCM VI in conjunction with APCOM'04, *Sept. 5-10, 2004, Beijing, China*

Modal Shape Control of Smart Composite Plates Using Piezoelectric Actuators

Quan Nguyen*, Liyong Tong

School of Aerospace, Mechanical & Mechatronic Engineering, Bldg. J07 The University of Sydney, NSW, 2006
e-mail: ltong@aeromech.usyd.edu.au

Abstract

Static shape control of a smart structure can be obtained through optimising the applied electric fields, their loci and sizes of the piezoelectric actuators attached to the structure. Shape control can be categorised as either static or dynamic shape control. Whether it is transient of gradually changing, static and dynamic shape control both aim to determine the size of piezoelectric actuators, their applied electric fields and loci such that the desired shape is achieved.

Dynamic shape control of smart composite structure is widely accepted. And the smart composite structure can be modelled as beams or plates or shells for investigating the disturbance that affect the shape of the structure as transient and associate with fixed deformation or slowly vary in time by using piezoelectric layer. The previous work in this field focussed on optimisation of the size, number and location of the piezoelectric to control structural vibration in the smart composite beams [4-5]. As a result, recent work focuses on the control of structural vibration in plate structure.

In this paper, Finite element software has been developed for analysing dynamic deformation of smart composite plate structures. The mechanical deformation is modelled using a 3rd order plate theory, whilst a layer-wise theory is used to model the electric field. Methods used to determine the rigid body mode shapes and their corresponding natural frequencies are also outlined. Using a recently developed Linear Least Square algorithm, numerical examples are executed to obtain the optimum values of the electric fields in actuators in order to achieve the required mode shape.

REFERENCES

[1] D. B. Koconis, L. P. Kollar, G. S. Springer, *Shape control of composite plates and shells with embedded actuators, II desired shape specified*, J. Composite Materials, 28(3), (1994), 262-285.

[2] C. Chee, L. Tong, G. Steven, *A mixed model for adaptive composite plates with piezoelectric for anisotropic actuation*, Computers and Structures, 77, (2000), 253-268.

[3] K. H. Lo, R. M. Christensen, E. M. Wu, *A high order theory of plate deformation, part 1 & 2: homogenous plates*, Journal of Applied Mechanics, 44, (1977), 663-676.

[4] A. Mackiewicz, J. Holnicki-Szulc, F. Lopez-Almansa, *Optimal sensor location in active control of flexible vibration*, American Institute of Aeronautics and Astronautics Journal, 34, (1996), 857-859.

[5] Y. K. Kang, H. C. Park, W. Hwangand, K. S. Han, *Optimum placement of piezoelectric sensor/actuator for vibration control of laminate beams*, American Institute of Aeronautics and Astronautics Journal, 34, (1996), 1921-1926.

COMPUTATIONAL MECHANICS (Abstracts)
WCCM VI in conjunction with APCOM'04, *Sept. 5-10, 2004, Beijing, China*

Asymptotic Analysis of Active Drive Shafts

Ajay Kumar Tamrakar, Pramod Kumar, Dineshkumar Harursampath *

Department of Aerospace Engineering, Indian Institute of Science, Bangalore-560012, India
e-mail: ajay@stc.iisc.ernet.in, pramod@aero.iisc.ernet.in, dinesh@aero.iisc.ernet.in

Abstract

In this paper, analytical modeling of active shafts made of Active Fiber Composites (AFC) for twist actuation as well as twist sensing is carried out by using the Variational Asymptotic Method (VAM). Closed form solutions for twist actuation and twist sensing problems are derived with the approach of generalizing all piezoelectric stress constants of AFC to be non-zero. For the material ([α^o] PZT-5H/Epoxy, where α is active fiber orientation angle, constituting the radially innermost and outermost layers of a generally anisotropic composite tube) under numerical study, it is found that when externally applied torque is zero, the twist generated per unit length κ_1 due to actuation electric field (in radially outward direction), is maximum at nearly $\alpha = 51^o$. There is a linear variation of twist actuated with the applied electric field. It has been proved here that antisymmetric layups for AFC in tubes cannot produce any twist actuation.

The second objective (i.e. sensing of twist) of the work is to study the case of active laminated thin-walled shaft with symmetric AFC layup loaded by external torque. For this problem, closed form expression for average charge, q_{avg}, generated in a lamina is derived in terms of laminae and laminate properties. q_{avg} and sensor voltage, V_{sensor}, across electrodes placed at inner and outer surfaces of the shaft are related by the effective capacitance, C_{eff}. Expression for q_{avg} is used to calculate the sensor voltage generated in the active shaft considered in numerical study of twist actuation problem. It is found that q_{avg} is maximum when $\alpha = 45^o$.There is a linear variation of average charge, q_{avg}, with twist-rate, κ_1. The variation of V_{sensor} with respect to κ_1 is also found to be linear. The procedure to obtain higher order approximations resulting in more accurate nonlinear results is outlined.

REFERENCES

[1] D. Harursampath, D. H. Hodges, *Asymptotic analysis of the non-linear behaviour of long anisotropic tubes,* Int. J. Non-linear Mechanics, 34, (1999), 1003-1018.

[2] A. K. Tamrakar, D. Harursampath, *Non-linear analysis of active pretwisted anisotropic strips,* 30th European Rotorcraft Forum, Marseilles, France. 14-16 September, (2004), (submitted).

[3] V. L. Berdichevskii, *Variational-asymptotic method of shell theory construction,* PMM., 34, (1979), 664-687.

[4] A. K. Tamrakar, D. Harursampath, *General constitutive relations for active fiber composites,* Third Int. Conf. on Advances in Structural Engineering and Mechanics (ASEM'04), Seoul, Korea, 2-4 September, (2004), (submitted).

[5] A. K. Tamrakar, D. Harursampath, *Application of active strip as twist sensor,* International Congress on Computational Mechanics and Simulation, Kanpur, India, 9-12 December (2004), (under preparation).

COMPUTATIONAL MECHANICS (Abstracts)
WCCM VI in conjunction with APCOM'04, *Sept. 5-10, 2004, Beijing, China*

Molecular Dynamics Simulations on Shape Memory Effect in Ni-Al Alloy

Takuya Uehara *, Takato Tamai

Department of Energy Conversion Science, Graduate School of Energy Science, Kyoto University, Yoshida Hon-machi, Sakyo-ku, Kyoto 606-8501, Japan
e-mail: uehara@energy.kyoto-u.ac.jp, tamai@force.energy.kyoto-u.ac.jp

Abstract

Molecular dynamics simulations on martensitic phase transformation and shape memory effect are carried out by using the EAM potential for Ni-Al alloy. The stable configuration of atoms for the potential parameters is first investigated by static calculations, and a body-centered cubic and a monoclinic structure, or martensite hereafter, are obtained. The potential energies of the two structures are calculated for various Ni contents, which reveal that the martensite is more stable for 60-70 % Ni, while bcc structure shows lower potential for less than 60 % of Ni. Also the temperature dependence of the stable structure is simulated: at any Ni content between 60 and 70 %, martensite is more stable at lower temperature, and bcc at higher temperature. Hence, the phase transformation from martensite to bcc is expected during heating, and vise versa in cooling. Therefore, we chose 68 % Ni model for the simulation of the martensitic transformation and related shape memory effects.

The shape memory effect is summarized as follows: A martensite specimen is generally composed of several variants, and two variants are connected with a twin boundary. When external shear force is imposed to the specimen, the twin deformation occurs under shear stress lower than yield stress. Even when the external force is removed, the inverse deformation at the twin boundary does not occur except the elastic recovery of the variant, and the specimen retains a macroscopic deformation. When this specimen is heated up, phase transformation to bcc structure occurs. Since the bcc structure is isotropic and independent of the orientation of martensite variant, the deformation disappears. When the specimen is cooled down to the room temperature, martensitic transformation occurs again. Since each martensite variant generated in this process has random orientation, which means the strain in any orientation is averaged over, the macroscopic shape of the specimen retains the original one.

Following the process mentioned above, loading, unloading, heating and cooling process is simulated by using the molecular dynamics method. A single crystal model with several variants in two orientations is prepared, and a shear load parallel to the twin boundary is imposed. Then the variants in unstable orientation against the load are shifted to the preferable orientation, and a macroscopic deformation is generated. Although this deformation is maintained after unloading, the original shape is recovered in the thermal process incorporating phase transformation to the bcc. The stress-strain curve and strain-temperature curve during the consecutive process show the typical hysteresis loop corresponding to the experimental one, if neglecting the differences originating from the far simplified modeling in atomistic scale. A larger model having grain boundaries are also applied to the simulation, which shows better correspondence to the experimentally obtained behavior of shape memory alloys.

COMPUTATIONAL MECHANICS (Abstracts)
WCCM VI in conjunction with APCOM'04, *Sept. 5-10, 2004, Beijing, China*

Algorithmic Implementation of an Elastic-Transformation Model for Shape Memory Alloys

Nunziante Valoroso *

Istituto per le Tecnologie della Costruzione, Consiglio Nazionale delle Ricerche, viale Marx 15 ,Roma, 00137, Italy
e-mail: nunziante.valoroso@itc.cnr.it

Abstract

The large body of literature dealing with modeling and implementation of shape memo ry alloys (SMAs) shows the growing interest in these materials for developing innovative applications, see e.g. [1] for a survey account. This has led to the formulation of a considerable number of constitutive models, many of which are however so demandin g from a computational standpoint that their application has been limited only to the simplest one -dimensional situation.
An approach that has gained a great popularity for the description of SMA pseudoelastic behaviour in a fully three-dimensional setting relies upon internal variables theories of inelasticity, see e.g. [2 -4], among others. Owing to the existing similarities with classical plasticity theories, this approach looks very attractive, mainly in view of numerical implementation via return mappin g algorithms [5]. Nonetheless, a major difference with respect to classical plasticity models arises from the fact that the elastic range of a SMA pseudoelastic model is not simply connected, so that a mere specialization of the classical closest-point projection scheme of elastoplasticity cannot be pursued.
In the present work return mapping algorithms along with the relevant implementation for a SMA elastic-transformation model are discussed in detail and the performances of the solution scheme are illustrated with reference to representative numerical examples.

REFERENCES

[1] F. Auricchio, L. Faravelli, G. Magonette, V. Torra, (eds.), Shape Memory Alloys. *Advances in Modelling and Applications*. CIMNE, Barcelona, (2001).
[2] B. Raniecki, C. Lexcellent, K. Tanaka, *Thermodynamics models of pseudoelastic behaviour of shape memory alloys*, Arch. Mech., 44, 3, (1992), 261-284.
[3] J. G. Boyd, D. C. Lagoudas, *A thermodynamic constitutive model for shape memory alloys materials. Part I. The monolithic shape memory alloy*, Int. J. Plasticity, 12, (1996), 805-842.
[4] A. Souza, E. Mamiya, N. Zouain, *Three-dimensional model for solids undergoing stress-induced phase transitions*, Eur. J. Mech. A/Solids, 17, (1998), 789-806.
[5] M. A. Qidwai, D. C. Lagoudas, *Numerical implementation of a shape memory alloy constitutive model using return mapping algorithms*, Int. J. Num. Meth. Eng., 47, (2000), 1123-1168.

COMPUTATIONAL MECHANICS (Abstracts)
WCCM VI in conjunction with APCOM'04, *Sept. 5-10, 2004, Beijing, China*

Optimum Design of a Novel Magnetoelectric Composite Structure

Y. X. Liu[1*], J. M. Liu[2], J. G. Wan[2], C. W. Nan[3]

[1] *Department of Civil Engineering, Logistical Engineering University, Chongqing , 400041, China*
[2] *Laboratory of Solid state Microstructure, Nanjing University, Nanjing, 210093, China*
[3] *Department of Materials Science and Engineering, Tsinghua University, Beijing, 100084, China*
e-mail: lyuanxue@vip.sina.com, liujm@nju.edu.cn, wanjg@nju.edu.cn, cwnan@tsinghua.edu.cn

Abstract

Magnetoelectric (ME) effect is a cross two -field effect, characterized by appearanc e of an electric polarization upon applying a magnetic field or with the change of magnetization on applying an electric field. As a potential magnetic field probe and transducer for magneto -electric conversion among many other applications, the ME effect has been attracting attention since its discovery half-century ago. Besides the ferroelectromagnets as the long -time objects of the related researches, ME-composite materials represent the systems where strong ME effect is expected based on the idea of product property of composite materials. Along this line, quite a number of ME composite materials were fabricated in order to reach the optimized ME effect. The great difference between the experimental data and theoretical prediction reminds us of the neces sity of real-scale modeling of the ME effect in those ME composite materials. One may argue that the mechanical coupling between the two components is highly structure -sensitive, whiles an experimental search for such a resonance is tedious and costive. On the other hand, a numerical model ing of the ME effect is definitely suggestive for pre-design and optimization of novel ME composite materials and structures of potential giant ME property.

In this paper, we focus on the numerical modeling of the giant M E effect in a ME composite structure which consists of Terfenol-D/epoxy composite (magnetostrictive component) bonded with PZT/ epoxy composite (piezoelectric component) . This ME composite structure was recently developed by the authors and a giant ME effe ct was recorded at the resonance mode. First, the ME coupling mechanism and numerical simulation procedure for th is structure has been presented. A numerical optimization algorithm is put forward based on the mechanical coupling mechanism between the two components. It will be shown that the numerical algorithm to be developed here can reproduce well the experimental results. The surprising optimum result allows us to argue that the ME output must be optimized according to the mechanical coupling mechanism by which the magnetostrictive component and piezoelectric component interacts with each other. This algorithm can also be used for optimum design of other ME composite structures in terms of the largest ME output.

REFERENCES

[1] G. Srinivasan, E. T. Rasmus sen, B. J. Levin, and R. Haye s, *Magnetoelectric effects in bilayers and multilayers of magnetostrictive and piezoelectric perovskite oxides*, Phys. Rev. B, 65, (2002), 134402.

[2] J. Ryu, A. V. Carazo, K. Uchino, and H. E. Kim, *Magnetoelectric properties in piezoelectric and magnetostrictive laminate composites*, Jpn. J. Appl. Phys., 40, (2001), 4948-4951.

COMPUTATIONAL MECHANICS (Abstracts)
WCCM VI in conjunction with APCOM'04, *Sept. 5-10, 2004, Beijing, China*

Adaptive Finite Element Simulation during Directional Solidification of a Binary Alloy using Phase-Field Method

T. Takaki [1*], T. Fukuoka [1], Y. Tomita [2]

[1] *Department of Maritime Sciences, Kobe University, Higashinada, Kobe, 658-0022, Japan*
[2] *Graduate School of Science and Technology, Kobe University, Nada, Kobe, 657-8501, Japan*
e-mail: takaki@maritime.kobe-u.ac.jp,fukuoka@maritime.kobe-u.ac.jp,tomita@mech.kobe-u.ac.jp

Abstract

The alloy phase-field method is a powerful approach for simulating the complicated microstructure, such as a dendrite, formed during solidification [1]. The phase field variable, which is newly introduced in the method, changes sharply but smoothly over the thin liquid-solid interface and is to be constant value in the bulk liquid (=0) and solid (=1). The phase-field simulation requires high computational cost to evaluate the microstructure with high accuracy, since the fine meshes are needed around the interface. In the usual solidification simulation, however, the total area of the interface extremely small compared to one of the bulk phases. We need not to solve the phase-field equation in the bulk phases. The adaptive mesh techniques which use fine mesh only in the vicinity of the interface and coarse one in the bulk phases are, therefore, successfully applied to the phase-field simulation. The adaptive method could be more fascinating for the directional solidification problem than for the undercooling one, since the wide computational region and the long calculating time are required to evaluate the primary arm spacing at the steady state after the competitive growth between dendritic or cellular arrays and break down of initial planer interface.

In this study, the adaptive finite element method using isoparametric quadrilateral elements is applied to the simulation of the directional solidification of a binary alloy [2]. Using the fixed node and the quadrilateral element, the presence of a node on the side of an element must be allowed in order to refine the mesh efficiently, in which the mesh size decreases inversely as 2^n with increasing the element level number *n*. The method in which the degree of freedom of the node is eliminated by conducting the matrix operation is adopted here, instead of the previous one which uses the additional triangular elements [3]. In present simulations, the almost same parameters as those in reference [2] are chosen and the effects of some parameters, such as the mesh size and the strength of anisotropy, on the interface morphology, such as the primary arm spacing, are evaluated.

REFERENCES

[1] J. A. Warren, W. J. Boettinger, *Prediction of dendritic growth and microsegregation patterns in a binary alloy using the phase-field method*, Acta Metall. Mater., 43, (1995), 689-703.

[2] C. W. Lan, Y. C. Chang, Efficient adaptive phase field simulation of directional solidification of a binary alloy, J. Crystal Growth, (2003), 525-537.

[3] N. Provatas, N. Goldenfeld, J. Dantzig, *Adaptive mesh refinement computation of solidification microstructures using dynamic data structures*, J. Computational Physics, 148, (1999), 265-290.

COMPUTATIONAL MECHANICS (Abstracts)
WCCM VI in conjunction with APCOM'04, *Sept. 5-10, 2004, Beijing, China*

Membrane Analogy Driven Selective Finite Element Refinement in Torsional Problems

N. Troyani [1*], A. Perez [1], P. Baiz [2], C. Gomes [3]

[1] *Centro de Métodos Numéricos en Ingeniería, Departamento de Mecánica, Universidad de Oriente, Puerto La Cruz, 6023, Venezuela*

[2] *Department of Engineering, Queen Mary University of London, Mile End Road, London, E1 4NS, UK*

[3] *Department of Mechanical Engineering, Carnegie Mellon University, 408 Scaife Hall, Carnegie Mellon University, Pittsburgh, PA, 15213, USA*

e-mail: ntroyani@cantv.net, serdna@telcel.net.ve, p.m.baiz@queenmaryul.ac.uk, cgomes@andrew.cmu.edu

Abstract

This study was aimed at establishing a computational strategy to selectively refine a given initial arbitrary finite element (FE) mesh in order to progressively obtain better estimates for the maximum shearing stresses in non-circular cross sectional torsional members. The strategy utilized is based on the mathematical analogy between the torsional problem and the deflection of a membrane problem. According to this analogy, there is a direct correspondence between the gradients of a deflected membrane and the magnitudes of the shearing torsional stresses acting on a cross section, provided that the membrane exhibits the same boundary shape as the one exhibited by the cross sectional area of the loaded torsional member. The analogy also states that the resultant shearing stresses at each point in the cross section are aligned with the level lines (LL) corresponding to the deflected membrane. If the LL's are picked at equal level intervals the regions where they become closer to one another are clear indications of larger membrane gradients and as a result providing a pointer to locations of larger shearing stresses as well, thus identifying the sub-regions where FE refinement is required to produce better approximations to the magnitudes of the shearing stresses. The proposed procedure provides a graphical visual tool to selectively refine a FE mesh to improve maximum shearing stress estimates for the problems treated here. A Delphi environment based graphical interface was developed to aid in visualizing all the geometric aspects described herein such as deflected membrane, FE mesh, LL, refined meshes etc. This visualization process requires 3D real number to 2D real number transformations, as well as 2D real number to 2D integer pixels numbers transformations. Comparisons between the results of applying the proposed scheme and known analytic solutions to benchmark problems exhibit the necessary convergence characteristics. Numerical results will be presented.

COMPUTATIONAL MECHANICS (Abstracts)
WCCM VI in conjunction with APCOM'04, *Sept. 5-10, 2004, Beijing, China*

A Self-Consistent Approach to the Dynamic Effective Properties of Composites Reinforced by Distributed Spherical Particles

P. J. Wei *

Department of Mathematics and Mechanics, University of Sciences and Technology Beijing, Beijing, 100083, China
e-mail: weipj@126.com

Abstract

A self-consistent approach to the dynamic effective properties of composites reinforced by randomly distributed spherical inclusions is studied. A composite sphere consisted of the inclusion, the matrix and the interphase between them is assumed to be embedded in the effective medium. The effective wavenumbers of the coherent plane waves propagating through the composites are obtained by the dynamic self-consistent conditions which require that the forward scattering amplitudes of such a composite sphere embedded in the effective medium equal to zero. The prediction of the dynamic effective properties (effective phase velocity, effective attenuation and effective elastic moduli) obtained by the present self-consistent approach at various volume concentrations for SiC-Al composites is compared numerically with that obtained by the independent scattering approach proposed by Gubernatis (1984). It is found that there is a good agreement between two approaches at a relatively low frequency and volume concentration. However, the numerical results obtained by two approaches deviate from each other at a relatively high frequency and volume concentration. The deviations increase with the increase of the frequency and the concentration of inclusions. Because that Gubernatis's equation is formulated based on independent scattering approach which clearly neglects the correlations among the individual scatterers and thus is valid only in a relatively low frequency and low concentration, the dynamic effective properties predicted by the present dynamic self-consistent approach is expected to be more close to the actual effective properties than that predicted by Gubernatis's equation, especially in a relatively high frequency and high volume concentration.

REFERENCES

[1] S. K. Data, H. M. Ledbetter, Y. Shindo, *Phase velocity and attenuation of plane elastic waves in a particle-reinforced composite medium*, Wave Motion, 10, (1988), 171-182.
[2] L. O. Foldy, *The multiple scattering of waves*, Phys. Rev., 67, (1945), 107-119.
[3] J. E. Gubernatis, E. Domany, *Effects of microstructure on speed and attenuation of elastic waves in porous materials*, Wave Motion, 6, (1984), 579-589.
[4] M. Lax, *Multiple scattering of waves*, Rev. Modern Phys., 23, (1951), 287-310.
[5] Y. Shindo, H. Nozaki, S. K. Datta, *Effect of interface layers on elastic wave propagation in a metal matrix composite reinforced by particles*, ASME, J. Appl. Mech., 62, (1995), 178-185.
[6] P. C. Waterman, R. Truell, *Multiple scattering of waves*, J. Math. Phys., 2, (1961), 512-537.

COMPUTATIONAL MECHANICS (Abstracts)
WCCM VI in conjunction with APCOM'04, *Sept. 5-10, 2004, Beijing, China*

Molecular Dynamics Study on Bending Single Crystal Silver at Room Temperature

Lin Yuan [*], Debin Shan, Bin Guo

School of Materials Science and Engineering, Harbin Institute of Technology, P. O. Box 435, Harbin, 150001, China
e-mail: yuanlin@hit.edu.cn, shandb@hit.edu.cn, bguo@hit.edu.cn

Abstract

Recent experiments by Kiritani have revealed no dislocations are seen during high speed plastic deformation of fcc metals. As a possible explanation, they propose a dislocation -free deformation mechanism. In order to research fcc metals micro-mechanism of bending and crack initiation and growth, molecular dynamics (MD) was employed to simulate the bending process of single crystal silver (Ag). A new EAM potential was used in the simulation, the temperature was controlled at 298K by Nose-Hoover method. The results show that the bending process is divided into three stages. The deformation is elastic in the initial stage. Then the deformation is in the flow of plasticity. Finally it is in the stage of crack initiation and growth. However dislocations are not found in the plastic stage of such high speed bending of single crystal silver. In the stage of crack initiation and growth, vacancies are produced, then dislocations are easily formed nea r the vacancies, and vacancies combine to form crack initiation and growth. Dislocations are more easily created in a with groove model than in a without groove model at the crack initiation and growth stage. Crack defect promotes crack initiation and growth. The stress curve is elastic in the initial stage of bending of single crystal silver, and then in the plastic stage, the stress is stable, finally in the crack initiation and growth, the stress decreases. The potential energy of single crystal silver r ises with line in the elastic and plastic stage, and the potential energy in the plastic stage rises with no fluctuation compared with the elastic stage, but the potential energy decreases in the stage of crack initiation and growth.

REFERENCES

[1] M. Kiritani, Y. Satoh, Y. Kizuka, *Anomalous production of vacancy clusters and the possibility of plastic deformation of crystalline metals without dislocations*, Philosophical Magazine Letters. 79(10), (1999), 797-804.

[2] M. Doyama, Y. Kogure, *Embedded atom potentials in fcc and bcc metals*, Computational Materials Science, 14(1-4), (1999), 80-83.

[3] F. Cleri, *Representation of mechanical loads in molecular dynamics simulation*, Physical Review B, 65(1), (2002), 014107.

[4] T. Kitamura, K. Yashiro, R. Ohtani, *Atomic simulation on deformation and fracture of nano-single crystal of nickel in tension*, JSME International A, 40(4), (1997), 430-435.

COMPUTATIONAL MECHANICS (Abstracts)
WCCM VI in conjunction with APCOM'04, *Sept. 5-10, 2004, Beijing, China*

Modeling of Inter-Fiber Sliding in Fabric Shaping Process

Yutaka Arimitsu [1*], Masayuki Ohkita [1], Yuji Sogabe [1], Zhiqiang Wu [1], Tsu-Wei Chou [2]

[1] *Department of Mechanical Engineering, Ehime University, Matsuyama, 790-8577, Japan*
[2] *Department of Mechanical Engineering, University of Delaware, Newark, Delaware, 19716-3140 U.S.A.*
e-mail: arimitsu@eng.ehime-u.ac.jp

Abstract

In fabric shaping process of resin transfer molding, which is widely employed to produce a composite shell, fabrics deform with inter-fiber sliding and buckling of yarns. Fabric deformation depends on the kind of fabrics, initial orientation of yarns and shape of tools. The orientation of yarns after shaping and shape of stitches of fabrics affect the properties of a composite shell and fluidity of injected resin, respectively. Buckling of yarns and inter-fiber sliding also affect strength of a composite shell and deformation of fabrics, respectively. To predict these properties, simulation of fabric shaping process is actually required in engineering fields.
Inter-fiber sliding is neglected in the conventional finite element simulation using two-dimensional elements [1] and pin-jointed network models [2]. Therefore numerical results differ from actual deformation of fabrics, in deep drawing. Lai et al. [3] proposed the modeling of fiber slippage using FEM. In the proposed model, the amount of fiber slippage is geometrically estimated from yarn curvature. But inter-fiber sliding should be mechanically estimated including yarn curvature and tension of yarns.
Since plastic deformation is related to microscopic slippage of materials, solid elements possessing elastic-plastic properties can be applied to modeling of intersections of yarns. We employ truss and cable elements for yarn segments. And we employ soft solid elements, which are reinforced with hard cable elements on upper and lower surfaces, for intersections of yarns. In the present model, mutual yarns are not connected directly, like actual fabrics. Since elements of yarns are connected with upper and lower nodes, yarns possess undulation properties also. We can evaluate the distribution of buckling of yarns from compressive force of truss elements.
The simulation is performed with LS-DYNA. The results of the finite element simulation are in good agreement with the experimental results, even in large punch traveling.

REFERENCES

[1] L. Dong, C. Lekakou, M. G. Bader, *Solid-mechanics finite element simulations of the draping of fabrics: a sensitive analysis*, Composites, Part A, 31, (2000), 639.
[2] Y. Arimitsu, T-W Chou, *Modeling of yarn orientation and stress concentration in fabric shaping process*, Proc. of 23rd International SAMPE Tech. Conf., 1, (1999), 551.
[3] C-L. Lai, W-B Young, *Modeling fiber slippage during the preforming process*, Polymer Composites, 20, (1999), 594.

COMPUTATIONAL MECHANICS (Abstracts)
WCCM VI in conjunction with APCOM'04, *Sept. 5-10, 2004, Beijing, China*

Periodical Distribution of Ellipsoidal Inclusion with Polynomial Expansion of Eigenstrain in Three Lattice Figures SC, BCC and FCC

K. Behfar [1*], Q. Behfar [2]

[1] *Department of Civil Engineering, Sharif University of Technology, Azadi, Tehran, Iran*
[2] *Department of Computer Engineering, University of Sheikh Bahaie, Esfahan, Iran*
e-mail: k_behfar@mehr.sharif.edu, Q_behfar@USB.edu

Abstract

In composite materials, if it is assumed that the size of inclusions is fairly small in comparison with the whole size of material, for mathematical simplicity, it seems possible to apply an assumption that material is infinitely extended or assumed to have a periodic structure. When a composite has a periodic structure, the strains can't be considered constant; they are period functions of the space variables, as has been shown by Nemat-Nasser and Taya (1981) in estimating effective moduli of an elastic body containing periodically distributed voids.
In this paper, the averaged total strain, elastic energy and effective moduli of a composite with periodical distribution of ellipsoidal inclusion in three lattice figures (simple cubic, body-centered cubic, face-centered cubic lattice) are determined. Using equivalent-inclusion method, in order to satisfy the continuity of tractions and displacements, a transformation strain inside the inclusion called eigenstrain is introduced and to reach more accuracy, it is considered to be non-uniform and to have a second order of polynomial expansion. For three cases of loading including hydrostatic, pure shearing and a two-axial load, systems of linear equations are solved using language C as a fruitful compiler for programming in this study also the software Matlab, the eigenstrain multipliers are calculated. The resulting second-order eigenstrains are entered into the formulation to obtain the desired values of the current study containing the averaged total strain, elastic energy and effective moduli for the three lattice figures. The results achieved for different loadings and lattice figures are illustrated in some diagrams and the differences are easily seen and compared with each other. It should be noted that the values obtained for the effective moduli feature a bit difference from those mentioned in Ref. [1]. Also, the elastic energy obtained in this article is in the range of what has been determined for spherical inclusion and constant eigenstrain in Ref. [4].

REFERENCES

[1] S. Nemat-Nasser, T. Iwakuma, M. Hejazi, *On composites with periodic structure*, Mechanics of Materials, 1, (1982), 239-267.
[2] H. M. Shodja, A. S. Sarvestani, *Elastic fields in double inhomogeneity by the equivalent inclusion method*, Journal of Applied Mechanics, 68, (2001).
[3] T. Iwakuma, S. Nemat-Nasser, *Composites with periodic microstructure*, Computers & Structures, 16 (1-4), (1983), 13-19.
[4] Toshio Mura, *Micromechanics of defects in solids*, Kluwer Academic Publishers, (1991).

COMPUTATIONAL MECHANICS (Abstracts)
WCCM VI in conjunction with APCOM'04, *Sept. 5-10, 2004, Beijing, China*

Modeling of Actuation of Ionic Polymer Metal Composite by Electro-Osmosis and Ionization-Dependent Swelling: Concept

Marie Le Guilly [1], Chunye Xu [1]*, Minoru Taya [1], Sung-Soo Kang [2], Yutaka Toi [2]

[1] *Center for Intelligent Materials and Systems, Department of Mechanical Engineering, University of Washington, PO Box 352600, Seattle, WA 98195-2600, USA*
[2] *Institute of Industrial Science, University of Tokyo, 4-6-1 Komaba, Meguro-ku, Tokyo, 153-8505, Japan*

e-mail: Le Guilly@u.washington.edu, chunye@u.washington.edu, tayam@u.washington.edu
kang@iis.u-tokyo.ac.jp, toi@iis.u-tokyo.ac.jp

Abstract

The actuation of Ionic Polymer Metal Composites is an electro -chemo-mechanical phenomenon with numerous coupling effects. We propos e that the main phenomenon for the initial fast response is electro-osmosis. Cations move to the cathode and bring water along, swelling the cathode side and depleting the anode side. Once the cations have redistributed, the E -field in the bulk of the membrane is small and electro-osmosis stops. The subsequent behavior of the composite is proposed to be governed by the "equilibrium hydration state" of the membrane under ion concentration gradients. In Nafion, this state will be mostly uniform hydration, exc ept for a small contribution by osmotic pressure. In Flemion, we propose that the ionic concentration gradients lead to ionization gradients which dramatically influence the equilibrium hydration state, yielding large permanent deformations.

We believe the above formulation will allow us to present a unified model for the description of actuation of Nafion and Flemion. The key parameter to predict the composite behaviors after the characteristic time is the equilibrium hydration H_{eq} which we propose is altered by the applied voltage in the case of Flemion. Being a weak acid, Flemion can be in different ionization states which will strongly influence its equilibrium hydration.

REFERENCES

[1] P. E. Grimshaw, J. H. Nussbaum, A. J. Grodzinsky, M. L.Yarmush, *Kinetics of electrically and chemically induced swelling in polyelectrolyte gels.* J. Chem. Phys., 93(6), (1990), 4462-4472.

[2] P. E. Grimshaw, A. J. Grodzinsky, M. L. Yarmush, D. M. Yarmush, *Dynamic membranes for protein transport: chemical and electrical control,* Chemical Engineering Science, 44(4), (1989), 827-840.

COMPUTATIONAL MECHANICS (Abstracts)
WCCM VI in conjunction with APCOM'04, *Sept. 5-10, 2004, Beijing, China*

Direct and Indirect Metis Element Methods for Edge Delamination in Laminates

Nguyen Tien Duong [1], Nguyen Dang Hung [2*]

[1] *Welding and Metals' Technology Department, Ha noi University of Technology, Hanoi, No 1 Dai Co Viet, Vietnam*
[2] *Department of LTAS – Fracture Mechanics, University of Liege, B-4000 Liège, Belgium*
e-mail: tienduong02@yahoo.com, H.NguyenDang@ulg.ac.be

Abstract

A special class of hybrid finite element named "metis element" (mongrel element) which assures the monotone convergence was proposed by Nguyen Dang Hung. This element permits to avoid the calculation of a contour integral in 3 dimensions or in 2 dimensions by using a volume integral for element in 3 dimensions or a surface integral for element in 2 dimensions, respectively.
This paper presents the metis finite element method to solve the edge delamination problem in composite laminate structures. At first, the formulae of generalized plane strain state for delamination problem are presented. And then Wang's singular solution is used to determine the stress singularity order and eigen coefficients C_m, essential quantities to determine the singular stress field. After that, formulation of metis displacement finite element in two-dimensional problem is established. These elements including regular and singular part of stress fields are used to solve the edge delamination problem in mixed fracture modes I, II and III. Direct and indirect methods ware used to calculate stress parameters. Finally, numerical results for symmetrical layers of composite laminates loaded the uniaxial tension will be examined and compared with other authors. Three stress intensity factors and the energy release rates for delamination crack are calculated. It appears that both two (direct and indirect) methods give close results, first and foremost dominant mode.

REFERENCES

[1] Nguyen Dang Hung, *On the monotony and the convergence of a special class of hybrid finite element: the mongrel element*, in S. Nemat-Nasser ed. *Variational methods in mechanics of solids,* Pergamon (1978).
[2] De Saxce Géry, *Conception d'un élément fini hybride métis monocouche pour la modé lisation du delaminage dans les matériaux composites*, rapport interne du service M.M.S. de la F.P.Ms, Projet Multimatériaux, Octobre (1992).
[3] Nguyen Tien Duong and Nguyen Dang Hung, *Stress singularity order in the delamination of a composite laminate*, The 7th National congress on Mechanics, Hanoi , December 18-21, (2002), pp. 151-159.
[4] S. S.Wang, *Edge delamination in angle-ply composite laminates*, A.I.A.A. Journal, 22, (1984), 256-264.

COMPUTATIONAL MECHANICS (Abstracts)
WCCM VI in conjunction with APCOM'04, *Sept. 5-10, 2004, Beijing, China*

Mechanical Behaviors Analysis for Aligned Discontinuous Fiber Reinforced Composite

Jie Jin *

College of Mechanical & Electronical Engineering, Zhejiang University of Technology Hangzhou 310014, China
e-mail: jinsan@mail.hz.zj.cn

Abstract

Long continuous fiber reinforced thermoplastic composite has very high stiffness and strength along its fiber direction, but lacking of extensibility needed by plastic process, resulting in cost increasing. In the early of 90 years, American company Du Pont and other cooperators [2, 3] studied and explored aligned discontinuous fiber reinforced thermoplastic composite. This composite has the following advantages: 1) Comparable low cost for manufacturing; 2) Good thermo-formability which could be used to shape making complex composite structure such as I shape structure, valve plane wing, plane body and helicopter; 3) Its mechanical properties are comparable to long continuous reinforced composite; 4) Structure making from it is repairable. This kind of composite has wide application prospect in plane making and sport racing. Aligned discontinuous fiber reinforced has taken as uniform in macroscopic view, and non-uniform from microscopic view. The geometric feature of fiber distribution (e.g., fiber arranging manner, fiber ratio of length to radius) will influence the mechanical interaction between fiber and base, fiber and fiber, furthermore influence strength, stiffness and thermo-distension factors. It's important to understand these factors relationship to exploring new composite. Some researchers established models to describe the relationship among local geometric parameter, material physics constant and macro mechanical properties. These studies use FEM to analysis the influence of fiber ratio of length to radius, volume percentage ratio of the fiber and mechanics properties of fiber and base themselves to the composite properties. But it is rarely reported about the study of relationship between macro property of the composite with geometric structure and fiber distribution feature (e.g. fiber overlay ratio) by use of FEM and theory model, which is very important for the development of the composite application and computer aided material structure design.

The paper discusses Mechanical behaviors for aligned discontinuous fiber reinforced composite from the point of micro measure, and establishes the stress-strain field theoretical model by the shear-lag method based on the dual-fibers model. Further it puts out the aided design mathematics model for structure and performance of aligned discontinuous fiber reinforced composite by use of virtual displacement principle. The mathematics relationship for geometric structure parameters and elastic modulus of the composite is discussed. By comparison of the theoretical model analytical solution with the result of FEA, two results are in good agreement.

REFERENCES

[1] Ike Y. Chang, James F. Pratte, LDF™, *Thermoplastic composites technology*, J. of Thermoplastic Composite Materiral, 4, (1991), 24-36.

[2] W. C. Xu, *Structural optimization of thermomechanical properties for short-fiber-reinforced composites*, Journal of Thermoplastic Composites Material, 8, (1995), 180-193.

[3] M. H. Piggott, Short fiber polymer composites: a fracture-based theory of fiber reinforcement, J. Composite Materials, 28, (1994), 588-606.

COMPUTATIONAL MECHANICS (Abstracts)
WCCM VI in conjunction with APCOM'04, *Sept. 5-10, 2004, Beijing, China*

Modelization of Delamination in High Performance Composites by Finite Element Method

M. R. Khoshravan*, J. Azizi Morovvat

Department of Mechanical Engineering, University of Tabriz, Iran
e-mail: rkhosh@tabrizu.ac.ir

Abstract

High performance composite materials have largely opened their place in aerospace industries and they are on the good way in the other fields. The Carbon Fiber Reinforced Plastic composites are useful in the panels and all aeronautical structures. In this paper, the failure of U.D. CFRP composites, by delamination is studied. Test bars with the shape of Double Cantilever Beam are s olicited under static monotonic loading in mode I of rupture. The analysis is based on finite element method using ANSYS software. 2D models, using plane 82 element with 8 nodes, and 3D models, using solid 95 element with 20 nodes, are realized. For these s models, different mesh generations, such as singular and non singular are done. With Irwin -Kies theory and laws of elasticity, the strain energy release rate is calculated. Several theories of the rupture such as Tsai -Hill, Maximum Stress, VCCT (Virtual Crack Closure Technique), Method of Raju are used. The obtained numerical results are compared with the existing experimental results, and the scatters between them are discussed.

COMPUTATIONAL MECHANICS (Abstracts)
WCCM VI in conjunction with APCOM'04, *Sept. 5-10, 2004, Beijing, China*

Finite Element Modeling of Distributed Damage in Laminated Polymer Matrix Composites

J. N. Reddy [1*], E. J. Barbero [2], P. Lonetti [3], F. Rostam-Abadi [4]

[1] *Dept. of Mechanical Engineering, Texas A&M University, Collge Station, Texas 77843-3123, USA*
[2] *Dept. of Mechanical and Aerospace Engineering, West Virginia University, Morgantown, WV, USA*
[3] *Structural Engineering Department, University of Calabria, Italy*
[4] *U. S. Army TACOM, AMSTA-TR-R, MS 263, 6501 East 11 Mile Road, Warren, MI 48397-5000, USA*
e-mail: jnreddy@shakti.tamu.edu

Abstract

A damage constitutive model for polymer matrix composite materials is developed and implemented into a commercial finite element package. Laminated composite structures are then modeled using standard finite elements. The formulation is based on Continuous Damage Mechanics (CDM) in a consistent thermodynamic framework using internal state variables to describe in-plane and inter-laminar damage as well as unrecoverable deformations (plasticity). The model uses a small number of adjustable parameters that are identified from available experimental data. Furthermore, the damage threshold and evolution parameters are identified from available experimental shear stress strain data. The finite element formulation and identified material parameters are then used to investigate the influence of material properties, aspect ratio, stacking sequence, and geometric non linearity on the damage evolution, deflections, and failure of laminated beams under four-point loading. As shown in Figure 1, neglecting inter laminar damage (d_3=0) leads to misleading predictions of in plane damage evolution and poor predictions of failure load. In case of Glass/Epoxy composite, inter-laminar damage is significant for all $[0/90]_s$ and $[90/0]_s$ beams with low aspect ratios. As the aspect ratio increases, the inplane damage becomes dominant. In the case of Carbon/Epoxy composite, the interlaminar damage is dominant for all aspect ratios of the beam.

COMPUTATIONAL MECHANICS (Abstracts)
WCCM VI in conjunction with APCOM'04, *Sept. 5-10, 2004, Beijing, China*

3D Elastic-Plastic Analysis and Experiment of Interlaminar Stresses and Deformation in Quasi-Isotropic Thermoplastic Composite Laminates

Min Shen*, Jingwei Tong, Shibin Wang, Hongqi Li

Department of Mechanics, Tianjin University, Tianjin 300072, China
e-mail: smshenmin@sina.com

Abstract

The fully three-dimensional orthotropic elastic-plastic analysis of interlaminar stresses and deformation for AS4/PEEK thermoplastic laminates is developed. Using Shih's theory of anisotropic plasticity [1], accounting for differences in yield strength in tension and compression in orthogonal directions, the elastic-plastic behavior of anisotropy in AS4/PEEK thermoplastic unidirectional laminate is described by the orthotropic linear hardening. Based on the mechanical-properties of AS4/PEEK tested by Martin etc. [2], the orthotropic parameters of linear hardening of AS4/PEEK were determined by means of analysis of its orthotropic property in this work [3]. A three-dimensional orthotropic elastic-plastic finite element model of AS4/PEEK laminates was developed. Under uniaxial tensile loading, the 3D orthotropic elastic-plastic FE analysis of interlaminar stresses and deformation are carried out for the quasi-isotropic thermoplastic composite $[0/\pm45/90]_{2S}$ laminates, and its stress-strain curves of tensile experiment was simulated. The quantitative local-field of interlaminar displacements on the free edge was measured using microscopic moiré interferometry. It is indicated that the numerical tensile stress-strain curves, interlaminar displacement u and shear strain γ_{yz} of $[0/\pm45/90]_{2S}$ laminates computed with 3D orthotropic elastic-plastic model are consistent with the experimental results. The comparison indicates that the 3D orthotropic elastic-plastic FE analysis for interlaminar stresses of AS4/PEEK composite is reliable, and fairly reflects influence of ductility of PEEK matrix on the elastic-plastic behavior of whole AS4/PEEK laminate.

The present work can be able to gain a better understanding of the influence of elastic-plastic deformation on interlaminar stresses and deformation in AS4/PEEK thermoplastic composite laminates, and can make a foundation for further optimize design and application of this class of composites.

The project is supported by the National Natural Science Foundation of China (No.10172065).

REFERENCES

[1] C. F. Shih, D. Lee, *Further developments in anisotropic plasticity*, Journal of Engineering Materials and Technology, 100, (1978), 294-302.

[2] R. J. Martin, R. S. Sandhu, A. N. Palazotto, *Experimental and analytical comparisons of failure in thermoplastic composite laminates*, Experimental Mechanics, 34, (1994), 53-65.

[3] Shen Min, *Elastic-plastic analyses and micromechanical experiments of interlaminar stresses and deformation in thermoplastic composite laminates*, PhD Thesis, Tianjin University, China, (2001), (in Chinese).

COMPUTATIONAL MECHANICS (Abstracts)
WCCM VI in conjunction with APCOM'04, *Sept. 5-10, 2004, Beijing, China*

A Computational Method of Filament Wound Composite Cylindrical Beam under General Loadings

J. Sun [1*], M. Růžička [2], O. Uher [3]

[1] *Department of Electo-Mechanical Eng., Jiaxing University, Zhejiang Jiaxing, 314001, China*
[2] *Department of Mechanics, Czech Technical University in Prague, Technická 4,166 07, Prague, Czech Republic*
[3] *ComptechPlus, Družestevní 159, 34201, Sušice, Czech Republic*
e-mail: sunjiang@mail.zjxu.edu.cn, ruzicka@fsid.cvut.cz, ondrej@compotech.com

Abstract

In this paper, a computational method as an expansion of the computation method in [1] is presented for filament wound composite cylindrical beam under general loadings.
The problem considered consists of a long hollow cylindrical beam made of filament wound composite which has elastic properties varying circumferentially and in the direction normal to the middle surface as well. Three coordinate systems, the global and the local Cartesian systems and the local curvilinear system, are used for computation.
In the analysis, first, the hollow beam is divided into several pieces of beam elements. The displacements in three directions of local curvilinear system are developed. Then stresses and strains with 3D elastic properties are analyzed in the local curvilinear system, transferred to the local Cartesian coordinates and the global Cartesian system and integrated in the global system. Thus the relationship of the displacements of beam and the applied loads in the global system can be found. Based on the relationship equation of displacements and loads and in conjunction with appropriate boundary conditions, the deformations of thin- or thick-walled composite beams can be determined when the applied forces and moments are specified. If the deformations are transformed back to local curvilinear coordinate, strains can be obtained.
The method mentioned above is straightforward and can be easily programmed for engineering application. The calculation results of computation have a good correlation with the results from Kim and White theory [2] and with the test results, which means the present method using 3D elastic effects can effectively predict the deformation behavior of thin- or thick-walled composite beams with coupled material behavior.

REFERENCES

[1] M. Ruzicka, J. Sun, O. Uher, *Development and application of computer program for design of filament wound composite beams with circular and noncircular cross-section*, Proc. 1st Asian-Pacific Congress on Computational Mechanics, Sydney, (2001), pp.1761-1766.

[2] C. Kim, S. R. White, *Analysis of thick hollow composite beams under general loadings*, Composite Structures, 34, (1996), 263-277.

COMPUTATIONAL MECHANICS (Abstracts)
WCCM VI in conjunction with APCOM'04, *Sept. 5-10, 2004, Beijing, China*

Bending Behaviour of Thick Composite Plates Subjected to Mechanical and Hygrothermal Loading

Tran Ich Thinh *

Department of Mechanics of Materials and Structures, Hanoi University of Technology, I Dai Co Viet, Vietnam
e-mail: tranichthinh@Yahoo.com

Abstract

This paper deals with hygrothermomechanical behaviour of thick rectangular simply supported composite plates. The present studies differentiate from previous studies in that:
1) The full third-order theory is used, in which both in-plane and out-of-plane displacement components are assumed to have cubic variations through the thickness of the plate.
2) The laminate moisture concentration distribution *m(z, t)* is a true representation of a practical situation. The temperature distribution in the z-direction is assumed to be constant.
3) The analysis of bending behaviour of thick composite plates is carried out employing a C^0 Quad-9 isoparametric higher-order finite element with twelve-degrees of freedom per node.
4) The loading, *q*, is the distributed force acting on the top or bottom surface of the plate.
To efficiently effect the calculations, a computer program was developed. The output gives deflections, rotations, stress resultants, strains and stresses in the plates.
The numerical examples are presented for symmetrically *on-axis*, ($0^0/90^0/90^0/0^0$) and *off-axis*, ($45^0/-45^0/-45^0/45^0$) graphite-epoxy laminated plates under combined mechanical and hygrothermal loads.
The obtained results show a good agreement of the present finite element solutions for thin-to-thick composite plates when compared with other finite element solutions and closed form solutions in the literature.
The numerical results show that the hygrothermal environment causes the increases in σ_1, σ_2 and σ_3 in all layers of the plate, but that shear stresses have no appreciable influence and the environment has significant influence on the limit loads, which were calculated by Tsai-Wu criterion.

REFERENCES

[1] J. N. Reddy, *A simple high-order theory for laminated composite plates*, J. Appl. Mech., 51, (1984), 745-752.
[2] B. N. Pandya, T. Kant, *Finite element analysis of laminated composite plates using a higher-order displacement model*, Composites Science and Technology, 32, (1998), 137-155.
[3] Tran Ich Thinh, Ngo Nhu Khoa, *On a bending problem of thick laminated composite plates,* Vietnam Journal of Mechanics, 24(1), (2002).
[4] G. J. Turvey, *Bending of laterally loaded, simply supported, moderately thick, antisymmetrically laminated plates*, Fibre Science and Technology, 10, (1997).

COMPUTATIONAL MECHANICS (Abstracts)
WCCM VI in conjunction with APCOM'04, Sept. 5-10, 2004, Beijing, China

Decay Effect of Laminated Beam under Bending Vibration

Z. R. Tian [1*], Z. S. Tian [2]

[1] *Department of Mechanics, Northwestern Polytechnical University, Xi'an, 710072, China*
[2] *Department of Mechanics, Graduate School, Chinese Academy of Sciences, P. O. Box 2706, Beijing, 100080, China*
e-mail: zrtian@nwpu.edu.cn

Abstract

The decay effect of laminated beam with middle layers made of soft decaying materials under bending vibration has been studied by using finite element method and analytical method.
The middle layers of classical laminated composites are usually made of pure resin. Its decay property needs to be improved. One of the possible approaches is of using soft decaying materials instead of pure resin. The decay property of these laminated beam under bending vibration has been analyzed.
The differential equation is derived for laminated beams with symmetric or unsymmetric decaying layers and different shear modulus under bending. By the use of exponential sum method, the expression of the mode shape is found. Four different boundary conditions are investigated. For hinged supported beam with twisting inertia moment along its length, the effect of increase in force on the decay of natural frequency and the relationship between them are studied for unsymmetric surface layer, thick and soft core and small thickness compared with span. And the bending vibration of laminated pillar with shear deformation is also studied.
Numerical solutions of 13 models show that the frequencies and the decay coefficients given by both present analytical method and the finite element method are very close to the experimental results. It is demonstrated that present analytical and approximate methods to be correct and effective in dealing with the problem of decay effect of laminated beam under bending vibration.
Meanwhile, the numerical examples also provide following result: 1) The decay is lowest when the beams possess unsymmetric decaying layers. 2) When the middle layers are made of soft decaying materials, if the decay of the middle layer increases then the shear deformations will also increase and the stiffness will decrease. 3) In order for natural type (type I) to possess the greatest decay the shear modulus of the decaying layers under bending and pressure are about 3.5 times of those of classical laminated composites with resin.

REFERENCES

[1] Z. R. Tian, *Mathematic and mechanic methods in composite materials*, National Defence Industry Press, Beijing, (2004).
[2] R. W. Clough, J. Penzien, *Dynamics of structures*, Mc Graw-Hill Book Company, New York, (1975).
[3] K. J. Bathe, *Finite element procedures in engineering analysis*, Prentice, Hall, Inc, Englewood Cliffs, New Jersey, (1982).
[4] G. P. Ostermeger, *Mechanik(I)*, Technische Uni, Berlin, Berlin, Deutschland, (1995).
[5] Kurt Moser, *Zur Berechnung Geschichteter Schalentragwerke Mittels der Methode der Finiten Elemente*, Innsbruck Uni, Bericht-4, Österreich, (1989).

COMPUTATIONAL MECHANICS (Abstracts)
WCCM VI in conjunction with APCOM'04, *Sept. 5-10, 2004, Beijing, China*

Geometrically Linear and Nonlinear Behavior of Local Loaded Laminated Rubber Toroidal Shells by a Mixed Finite Element Method

Yu Ching Wu [*], Xiaoxiang Xue, Zhenhan Yao

Department of Engineering Mechanics, Tsinghua University, Beijing, 100084, China
e-mail: wuyuq@mail.tsinghua.edu.cn,xuexiaoxiang99@mails.tsinghua.edu.cn,demyz@tsinghua.edu.cn

Abstract

Based on mixed finite-element approximations, Kulikov and Plotnikova developed a numerical algorithm for solving linear static problems of pre-stressed multilayer composite shells subjected to large displacements and arbitrarily large rotations. These equations [1-3] are investigated and extended to analyze geometrically linear and nonlinear behavior of laminated toroidal rubber shells of non-uniform wall thickness. The first-order discrete-layer theory has been discussed in detail. The transverse shear and transverse normal effects are included. The equation of motion, the kinematical equation, the constitutive law and the Hu-Washizu functional of layered plates and shells are generated in Cartesian, cylindrical, spherical and toroidal coordinate systems. Danielson and Noor [4] generated finite elements in cylindrical coordinates for three-dimensional analysis of tires in 1997. The exact modeling of circular geometries can provide better finite element predictions and reduce the number of elements needed around the tire circumference. The cylindrical finite elements have been extended and modified as ones in toroidal coordinate. The solutions of displacement are calculated using a discrete layer finite element method and compared with the analytical solutions from previous papers to demonstrate the accuracy of the numerical model. The relevant algorithm is based on a differential approach using the method of discrete orthogonalization, and the finite element method with linear local approximation in the meridional direction. The results of study have a variety of applications in tire technology and rubber production to the analysis of a radial automobile tire subjected to rim mounting, non-conservative inflation pressure, and rigid pavement contact. The method will be used to analyze some more complex structures, such as plates and shells of arbitrary wall thickness, toroidal piezoelectric shells, material nonlinear rubber-cord composite shells, etc.

REFERENCES

[1] G. M. Kulikov, S. V. Plotnikova, *Simple and effective elements based upon Timoshenko-Mindlin shell theory*. Computer Methods in Applied Mechanics and Engineering. 191, (2002), 1173-1187.

[2] G. M. Kulikov, S. V. Plotnikova, *Investigation of locally loaded multilayer shells by a mixed finite-element method, 1. geometrically linear statement*, Mechanics of Composite Materials. 38(5), (2002), 397-406.

[3] G. M. Kulikov, S. V. Plotnikova, *Investigation of locally loaded multilayer shells by a mixed finite-element method, 2. geometrically nonlinear statement*. Mechanics of Composite Materials, 38(6), (2002), 539-546.

[4] K. T. Danielson, A. K. Noor, *Finite Elements Developed in Cylindrical Coordinates for Three-Dimensional Tire Analysis*. Tire Science and Technology, TSTCA, 25(1), (1997), 2-28.

COMPUTATIONAL MECHANICS (Abstracts)
WCCM VI in conjunction with APCOM'04, *Sept. 5-10, 2004, Beijing, China*

A Mixed Lagrangian Eulerian Finite Element Method with ALE Formulation for Large Deformation Fluid-Structure Interaction Simulations

Hsien-Jan Chang*, H. Theodore Lin

AMPS Technologies Company, 1138 Fox Chapel Road, Pittsburgh, PA 15238, USA
e-mail: tedlin@ampstech.com, hjchang@ampstech.com

Abstract

The fluid-structure interaction (FSI) effects have become an important design analysis consideration in the wide applications of finite element method to industrial problems. The effects are especially important in the microelectronics device simulations due to the relatively soft material and the increasingly dominating momentum effect of the surrounding fluid or air. The paper presents a direct-coupled mixed Lagrangian Eulerian finite element formulation for such FSI applications.
In the Eulerian flow field, an optimal Least-squares finite element formulation [1, 2] is adopted due to its numerical stability, theoretical completeness, and the direct compatible DOF's with the velocity based Lagrangian formulation for the continuum field. By adopting a velocity based Lagrangian formulation, the fluid-structure interface condition of the Eulerian flow field and the Lagrangian deformation are automatically preserved due to the direct-coupled velocity variable. Numerical results of this direct strong-coupled formulation are compared with the existing literatures to demonstrate the straightforward numerical applicability and accuracy.
To solve most of the industrial problems involving relative large deformation in the continuum field, we adopt an updated Lagrangian formulation for the continuum motion description. In the surrounding Eulerian field, an Arbitrary Lagrangian Eulerian (ALE) formulation is applied to accommodate the large deformation. The common Eulerian-Lagrangian interface becomes the ALE remeshing constraint, and the direct coupling of the Eulerian flow field ensures the compatibility of the continuum velocity and deformation compatibility. Numerical results are presented with demonstration to general industrial problems.

REFERENCES

[1] H. T. Lin, T. L. Lin, *An optimal least-squares finite element method with h-p adaptivity for CFD and conjugate heat transfer*, 4th National Congress on Computational Mechanics, August, (1997).

[2] H. T. Lin, A weighted optimal least-squares finite element method for CFD in multi-physics simulations", WCCM VI/APCOM '04, September, (2004), (submitted).

COMPUTATIONAL MECHANICS (Abstracts)
WCCM VI in conjunction with APCOM'04, *Sept. 5-10, 2004, Beijing, China*

Lattice Boltzmann Simulation of Flow past a Sphere up to Reynolds Number of 300

Z. X. Cui, Y. Liu *, R. M. C. So

Department of Mechanical Engineering, The Hong Kong Polytechnic University Hung Hom, Kowloon, Hong Kong, China
e-mail: mmyliu@polyu.edu.hk

Abstract

A sphere in a cross flow is one of the most basic and revealing cases in the three dimensional fluid-structural interaction problems. This problem has been both experimentally and numerically investigated in recent years for its widespread applications as well as complicated mechanism of fluid-structural interactions. Among different CFD approaches, lattice Boltzmann method (LBM) as a relatively new numerical scheme has been demonstrated to be an effective tool in simulating fluid flows. In this paper, a three-dimensional LBM model is used to numerically simulate the flow past a sphere at low Reynolds numbers (Re) of 100, 150, 200, 250 and 300, respectively. For Re below 200, the flow is separated, steady, axisymmetric and topologically similar, and the wake is laminar; for Re between 210 and 270, steady and planar-symmetric flow is observed; and for Re between 280 and 375, the flow becomes unsteady but planar-symmetry is maintained. The above Reynolds numbers are chosen to stand for these different flow regimes. To get better accuracy, a local mesh refinement technique is adopted.

To correctly simulate the fluid-structure interaction, the most important is the boundary treatment. The conventional treatment is bounce back method which is the first order accuracy. In this paper, we use a new boundary extrapolation treatment which has been proved to be of second order accuracy and has improved numerical stability. The objective of this study is to solve this three dimensional problem and establish the confidence of three dimensional simulation using LBM, also through this three dimensional simulation, to gain better understanding of the kinematic and vortical interactions.

The calculation has been validated against measurement and other available numerical simulation. The present calculation compares well with available results in mean drag coefficient, separation angle, and vortex position.

Fig. 1 shows clearly the hairpin like vortex structure at Re=300.

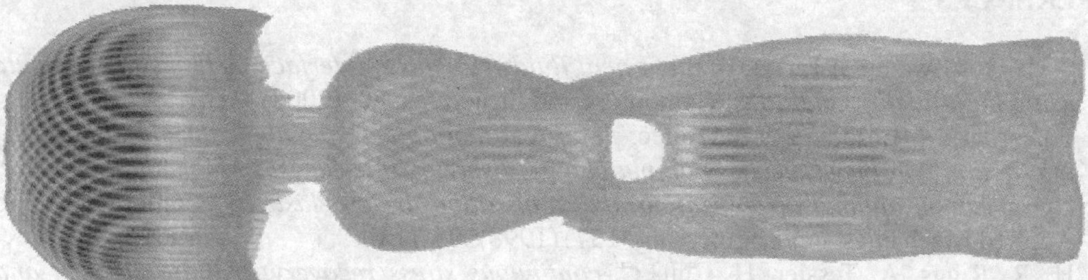

Fig. 1 Vortex structure at Re = 300

COMPUTATIONAL MECHANICS (Abstracts)
WCCM VI in conjunction with APCOM'04, *Sept. 5-10, 2004, Beijing, China*

An Interfacing Strategy for Fluid-Structure Interaction

Linda Huang, H. R. Riggs *

Department of Civil & Environmental Engineering, University of Hawaii at Manoa, 2540 Dole St. Holmes Hall 383, Honolulu, HI, 96822, USA
e-mail: linlin@hawaii.edu, riggs@hawaii.edu

Abstract

In nonlinear, time domain, fluid-structure analysis based on a staggered solution method, the displacements from the structural mesh must be transferred to the fluid mesh, and the pressures (or equivalent forces) must be transferred from the fluid mesh to the structural mesh. When the meshes describe two distinct surfaces in three-dimensional space (common when the wetted surface is curved) this coupling of the meshes is difficult and does not have a unique mathematical solution. Many strategies have been proposed, only some of which specifically ensure conservation of energy. See [1] for a discussion of methods and [2] for a discussion of energy conservation.

In this paper we propose a new interfacing strategy that is based on the smoothing element analysis (SEA) method developed to recover stresses in finite element analysis [3-4]. A transfer strategy has two components: a mapping component to relate points on one mesh to points on the other mesh, and an interpolation component so that the data can be obtained at the required locations. We assume that the wetted surface is defined (piecewise) by analytical functions, e.g., by a parametric-based CAD description of the structure. Hence, an analytical surface can be represented by 2-D parametric coordinates. Interpolation is then done in the 2-D parametric space.

SEA is a variationally-based, penalized discrete least squares technique that is applicable to recovery of smoothed data fields based on discrete input data. SEA is directly applicable to smoothing and then transferring displacements from the structure to the fluid. To ensure energy conservation, the variational principle is modified to smooth and then transfer pressures to the structural mesh. The method is unique in that pressures and not nodal loads are transferred.

The method is evaluated numerically with several examples from linear hydrodynamics and linear hydroelasticity. The advantage of choosing these examples is that the different components of the strategy can be evaluated separately, and the performance of the method can be more clearly examined. Numerical results show the method is quite promising.

REFERENCES

[1] G. P. Guruswamy, *A review of numerical fluids/structures interface methods for computations using high-fidelity equations*, Computers and Structures, 80, (2002), 31-41.

[2] C. Farhat, M. Lesoinne, P. LeTallec, *Load and motion transfer algorithms for fluid/structure interaction problems with non-matching discrete interfaces: Momentum and energy conservation, optimal discretization and application to aeroelasticity*, Computer Methods in Applied Mechanics and Engineering, 157, (1998), 95-114.

[3] H. R. Riggs, A. Tessler, H. Chu, *C^1-continuous stress recovery in finite element analysis*, Computer Methods in Applied Mechanics and Engineering, 143, (1997), 299-316.

[4] A. A. Yazdani, H. R. Riggs, A. Tessler, *Stress recovery and error estimation for shell structures*, International Journal for Numerical Methods in Engineering, 47, (2000), 1825-1840.

COMPUTATIONAL MECHANICS (Abstracts)
WCCM VI in conjunction with APCOM'04, *Sept. 5-10, 2004, Beijing, China*

Simulation of the Interaction among Pressure Wave Propagation, Wall Deformation and Bubble Dynamics in the Mercury Target

Ryuta Imai [1], Shuichi Ishikura [2], Hiroyuki Kogawa [2], H. Sato [2], R. Hino [2], Chuichi Arakawa [3*]

[1] *Center for Computational Science and Engineering, FUJI Research Institute Corporation, 2-3, Kanda-nishiki-cho, Chiyoda-ku, Tokyo, 101-8443, Japan*
[2] *Center for Neutron Science, Japan Atomic Energy Research Institute, Tokai-mura, Naka-gun, Ibaraki-ken, 319-1195, Japan*
[3] *Graduate School of Interdisciplinary Information Studies, the University of Tokyo, Hongo, Bunkyo-ku, Tokyo, 113-8656, Japan*
e-mail: rimai@ccse.fuji-ric.co.jp, {ishikura, kagawa, hsato, hino}@cat.tokai.jaeri.go.jp, arakawa@iii.u-tokyo.ac.jp

Abstract

Japan Atomic Energy Research Institute (JAERI) is promoting to construct the mercury target as a neutron source by the nuclear spallation reaction. In the spallation reaction process, high-density heat is released and strong pressure wave is generated consequently. Then, in some experiments, it is shown that some pits type scratch is created on the wall when strong pressure wave reflects on it. And it is considered that an erosion of the wall might be caused by the cavitation. In this paper a simulation of the interaction among the pressure wave propagation of the liquid mercury, the wall deformation and the bubble dynamics in the mercury target is carried out in order to investigate the possibility of the cavitation in the mercury target.
In particular, weakly coupled simulation between compressible fluid dynamics and elastic solid dynamics is adopted and the interaction between the pressure wave propagation and the bubble dynamics is introduced through the local fraction of volume occupied by the bubble at every cell. Thus our simulations show that the wall deformation causes a rapid reduction of the pressure near the target wall and that a negative pressure might be reached there. Moreover the bubble dynamics simulation in the negative pressure mercury near the wall is carried out to advance the coupling simulation.

REFERENCES

[1] A. Prosperetti, *The thermal behavior of oscillating gas bubbles*, J. Fluid Mech., 222, (1991), 587-616.
[2] S. Ishikura, H. Kogawa, M. Futakawa, K. Kikuchi, R. Hino, C. Arakawa, *Bubble dynamics in the thermal shock problem of the liquid metal target*, J. Nucl. Mater., 318, (2003), 113-121.
[3] R. Imai, S. Ishikura, H. Kogawa, R. Hino, C. Arakawa, *Development of a fluid-structure interaction code for a mercury target*, Transactions of JSCES, Paper No. 20030014, (2003).

COMPUTATIONAL MECHANICS (Abstracts)
WCCM VI in conjunction with APCOM'04, *Sept. 5-10, 2004, Beijing, China*

Numerical Analysis of Fluid-Solid Coupled Seepage in Compacted Overbroken Rock Mass

W. Q. Liu*, X. X. Miao

School of Sciences, China University of Mining & Technology, Xuzhou, 221008, China
e-mail: wqliu@cumt.edu.cn, xxmiao@cumt.edu.cn

Abstract

In mining engineering, effectively solving the seepage problems of fractured or broken rock mass is regarded as a most significant challenge. Statistical reports show that the coal-mining incidences in China have killed more than 10,000 workers during the past 10 years, and these incidences include about 700 serious disasters, which killed more than 10 workers at a time. In all the incidents, those happening because of methane explosion and water burst almost take over 80% [1]. As we know, mining activity probably brings about the fracturing and collapsing of surrounding rock mass, and since the fracture permeability is far larger than the pore one [2, 3], majority of the seepage disasters naturally take place in fractured and overbroken rock mass. By now, the numerical techniques for the seepage calculation of thoroughly integrated and generally fractured rock masses have gotten abundant introduction in references, but they are impractical for overbroken rock mass. Therefore, it is quite necessary to develop a numerical method to calculate overbroken rock seepage.
Based on Eulerian control-volume principle, this paper puts forward a set of procedures to describe the plastic finite deformation of overbroken rock mass, and then employs them in construction of corresponding seepage models joined with the program RFPA, which possesses the characteristics of parameter heterogeneity and fluid-solid coupling. The calculated results meet the testing data very well, and also show some other seepage properties difficult to be tracked in testing, such as flux and pressure field distributions. The given procedures and conclusion can provide valuable reference in the study of overbroken rock seepage.
The supports of the key National Natural Science Foundation (50134040), the National Outstanding Youth Foundation (50225414), and the PhD Foundation of Educational Ministry of China (20010290003) are gratefully acknowledged.

REFERENCES

[1] S. N. Zhou, X. F. Xian, W. X. Zhu, *Strategy of prevention and control of coal methane*, CUMT Press, Xuzhou, China (2001).
[2] G. I. Barenblatt et al, *Basic concepts in the theory of seepage of homogeneou s liquids in fissured rocks*, Prikl. Mat. Mekh., 24, (1960), 852-864.
[3] W. Q. Liu, *Studies on theory and application of seepage in overbroken rock mass*, PhD Dissertation, China Univ. of Min. & Tech., Xuzhou, China, (2002).

COMPUTATIONAL MECHANICS (Abstracts)
WCCM VI in conjunction with APCOM'04, *Sept. 5-10, 2004, Beijing, China*

Computational Modelling of Strain Localization in Elastoplastic Saturated Porous Media

X. Liu*, A. Scarpas

Section of Structural Mechanics, Delft University of Technology, Stevinweg 1, Delft, 2628CN, The Netherlands
e-mail: X.Liu@ct.tudelft.nl, A.Scarpas@ct.tudelft.nl

Abstract

The analysis of strain localization is of importance in engineering practice because localization is a precursor to sudden failure. Localized deformations in the form of narrow shear bands are often observed to develop after large inelastic deformation in materials. Within this shear band, the material behaviour is inelastic. The width and orientation of these shear bands depend on material parameters, geometry, boundary conditions and loading conditions. In other words, knowledge concerning the development of the strain localization phenomenon will also lead to insight into failure mechanisms.
In order to properly characterize the mechanisms of strain localization, the investigations of strain localization are needed both numerically and experimentally. Strain localization has been extensively studied in recent years, in particular in connection with single phase solids. Due to the complexities associated with solid fluid phase interactions, the mechanisms responsible for strain localization in multiphase porous media can vary significantly from case to case. Several problems need to be addressed such as: the role of the fluid components in localization, the significant influencing factors in the development of strain localization, the question of why localization pattern vary between different tests.
The aim of this investigation is to provide insight into the phenomenon of strain localization in fluid saturated sands by means of numerical modelling. In order to achieve this goal, a constitutive model capable of describing the elastoplastic response of porous media including model nonassociativity is developed.
A fully saturated 3D sand specimen subjected to true triaxial loads is chosen for numerical investigation. The importance of the presence of fluid on the formation and evolution of strain localization are identified. The role of influencing factors, such as soil permeability, water suction, and the draining conditions is presented. Also, the influence of specimen geometry is illustrated. Various combinations of specimen geometry and material characteristics are examined.

COMPUTATIONAL MECHANICS (Abstracts)
WCCM VI in conjunction with APCOM'04, *Sept. 5-10, 2004, Beijing, China*

Coupling of Mechanical Models and Application to Vehicle Guide Way Interaction Simulation

S. Lutzenberger *

Department of structural mechanics, Munich University of Technology, Arcisstr 21, München, 80333, Germany
e-mail: s.lutzenberger@bv.tum.de

Abstract

The coupling of dynamic models is the basic principle in order to describe the vehicle guide way interaction, which simulates the movement of a vehicle along a guide way (bridge, rail system). A difficulty herby is the movement of the vehicle, which causes a time dependent coupling.
The algorithm [3] provides the coupling of different models (Multibody, Finite Elements, Boundary elements). The coupling conditions are included in the system of equations by means of Lagrange constraints. The equations of motion are formulated with respect to Euler/Lagrange whereas the dynamic wheel loads are included in the vector of unknowns. The movement of the vehicle along the guide way is described by time dependent interpolation functions.
This formulation leads to a system of differential algebraic equations (DAE). The index of the DAE is a measure for the degree of difficulty during the numerical solution of the system [1]. The time integration scheme depends of the index of the DAE. Coupled structural dynamic systems can have an index from 1 to 3. As structural dynamic systems possess a great many of degrees of freedom, time integration schemes that integrate the DAE 2. order are preferable (Newmark-β, Generalized-α, DAISY). The results of the interaction simulation are compared to other solution strategies ([2]) and show a very good agreement.
The algorithm is used to simulate the crossing of a three axle truck over a two span composite bridge. The vehicle is modelled as a vertical dynamic Finite Element model, by taking the most important components as tires, axles, leaf springs, the frame and the vehicle masses into account. The bridge is modeled by a Finite Element model with 7000 Dof. Dynamic measurements were taken of the vehicle and the bridge separately and used for verification purposes of the models. The unevenness of the road, which is the major influence for the excitation of the vehicle is included in the simulation. Results of the interaction calculation will be shown. The verification of the interaction simulation is done by comparing the calculation results to measurements, taken during the crossing of the truck over the bride.

REFERENCES

[1] K. E. Brenan, S. L. Campbell, L. R Petzold, *Numerical solution of initial-value problems in differential-algebraic equations*, North-Holland, New York, (1989).
[2] L. Fryba, *Vibration of solids and structures under moving loads*, Noordhoff International Publishing, Groningen, (1972).
[3] S. Lutzenberger, *Ein differentiell – algebraisches Simulationsmodell zur Fahrzeug – Fahrweg Interaktion*, Dissertation, TU München, (2003).

COMPUTATIONAL MECHANICS (Abstracts)
WCCM VI in conjunction with APCOM'04, *Sept. 5-10, 2004, Beijing, China*

Simulation of Offshore Wind Turbines with a Coupled Multi-Physics Model

T. Srisupattarawanit [1*], L. Lehmann [2], H. G. Matthies [1]

[1] *Inst. of Scientific Computing, TU Braunschweig, Hans-Sommer Str.65, Brunswick, 38106, Germany*
[2] *Inst. of Applied Mechanics, TU Braunschweig, Spielmann Str.11, Brunswick, 38106, Germany*
e-mail: T.Srisupattarawanit@tu-bs.de, lutz-o.lehmann@tu-bs.de, H.Matthies@tu-bs.de

Abstract

Presently there are quite a few projects in various stages of planning and construction which aim at placing wind turbines offshore for electricity generation. There the usually higher wind speeds can be exploited, but the turbines are also subjected to the offshore environment, i.e. especially wave loading. Fatigue calculations of wind turbine structures usually involve Monte Carlo Simulations of their operation in turbulent wind. In the offshore environment, we have to consider a coupled system, consisting of the structure, foundation, surrounding soil, the aeroelastic system for the wind model, and the hydroelastic system for the wave model.

In this work, the models have been developed initially separately, and are consistently coupled here to give the whole system. They consist of:

1) the structural subsystem, which is a nonlinear large displacement finite element model (FEM);
2) the aerodynamic subsystem, with prescribed stochastic wind characteristics and an instationary dynamic stall model for the aeroelastic blade loading;
3) the hydrodynamic subsystem, modeled mainly as potential flow and coupled to the stochastic wave field to describe the real sea state, and take the far-field into account. The computation on this part is based on the boundary element method (BEM);
4) the soil dynamic subsystem, modeled as the near-field soil close to the underground structure and discretised by the FEM, and the far-field effect which is described by the scaled boundary finite element method (SBFEM).

All of these models are coupled in the time domain, and the consistent simulation of the offshore wind turbine is presented. As the computational cost is high, reduction methods are used to reduce the amount of computation.

REFERENCES

[1] H. G. Matthies, C. Nath, T. E. Schellin, A. D. Garrad, M. A. Wastling, D. C. Quarton, J. Wei, M. Scherweit, T. Siebers, *Study of Offshore Wind Energy in the EC, JOULE I. Verlag natüliche Energie*, Germanischer Lloyd, Germany, (1995).

[2] Marcus Meyer, *Reduktionsmethoden zur aeroelastischen Simulation von Windkraftanlagen*, Dissertation, TU Braunschweig, Germany, (2002).

[3] John P. Wolf, *The Scaled Boundary Finite Element Method*, John Wiley & Sons Ltd., Chichester, England, (2003).

COMPUTATIONAL MECHANICS (Abstracts)
WCCM VI in conjunction with APCOM'04, *Sept. 5-10, 2004, Beijing, China*

The Volume-Surface Integrated Finite Difference Method in the Modelling of Complex Multiple Porosity Poroelastic Reservoirs

Mario-César Suárez-Arriaga [1*], Fernando Samaniego-Verduzco [2]

[1] School of Sciences, Michoacan University, Patzimba 438, Vista Bella, Morelia, Mich., 58090 Mexico
[2] Faculty of Engineering, National University of Mexico, Ciudad Univ ersitaria, 04250 Mexico, D.F.
e-mail: msuarez@zeus.umich.mx, ance@servidor.unam.mx

Abstract

Under the general concept of multiple porosity reservoirs, N (> 2) continuous poroelastic media interact with each other. Every medium has its own parameters and its own interporosity flow. Double and triple porosity models can be classified as special cases of this general theoretical concept, applicable to all-class reservoirs, including many productive fields in the world (oil, gas, water, geothermal). Petrophysical parameters in this kind of reservoirs are heterogeneous and randomly distributed, conforming extremely complex natural systems. This heterogeneity affects both mass and energy flows and the thermodynamic evolution and mechanical behavior of the whole system. The simultaneous heat and mass flows are coupled to the deformation of thermoporoelastic rocks. The fundamental equation of the model is the total flow of mass, solid plus fluid, in a porous rock. This conduces to a natural generalization (Biot, 1941) of Darcy's Law for a deformable medium. The sum of the tensor acting on the solid rock plus the tensor acting on the fluid is the total stress tensor satisfying classical equilibrium conditions. Assuming that the geothermal rock is only subjected to small deformations, a linear Hooke's Law can be used to represent the relationship between strains and stresses. The elastic matrix coefficients could be functions of both pressure and temperature and must be determined experimentally. Modeling these coupled phenomena requires averaging highly contrasting physical properties, independently of the method used in the solution of equations. The treatment of heterogeneity in complex porous systems opens the door to new and challenging mathematical problems related to modeling and numerical simulation: optimal construction of deformable meshes, efficient and accurate calculation of derivatives, multiscaling, discontinuities, boundary interaction between different media, etc. Because of both practical and theoretical reasons, we introduce the Volume-Surface Integrated Finite Difference technique (Pruess and Narasimhan, 1985) as the most appropriate and intuitive method to model the thermo-mechanical behavior of a deformable porous continuum. The reservoir is divided into N interacting continua of volume V_n each, the mass and energy integral flows occurring through its surface S_n. The thermoporoelasticity of the rock is coupled through the deformable volume V_n and its porosity. The model is useful in the study of enhanced hydrothermal reservoirs, in hot dry rock systems, in petroleum reservoirs and in the interpretation of microseismicity data.

REFERENCES

[1] M. A. Biot, *A general theory of three-dimensional consolidation*, J. of Applied Physics, 12, (1941), 155-164.

[2] K. Pruess, and T. N. Narasimhan, *A Practical Method for Modeling Fluid and Heat Flow in Fractured Porous Media*, SPE Journal, (1985), 14-26.

COMPUTATIONAL MECHANICS (Abstracts)
WCCM VI in conjunction with APCOM'04, *Sept. 5-10, 2004, Beijing, China*

On the Parametric and Geometric Dependence of the Second-Order Sum-Frequency Forces on Deep-Water Offshore Structures

Longbin Tao [1*], Stig Maeland [2], Shunqing Cai [3]

[1] *School of Engineering, Griffith University, PMB50, Gold Coast Mail Centre, QLD 9726, Australia*
[2] *TNCNO733, Offshore Structures, Det Norske Veritas AS, Veritasveien 1, 1322 Høvik, Oslo, N orway*
[2] *School of Civil and Resource Engineering, The University of Western Australia, WA6009, Australia*
e-mail: l.tao@griffith.edu.au, stig.meland@dnv.com, shunqingcai@hotmail.com

Abstract

The nonlinear sum-frequency wave excitation is identified as a pri mary source to cause high-frequency resonance of a Tension Leg Platform (TLP) in vertical planes (heave, pitch and roll), so called "springing" vibration. Due to slightly damped system of a TLP, 'springing' persists very long period of time and causing con siderable fatigue stress in the risers and tendons. It has been demonstrated that the second or higher -order wave forces due to sum -frequency effects in the wave spectrum are the main contributor to the oscillation. In addition to the coupling effect of structure and mooring system, the hydrodynamic response of a compliant floating offshore structure in the ocean is primarily governed by wave exciting force and damping. Previous study showed that the viscous damping is a crucial effect on the nonlinear reso nance of a TLP. This study focuses on the parametric and geometric dependence of the springing forces of a TLP using numerical simulation. Several issues closely related to the numerical computation of the second-order sum-frequency forces such as mesh par ameters including panel numbers on structural surface and free surface, and the partition circle of the free surface discretization are examined based on convergence tests. The interaction between the structural members on the springing force estimation is investigated. The effects of the major geometrical dimensions of a TLP on the springing forces are examined. Other effects including the incident wave direction and different wave spectrum on the springing forces are also investigated. The recommendations based on the results obtained in this study are of great interest in the TLP design.

REFERENCES

[1] S. K. Chakrabarti, *Hydrodynamics of offshore structures*, WIT Press, Southampton, U. K., (1987).

[2] DNV, *The implementation of second-order force computation in WADAM*, DNV SESAM Report No. 93-7081, Rev. 0, Oct. Det Norske Veritas, (1993).

[3] M. H. Kim, *Second-order sum-frequency wave loads on large-volume structures*, Applied Ocean Research, 13(6), (1991), 287-296.

COMPUTATIONAL MECHANICS (Abstracts)
WCCM VI in conjunction with APCOM'04, *Sept. 5-10, 2004, Beijing, China*

Method of Fluid-Structure Coupling for Turbomachinery with Multi-Parameter Modeling of Generalized Aerodynamic Forces

Duc-Minh Tran *, Fabrice Poirion, Cédric Liauzun

Structural Dynamics and Coupled Systems Department, Office National d'Études et de Recherches Aérospatiales, 29 avenue de la Division Leclerc, B.P. 72, Châtillon cedex, 92322, France
e-mail: minh.tran_duc@onera.fr, fabrice.poirion@onera.fr, cedric.liauzun@onera.fr

Abstract

This paper is concerned with the coupled fluid-structure dynamic analysis of turbomachinery. The structure consists of a rotating bladed disk submitted to the unsteady aerodynamic forces exerted by the fluid, which are themselves generated by the structural motion. The structure and the fluid are assumed to have a perfect cyclic symmetry, so that the reduction of the analysis to only one structural reference sector and one fluid channel can be applied. For each phase number, the structural displacements of the reference sector in the traveling wave coordinates are expressed as a linear combination of the complex modes on which the motion equations are projected to obtain a reduced system.

The unsteady aerodynamic forces are assumed to depend linearly on the structural displacements and velocities. They are computed prior to the coupling calculations by solving the Euler equations and by imposing an harmonic motion to the modes, for an inter-blade phase angle and for several reduced oscillation frequencies. Karpel's minimum state smoothing method [1] provides a rational function approximation of the generalized aerodynamic forces and the reduced coupled system is then solved in both frequency and time domains to obtain either the flutter diagram or the time response [3].

In order to detect the instability areas, the aerodynamic and coupling calculations should be performed at several points in the turbomachinery operation map, depending on parameters such as the mass flow, the pressure ratio, the rotation speed etc. As the aerodynamic computations are very time consuming, a multi-parameter modeling of the generalized aerodynamic forces is proposed in order to reduce their number and their cost. This method, which has already been applied to aircraft applications [2], uses at first a spline function interpolation and then Karpel's minimum state smoothing of the spline coefficient matrices. With the generalized aerodynamic forces computed for a few values of a parameter, this method allows the coupling calculations for any intermediate values of the parameter in both frequency and time domains, without any additionnal aerodynamic computations. This multi-parameter method is applied to a compressor blade with the rotation speed as parameter.

REFERENCES

[1] M. Karpel, *Design for active flutter suppression and gust alleviation using state-space aeroelastic modeling*, Journal of Aircraft, 19, (1982), 221-227.

[2] F. Poirion, *Multi-Mach rational approximation to generalized aerodynamic forces*, Journal of Aircraft, 33, (1996), 1199-1201.

[3] D.-M. Tran, C. Labaste, C. Liauzun, *Methods of fluid-structure coupling in frequency and time domains using linearized aerodynamics for turbomachinery*, Journal of Fluids and Structures, 17, (2003), 1161-1180.

COMPUTATIONAL MECHANICS (Abstracts)
WCCM VI in conjunction with APCOM'04, *Sept. 5-10, 2004, Beijing, China*

Numerical Analyses of Transonic Fluid/Structure Interaction in Aircraft Engineering

Guowei Yang [1*], Shigeru Obayashi [2]

[1] *Institute of Mechanics, Chinese Academy of Sciences, Beijing, 100080, China*
[2] *Institute of Fluid Science, Tohoku University, Sendai, 980-8577, Japan*
e-mail: gwyang@imech.ac.cn, obayashi@ieee.org

Abstract

Through the coupling between aerodynamic and structural governing equations, a fully implicit multiblock aeroelastic solver was developed for transonic fluid/structure interaction. The Navier-Stokes fluid equations are solved based on LU-SGS (lower-upper symmetric Gauss-Seidel) time-marching subiteration scheme and HLLEW (Harten-Lax-van Leer-Einfeldt-Wada) spacing discretization scheme and the same subiteration formulation is applied directly to the structural equations of motion in generalized coordinates. Transfinite interpolation (TFI) is used for the grid deformation of blocks neighboring the flexible surfaces. The infinite plate spline (IPS) and the principal of virtual work are utilized for the data transformation between fluid and structure.
The developed code was first validated through the comparison of experimental and computational results for the AGARD 445.6 standard aeroelastic wing. In the subsonic and transonic range, the calculated flutter speeds and frequencies agree well with experimental data; however, in the supersonic range, the present calculation overpredicts the experimental flutter points similar to other computations. Then the flutter character of a complete aircraft configuration is analyzed through the calculation of the change of structural stiffness. Finally, the phenomenon of aileron buzz is simulated for the weakened model of a supersonic transport wing/body model at Mach numbers of 0.98 and 1.05. The calculated unsteady flow shows, on the upper surface, the shock wave becomes stronger as the aileron deflects downward, and the flow behaves just contrary on the lower surface of the wing. Corresponding to general theoretical analysis, the flow instability referred to as aileron buzz is induced by a stronger shock alternately moving on the upper and lower surfaces of wing. For the rigid structural model, the flow is stable at all calculated Mach numbers as observed in experiment.

REFERENCES

[1] R. E. Gordiner, R. B. Melville, *Transonic flutter simulations using an implicit aeroelastic solver*, Journal of Aircraft, 37, (2000), 872-879.
[2] A. S. F. Wong, H. M. Tsai, J. Cai, Y. Zhu, F. Liu, *Unsteady flow calculations with a multi-block moving mesh algorithm*, AIAA Paper 00-1002, (2000).
[3] S. Obayashi, G. P. Guruswamy, *Convergence acceleration of a Navier-Stokes solver for efficient static aeroelastic computations*, AIAA Journal, 33, (1995), 1134-1141.
[4] Guowei Yang, Shigeru Obayashi, Jiro Nakamichi, *Aileron buzz simulation using an implicit multibolk aeroelastic solver*, Journal of Aircraft, 40, (2003), 580-589.

COMPUTATIONAL MECHANICS (Abstracts)
WCCM VI in conjunction with APCOM'04, *Sept. 5-10, 2004, Beijing, China*

The Numerical Simulation of Non-linear Aeroservoelasticity

Weiwei Zhang, Zhengyin Ye*

College of Aeronautics, Northwestern Polytechnical University, Xi'an, 710072, China
Email: zww12345@sina.com yezy@nwpu.edu.cn

Abstract

Aeroservoelasticity studies the problems coupled with aerodynamics, structure and control system. The classical aeroservoelasticity is often analyzed infrequency-domain. The unsteady aerodynamic loads are computed by frequency aerodynamic model, the aeroelastic stability is analyzed by VG-method *or* PK-method [*1-2*]. *Because these methods are basing on many linear assumptions, they are very difficult to deal with the non-linear aeroelasticity, such asaerodynamic non-linearity, structure non-linearity or control non-linearity. The flow of high angle of attack, transonic or hypersonic is strongly non-linear.* Unsteady CFD *technology is usually used to compute the non-linear aerodynamicforce* [3]. *Aeroservoelasticity with aerodynamic non-linearity is simulated in this paper.*

By using unsteady aerodynamic solver, the wing's unsteady loads can be solved directly in time-domain. The transfer function of the servo system in Laplace-domain is transformed to the time-domain to produce the servo state equations. By solving the time-domain aerodynamic equations, structure state equations and servo state equations in every step, the aeroservoelasticity is simulated in time-domain.

Considering a supersonic control surface, the calculated flutter speeds without servo system are within 10% of experimentally determined ones. The states of the free steam is: M = 3, $\alpha = 0^\circ$, $\rho = 0.11954\text{kg/m}^3$. *By numerical simulation, the servo system greatly reduces the aeroelastic stability* (*reduced from* 995m/*s to* 750m/*s*). *The structural flitter added to the system can reduce the coupling effect of servo system and structure mode, and increase the flutter velocity (increased to 900m/s). The calculated flutter speed of the aeroservoelastic system with increasing angle of attack at 7* Mach *number is shown in paper.* We *compute aeroservoelastic flutter boundary over a range of* Mach *number from 2 to 12 at* 3° *angle of attack.* We *find the aeroservoelastic system is non-linear, and the flutter speed decreases greatly with the increasing angle of attack. Flutter speed increases with the increasing of* Mach *number, but the increasing amplitude becomes smaller and the flutter speed tends to constant number at high* Mach *number. This shows the stability of aeroservoelasticity also conforms to the hypersonic* Mach *number independence principle.*

REFERENCES

[1] H. T. Lee, I. M. Kroo, S. Bieniawski, *Flutter suppression for high aspect ratio flexible wings using microflap,* AIAA, (2002), 1717.

[2] W. P. Rodden, *MSC/NASTRAN aeroelastic aanlysis – user's guide*, The Macneal-Schwendler Corporation, USA, (1994).

[3] K. K. Gupta, L. S. Voelker, *CFD-based aeroelastic analysis of the X-43 hypersonic flight vehicle, AIAA , (2001), 0712.*

COMPUTATIONAL MECHANICS (Abstracts)
WCCM VI in conjunction with APCOM'04, *Sept. 5-10, 2004, Beijing, China*

Selection with Reason to Interface Element of Structure-Soil Interaction

Tongchun Li [1*], Yu Li [1], Miao Li [1], Xianglai Wang [2]

[1] *College of Water Conservancy and Hydropower Engineering, Hohai University, Nanjing, 210098 China*

[2] *Water Conservancy Investment and Construction Company of Shanghai, Ltd, Shanghai, 200012 China*

e-mail: xixi9118@sina.com

Abstract

According to material characteristic of contact bodies, contact problems are divided two types. One is that material characteristic of contact bodies are equivalent, for example concrete dam and bedrock surface, construction joint in the concrete dam. The other is that material characteristic of contact bodies has extremely discrepancy, for example contact between pill and soil, between soil and central core or seepage prevention wall in the earth dam. This ki nd of question is manifested mainly soil-structure contact problem in the civil engineering and hydraulic engineering. Its characteristics are that concrete structure behaves elasticity, and there is no crack in the condition of small deformation, and shea ring slip of interface depends on the material characteristics of soil. Generally the structure is known as the study object in this problem. In order to offer bases to reasonable evaluation of safety and structure design, its main idea is to make sure acc urately the distribution of deformation and internal force. Deformation of structure is made of structure deformation, soil deformation and relative deformation of soil -structure. Internal force of structure is decided by external force of structure and st ress of interface. So it is important that structure, soil and interface are described reasonably. In this paper, a mechanical behavior of interface is simulated reasonably in the analyses of soil –structure interaction.

The study of mechanical behaviors of interface in soil-structure interaction includes two aspects. One is soil behavior of interface, in other words, the relation of contact stress and contact strain. The other is the form of interface element, which is corresponding to behaviors of interface. Two types of interface element are introduced, which are spring element and thin-layer element. The two types elements are introduced and theirs accessible are analyzed in this paper. Rectangular (plane) or cubic (solid) thin layer element with improved yield criteria is recommended to simulate structure-soil interaction. One numerical example is presented in the paper in order to valid this element numerical stability.

COMPUTATIONAL MECHANICS (Abstracts)
WCCM VI in conjunction with APCOM'04, *Sept. 5-10, 2004, Beijing, China*

Analysis of Behavior of Cement Stabilized Soil-Wall for Pavement in Wet Soft Soil

Y. Zhao[1*], X. Liu[1], A. de Bondt[2], A. Scarpas[1]

[1] *Section of structural Mechanics, Delft University of Technology, Stevinweg 1, 2628 CN, Delft, The Netherlands*
[2] *Ooms Avenhorn Holding bv, Scharwoude, The Netherlands. http://www.ooms.nl*
e-mail: Y.Zhao@citg.tudelft.nl

Abstract

The problems caused by weak and wet soil to road/rail pavement construction demand an elaborate knowledge of the interaction between the foundation and the overlaying pavement. This paper presents 3D non-linear finite element analyses of an asphalt pavement constructed on a cement stabilized soil-wall system, which is widely employed technique in road/rail construction. The emphasis is placed on the interplay between the pavement and soil foundation via soil-walls, with the goal of optimizing overall pavement performance. The finite element system INSAP-PM [1], developed at TU-Delft, was utilized.
A FEM mesh is built up, simulating the foundation system which is assumed to consist of certain number of regularly spaced cement stabilized soil-walls of equal length installed in a weak soil layer and supporting the asphalt pavement. Numerical modeling of the interaction between the mechanical behavior of the solid phase and the flow characteristics of the liquid phase has been formulated on the basis of Biot's consolidation theory. Constitutive material models capable of describing the inelastic, short/long-term response of the materials were utilized [2]. The mechanical characteristics of geotechnical materials such as isotropic and anisotropic hardening, softening, pressure sensitivity, associated and non-associated plasticity can be captured through this approach.
The influence of interaction between the soil-wall and the wet soil on the mechanical response of the overall structure is investigated for various combinations of material properties and traffic loading situations. Such factors were examined, as the depth and spacing of the soil-wall, the weak soil properties, properties of the soil-wall, the depth of the bedrock relative to the tip of the soil-walls and the magnitude and position of the load carried by the asphalt pavement (expressed as cyclic moving loads). In light of the study, the role and the contribution of the soil-wall to the load carrying capacity were interpreted and quantified. In addition, the analyses will allow clarification of the long-term phenomenon like consolidation, degradation of foundation material, repetitive loading etc. The findings of this investigation will provide to the manufacturers/suppliers/construction industry the necessary numerical information for improvement of existing designs and the development of design specifications.

REFERENCES

[1] X. Liu, *Numerical modelling of porous media response*, Ph. D. Thesis, Faculty of Civil Engineering, TU-Delft, The Netherlands, (2003).

[2] Desai, C.S., Somasundaram, S. and Frantziskois, G.N. (1986). A hierarchical approach for constitutive modeling of geologic materials. Int. J. Num. Analyt. Meth. In Geomech., 10, pp,.225-257.

COMPUTATIONAL MECHANICS (Abstracts)
WCCM VI in conjunction with APCOM'04, *Sept. 5-10, 2004, Beijing, China*

A Study on Microwave Heating

A. K. Haghi *

The University of Guilan, P.O.Box 3756, Rasht, Iran
e-mail: Haghi@guilan.ac.ir

Abstract

The focus of this investigation is the development of a mathematical model to simulate the microwave drying of wool. In the first part of this study, the yarn sample is considered to be a porous circular cylinder. The yarn model is developed by considering mass and heat transfer within the wet and sorption regions of the differential control volume in a given slice of yarn. Heat and moisture transport are assumed to vary only in the radial direction. The model prediction and experimental results for the average moisture is presented. The numerical and experimental results found to be in good agreement. In the second part of this study, the sample was heated in a conventional oven. The results of the strength tests were then compared for the two heating method. These reveled that microwave heating reduced the strength of the dried sample.

COMPUTATIONAL MECHANICS (Abstracts)
WCCM VI in conjunction with APCOM'04, *Sept. 5-10, 2004, Beijing, China*

Effect of Rotation and Surface Roughness on Heat Transfer Rate to Flow through Vertical Cylinders

Ahmed Mohamed Hany *

Department of Mechanical Engineering, Faculty of Engineering, Assiut University, Assiut, Egypt
E-mail: hah@aun.eun.eg

Abstract

The enhancement in the rate of the heat transfer resulted from rotating smooth and rough cylinder surfaces are experimentally studied. Experiments were carried out for constant cooling fluid Reynolds numbers from 3300 to 7800 with varying the rotational speed up to 280 rpm. In addition, the experiments were run at the stationary case (no rotation) and the results show an agreement with the theoretical values involved small deviations. The Nusselt number values, obtained from the experimental results at various rotational speeds are correlated as a function of Reynolds, Weber, and Prandtl numbers for smooth and rough surfaces. The correlated equations were compared with the correlation obtained by another author.

The results show that the enhancement on the heat transfer rate becomes more appreciable for low cooling flow Reynolds numbers at high rotational speeds and for high Reynolds numbers at low rotational speeds. The rotation causes an enhancement in the overall heat transfer coefficient with percentage values of about 89 % at Re = 7800, We = 1084, and Pr = 1.48 for smooth surface and about 13.7 % at Re = 4700, We = 4891, and Pr = 1.696 for rough surface. Also, the enhancement in the heat transfer rates utilizing rotary surface becomes more pronounced for the smooth surface compared with the rough one, therefore the choice of the heat transfer surface is very important. The present work shows a reduction in the heat transfer rate below its peak value depends on the type of the heat transfer surface. It is shown that the enhancement in the heat transfer, i.e. enhancement in the Nusselt number, depends on the Weber number value and the surface type while the Nusselt number value mainly depends on the Reynolds and Prandtl numbers. Present correlated equation could be used to represent the Nusselt number values as function of the Weber and Reynolds numbers within the stated ranges of the parameters.

COMPUTATIONAL MECHANICS (Abstracts)
WCCM VI in conjunction with APCOM'04, *Sept. 5-10, 2004, Beijing, China*

Monte Carlo Solution of Transient Heat Conduction in Anisotropic Media

F. Kowsary *, Saeed Irano

Department of Mechanical Engineering, Faculty of Engineering, University of Tehran, Tehran, Iran
e-mail: fkowsari@ut.ac.ir,

Abstract

Solution of transient heat conduction by the Monte Carlo method is presented for anisotropic media. Although the utility of the Monte Carlo method in radiation heat transfer has been well established, not as much attention is given to its use in the conduction heat transfer problems [1]. This is despite the fact that the use of the Monte Carlo method has shown to be very convenient for computer programming multi-dimensional bodies with complicated geometries and to have this very attractive feature that the solution at a specific point may be obtained without the need for solving the whole domain [1]. This latter feature of the Monte Carlo method is of great value in inverse heat conduction problems [2]. Monte Carlo method is a perfect for parallel computations as processors do not need to communicate until the random walk steps are complete. Recently Deng and Liu [3] utilized the Monte Carlo method in bio-heat transfer applications. The use of the Monte Carlo method for solving the steady heat conduction problem in an anisotropic medium was addressed by Kowsary and Arabi [4]. However, its use for solving the unsteady heat conduction problems in anisotropic media has not yet been discussed. The present paper addresses this issue. The probability coefficients derived from finite differencing of transient heat conduction equation are presented and are used subsequently in a fixed-step random walk procedure. Results for a test case having Robin and adiabatic boundary condition types are presented and validated against results of the existing numerical methods.

REFERENCES

[1] A. Haji-Sheikh, *Monte Carlo methods*, in W. J. Minkowycz et al eds. Handbook of Numerical Heat Transfer, Chap. 16, Wiley, NY, (1989).

[2] A. Haji-Sheikh, F. P. Buckingham, *Multidimensional inverse heat conduction using the Monte Carlo method*, Journal of Heat Transfer, 115, (1993), 26-33.

[3] Zhong-Shan Deng, Jing Liu, *Monte Carlo method to solve multidimensional bioheat transfer problems*, Numerical Heat Transfer, Part B, 42, (2002), 543-567.

[4] F. Kowsary, M. Arabi, Monte Carlo solution of anisotropic heat conduction, International Communications in Heat and Mass Transfer, 26, (1999), 1163-1173.

COMPUTATIONAL MECHANICS (Abstracts)
WCCM VI in conjunction with APCOM'04, *Sept. 5-10, 2004, Beijing, China*

A Model of Multiphase Flow Transfer in Porous Textiles

Yi Li, Fengzhi Li*

Institute of Textiles and Clothing, The Hong Kong Polytechnic University, Hung Hom, Kowloon, Hong Kong, China
e-mail: tcliyi@polyu.edu.hk, lfz7553.student@sina.com

Abstract

The recent SARS (severe acute respiratory syndrome) outbreak has spawned a major controversy concerning the protective performance of facemasks. The structure and physical property of the mask materials significantly impact the protective performance. For describing the mechanism of virus transfer in the facemask with different structure and material, a model of multiphase flow transfer in porous textiles is established.
The virus transfers in masks being taken by liquid water. So controlling liquid water transfer in mask is a key to restrain the virus transfer. In this paper, the liquid water transfer mechanisms are based on capillary pressure model proposed by Yi Li and Q. Zhu [1]. Meanwhile, breathing effect on it is considered by adding a pressure gradient term. The intrinsic permeability of the Darcy's Law in porous textiles is obtained from theoretically eduction. It is a function of material capillary average angle, pore effective radius and material porosity. Another important factor for virus transfer is the diffusion in liquid water, deposition and release on fibers surface. The deposition/release rate in this paper is obtained by modification the two-parameters rate expression for spherical collectors proposed by Gruesbeck and Collins [2]. For the fiber collectors, the specific area of fabric is introduced in the model. Furthermore, the water vapor condensation and liquid water evaporation, moisture sorption by fibers, phase change and latent heat generation are considered in the model. The classical multi-phase flow model is extended to the area of fabric. Also on the basis of finite volume method, the mathematical equations are solved. As an example, a KCL (potassium chloride) filtration experiment in N95 mask during breathing cycles is simulated. The predictions given by the model agree well with that of the experiment, indicating that the model has good prediction ability. The conclusion shows that:
For material with large contact angle and small pore size such as N95 mask, the effect of breathing cycles on liquid water diffusion is small, but that on water vapor concentration and gas pressure is large. The viruses transfer in the mask mainly by diffusion in liquid water, and the evaporation of liquid water increases the virus concentration in liquid water and impacts the virus diffusion amount. The virus mass in liquid water is larger than that on the surface of fibers in deposition because of small penetration velocity of liquid water.

REFERENCES

[1] Yi Li, Qingyong Zhu, *A model of coupled liquid moisture and heat transfer in porous textiles with consideration of gravity*, Numerical heat transfer, 43, (2003), 501-523.

[2] C. Gruesbeck, R. E. Collins, *Entrainment and deposition of fine particles in porous media*, Soc. Pet. Engin. J., (1982), 847-856.

COMPUTATIONAL MECHANICS (Abstracts)
WCCM VI in conjunction with APCOM'04, *Sept. 5-10, 2004, Beijing, China*

Multi-Scale Analysis Method for Thermal Conductivity of Porous Material with Radiation

Shutian Liu *, Yongcun Zhang

State Key Laboratory of Structural Analysis for Industrial Equipment and Department of Engineering Mechanics, Dalian, 116024, China
e-mail: stliu@dlut.edu.cn,yongcunzhang@126.com

Abstract

Porous material is widely used in weight-sensitive structures as core material or heat-insulate material. Besides the mechanical properties, the determination of thermal properties of porous materials is an important issue. Radiation is a way of heat transmission and sometimes plays an important role in heat transfer, especially at high temperature environment. Owing to the complexity of the microstructure and the multiple ways of heat transfer, it is almost impossible to determine exactly the temperature distribution in microstructure. The conventional method is to assume an approximate distribution of temperature. For example, in paper [1], the temperature distribution is assumed to be linear along the heat flux over a unit cell. This linear distribution assumption is obvious inaccurate because the cavities have great influence on the temperature fields. In this paper, a homogenization-based multi-scale method with a mathematical rigorous methodology is used to predict the effective thermal conductivity and a temperature field in a local domain of the unit cell without considering radiation. This temperature field is taken as the approximation of the real temperature distribution, and the radiative thermal conductivity is obtained. The effect of the microstructure, the distribution and geometry of cavities on heat transfer of the porous material is discussed. It is concluded that the radiative thermal conductivity of the porous material with cylindrical cavities is proportional to cavity's diameter; the radiative thermal conductivity of the porous material with spherical cavities is proportional not only to the diameter, but also to the porosity. The effective thermal conductivity without radiation and temperature distribution in microstructure domain are important factors influenced the effective thermal conductivity if the heat transfer of radiation is considered.

This project is supported by the Major Research Plan (90205029) and the Major Program (10332010) of the National Nature Science Foundation of China and the Excellent Young Teacher's Program of MOE of China. Their financial contributions are gratefully acknowledged.

REFERENCES

[1] X. G. Liang, W. Qu. *Effective thermal conductivity of gas-solid composite materials and the temperature difference effect at high temperature*, Int. J. Heat Mass Transfer 42, (1999), 1885-1893.

[2] Shutian, Liu, *Prediction of properties of composites and optimal design of functionally gradient materials*, PhD Dissertation, Dalian: Dalian University of Technology, China, (in Chinese).

COMPUTATIONAL MECHANICS (Abstracts)
WCCM VI in conjunction with APCOM'04, *Sept. 5-10, 2004, Beijing, China*

Lattice Boltzmann Method for Natural-Convection with High Rayleigh Number in a Triangular Enclosure

Baochang Shi, Rui Du *

Department of Mathematics, Huazhong University of Science & Technology, Wuhan, 430074, China
e-mail: sbchust0382@sina.com, karen_du@sohu.com

Abstract

The problem of natural convection heat transfer and fluid flow in enclosures, especially in a triangular enclosure, has received considerable attention from researchers in many diverse fields of application. But the natural convection flows with high Rayleigh number have not been solved efficiently by traditional ways up to now because of their complex behaviors. Lattice Boltzmann method is a new approach using simple microscopic kinetic models to simulate complicated macroscopic behaviors of transport phenomena. A most popular LB model is lattice BGK one.

In this paper, an improved coupled thermal lattice BGK model with a robust boundary scheme for complex boundary conditions has been proposed on basis of the coupled LBGK (CLBGK) approach [1-2]. For temperature field a D2Q9 scheme is used which can increase the numerical accuracy and stability especially for complex physical boundary conditions. The basic idea of the CLBGK model is to use two independent LBGK equations to solve the velocity and temperature fields respectively, and then combines them into one coupled model for the whole system. Because the computing grid is regular, it is important about how to treat the curved boundary in the model. Here an extrapolation method [3] has been applied in both velocity and temperature fields accordingly.

We have simulated a natural convection flow in a triangular enclosure with a heated transfer on the bottom and upper cool walls with different *Ra* number from 10^3 up to 10^8 and Pr=0.71. In order to improve the numerical accuracy and stability, one of the most important ways is to refine the computing grids which may lead to a large numerical scale. Due to the intrinsic parallel character of the method it can be operated on a parallel computer easily. We implemented the parallel version using Message Passing Interface (MPI) on our "DeepComp 1800" with 24 computing nodes. From the numerical results the model proposed in this paper is of better numerical accuracy and stability. And it has been predicted in Ref [4].

REFERENCES

[1] Z. L. Guo, B. C. Shi, C. G. Zheng, *A coupled lattice BGK model for the Bouessinesq equation*, Int. J. Num. Fluids., 39, (2002), 325-342.

[2] Baochang Shi, Zhaoli Guo, *Thermal lattice BGK simulation of turbulent natural-convection due to internal heat*, Int. J. of Modern Physics B, 17, (2003), 173-177.

[3] Zhaoli Guo, Chuguang Zheng, Baochang Shi, *An extrapolation method for boundary conditions in lattice Boltzmann method*, Phys. Of Fluids, 14, (2002), 2007-2010.

[4] S. Völker, T. Burton, S. P. Vanka, *Finite-volume multigrid calculation of natural-convection flows on unstructured grids*, Num. Heat Transfer, Part B, 30, (1996), 1-22.

COMPUTATIONAL MECHANICS (Abstracts)
WCCM VI in conjunction with APCOM'04, *Sept. 5-10, 2004, Beijing, China*

Transient Temperature Distribution in the Sheet Metal by Low-Energy Laser Scanning

Yinchun Wang [*], Yuren Wang

Institute of Mechanics, Chinese Academy of Sciences, 15 Beisihuanxi Road, Beijing 100080, China
e-mail: weasky@hotmail.com

Abstract

Low-energy laser-heating techniques are widely used in engineering applications such as, thin film deposition, surface treatment, metal forming and micro-structural pattern formation [1]. To have a comprehensive understanding of the laser-material interaction and control the manufacturing processes in precise, it is necessary to correctly analyze the transient temperature field in a material produced by the scanning laser beam heating. However, the problem is far from to be resolved due to the complexity of the laser-material interaction process such as the nonlinearity of the material properties, although many models have been put forward [2]. A numerical simulation is a powerful tool for the analysis of the transient temperature field and can provide useful data for the model set-up.

In this paper, under the conditions of ignoring the thermo-mechanical coupling, we presented a numerical simulation on the spatial and temporal temperature distribution in sheet metal produced by the laser beam scanning in virtue of the finite element method. Both the three-dimensional transient temperature field and the temperature evolution as a function of heat penetrating depth in the metal sheet are calculated. The results are compared with the previous semi-analytical and numerical results. The temperature dependence of material properties was taken into account. It was shown that, after taking the temperature dependence of the material absorbance effect into consideration, the temperature change rate along the scanning direction was reduced, while the temperature maximum was increased.

The influences of the processing parameters such as, the scanning velocity, the sheet thickness, the spot size of the laser beam and the laser power, on the temperature field distribution are also studied. The results indicated that the peak values of the temperature distribution increased with the increase of laser power and the beam-spot size, but decreased with the increase of the scanning velocity and the sheet thickness. Moreover, it was found that the temperature difference along the thickness direction increased with the increase of laser power and the thickness, and the decrease of the scanning velocity.

REFERENCES

[1] T. W. Murray, S. Krishnaswamy, J. D. Achenbach, *Laser generation of ultrasound in films and coatings*, Appl. Phys. Lett., 74, (1999), 3561-3564.

[2] Z. Ji, S. Wu, *FEM simulation of the temperature field during the laser forming of sheet metal*, J. Mater. Process. Technol, 74, (1998), 89-95.

COMPUTATIONAL MECHANICS (Abstracts)
WCCM VI in conjunction with APCOM'04, *Sept. 5-10, 2004, Beijing, China*

Numerical Study on Thermal and Chemical Nonequilibrium Aerothermodynamics of an Inflatable Re-entry Vehicle

Gang Xia*, Zhongxi Hou, Yang Li

College of Aerospace and Material Engineering, National University of Defense Technology, Changsha, 410073, China
e-mail: arhgang@yahoo.com

Abstract

Inflatable Re-entry Vehicle (IRV) is an innovative aerospace recovery concept combining the three main re-entry subsystem elements, i.e. 1) Thermal Protection System (TPS) for re-entry, 2) Parachute system for descent and 3) Damping/floatation system for landing [1]. This results in simple and cost effective design of the re-entry system, which will provide sufficient aerodynamic and aerothermodynamic behavior during the critical reentry phase and will allow for touch-down of the vehicle in the predefined landing area. Multi-Layer Insulation (MLI) materials are designed to cope with the expected atmospheric-entry heat loadings which should be well predicted by numerical study. Thermal and chemical nonequilibrium aerothermodynamics of a new type of IRV named Spacelab Inflatable Download Transporter (SLiDT) [2] was simulated in this work. Firstly, a pre-solver analysis method of aeroheating prediction was developed and numerical solutions for hypersonic flows of air around a 70-deg sphere-cone Mars entry vehicle configuration [3] were computed to validate it. Using a turbulent, 3-D Navier-Stokes solver, the computations showed the distribution and accuracy of heat flux are sensitive to grid resolution, near-wall grid adaptation, turbulence models and their near-wall treatment. Spalart-Allmaras, k-ω and k-ε turbulence models were implemented with both fine and coarse grids. Comparisons between numerical and experimental surface heating distributions showed when near-wall grids satisfy $y^+ \approx 5$, the difference were well within the estimated experimental uncertainty [4], and the SA model was found to be much more suitable for these slightly separated flow fields. The pre-solver analysis method evaluated effects of the factors mentioned above as well as gas models and numerical scheme accuracy, then established guidelines to improve the precision of computed heat flux. Secondly, the high-enthalpy 3-D aerothermodynamics of the SLiDT during re-entry phase with large attack angle were computed and analyzed for the thermal protection system design purpose.

REFERENCES

[1] Detlef Wilde, Stephan Walther. *Inflatable reentry and descent technology (IRDT) -further developments*, 2nd International Symposium of Atmospheric Reentry Vehicles and Systems, Arcachon, France, March, (2001).

[2] Gang Xia, *Theoretical, numerical simulation investigation of spacelab inflatable download transporter*, PhD. Thesis, National University of Defense University, China, (2003).

[3] Brian R Hollis, John N Perkins, *High-enthalpy aerothermodynamics of a Mars entry vehicle part 2: computational results*, Journal of Spacecraft and Rockets, 34, (1997), 457-463.

[4] Brian R Hollis, John N Perkins, *High-enthalpy aerothermodynamics of a Mars entry vehicle part 1: Experimental results*. Journal of Spacecraft and Rockets, 34, (1997), 449-456.

MECHANICS (Abstracts)
WCCM VI in conjunction with APCOM'04, *Sept. 5-10, 2004, Beijing, China*

Multiple Scale Method for Numerical Simulation of Nonclassical Heat Conduction Problems

Hongwu Zhang, Sheng Zhang*, Xu Guo

State Key Laboratory of Structural Analysis for Industrial Equipment, Department of Engineering Mechanics, Dalian University of Technology, Dalian 116024, China
e-mail: zhanghw@dlut.edu.cn, zhangs@student.dlut.edu.cn

Abstract

A multiple spatial and temporal scale method is studied to simulate numerically the phenomenon of non-Fourier heat conduction in periodic heterogeneous materials. The model developed is based on the higher-order homogenization theory with multiple spatial and temporal scales in one dimensional case. The amplified spatial scale and the reduced temporal scale are introduced to account for the fluctuations of non-Fourier heat conduction due to material heterogeneity and nonlocal effect of the homogenized solution, respectively. By separating various scales the different orders of homogenized non-Fourier heat conduction equations are obtained, and then combining these equations the reduced time dependence is eliminated and the fourth-order governing differential equations are derived. To avoid the necessity of C^1 continuous finite element implementation, a C^0 continuous mixed finite element approximation scheme is put forward. Numerical results are shown to demonstrate the efficiency and validity of the proposed method.

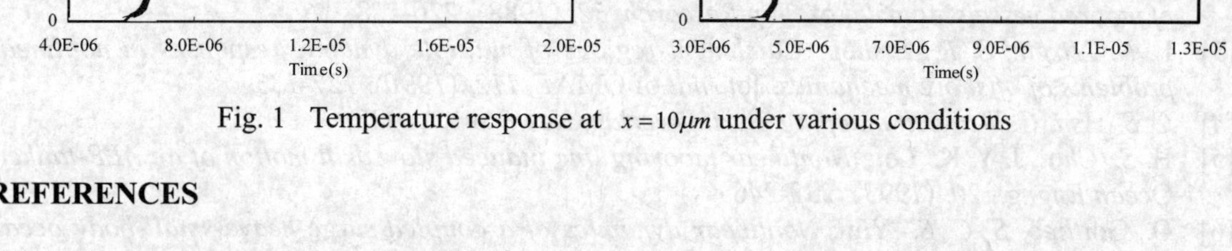

Fig. 1 Temperature response at $x = 10\mu m$ under various conditions

REFERENCES

[1] C. Boutin. *Microstructural influence on heat conduction*, Int. J. Heat Mass Transfer, 38(17), (1995), 3181-3195.

[2] J. Fish, W. Chen, G. Nagai, *Non-local dispersive model for wave propagation in heterogeneous media: one-dimensional case*, International Journal for Numerical Methods in Engineering, 54, (2002), 331-346.

[3] H.W. Zhang, B.A. Schrefler, *Global constitutive behaviour of periodic assemblies of inelastic bodies in contact*, Mechanics of Composite Materials and Structures, 7(4), (2000), 355-382.

COMPUTATIONAL MECHANICS (Abstracts)
WCCM VI in conjunction with APCOM'04, *Sept. 5-10, 2004, Beijing, China*

Research on the Periodic Motions and Period-doubling Bifurcations of a Moored Ocean Platform Using Cell-to-Cell Mapping Method

Du Du [1*], Weikang Zhang [2], Cheng Ma [1], Yi Bi [2]

[1] *Naval Research Center, Beijing, 100073, China*
[2] *Department of Naval Architecture & Ocean Eng., Naval University of Engineering, Wuhan, 430033, China*
e-mail: dudu0736@yahoo.com.cn

Abstract

Cell-to-Cell Mapping Method was *used to study the periodic orbits and period-doubling bifurcations* of *a* moored *ocean production plat*form *subjected* to *waves.* The mathematic model *was a nonlinear oscillator driven* by *wave exciting* force *while the restoring* force *was exponential.* Investigation *on period*-doubling *bifurcations was presented* by *defining the period* of *wave exciting* force *as a bifurcation control parameter. Cell-to-Cell* Mapping Method *was used* to *find periodic orbits* of *the system and* domains *of attraction* of *different attractors.* For *every attractor, time history, phase plane portrait, power spectrum* and Poincaré *section were presented* and *then used* to *identify its type.* It *is shown that when the value* of wave period *is smaller than the critical value, there are* competing *attractors in this system* which *domains of attraction are twining each other.* There *are different period*-doubling *bifurcation characteristics in this system when the* bifurcation *parameter is in different* parts of *parameter space.* Onset *and abrupt disappearance* of *a period-doubling bifurcation sequence is presented, while a* bifurcation *sequence converging to a chaotic attractor is also* observed, *in which each attractor is global attracting.* This *phenomenon is clarified using the results* of *Cell-to-Cell* Mapping *analysis.*

REFERENCES

[1] J. M. T. Thompson, H. B. Stewart, *Nonlinear dynamics and chaos, Wiley & Sons, Chichester* (1986).

[2] F. A. Papoulias, M. M. *Bernitsas, Autonomous oscillations, bifurcations, and chaotic response of moored vessels,* Journal of *Ship Research, 32,* (1988), 220-228.

[3] L. N. Virgin, S. R. Bishop, *Catchment regions of multiple dynamic responses in nonlinear problems of offshore mechanics,* Journal of OMAE, *112,* (1990), 127-133.

[4] *C. S. Hsu, Cell-to-cell mapping, Spring-verlag,* New York, (1987).

[5] H. S. Choi, J. Y. K. Lou, *Nonlinear mooring line induced slow drift motion of an ALP-tanker,* Ocean Engng., 20, (1993), 233-246.

[6] O. Gottlieb, S. C. S. Yim, *Nonlinear dynamics of a coupled surge-heave small-body ocean mooring system,* Ocean Engng. 24, (1997), 479-495.

[7] *R.* Ouyang, J. Zhu, *Computation of attracting basin in multi-point mooring system,* The Ocean *Engineering, 17,* (1999), 1-7, (in Chinese).

COMPUTATIONAL MECHANICS (Abstracts)
WCCM VI in conjunction with APCOM'04, *Sept. 5-10, 2004, Beijing, China*

Robust Modeling of Intermittent Contact in Multibody Dynamics

J.C. Garcia Orden *

Departamento de Mecanica de Medios Continuos *y Teoria de Estructuras, Universidad Politecnica* de Madrid, E.T.S.I. Caminos, c/ Profesor Aranguren s/n, Madrid, 28040, Spain
e-mail: jcgarcia@mecanica.upm.es

Abstract

A great variety of engineering applications involving mult ibody systems demand an effective numerical representation of contacts [3]. These situations often contain interactions between rigid-deformable and rigid-rigid bodies, and typically involve a very high number of contact events during the characteristic ti me scale of rigid body motions. An example is the analysis of joint clearances in machinery components, which could be required by design criteria or can be present due to geometrical imperfections caused by limitations of manufacturing processes or wear.
From the numerical point of view, a feasible analysis option is to employ time steps small enough to capture with precision every contact taking place during the interaction, and to use explicit schemes. However, this approach can lead to large computing t imes for typical motions of multibody systems. On the other hand, standard implicit schemes allow for larger time steps, but they still exhibit serious stability problems due to the stiff character of the equations.
As an alternative, the approach presen ted in this work is able to model high number of impact events with rather large time steps, without demanding great detail for local effects of the individual contacts, which is precisely the case in typical multibody simulations. Three are the key ingred ients of the proposed methodology; firstly, it takes advantage of the analytical definition of one of the contacting surfaces. Secondly, it regularizes the contact problem with a penalty technique and, in contrast with standard approaches [3], it avoids the use of the gap (minimum distance between the colliding surfaces) as a primary variable. Thirdly, the contact force is formulated in the context of a conservative time integration method [1 -2], which possesses extraordinary stability properties without introducing artificial numerical damping.
Some representative numerical simulations involving the dynamics of mechanical systems experiencing high number of intermittent contacts (with and without friction) are presented, and evaluated in terms of accuracy a nd time step size. As a conclusion, the proposed methodology appears to be promising in terms of numerical efficiency and robustness in the treatment the dynamics of complex multibody systems undergoing high number of intermittent contacts.

REFERENCES

[1] J. C. Simo, N. Tarnow, *The discrete energy-momentum method, conserving algorithms for nonlinear elastodynamics*, ZAMP, (1992).
[2] J. M. Goicolea, J. C. Garcia Orden, *Dynamic analysis of rigid and deformable multibody systems with penalty methods and energy-momentum schemes*, Computer Methods in Applied Mechanics and Engineering, 188(4), (2000), 789-804.
[3] P. Wriggers, *Computational contact mechanics*, Wiley, (2002).

COMPUTATIONAL MECHANICS (Abstracts)
WCCM VI in conjunction with APCOM'04, *Sept. 5-10, 2004, Beijing, China*

A Mechanical Model to Simulate Interactively a Bending Actuator Composed of Three Parallel Bellows

P. Joli [1*], N. Seguy [1], C. François [2]

[1] *Laboratoire Systèmes Complexes, University of Evry, 40 rue du Pelvoux, CE 1455, Courcouronnes, 91020 Evry Cedex, France*
[2] *I.U.T. de Cachan, 9 av. de la division Leclerc, BP 140 94234 Cachan, France*
e-mail: pjoli@lsc.univ-evry.fr, nseguy@lsc.univ-evry.fr, chris.francois@noos.fr

Abstract

In response to a project on the risks related to the endovascular method stent placement in an aorta with a flexible catheter, it has been decided to place inside the distal part of the catheter, a "bending actuator " which could be remote controlled by the operator to prevent excessive contact forces and to open the way through the complex path of the vessels. A prototype of a "bending actuator" has been designed and presented in a previous paper [1]. The actuator is composed of three cylindrical nickel bellows, each one can be deformed due to internal hydraulic pressure of physiological water (see figure below).

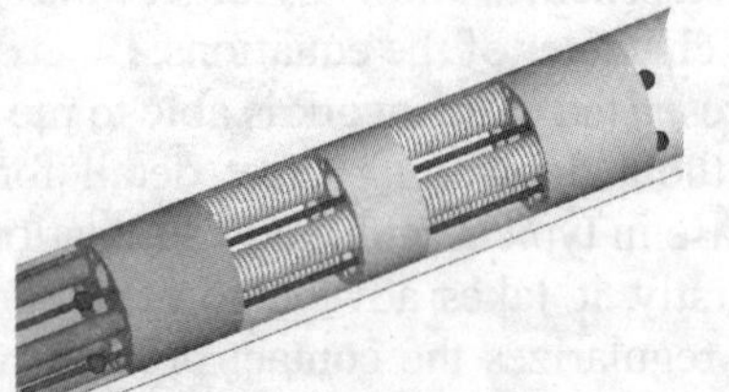

In this paper we present a suitable numerical model for such a syste m to be simulated interactively. The challenge is to have a software tool to build a first draft virtual prototype of the "bending actuator" with the possibility to easily change geometrical parameters or internal pressure. In order to reduce the computat ional time, each bellow is modeled individually as an articulated multi -body system with elastic joints, and the relative joint coordinates are calculated by a backward formulation [3]. The difficulties are to calculate the tangent stiffness matrix of each bellow by a forward formulation and then to satisfy all the relative geometrical constraints between each bellow by a "gluing algorithm" [2].

REFERENCES

[1] P. Joli, Ch. François, T. Gagarina, F. Boudghène, *Modeling and process design of a new type of catheter for special endovascular treatment of abdominal aortic*, Proc 16th Int. Conf. CARS, Paris, (2002), Springer, pp. 849-854.
[2] F.-C. Tseng, G. M. Hulbert, *A gluing algorithm for network-distributed dynamics simulation*, Multibody System Dynamics, 6, (2003), 377-396.
[3] R. Featherstone, *A divide-and-conquer articulated body algorithm for parallel calculation of rigid body dynamics, part 1: basic algorithm*, International Journal of Robotic Research, 18(9), (1999), 867-865.

COMPUTATIONAL MECHANICS (Abstracts)
WCCM VI in conjunction with APCOM'04, *Sept. 5-10, 2004, Beijing, China*

Multi-Body Nonlinear Analysis for Engine Vibration Simulation

Jun Lan *

AVL-List Shanghai Office, Shanghai, 201206, China
e-mail: jun.lan@avl.com

Abstract

NVH (noise, vibration and harshness) issues are of increasing significance for product differentiation and for noise and vibration regulations in the vehicle industry. The simulation of engine vibrations and noise radiation becomes more frequently applied in the engine development process together with the well established experimental engineering approach. The paper summarizes the state-of-the-art in computational engine vibrations and tries to give an outlook on new methods and techniques for the engine and vehicle development process.

AVL EXCITE has been developed especially for these applications using an outstanding simulation technology, which enables to produce realistic results of engine dynamics and noise that are very close to the absolute ones. Up to 1 kHz engine block and mount vibrations can be analyzed, and have good agreement with the experiments.

AVL EXCITE is a multi-body nonlinear dynamics program for engine dynamic simulation. The simulation concept is based on dividing the non-linear mechanical system into sub-systems with linear elastic behavior and non-linearities occurring only at the connections between these sub-systems. In the simulation model linear elastic bodies interact through highly non-linear connections (e.g. oil film in the slider bearings or between piston and liner, and engine mounts). The main individual bodies consist of power unit, crankshaft, connecting rods. These components are reduced substructures that are obtained by finite element program after defining suitable master degree of freedom. Nonlinear stiffness and damping properties of joint are considered. AVL EXCITE determines initial conditions and provides an effective time integration procedure with adjusted step size. The external loads must be given in time domain and the joint loads and additional inertia forces are calculated iteratively in each time step.

REFERENCES

[1] Loibnegger, B.; Resch, T.; Zmire, B., *Enhanced Crankshaft Vibration and Stress Analysis*, 19th CAD-FEM Users' Meeting 2001, International Congress on FEM Technology, Hotel Dorint Sanssouci, Berlin, Potsdam, October 17-19, 2001.

[2] Product Description AVL-EXCITE v6.0, *Dynamics and Acoustics of Power Units*, April 2003.

[3] Priebsch, H.H.; Loibnegger, B.; Pramberger, H., *Simulation of Engine Noise – Influence of Design Parameters*, JSAE 1998, 20.5. - 22.5. Pacific Convention Plaza Yokohama, Japan.

COMPUTATIONAL MECHANICS (Abstracts)
WCCM VI in conjunction with APCOM'04, *Sept. 5-10, 2004, Beijing, China*

Distributed Simulation and Visualization of Interactive Multibody Systems

Zhuo Li [1*], Peter Eberhard[2]

[1] *Institute of Applied Mechanics, University of Erlangen-Nuremberg, Egerlandstr. 5, Erlangen 91058, Germany*
[2] *Institute B of Mechanics, University of Stuttgart, Pfaffenwaldring 9, Stuttgart 70550, Germany*
e-mail: li@ltm.uni-erlangen.de, eberhard@mechb.uni-stuttgart.de

Abstract

The technology of immersive Virtual Reality (VR) introduces many additional possibilities of the interaction for the numerical simulation of Multibody Systems (MBS). In an immersive VR-Environment the user can manipulate bodies of the concurrent ly running simulation of MBS with an input device which usually is a tracked pen. Since the simulated MBS is stereoscopic ally presented in the same 3D room where the user works, the user can manipulate the system intuitively and observe the effect of his manipulation immediately.

The server/client architecture is employed to construct the programs of the simulation and the visualization of MBS. The server is used to coordinate the clients and save temporal data. One client is used for the numerical simulation of MBS while two more clients are used for the stereoscopic visualization of MBS. We use CORBA to build the server/client architecture.

Compared to the manipulation in the real world, the user's manipulation in VR-Environment is not restricted. Sometimes this is convenient. However, it may cause the kinematical inconsistence of MBS. Two categories of approaches have been developed to deal with this inconsistency and model the interaction in view of MBS. The approach based on kinematics tries to transform the user's hand motion in consistent positions, while the approach based on dynamics introduces driving forces into the system for the simulation of the hand motion. By both approaches the user's demands can be satisfied.

REFERENCES

[1] A. A. Shabana, *Dynamics of multibody systems, 2nd Edition*, Cambridge University Press, (1998).

[2] R. G. Belleman, B. Stolk, R. de Vries, *Immersive virtual reality on commodity hardware,* ASCI 2001 Conference, The Netherlands, (2001).

[3] E. Hairer, G. Wanner, *Solving ordinary differential equations, II, stiff and differential-algebraic problems, 2nd Edition*, Springer, Berlin, (1996).

[4] J. J. Craig, *Introduction to robotics,* Mechanics & Control, Addision-Wesley, (1986).

COMPUTATIONAL MECHANICS (Abstracts)
WCCM VI in conjunction with APCOM'04, *Sept. 5-10, 2004, Beijing, China*

Numerical Study on the Dynamical Characteristics of Nonlinear System

Shuyong Liu*, Shijian Zhu, Jianbo Xie

Institute of Noise and Vibration, Naval University of Engineering, Wuhan 430033, China
e-mail: lsydh@sina.com

Abstract

The paper analyzes the dynamical characteristics of the nonlinear system. In general, the autonomous and nonautonomous system are often rewritten as first order state equations, and they can be considered as vector field, where $f(t,x)$ relates not only to location but also to time, so that it is known as time-varying field or nonautonomous field [1]. The vectors passing through any point have many, even infinite directions, and they may be different at any instant. For vector field $f(x)$, it is only determined by location, and thus it is called constant vector field. If the theorem of the unique solution is satisfied, the trajectories of the autonomous system will not intersect each other in the phase space. However, the trajectories of the nonautonomous system may intersect. Although the approximate profile of the phase trajectories is analyzed with equal slop method and Li nard drawing method, these methods are not only complex but also imprecise. Especially, the variation of the parameters makes the analysis more difficult and even some important characteristics of the system are lost. Here, the combination of linearization method and numerical algorithm is applied to reveal the interest features of the system. Linearization reveals the dynamical characteristics of the nonlinear system at the equilibrium point, so that the behavior of the system can be predicted in the vicinity of that point. Four cases are discussed depending on the eigenvalues of matrix a. Thus, the behavior of a nonlinear system near the original point can be deduced. Numerical analysis is used to research the trajectory of the system in the phase plane. Results show that the vector field and phase trajectory can reveal the effect of different parameters, where the sparseness and denseness degree of the trajectories reflects the speed of system approaching the equilibrium point, and the direction of the vector field illustrates whether the system's behavior is stable or not. When the parameters are in a special range, the nonlinear system is in a chaotic state [2] and its response contains rich dynamic information. As examples, a pendulum system with damp and a nonlinear vibration isolation system are studied.

REFERENCES

[1] F. Tankens, *Singularities of vector fields*, Publ. Math. I. H. E. S., 43, (1973), 47-50.
[2] Francis C. Moon, *Chaotic vibrations*, John Wiley & Sons, Inc., (1987), pp. 78-79.

COMPUTATIONAL MECHANICS (Abstracts)
WCCM VI in conjunction with APCOM'04, *Sept. 5-10, 2004, Beijing, China*

The Predictor-Corrector of Exponential Time Differencing for Strongly Nonlinear Autonomous Oscillators with Many DOFs

Haiqing Yan [1*], Chen Tang [1,2], Zhanqing Chen [3], Hao Zhang [1], Ming Liu [1], Guimin Zhang [1]

[1] *Deptartmeny of Applied Physics, University of Tianjin, Tianjin, 300072, China*
[2] *State Key Lab. of Nonlinear Mechanics, Institute of Mechanics, Chinese Academy of Sciences, Beijing, 100080, China*
[3] *Department of Mechanical Engineering and Automation, Harbin Institute of Technology, Harbin, 150086, China*
e-mail: yanhg@eyou.com, tangen@eyou.com

Abstract

The predictor-corrector methods of exponential time differencing are extended to determine the bifurcations and limit cycles of strongly nonlinear autonomous oscillators with many degree s of freedom.

The numerical method treated in this paper is the so-called "exponential time differencing" (ETD) scheme [1, 2, 3], or called "exponential fitting"(EF) [4, 5]. In [6] the various order explicit multistep exponential time differencing was developed which works extremely well for nonlinear systems. In this paper, we use explicit multistep exponential time differencing to obtain the predicted solution value, then use these predicted values in the corresponding implicit method to obtain the corrected solution value. This combination of an explicit ETD and implicit ETD scheme is called predictor-corrector ETD method. We test the predictor-corrector ETD method on coupled van der Pol oscillators and coupled Rayleigh oscillators, and compare with the well studied and widely used four-order Runge-Kutta method. Limit cycles of the oscillators can be calculated to and desired degree of accuracy. The numerical results show that the scheme s are highly accurate and computationally efficient.

REFERENCES

[1] R. Holland, IEEE Trans. Electromagn. Compat., 36, (1994), 32.
[2] P. G. Petropoulos, IEEE Trans. Antennas Propaga tion, 45, (1997), 1054.
[3] C. Schuster, A. Christ, W. Fichtner, Microwave Optical Technol. Lett., 25, (2000), 16.
[4] R. J. McCartin, et al, Proceedings of the Fourth Conference on Mathematical Modeling in the Undergraduate Curriculum, University of Wisconsin-La Crosse, (2000).
[5] R. J. McCartin, *Exponential fitting of the delayed recruitment/renewal equation*, J. Comput. Appl. Math., 136, (2001), 343.
[6] C. Tang, H. Q. Yan, H. Zhang, W. R. Li, *The various order explicit multistep exponential fitting for systems of ordinary differential equations*, J. Comput. Appl. Math., (accepted).

COMPUTATIONAL MECHANICS (Abstracts)
WCCM VI in conjunction with APCOM'04, *Sept. 5-10, 2004, Beijing, China*

The Computation for Active Vibration Control of Structure in an Inverse System Method

Jianhua Yu [1*], Qingshui Qiu [2], Yong Liu [1]

[1] *Dept. of Civil Engineering & Applied Mechanics, Sichuan University, Chengdu, 610065 , China*
[2] *Dept. of Architecture & Civil Engineering, University of West China, Chengdu, 610039 , China*
e-mail: hemeiyu8@mail.sc.cninfo.net

Abstract

The inverse system method of multiple-variable control is an emerging approach for nonlinear control. It has very important application on the theory and engineering of the automation control, but its application on the vibration active control is seldom found. This article deals with our studies of the inverse system method for the active vibration control of structure in the recent years, the related theoretical and computational aspects are discussed.

The basic conception of the inverse system method is firstly presented. By means of an inverse system of an original system, an inverse system in α-order integral can be built by the feedback method, the original system can then compensated as a mapping decouples linear system operating liner feedback. The control theory based on the linear system can be easily used for realization of the control of a system.

The inverse system method is introduced to the active vibration control here. A common system can be transformed to a decoupled system if we use the modal coordinate in replace of the physics coordinate system. By Laplace transformation the transmission matrix for a closed loop system can be introduced. A discussion about the foundation of the mechanical-mathematical model for the inverse system method of active vibration control is given. From this study, for the reference input, the given system can feedback smoothly, the force vector excerting the structure for active vibration control in the process of the vibration control can be obtained by the feedback.

In the vibration-centrifugal compound environment, the active vibration control is still difficult. The simulation of a vibration-centrifugal system can be realized by a multiple-rigid bodies system. Their governing differential equations are introduced by Hamilton principal and Lagrangian motion equations. A corresponding inverse system is then treated. The related mathematical model and computation is focused upon for this kind system.

The computation approach adopted is presented here. The realization of the computation for the inverse control is discussed. The efficiency of the computation arithmetic is concerned.

Two numerical examples are given to show the feasibility of the suggested mechanical-mathematical model of the inverse system method for vibration control and the suggested computation approach. An active vibration for frame of three storeys and a system of two rigid bodies are concerned. The computation data treatment and the results are presented. The main numerical results are expressed by a series of figures and tables.

The feasibility and efficiency of the suggested inverse system method using the suggested computational approach are discussed; their comparative studies with other methods are also mentioned. Further considerations for future studies for the inverse system method both in theory and computation fields are offered.

The project is supported by the National Natural Science Foundation of China (Grant No.10276030).

COMPUTATIONAL MECHANICS (Abstracts)
WCCM VI in conjunction with APCOM'04, *Sept. 5-10, 2004, Beijing, China*

Numerical Solution for DAEs of Constraint Multibody Dynamics

Yubo Zhao [1*], Jun Zou [1], Guoping Liang [2]

[1] *Department of Mathematics, The Chinese University of Hong Kong, Shatin N.T., Hong Kong, China*
[2] *Beijing FEGEN Software Co., Ltd, China Offshore Enterprise Mansion, No. 56 Zhichun Rd., Haidian District, Beijing, 100086, China*
e-mail: ybzhao@math.cuhk.edu.hk, zou@math.cuhk.edu.hk, ling@fegensoft.com

Abstract

When performing dynamic analysis of a constrained mechanical system, a set of index-3 Differential Algebraic equations (DAEs) describes the time evolution of the model. In this paper, we construct a second-order one-leg multistep method to integrate the DAEs directly while the formulae to treat the acceleration and velocity are different. Velocity elimination (VE) technique is applied to present a framework that the position and velocity vectors can be obtained in different stages, i.e. only the position variables and Lagrange multipliers take part in the nonlinear iterations at each time step and the constraint equations on the position level can be satisfied exactly. Numerical experiments for three mechanical systems including a seven-body mechanism which is used as the benchmark problem to test the multibody program are demonstrated. The test problems are used to validate the proposed algorithm. Good stabilization and high efficiency can be achieved.
The work is partial supported by the CUHK postgraduate student grants for overseas academic activities. We also thank a lot for the support of the center of R&D, China academy of railway sciences.

REFERENCES

[1] A. A. Shabana, *Dynamics of Multibody Systems* (Second Edition), Cambridge University Press, (1998).
[2] C. W. Gear, G. K. Gupta, B. J. Leihmkuhler, *Automatic integration of the Euler-Lagrange equations with constraints*, J. Comput. Appl. Math. 12-13, (1985), 77-90.
[3] Dan Negrut, E. J. Haug, H. C. German, *An implicit Runge-Kutta method for integration of differential algebraic equations of multibody dynamics*, Multibody System Dynamics, 9, (2003), 121-142.
[4] E. J. Haug, *Computer Aided Kinematics and Dynamics of Mechanical Systems, Volume 1: Basic Methods*, Allyn and Bacon, New York, (1989).
[5] K. Brenan, S. Campbell, L. Petzold, *Numerical Solution of Initial-Value Problems in Differential-Algebraic Equations*, North-Holland, (1989).
[6] Uri M. Ascher, Hongsheng Chin, Sebastian Reich, *Stabilization of DAEs and invariant manifolds*, Numer. Math., 67, (1994), 131-149.
[7] W. Schielen ed., *Multibody Systems Handbook*, Springer, (1990).

COMPUTATIONAL MECHANICS (Abstracts)
WCCM VI in conjunction with APCOM'04, *Sept. 5-10, 2004, Beijing, China*

A Self-Adaptive Arc-Length Method for Computing Accurate Poincaré Map

Jingwu Gao*, Zhongmin Cai

Institute of Applied Mechanics, Taiyuan University of Technology, Taiyuan 030024, China
e-mail: gaojingwu@sina.com, caizhongmin_3@hotmail.com

Abstract

In many physical applications and particularly in the theory of dynamical systems, one is often interested in computing a Poincaré map of a system. This map is produced by considering successive intersections of a trajectory with a codimension-one surface Σ — a Poincaré section of the phase space. In Classical numeric methods, when such a crossing of Σ has been detected, the actual intersection can be approximated by an interpolation scheme, e.g. a bisection- or a Newton algorithm. Such an iterative ending may require many vector field evaluations, as well as a diminishing step-size t. This complicates any attempt to make a rigorous, global error estimate, and the largest computational error usually occurs at this final stage. An improving procedure was described by Hénon [1] through transforming one of the phase variables into an independent one, which allow us to make the final approach to Σ in one single step, thereby avoiding the error accumulation associated with classical methods. As a generalization of such an approach, Warwick Tucker [2] applied the transformation not only at the last stage, but for whole integration process, where independent variable need to be changed frequently at each integration step according to the dominating coordinate in the vector field. In this paper, an arc-length method, firstly presented by Riks for a snapping or buckling question [3], and developed by Jike Wu et al [4, 5] for solving the differential equations with singularity, was introduced to compute Poincaré Map. Furthermore, a self-adaptive step-length control was developed for avoiding the error accumulation caused by a sharp turn of trajectory within one integration step. However, in our approach, arc-length variable is always the dominating one, so it is easier for programming and more efficient than that before. An example is also given, and numeric results demonstrate its accuracy.

REFERENCES

[1] M. Hénon, *On the numerical computation of Poincaré maps*, Physica, D 5, (1982), 412–414.
[2] Warwick Tucker, *Computing accurate Poincaré maps*, Physica, D 171, (2002), 127–137.
[3] E. Riks, *An incremental approach to the solution of snapping and buckling problems*, Int. J. Solid Structures, 15(4), (1979), 529-551.
[4] Jike Wu, W. H. Hui, Hongli Ding, *Arc-length method for differential equations*, Applied Mathmatics and Mechanics, 20, (1999), 875-880.
[5] Zuochang Zheng, Jike Wu, *Stability analysis of periodic solutions of dynamical systems in arclength parametrization*, Acta Scientiarum Naturalium Universitatis Pekinensis, 33(3), (1997), 315-318.

COMPUTATIONAL MECHANICS (Abstracts)
WCCM VI in conjunction with APCOM'04, *Sept. 5-10, 2004, Beijing, China*

Series Solution of Avoiding Calculating the Inversion of the State Matrix for Nonlinear Dynamics Equation

Jinqiao Li [1*], Jianhua Yu [2]

[1] *Shenzhen Terra Co., LTD, Shenzhen, 518040, China*
[2]*Dept. of Civil Engineering & Applied Mechanics, Sichuan University, Chengdu, 610065, China*
e-mail: ljqjfx@163.net, hemeiyu8@mail.sc.cninfo.net

Abstract

Based on the precise integration method of the exponential matrix developed by Wanxie Zhong, we discuss a general dynamics system governed by the state equation $\dot{\boldsymbol{v}} = \boldsymbol{H} \cdot \boldsymbol{v} + \boldsymbol{f}(\boldsymbol{v},t)$, in which $\boldsymbol{v}$ is an unknown n-dimensional vector, $\boldsymbol{H}$ is a state coefficient matrix, $\boldsymbol{H} \cdot \boldsymbol{v}$ and $\boldsymbol{f}(\boldsymbol{v},t)$ are the linear homogeneous part and nonlinear part in the right members of the equation respectively.

It is specially necessary to pay high attentions to the numerical method of nonlinear dynamics equation in that the analytical solution are difficult to be found out for the most cases. The high precise solution and calculation stabilization have been two extremely important problems for the numerical method.

For nonlinear dynamics equation, the existing numerical methods have some disadvantages, such as the precision is sensitivity to the time-step or some components of the high frequency vibration can't be reflected for the large time-step, etc. Fortunately, the precise integration method overcomes these disadvantages for its striking high precision and has been applied in many engineer fields. In some cases it is more or less difficult due to the inversion matrix calculation, in a sense this limits the exercise of precise integration. Two kinds of explicit series solution under precise integration for the nonlinear dynamic equation is provided in this article.

In this paper, we deal with the nonlinear dynamics state equation $\dot{\boldsymbol{v}} = \boldsymbol{H} \cdot \boldsymbol{v} + \boldsymbol{f}(\boldsymbol{v},t)$, the nonlinear part $\boldsymbol{f}(\boldsymbol{v},t)$ is expressed in Taylor series at the time t_k. Then the authors suggest that the integral of the state equation can be evaluated by different numerical method combined with the precise integration.

By the calculative strategy elaborated above, two classes of series solution can be successively obtained. In the first one, the exponential matrix in the integral part is expressed in Taylor series, so the state equation is evaluated by numerical method directly through exponential matrix and its precise algorithm. In the other one, the high precise numerical integral technique such as Gauss-Legendre class can be applied directly in treating with the integral term, and then a high precision and convenient solution can be attained satisfactorily. The precision of the two kinds of series solution can be controlled easily in practical utilization.

This algorithm acquired avoids calculating the inversion of the $\boldsymbol{H}$ matrix and embodies much better advantage when the inversion of matrix $\boldsymbol{H}$ does not exist or is of oddity, meanwhile the stabilization and efficiency of the computation can be ensured. The algorithm discussed is more benefit to the large-scale problems especially. It is not only very convenient for programming implementation, but also more efficient to parallel computer, on the other hand the satisfied solution can be achieved in relatively great time-step.

Finally, the numerical examples presented demonstrate the validity and efficiency of this algorithm, and some useful conclusions are drawn.

COMPUTATIONAL MECHANICS (Abstracts)
WCCM VI in conjunction with APCOM'04, *Sept. 5-10, 2004, Beijing, China*

Identification of Nonlinear Regions of System Responses by Using Box Dimensions of Poincaré Maps

F. S. Lin [1,2*], T. Zhang [2], G. Meng [2]

[1] *Department of Electromechanical Engineering, Wuhan University of Science & Engineering, 1 Fangzhi Rd, Wuhan, 430073, China*
[2] *State Key Laboratory of Vibration, Shock & Noise, Shanghai Jiaotong University, 1954 Huanshan Rd, Shanghai, 200030, China*
e-mail: Lfsfs@163.com, zt000@263.net, gmeng@sjtu.edu.cn

Abstract

The fractal theory was previously based on systems with self-similarity; however, this theory has offered a new approach of obtaining the features of faults of rotary machinery. Studies have shown that nonlinear dynamic systems also have fractal characteristics and that the fractal dimension has some corresponding relations with chaos.

In the literature, to calculate the fractal dimension of a nonlinear system, most methods, such as the G-P algorithm and the box-counting approach, usually use the time series of the response. The authors analyzed the problem of these algorithms used in practical engineering, then proposed a modified box-counting measure to calculate the box dimension of the Poincaré map of the response. The measure was proved using the fractal theory. Using this technique we can identify periodic, quasi-periodic and chaotic response regions of the system and farther learn the fault characteristics of the system. Compared with the traditional G-P algorithm, the improved G-P algorithm and the box-counting approach, the computation time required for the proposed method was drastically reduced. The mode shapes of nonlinear vibratory systems are found to be associated with the box dimensions of Poincaré maps. The response is chaotic if the box dimension is less than 1.0, quasi-periodic if equal to 1.0, and periodic if equal to 2.0.

Squeezed film damper (SFD) have been installed on the rotor systems of many aerial engines to reduce the vibration amplitude. However, the strong nonlinear forces of the squeezed film probably result in strong nonlinear responses of the rotor system. Besides, the impact-rub between the rotor and the stator of rotary machinery also brings on nonlinear responses. There exist a numerous nonlinear bifurcation and chaotic phenomenon in these systems. The proposed method is demonstrated and validated by a numerical calculation of an impact-rub rotor system supported on squeezed film dampers. The bifurcation diagram of the rotor system was compared with the plot of the box dimension of Poincaré maps versus the corresponding control parameter.

The great advantage of the proposed algorithm is the convenience of identifying the type of responses, in addition to other advantages such as the possibility of identifying route to chaos and the saving in the execution time.

COMPUTATIONAL MECHANICS (Abstracts)
WCCM VI in conjunction with APCOM'04, *Sept. 5-10, 2004, Beijing, China*

Analysis of Fractional-Order Numerical Method for the Fractional Relaxation Equation

Ran Lin [1*], Fawang Liu [1, 2]

[1] *School of Mathematical Sciences, Xiamen University, Xiamen 361005, China*
[2] *School of Mathematical Sciences, Queensland University of Technology, GPO Box 2434, Brisbane, Qld. 4001, Australia*
e-mail: 3425linran@sohu.com, fwliu@xmu.edu.cn

Abstract

Various fields of science and engineering deal with dynamical systems described by fractional-order equations. This topic has received a great deal of attention in the last decade [1 -7], especially in the fields of viscoelastic materials, electrochemical processes, dielectric polarization, colored noise, anomalous diffusion and chaos. The main reason for the success of the theory in these cases is that these new fractional-order models are more accurate than integer-order models.
Numerical methods associated with integral order ordinary differential equation, for example, Euler method, linear multiple step method, and so on, have been treated extensively in the literature. On the other hand, theoretical studies of the fractional-order numerical methods and the error estimate of fractional order differential equation are quite limited, because theoretical analysis of fractional order numerical methods is very difficult [3-5].
In this paper, we consider the fractional-order relaxation equation:
$_0D_t^{\alpha} y(t) + Ay(t) = f(t), \ (t > 0), \quad y(0) = 0, \ 0 < \alpha \le 1.$
A fractional-order numerical method of the fractional-order relaxation equation is presented. The consistence, convergence and stability of the fractional-order numerical method are proven. Finally, some numerical examples are provided to show that the fractional order numerical method for solving the fractional-order relaxation equation is an effective solution method.

REFERENCES

[1] K. B. Oldham, J. C. Spanier, *The fractional calculus*, Academic Press, San Diego, CA, (1974).
[2] K. S. Miller, B. Ross, *An introduction to the fractional calculus and fractional differential equations*, Wiley, New York, (1993).
[3] I. Podlubny, *Fractional differential equations*, Academic Press, San Diego, CA, (1999).
[4] F. Liu, V. Anh, I. Turner, *Numerical solution of space fractional Fokker-Planck equation*, J. Comp. and Appl. Math., 166, (2004), 209-219.
[5] Ch. Lubich, *Discretized fractional calculus*, SIAM J. Math. Anal., 17(3), (1986), 704-719.
[6] R. L. Bagley, R. A. Calico, *Fractional order state equations for the control of viscoelastic structures*, Journal of Guidance, Control, and Dynamics, 14(2), (1999), 304-311.
[7] R. Lin, F. Liu, *High order approximations for the fractional ordinary differential equation with initial value problem*, Journal of Xiamen University (Natural Science), 43(1), (2004), 21-25, (in Chinese).

COMPUTATIONAL MECHANICS (Abstracts)
WCCM VI in conjunction with APCOM'04, *Sept. 5-10, 2004, Beijing, China*

Responses' Convergence for Time Integration Analyses Involved in Linearly-Elastic/Perfectly-Plastic Behavior and Impact

Aram Soroushian *

Department of Civil Engineering, University of Tehran, Teranpars, Rashid Street, 136 alley, No. 35, Tehran, 16517, Iran
e-mail: asoroush@ut.ac.ir, aramsoro@yahoo.com

Abstract

In order to analyze structural dynamic models with nonlinear behavior, time integration is a broadly accepted analysis method that generates approximate responses for the general case of MDOF models. Hence as a numerical method, time integration is successful only when the computed approximate responses converge to exact responses. Though convergence can be simply studied and maintained in the range of linear analyses, in the presence of nonlinearity, convergence is not unfailing; to say better when repeating time integration analyses with smaller time steps, there is no guarantee to arrive at more accuracy, and even errors in order of instability are reported in the literature. The phenomenon of responses' probable improper convergence is discussed in the literature by times. Nevertheless, as yet there is no reliable method to overcome the general case of the shortcoming. The author(s) of this abstract has recently carried out a detailed and mathematical study on this subject, and among several consequences, appropriately controlling the iterations of nonlinearity solutions is sought as a tool that can result responses' convergence for nonlinear problems. Methods to overcome the shortcomings are thereafter developed, and recently applied to dynamic problems involved in linearly-elastic/perfectly-plastic behavior and problems involved in impact, separately. In this paper, in order to display the efficacy of the proposed method in presence of the complicated nonlinearities of real structural systems, after a review on the formulation, the resulting method is applied to linearly-elastic/perfectly-plastic models that are also involved in impact. Considering different parameters and time integration methods, it is clearly illustrated that the proposed method can maintain appropriate responses' convergence with respect to time step size even in complicated nonlinear problems. Some discussions about the computational cost will be also presented, and finally the paper will be closed with a set of brief conclusions.

REFERENCES

[1] P. Henrici, *Discrete Variable Methods in ODEs*, John Wiley and Sons, USA, (1962).

[2] T. Belytschko, W. K. Liu, B. Moran, *Nonlinear Finite Elements for Continua and Structures*, John Wiley & Sons Inc., USA, (2001).

[3] A. Soroushian, J. Farjoodi, P. Wriggers, *Reliable convergence for dynamic linearly-elastic/perfectly-plastic systems analyzed with different integration methods*, in A. Nilsson, H. Boden eds. *Sound and vibration*, 10th Int. Cong., Stockholm, 2003, IIAV, Stockholm (2003), (CDROM).

[4] A. Soroushian, J. Farjoodi, *Responses convergence in time integration of nonlinear semi-discrete equations of motion*, In M. Brennan ed. *Advances in Structural Dynamics*, 8th Int. Conf., Southampton, 2003, ISVR, Southampton, (2003), (CDROM).

COMPUTATIONAL MECHANICS (Abstracts)
WCCM VI in conjunction with APCOM'04, *Sept. 5-10, 2004, Beijing, China*

Study of Two Dimensional Magneto Thermoelastic Problems in Rotating Medium by Eigenfunction Expansion Method

Arup Baksi [1*], Rasajit Kumar Bera [2]

[1] *Umeschandra Collge, 13, Surya Sen Street, Kolkata, 700 012, India*
[2] *Technical Teachers' Taining Institute, Block-FC, Sector-III, Salt Lake City, Kolkata, 700 106, India*
e-mail: rasajit@yahoo.com

Abstract

The governing equations for displacement and temperature fields in the linear dynamical theory of classical thermo-elasticity consist of the coupled partial differential equation of motion and the Fourier 抯 law of heat conduction equation. The equation for di splacement field is governed by a wave type hyperbolic equation, whereas, that for the temperature field is a parabolic diffusion type equation. This amounts to the remark that the classical thermoelasticity predicts a finite speed for predominantly elasti c disturbances but as infinite speed for predominantly thermal disturbances, which are coupled together. This means that a part of every solution of the equations extends to infinity, vide, Lord and Shulman. Experimental investigations by Ackerman et al, v on Gutfeld and Nethercot, Taylor et al, Jackson and Walker and many others, conducted on different solids have shown that heat pulses do not propagate with infinite speeds. In order to overcome this paradox, efforts were made to modify classical thermoelas ticity, on different grounds, for obtaining a wave type heat conduction equation, vide, Kaliski, Norwood and Warren, Green and Lindsay, Suhubi, Lebon, Dhaliwal and Singh and others. A comprehensive list on this generalization for the last two decades can b e available in the work of Chandrasekharaiah [1].
At present there are two generalizations of the classical theory of thermoelasticity. The first proposed by Lord and Shulman [2] (L-S theory) which involves one relaxation time for a thermo -elastic process, and the second by Green and Lindsay [3] (G-L theory), which takes into account two parameters on relaxation times. Owing to the mathematical complexities encountered in coupled thermo -elasticity, mainly due to inertia and coupling terms in governing equat ions, these types of problems are mostly confined to one dimensional problems, vide, Noda et al, Furukawa et al, Anwar et al, Chandrasekharaiah et al, Roy Choudhari, Mukhopadhyay and Bera and many others. However, an attempt has been made to investigate a two dimensional problem in generalized thermoelasticity with the help of eigenvalue approach in the preset analysis. However, in the present paper following one parameter L -S theory, the authors have considered a problem of rotation in two dimensions, in g eneralized theory of thermo-elasticity by taking into account of both the dynamic effect and the influence of coupling terms. Although this is a two dimensional problem under the influence of instantaneous heat sources distributed over a plane area in an infinite, transversely isotropic elastic medium, it involves two displacement components due to rotation unlike that in the one dimensional case where the number of displacement components increase due to rotation . The new eigen-value approach of Das et al has been applied for the solution of the problem. The solution has been achieved in closed form in Laplace -Fourier transform domain and finally numerical inversions in space-time domain have been made with the help of MATLAB Software and some of the results have been shown graphically.

REFERENCES

[1] H. W. Lord, Y. Shulman, J. Mech. Phys. Solids, 15, (1967), 299-309.
[2] A. E. Green, K. A. Lindsay, *Thermoelasticity*, J. Elasticity, 2, (1972), 1-7.
[3] D. S. Chandrasekharaiah, Appl. Mech. Rev., 39(3), (1986), 355 -376.

COMPUTATIONAL MECHANICS (Abstracts)
WCCM VI in conjunction with APCOM'04, *Sept. 5-10, 2004, Beijing, China*

Research on Magnus Force of Common Ammunition Bodies with Turbulent Layers

Jun Chen *

School of Mechanical Engineering, Nanjing University of Science & Technology, Xiaolingwei 200#, Nanjing City, 210014, China
e-mail: cjsky@public1.ptt.js.cn, cjsky@sina.com

Abstract

Usually most researches on Magnus force of an ammunition body only gave calculating formulae of laminar layers [1,2]. In fact, modern ammunitions often worked with high velocity. So it's necessary that calculating model of Magnus force of turbulent layers be built. Calculating model of Magnus force on common ammunition bodies (cone-shaped and parabola-shaped) with turbulent layers was built based on Magnus theory [3]. The force coefficients and moment coefficients of Magnus force could be achieved by the model. The research considered effect of temperature exponential, adopted curve-fitting approaches which could give more exact result data. The format of the calculation formulae of turbulent layers is similar with that of laminar layers. Similarity constants in Magnus force formulae respectively called flow layer constants and shape constants were presented which were useful to evaluate Magnus force in different states. Referring to other references and the formulae derivng from the model, we can make a conclusion that Magnus force coefficients may be expressed follow formats

$$c_{z\overline{\omega}_x\overline{\alpha}} = \frac{\Theta_c H}{\mathrm{Re}_L^{\Theta_{\mathrm{Re}}}}(1-1.52\overline{\alpha})\lambda^2\left[1-\left(1-\frac{\Theta_{c\alpha}}{1-1.52\overline{\alpha}}\right)\left(\frac{\lambda_n}{\lambda}\right)^{1.8}\right], \quad m_{y\overline{\omega}_x\overline{\alpha}} = \frac{\Theta_m H}{\mathrm{Re}_L^{\Theta_{\mathrm{Re}}}}(1-1.52\overline{\alpha})\lambda^3\left[1-\left(1-\frac{\Theta_{m\alpha}}{1-1.52\overline{\alpha}}\right)\left(\frac{\lambda_n}{\lambda}\right)^{2.8}\right]$$

It's obvious that constants Θ differ with different flow state (i.e. laminar or turbulent layer) and different shape of ammunition bodies. The data in Table 1 show that Θ_{Re}, Θ_c and Θ_m (called flow layer constants) are only related to flow state (i.e. laminar or turbulent layer), but $\Theta_{c\alpha}$ and $\Theta_{m\alpha}$ (called shape constants) almost have no relation to flow state but to shapes of bodies.

Table 1. Constants Θ changed with different states

flow state / shape	laminar layer					turbulent layer				
	Θ_{Re}	Θ_c	$\Theta_{c\alpha}$	Θ_m	$\Theta_{m\alpha}$	Θ_{Re}	Θ_c	$\Theta_{c\alpha}$	Θ_m	$\Theta_{m\alpha}$
cone-shaped	0.5	20.8	0.231	12.48	0.275	0.2	1.23333	0.24744	0.793	0.28350
parabola-shaped	0.5	20.8	0.365	12.48	0.407	0.2	1.23333	0.36840	0.793	0.40873

REFERENCES

[1] H. J. Luckert, *Static stability and drag studies for bodies of revolution in super sonic flow*. AIAA 73295.

[2] E. R. Van Driest, *Turbulent boundary layer in compressible fluids*, JAS, 18(3), (1951).

[3] H. R. Vaughn, G. E. Reis, *A Magnus theory*. AIAA Journal, 11(10), (1973).

COMPUTATIONAL MECHANICS (Abstracts)
WCCM VI in conjunction with APCOM'04, *Sept. 5-10, 2004, Beijing, China*

Stabilization of Staggered Coupled Analysis by Numerical Filter for Electromagnetic and Structural Coupled Problem

T. Horie [*], T. Niho

Department of Mechanical Information Science and Technology, Kyushu Institute of Technology
680-4 Kawazu, Iizuka, Fukuoka, 820-8502, Japan
e-mail: horie@mse.kyutech.ac.jp, niho@mse.kyutech.ac.jp

Abstract

In analyses of an electromagnetic and structural coupled problem or magnetic damping problem, two types of coupled analysis methods have been proposed [1]; one is the simultaneous method and the other is the staggered method. The staggered method, in which eddy current analysis and structural analysis are performed alternately at each time step, is suitable for a large scale coupled analysis; however, the solution diverges as the result of numerical instability under some analysis conditions of magnetic field and time increment. Studies on the numerical instability of the electromagnetic and structural coupled analysis have been performed, such as a proposal of numerical instability condition [2], a comparison of stability of some staggered methods [3] and a proposal of an augmented staggered method with reformulation [4]; but little attention has been given to the stabilization technique which is easily applied to the staggered coupled analysis methods.

In this study, a stabilization technique of numerical filter, which is widely used for the signal processing, is proposed for the staggered coupled analysis. At each time step, vibration components caused by the numerical instability is eliminated from the analysis results by the numerical filter between two analyses such as the eddy current analysis and the structural analysis.

Analyses of a plate with electromagnetic and structural coupled effect were carried out to examine frequency characteristics of the vibration including numerical oscillation. Since the frequency of numerical oscillation is higher than that of the plate vibration modes, numerical filters including low pass filter or the moving average method may be effective for the staggered coupled analysis. Analyses with numerical filter were also performed. According to the results with low pass filter, accuracy of solution is not satisfactory although the numerical oscillation mode is eliminated. Stable and accurate solution is obtained using the moving average method with recalculation. Stabilization methods using other numerical filters were also examined from the viewpoints of stability, accuracy and computing cost.

REFERENCES

[1] T. Takagi, *Summary of the results for magnetic damping in torsional mode (TEAM problem 16)*, COMPEL, 14, (1995), 77-89.

[2] T. Niho, T. Horie, Y. Tanaka, *Numerical instability of magnetic damping problem of elastic plate*, IEEE Trans. on Magn., 36(4), (2000), 1373-1376.

[3] T. Horie, Y. Tanaka, T. Niho, in Proceedings of the First Asian-Pacific Congress on Computational Mechanics, vol. 2, (2001), pp. 1335-1340.

[4] Y. Tanaka, S. Furuki, E. Shintaku, Y. Fujimoto, in Record of the 14th COMPUMAG Conference on the Computation of Electromagnetic Fields, (2003), pp. 46-47.

COMPUTATIONAL MECHANICS (Abstracts)
WCCM VI in conjunction with APCOM'04, *Sept. 5-10, 2004, Beijing, China*

Molecular Simulation of the Adsorption Behavior of Chemical Admixtures on Cement Surfaces

Li-Chen Lee, Kung-Chung Hsu *

Department of Chemistry, National Taiwan Normal University, 88, Sec. 4, Tingchow Road, Taipei, 116, Taiwan, China
e-mail: kchsu@cc.ntnu.edu.tw

Abstract

Chemical admixture, after been adsorbed onto cement particles, can enhance the fluidity or workability of cementitious materials. This paper has applied a molecular dynamics (MD) technique to investigate the adsorption behaviors of chemical admixtures onto cement surfaces. MD simulations were performed according to Cerius^2 3.5 molecular modeling software from Molecular Simulations Incorporated. Organic chemicals with different functional groups or degree of polymerization (DP) were constructed. The crystal structures of C_3A and C_3S were built. During simulation, the organic molecule was allowed to move freely onto a fixed inorganic material surface until an equilibrium condition was established, as evidenced by the calculated system energy reaching a stable minimum value. The factors such as the functional group and molecular weight of organic molecules, and the composition of cement on the adsorption of chemical admixtures were examined.

The simulation results indicate that chemical admixtures are adsorbed stronger on C_3A, which contains aluminum ions with greater positive charges, than on C_3S, which contains calcium ions. The ionic functional groups of admixture molecules exhibit greater adsorption capability than the polar and nonpolar groups. The adsorption forces increase with the degree of polymerization (DP) of admixtures up to a critical DP value.

REFERENCES

[1] M. R. Rixom, N. P. Mailvaganam, *Chemical admixture for concrete*, 2nd Ed., E & FN Spon, New York, (1986).
[2] C. Jolicoeur, M. A. Simard, *Chemical admixture-cement interactions: phenomenology and physico-chemical concepts*, Cem. Conc. Compos., 20, (1988), 87-101.
[3] P. J. Anderson, D. M. Roy, J. M. Gaidis, *The effect of superlasticizer molecular weight on its adsorption and dispersion of cemen*t, Cem. Conc. Res., 18(6), (1988) 980-986.
[4] *Cerius^2 3.5*, Molecular Simulations Inc., San Diego, (1998).

COMPUTATIONAL MECHANICS (Abstracts)
WCCM VI in conjunction with APCOM'04, *Sept. 5-10, 2004, Beijing, China*

Studies of Alumina Nanoparticles Using the First Principles Density Functional Theory Calculations

X. Song [1*], Q. Ge [2], S. C. Yen [1]

[1] *Materials Technology Center, Southern Illinois University, Carbondale, IL 62901, USA*
[2] *Department of Chemistry and Biochemistry, Southern Illinois University, Carbondale, IL 62901, USA*

e-mail: xiaolan@siu.edu, Qge@chem.siu.edu, myen@siu.edu

Abstract

Alumina nanoparticle is one of the most commercially important nanoparticle materials with broad application in ceramics, microelectronics, coatings, reinforcing materials, and catalysis. Introduction of the alumina nanoparticles into polymer is shown to make new composite materials with enhanced mechanical, tribological, thermal, electronic, and optical properties. An accurate characterization of the structure, stability, and property of alumina nanoparticles is very important to control and optimize the properties of new nanocomposite materials.

In this work, fifteen structures consisting of 5-32 atoms were investigated. Both stoichiometric and nonstoichiometric clusters were included in our research. The largest particle studied here is 1.4 nm in one dimension. The initial geometries of the clusters were fragments of the Al_2O_3 bulk crystalline structure. Once the initial geometry is selected, the structure is allowed to relax according to the Hellmann-Feynman forces on the atoms and the total energy. The calculations were conducted using the Vienna Ab-initio Simulation Package (VASP) [1], which is based on DFT with a plane wave basis set. A plane wave cutoff energy of 500 eV was used in the calculations. The bond length, formation energy, HOMO-LUMO band gap and the magnetic moment were computed for all clusters.

The results show a trend of increasing stability with increasing size for the clusters. The O/Al atomic ratio also strongly influences the stability of the cluster. Stoichiometric clusters are more stable than the under-oxidized counterparts. The clusters with the composition closer to the stoichiometric ratio are energetically more stable. The HOMO-LUMO band gaps of the clusters are in the range of 0.015-3.791 eV. These are significantly lower than the calculated bulk band gap of the 5.936 eV. With a narrower band gap, alumina nanoparticles have the potential to be used as photocatalyst [2] and explored for other applications. Some clusters have finite magnetic moments although the bulk alumina material is known as non-magnetic.

REFERENCES

[1] G. Kresse, J. Hafner, *Ab initio molecular dynamics for liquid metals*, Phys. Rev. B, 47, (1993), 558-561.

[2] T. A. Yamamoto, S. Seino, M. Katsura, R. Oshima, Y. Nagata, *Hydrogen gas evolution from alumina nanoparticles dispersed in water irradiated with γ-ray*, Nanostructured Materials, 12, (1999), 1045-1048.

COMPUTATIONAL MECHANICS (Abstracts)
WCCM VI in conjunction with APCOM'04, *Sept. 5-10, 2004, Beijing, China*

The Non-Stationary 3D Coupled Problems of the Theory of Thermo-Plasticity

Pavlo Steblyanko *

Department of Higher Mathematics, State Technical University, Dneprostroevskaya 2, Dneprodzerzhinsk, 51918, Ukraine
e-mail: caf-vmi@ukr.net

Abstract

A new effective method for solution of three-dimensional coupled problems of the theory of thermo-plasticity, which proposed in work [3], for case non-stationary loading of solids with concentrators, is proposed.
The numerical solution is based on the application to complete simultaneous differential equations (equation of motion, geometrical correlation, physical equations, equation of heat-conduction, where the heat source is stipulated cyclically by varied load and that depends of a stress tensor and deformations) the method of decomposition which the unknown values (displacement velocity vector, stress tensor, deformation tensor and temperature) given at each step in the form of spline-functions.
The simultaneous equations, which above mentioned may be to reduce to vector form

$$\frac{\partial \vec{W}}{\partial t} = \sum_{i-1}^{3}\left(A_i \frac{\partial \vec{W}}{\partial \alpha_i} + B_i \frac{\partial^2 \vec{W}}{\partial \alpha_i^2} \right) + \vec{F}_i ,$$

that is let to apply for its decision the method of on component slitting in combination with by notion of unknown values in form of strengthening splines [1-2]

$$W_m = \sum_{l=-1}^{2} b_{j+l} \cdot H_N^{j+l}(\alpha_i).$$

Here factors b_{-1}, b_0, b_1, b_2 are determined from conditions of collocation and boundary conditions.
In the capacity as example the problem about definition strained-deforming state of preliminary stretched tube and with circular-fronted notch by cyclic turn of small amplitude are presented. The problem is deciding take into consideration evolution of near section for deformation process.
This method gives the third order of accuracy on coordinates.

REFERENCES

[1] P. A. Steblyanko, *Methods of decomposition in space problems of the theory of plasticity*, Naukova dumka, Kyiv, Ukraine, (1998).

[2] P. A. Steblyanko, *The schemes of abnormally high accuracy solution of nonstationary problems of theory of the thermo-elastic-plasticity for plates and shells – thermal stresses and related topics*. Proc. 5th Int. Conf., Blacksburg, 2003, Virginia, (2003), pp. 231-234.

[3] U. N. Shevchenko, V. G. Savchenko, *The mechanics of coupled fields in elements of constructions, vol. 2: the thermo-viscous plasticity*, Naukova dumka, Kyiv, Ukraine, (1987).

COMPUTATIONAL MECHANICS (Abstracts)
WCCM VI in conjunction with APCOM'04, *Sept. 5-10, 2004, Beijing, China*

Numerical Simulation on Electrohydrodynamics by Least-Squares Meshfree Method

X. K. Zhang [1*], K. M. Lim [1, 2], B. C. Khoo [1, 2], S. P. Lim [1, 2], R. F. Carlos [1], Z. C. Xuan [1]

[1] *Singapore-MIT Alliance, National University of Singapore, 117576, Singapore*
[2] *Department of Mechanical & Production Engineering, National University of Singapore, 119260, Singapore*
email: {smazxk, mpelimkm, mpekbc, mpelimsp, smarfc, smaxzc}@nus.edu.sg

Abstract

Microelectrode structures in AC electrokinetics can generate high electric field strength to manipulate, characterize and separate particles in suspending medium. It has been widely used in biological, pharmaceutical and medical fields. Joule heat generated by electric field will have effect on the temperature and velocity of suspending medium. This paper presents the numerical study on two-dimensional electrohydrodynamics, electrothermally induced fluid flow, by the least-squares meshfree method. In the present study, Maxwell equation will be used to compute electric field. Time-dependent Navier-Stokes equations in velocity-pressure-vorticity form are employed to describe incompressible viscous flow of the suspending medium. The heat flux is introduced as an independent variable in the heat conduction equation to obtain the linear first-order equation for temperature field. The discretization of all governing equations is completed by the least-squares method. In the least-squares formulation, equal-order local approximation is used with Gaussian quadrature in the background cells. An element-by-element (EBE) Jacobi preconditioned conjugate method (JCG) is applied to solve the resulting systems. Numerical solutions of electric field, temperature field and fluid flow are presented.

REFERENCES

[1] B. N. Jiang, *The least-squares finite element method — theory and Applications in Computational Fluid Dynamics and Electromagnetics*, Springer: Berlin, (1998).

[2] S. H. Park, S. K. Youn, *The least-squares meshfree method*, Int. J. Numer. Meth. Eng., 52, (2001), 997-1012.

[3] T. Belytschko, Y. Krongauz, D. Organ, M. Fleming, P. Krysl, *Meshless methods: an overview and recent developments*, Comp. Meth. Appl. Mech. Eng, 139, (1996), 3-47.

[4] N. G. Green, A. Ramos, A. Gonzalez, A. Castellanos, H. Morgan, *Electrothermally induced fluid flow on microelectrodes*, J. Electrostatics, 53, (2001),71-87.

COMPUTATIONAL MECHANICS (Abstracts)
WCCM VI in conjunction with APCOM'04, *Sept. 5-10, 2004, Beijing, China*

Biomechanical Evaluation of PMMA Injection in Vertebroplasty

Soo-Won Chae [1*], Jun Hyoung Lee [1], Tae Soo Lee [2], Jung Gun Suh [3], Jung Yul Park [3]

[1] *Department of Mechanical Engineering, Korea University, 1,5-ka, Anam-dong, Sungbuk-ku, Seoul, 136-701, Korea*

[2] *Department of Mechanical Engineering, Sogang University, 1, Sinsu-dong, Mapo-ku, Seoul, 121-742, Korea*

[3] *Neurosurgery, Korea University Medical Center, 1,5-ka, Anam-dong, Sungbuk-ku, Seoul, 136-701, Korea*

e-mail: {swchae, hyoung944}@korea.ac.kr, tslee@sogang.ac.kr, jcjks@unitel.co.kr, jypark@kumc.or.kr

Abstract

The compressed fracture of spine caused by osteoporosis is one of the most frequent disease in bone fracture. Recently the vertebroplasty has drawn much attention as a medical treatment for the compressed fracture of spine, which strengthens the vertebral body and corrects deformity, and relieves pain in patients by injecting bone cement. Although this method is considered as the most ideal treatment available so far, most of the research efforts made on this method are nothing else than qualitative investigations or clinical statistics. In this study, method of PMMA injection is suggested for vertebroplasty in patients with osteoporotic compression fracture. The finite element analysis is used to investigate the vertebroplasty quantitatively. In order to improve previous works with simplified geometry of vertebral body more exact geometry has been constructed from CT image data with 1mm thickness. An ideal method of PMMA delivery, with respect to location and amount of injectate, into vertebral body has been suggested based on evaluation of the insert positions and the insert shapes of injected PMMA. It is shown that vertebral body can be compensated most efficiently when PMMA is highly concentrated on the top-front of trabecular bone of compressed vertebra.

REFERENCES

[1] J. S. Thomsen, E. N. Ebbesen, L. Mosekilde, *Zone dependent changes in human vertebral trabecular bone: clinical implications, bone*, 30(5), (2002), 664-669.

[2] A. Cotton, N. Boutry, B. Cortet, et al, *Percutaneous vertebroplasty, state of the art*, Radio Graphics, 18, (1998), 311-320.

[3] M. Liebschner, W. Rosenberg, T. Keaveny, *Effect of bone cement volume and distribution on vertebral stiffness after vertebroplasty*, SPINE, 26(14), (2001), 1547-1554.

COMPUTATIONAL MECHANICS (Abstracts)
WCCM VI in conjunction with APCOM'04, *Sept. 5-10, 2004, Beijing, China*

Triphasic Finite Element Simulation of Articular Cartilage Swelling and Curling Behaviors

Yuan Chen*, Xian Chen, Toshiaki Hisada

School of Frontier Sciences, The University of Tokyo, 7-3-1 Hongo, Bunkyo-ku, Tokyo 113-8654, Japan
e-mail: chenyuan@sml.k.u-tokyo.ac.jp, xchen@ sml.k.u-tokyo.ac.jp, hisada@ingram.t.u-tokyo.ac.jp

Abstract

Swelling of articular cartilage originates from osmotic pressure caused by fixed charges on its collagen-proteoglycan solid matrix. The swelling pressure is balanced by the stress due to matrix swelling and thus results in residual stress in articular cartilage *in vivo* in the absence of external load. The inhomogeneous fixed charge density (FCD) causes nonuniform distributions of swelling pressure and hence residual stress in the cartilage, and as a result the cartilage exhibits curling deformation after excision from the subchondral bone. Such curling phenomena reflecting the existence of residual stress have been observed in experiments. In this study, an articular cartilage is modeled as a material governed by triphasic mechano-electrochemical theory [1]. In order to describe large deformation of cartilage curling, the triphasic theory is reformulated by introducing underformed solid configuration as reference configuration for the whole tissue. A nonlinear finite element formulation based on the reformulated theory was applied to simulate three-dimensional free swelling phenomena of cartilage. The deformation dependent permeability, and the distributions of solidity and fixed charge density reported in [2, 3] were utilized in the present analysis. The numerical results reproduced the curling of cartilage strip *ex situ* when submerged in solution bathes with various concentrations. And the swelling parameters such as swelling-induced stretch and a real change obtained by finite element analysis are in agreement with those measured by experiments [4]. Also the changes of fixed charge density, osmotic pressure distribution and interstitial fluid flow during warping were investigated to explain the occurrence of curling.

REFERENCES

[1] W. M. Lai, J. S. Hou, V. C. Mow, A *triphasic theory for the swelling and deformation behaviors of articular cartilage*, ASME J. Biomech. Eng., 113, (1991), 245-258.
[2] V. C. Mow, M. H. Holmes, W. M. Lai, *Fluid transport and mechanical properties of articular cartilage: a review*, J. Biomech., 17, (1984), 377-394.
[3] A. Maroudas, *Mechanisms of fluid transport in cartilaginous tissue*, in A. R. Hargens ed. *Tissue nutrition and viability*, Springer, NY, USA, (1986).
[4] L. A. Setton, H. Tohyama, V. C. Mow, *Swelling and curling behaviors of articular cartilage*, ASME J. Biomech. Eng., 120, (1998), 355-361.

COMPUTATIONAL MECHANICS (Abstracts)
WCCM VI in conjunction with APCOM'04, *Sept. 5-10, 2004, Beijing, China*

Detection of HIV Cellular Evidence by Pattern Recognition Analysis

A. Conci [1*], J. L. Fernandez [2], E. O. Nunes [2]

[1] *Mechanical Engineering Postgraduate Program, Federal Fluminense University, R. Passo da Patria 156, Niteroi, 24210 -240, Brazil*
[2] *Department of Computer Science, Federal Fluminense University, R. Passo da Patria 156, Niteroi, 24210 -240, Brazil*
e-mail: aconci@ic.uff.br, jlf@vm.uff.br, enunes@ic.uff.br

Abstract

More than 80% of AIDS (seropositive) diagnoses during 2001 were among residents of the poorest counties, although these counties represented only one quarter of the population living with diagnosed HIV due to the highly active anti-retroviral therapy (HAART). That is the declines in deaths among persons with AIDS since 2001 were caused primarily by the slower progression of HIV -associated immune deficiency among persons who used HAART. However, AIDS diagnoses and deaths were relatively constant among the poor. The populations in which the HIV epidemic is becoming concentrated (racial, ethnic) need increased access to prevention programs and health care services so the importance of this low cost HIV and AIDS estimator. During the last 10 years, groups with less access to medical care have been affected more and more by the HIV epidemic. To monitoring the status of the epidemic and evaluating the effectiveness of prevention, cheaper HIV diagnoses will be essential. This work presents a novel numerical method of analysis for HIV diagnoses in possible contaminated blood laboratory tests. It turns possible quick estimation in early stages where the numbers of HIV cells are small. The implemented system of pattern recognition is based not only on conventional feature extraction but use also chaotic measurements (fractal dimension and lacunarity estimators). The presented methodology estimates AIDS using secure and relatively low in price techn iques. It is very adequate for underdeveloped or developing nations of Africa, Asia, and Latin America where greater the proportion of the population below the poverty level.

REFERENCES

[1] R. S. Janssen, G. A. Satten, S. L. Stramer et al, *New testing strategy to detect early HIV-1 infection for use in incidence estimates and for clinical and prevention purposes,* JAMA, 280, (1998), 42-48.

[2] Centers for Disease Control and Prevention, *CDC guidelines for national human immunodeficiency virus case surveillance, including monitoring for human immunodeficiency virus infection and acquired immunodeficiency syndrome,* MMWR Morb Mortal Wkly Rep. 48(RR-13), (1999), 1-31.

[3] J. M. Karon, P. L. Fleming, R. W. Steketee, Kevin M. De Cock, *HIV in the united states at the turn of the century: an epidemic,* Am. J. of Public Health., 91(7), (2001), 1060-1068.

[4] E. O. Nunes, A. Conci, *Segmenting multi bands images by color and texture analysis,* in B. Simak, P. Zahradnik eds. *Recent trends in multimedia information processing*, Proceedings 10th International Workshop on Systems, Signals and Image Processing - IWSSIP'03, Prague, Czech Republic, (2003), pp. 231-234.

[5] E. Azevedo, A. Conci, *Computação gráfica: teoria e prática,* Editora Campus, Elsevier, Rio de Janeiro, Brazil, (2003), (in Portuguese).

COMPUTATIONAL MECHANICS (Abstracts)
WCCM VI in conjunction with APCOM'04, *Sept. 5-10, 2004, Beijing, China*

Malignant Glioma Progression and Response to Therapy In Silico

Vittorio Cristini [1,2], J. Sinek [2], H. Frieboes [1], Steven Wise [1,2*], Xiaoming Zheng [2]

[1] *Department of Biomedical Engineering, University of California, Irvine, REC 204, Irvine, CA 92697-2715, USA*

[2] *Department of Mathematics, University of California, Irvine, REC 204, Irvine, CA 92697 -2715, USA*
e-mail: cristini@math.uci.edu

Abstract

We present computer simulations of a mathematical model of glioma [1] (malignant brain tumor) progression. The model describes the coupling of cell proliferation and tumor mass growth with the development of a network of new capillaries (angiogenesis) that provides glucose and oxygen through diffusion into the cancerous tissue. The simulations are based on a novel adaptive finite-element/level-set method. The microphysical parameters of input to the simulations w ere determined by fitting model predictions to *in vitro* experiments on glioblastoma tumor spheroids and to *in vivo* data of vascular tumor growth in brain cancer patients.

The simulations exploit the dependence of the morphology (phenotype) of these carcin omas on the genotype and in particular on the strength of cell adhesive forces and on mitosis rates that are affected by genetic instability. We demonstrate that the transition to malignancy and the development of invasive morphologies can be promoted by w eak adhesive forces through a diffusional growth instability coupled to local degradation of extra -cellular matrix by the tumor cells. This diffusional instability is analogous to that occurring in complex materials during phase transformations, and leading for example to complex morphologies during solidification and crystal growth. In tumors, this diffusional instability occurs, in the presence of genetically driven weakened adhesive forces, from the competition between cell mitosis and death (also affect ed by mutations), and leads to the growth of low-wave-number perturbations (analogous to dendritic shapes during solidification). These results help explain the experimental observation that cancer cells, unlike normal tissue cells, tend to migrate and invade in large clusters rather than individually through the host tissue. We demonstrate that this diffusional instability can be a powerful mechanism for tumor invasion: a colony of highly proliferative and malignant cells would not be able to grow to a cli nically relevant size otherwise. The model captures the transition to malignancy and the development of an invasive morphology with spread of tumor cells into the external tissue and separation of infiltrating tumor fragments. Interestingly, we demonstrate that it is possible for malignant glioma to grow to centimeter size without significant expression of angiogenic factors and without inducing angiogenesis by using the pre -existing brain vasculature. This is in contrast to Folkman's hypothesis that cancer requires neo-vascularization of the tumor. Therapy strategies *via* simulations are also illustrated.

REFERENCES

[1] E. A. Maher et al, *Malignant glioma: genetics and biology of a grave matter*, Genes Develop., 15, (2001), 1311-1333.

COMPUTATIONAL MECHANICS (Abstracts)
WCCM VI in conjunction with APCOM'04, *Sept. 5-10, 2004, Beijing, China*

A Mechanics Model of Plant and Vegetative Cell for the Compression Loading

Zhiyi Fu [*], Qunying Jiao, Jianli Wang

Department of Application Mechanics, College of Science' China Agricultural University, China Agricultural University, east campus, Mailbox 75, Beijing, 100083, China
e-mail: fzy@cau.edu.cn, jiaoqy@cau.edu.cn

Abstract

The rigidity of plants and vegetables has been important subject for agricultural and food processing engineering. In order to study mechanical behavior of a single cell of plant, an axisymmetric parenchyma fluid-filled spherical cell model was developed in this paper. The model has a spherical soft thin cell wall. The cytoplasm was simulated as static incompressible liquid. The wall of the cell was assumed as an incompressible super-elastic thin membrane with penetrability for inside fluid. The contributive of the wall was chosen as Moony-Rivlin's strain-energy function, which is widely used in showing the stress-strain relation of such materials as natural rubber and plant tissue [1-2]. In this paper, the osmosis relation that the fluid pass through the wall was assumed at a volumetric rate proportional to the cell turgor (pressures difference between inside and outside of the cell wall). The cell was compressed by contacting with two parallel rigid planes. Consulting the mathematic analysis given by Feng and Yang in 1973[1], a group of first-order nonlinear control differential equations in contacted and non-contact areas were formed. In these equations, the geometric relation for large deformation denoting was used to describe the volumetric rate inside the cell during deforming. The solutions were obtained by numerical simulation such as Runge-Kutta etc. Using the model, the main stresses (Cauchy) and main tensions in the wall, the deformations of the wall and the relation between turgor pressure and compressional load were obtained through simulation during loading. From the results above, following significant points were found.

1) Cell rigidity increases along with the climbing of the external compressive force. This fact is in accordance with the result of experimental research and other modeling simulation [3]. The nonlinear character of the rigidity comes from the enhancement of the turgor pressure.

2) Under the constant external compressive force, because of the effect of pervasion together with the elasticity of the wall, the contacted area decreases slightly, so the turgor pressure increases accordingly.

3) When the compressive displacement of the two planes remains constant, due to the effect of pervasion, the turgor pressure gradually declines along with time, inclines towards the initial value.

REFERENCES

[1] W. W. Feng, W. H. Yong, *On the contact problem of an inflated spherical nonlinear membrane*, Journal of Applied Mechanics, 40, (1973), 209-214.

[2] H. X. Zhu, J. R. Melrose, *A mechanics model for the compression of plant and vegetative tissues*. J. theor. Biol., 211, (2003), 89-101.

[3] R. E. Pitt, *Models for rheology and statistical strength of uniformly stressed vegetative tissue*, Trans. ASAE, 25(6), (1982), 1776-1784.

COMPUTATIONAL MECHANICS (Abstracts)
WCCM VI in conjunction with APCOM'04, *Sept. 5-10, 2004, Beijing, China*

The Effect of a Longitudinal Defect in a Long Bone on Its Torsional Stiffness and Stress Distribution

Y. X. Liu[1*], W. Z. Zhao[1], J. Zhang[1], P. Lv[2], L. Yang[2]

[1] *Department of Engineering Mechanics, Dalian University of Technology, Dalian, 116024, China*
[2] *Department of orthopaedics, the Second Affiliated Hospital of Dalian Medical University, Dalian, 116025, China*
e-mail: yxliu@dlut.edu.cn, drzwz@163.com, armyzhang@sohu.com

Abstract

A mechanical model of a femur mid-diaphysis with a rectangular defect in the exterior cortex was developed with computer technique and finite element method to calculate and analyze the influence of a longitudinal defect in a long bone on its torsional stiffness and the stress distribution. The defect widths range from one-tenth of the femur outer diameter (0.1OD) to 0.3 OD, and defect lengths range from 1 to 5OD. The result shows that defects with a length of 1OD had little influence on the femur torsional stiffness or stress distribution. The torsional stiffness decreased most dramatically as the defect length increased from 2 to 4OD. Shear stress peaked at the center of the defect. This conclusion provides an important theoretical gist to the operative methods and the dimension control of the defect size in clinical orthopaedics as follows: 1) If the treatment effect will not be influenced the length of the bone defect should be made shorter than 1OD as can as possible; 2) Orthopaedic surgeons should try their best to make the length of bone defect shorter than 2 -3OD. A long defect had better be made into two shorter defects and there is a 0.5cm -long normal cortex in the middle of them if patient's condition is not allowed. 3) If the width of the defect exceeds 0.3OD, the length of defect should be controlled within 2 OD. 4) In order to get rid of the stress concentration the shape of defect should be made curved instead of rectangular. 5) This conclusion prompt that a longitudinal defect in a long bone influences its torsional stiffness dramatically, so the affected limb should not bear torsional stress after operation for fear bone fracture.

REFERENCES

[1] M. L. Desouza, K. N. An, B. F. Morrey, E. Y. Chao, *Strength reduction of rectangular cortical defects in diaphyseal bone*, in: R. D. Coutts, Ed., *Proceedings of the 35th Annual Meeting of the Orthopaedic Research Society*, Las Vegas, NV, (1989).

[2] B. C. Edgerton, K. N. An, B. F. Morrey, *Torsional strength reduction due to cortical defects in bone*, Journal of Orthopedic Research, 8, (1990), 851- 855.

[3] D. T. Reilly, A. H. Burstein, *The elastic and ultimate properties of compact bone tissue*, Journal of Biomechanics, 8, (1975), 393-405.

[4] J. H. Keyak, S. A. Rossi, S. A. Jones, et al., *Prediction of femoral fracture load using automated finite element modeling*, Journal of Biomechanics, 31, (1998), 125-133.

COMPUTATIONAL MECHANICS (Abstracts)
WCCM VI in conjunction with APCOM'04, *Sept. 5-10, 2004, Beijing, China*

Modeling and Simulation of Biofilm Growth

B. Moran *, S. Hao, D. L. Chopp

Department of Mechanical Engineering, Northwestern University, Evanston, IL 60208, USA
e-mail: b-moran@northwestern.edu, suhao@northwestern.edu, d-chopp@northwestern.edu

Abstract

The motion of biofilm can be modeled as the two-phase flow where the biofilm phase contains two parts, the active species bacteria and the deactive biomass water [2]. The evolution and decay of biofilm is actually the processes that bacteria cell produces and emits synthase signal; at high concentration of cell density and signals, the signals interact with certain transcriptional regulators which modulates expression of quorum sensing-regulated genes and catalyzes the synthesis of the bacteria cell. This dynamics can be expressed as the diffusion-like process governed by three key-variables: living cell volume fraction f, quorum sensing concentration a, and the limiting growth substrates S [1-2]. In the framework of continuum computational mechanics, a biofilm growth can be represented by Navies-Stokes equations. As an extension of 1-dimensional bacteria evolution model introduced in [2], a simulation of a growing "Pseudomonas aeruginosa" biofilm has been performed using a MPFEM-based Level Set with E-FEM enrichment [3-6]). An example of simulation is given in Fig.2 with a qualitative comparison with the experimental observation shown in Fig. 1 [7].

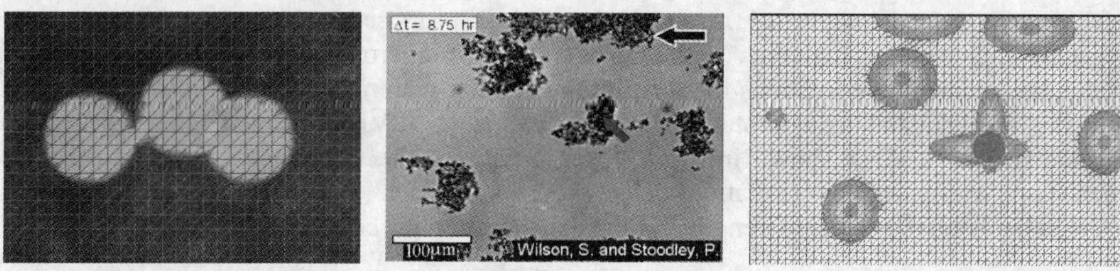

(a) (b) (c)

Fig. 1 (a) merging of three pieces of biofilm, simulated as level-sets conjunction; (b) an experimental observation of the evolution of biofilm [7]; (c) a simulation of biofilm evolution for the case in (b); where the contours represent the volume fraction of the active biomass.

REFERENCES

[1] E. Bruce, P. L. M. Rittmann, *Environmental biotechnology*, Biological Science Ser., McGraw-Hill, (2000).
[2] D. L. Chopp, M. J. Kirisits, B. Moran, M. R. Parsek, Bulletin of Mathematical Biology, (1), (2003), pp. 1-34.
[3] S. Hao, W. K. Liu, T. Belytschko, Int. J. Numer. Meth. Engr., 59, (2004), 1007.
[4] J. A. Sethian, *Level set methods and fast marching methods*, Cambridge, Cambridge University Press, UK, (1999).
[5] T. Belytschko, T. Black, Int. J. Numer. Meth. Engrg, 45(5), (1999), 601 -620.
[6] N. Moes, J. Dolbow, T. Belytschko, Int. J. Numer. Meth. Engrg, 46, (1999), 131 -150.
[7] P. Stoodley, J. D. Boyle, D. DeBeer, H. M. Lappin -Scott, Biofouling, 14(1), (1999), 75-90.

COMPUTATIONAL MECHANICS (Abstracts)
WCCM VI in conjunction with APCOM'04, *Sept. 5-10, 2004, Beijing, China*

Computational Mechanics Model of 2D-Cellular Solid in Compression with Consideration of Microstructural Development

Yoshihiro Suwa [1*], Toshiji Mukai [1], Tatsuhiko Aizawa [2]

[1] *Research Center for Advanced Science and Technology*, The *University of Tokyo, 4-6-1 Komaba, Meguro-Ku, Tokyo, 153-8904, Japan*
[2] *Center for Collaborative Research, The University of Tokyo, 4-6-1 Komaba, Meguro-Ku, Tokyo, 153-8904, Japan*
e-mail: suwa@hpm.rcast.u-tokyo.ac.jp,mukai@hpm.rcast.u-tokyo.ac.jp, aizawa@hpm.rcast.u-tokyo.ac.jp

Abstract

Cellular materials have much potential for absorbing impact energy during the crashing. In the efficient design for impact energy absorption, it is required to exhibit an extended stress plateau in its stress-strain curve. The cell size and/or shape distributions have much effect on that. Because of high-complexity of cellular structures, it is difficult to determine the optimum solution of the size and/or shape distributions by only experiments. Many researchers reported the compressive mechanical response of the ideal cellular materials with periodic boundary conditions. Irregularity in cellular structure is only considered from the computational geometry with use of Voronoi tessellation.

In this paper, new computational mechanics model is proposed to describe the compression response of 2-D cellular materials with consideration of microstructural development. In this modeling, the processing conditions to fabricate the cellular materials are taken into account as the cell-growth mechanism. Representative volume elements (RVE) are generated with use of phase-filed model [1, 2] as shown in Fig.1, Fig.2 compares the calculated stress-strain relations between two configurations (Relative density=0.1). The cell size and/or shape distribution especially affects on the critical strength and the densification process.

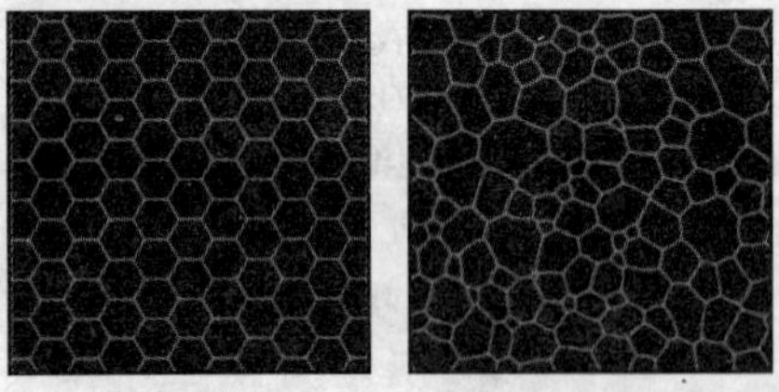

(a) (b)

Fig.1 Examples of 2-D cellular structures

Fig. 2 Comparison of stress-strain relations

REFERENCES

[1] D. Fan, L.-Q. Chen, *Computer simulation of grain growth using a continuum filed model*, Acta. Mater., 45, (1997), 611-622.

[2] Y. Suwa, Y. Saito, *Computer simulation of grain growth by the phase field model, effect of interfacial energy on kinetics of grain growth*, Mat. Trans., 44, (2003), 2245-2251.

COMPUTATIONAL MECHANICS (Abstracts)
WCCM VI in conjunction with APCOM'04, *Sept. 5-10, 2004, Beijing, China*

Inertia and Aerodynamic Forces in Free Flight Insect

Hao Wang [1*], Hao Liu [2]

[1] *PRESTO, Japan Science and Technology Agency (JST), Department of Electronics & Mechanical Engineering, Faculty of Engineering, Chiba University, Chiba 263-8522, Japan*
[2] *Department of Electronics & Mechanical Engineering, Faculty of Engineering, Chiba University, Chiba 263-8522, Japan*
e-mail: hwang@meneth.tm.chiba-u.ac.jp

Abstract

Flapping flight is the significant characteristic of insects' locomotion. The periodic flapping motion induces the periodic inertia forces, which is the marked difference from the traditional airfoil s. Insects have the excellent skill on controlling the flight attitude maneuvers. And the inertia and aerodynamic forces are two critical factors affecting the instantaneous attitude of body and wings. At the same time, the instantaneous attitude of insect has a profound effect on the aerodynamic forces it can generate. And the relationship is non -monotonic, between the unsteady airflow passing along the body and wings and the lift and thrust developed.

In this paper, a numerical rigid multibody modeling of free flight insect is addressed, which is core to understanding the contribution of inertia and aerodynamic forces in insect flight control. The rigid multibody structure of whole insect is studied by using third -order method with the energy correction applied to preserve angular momentum and energy and to improve the accuracy and efficiency of simulation. And the unsteady aerodynamics around flapping wings and vibrating body is studied by using a time accurate solution of the three -dimensional, incompressible, laminar Navier–Stokes equations. The instantaneous attitude of insect is expressed by Euler angles. The changes of Euler angles arise, in part, from the passive mechanics of flapping wings. And meanwhile, their instantaneous values may well emerge from details of the airflow loading. We also discussed what change attitude of insect is controlled by structural mechanics versus aerodynamic loading. And that how changing attitude arising from inertia mechanisms compare to those Euler angles changing that arise from air pressure forces was explored using a combination of computational and analytic methods. Additionally, the metabolic forces, which are the active forces of flight drive and approximate equivalent to the nonideal restraint elemental forces b etween body and wings, were calculated and compared based on the analysis of inertia and aerodynamic forces . The work of metabolic forces, or the energy transformation between metabolizable energy and mechanical energy, was also compared among the different flight patterns, mainly including hovering, forward flight and turning maneuvers.

REFERENCES

[1] J. C. Simo, K. K. Wong, *Unconditionally stable algorithms for rigid body dynamics that exactly preserve energy and momentum*, Int. J. Num. Meth. Eng., 31, (1991), 19-52.
[2] M. H. Dickinson, F. O. Lehmann, S. P. Sane, *Wing rotation and the aerodynamic basis of insect flight*, Science, 184, (1999), 1954-1960.
[3] S. P. Sane, M. H. Dickinson, *The aerodynamic effects of wing rotation and a revised quasi-steady model of flapping flight*, J. Exp. Biol., 205, (2002), 1087-1096.

COMPUTATIONAL MECHANICS (Abstracts)
WCCM VI in conjunction with APCOM'04, *Sept. 5-10, 2004, Beijing, China*

On the Mechanical Model and the Behaviors of the Fruit Tissue under Compression and Shear Forces

Rong Wang [1], Yongyi Gao [2], Qunying Jiao [1*]

[1] *Department of Applied Mechanics, China Agricultural University, Qinghua Donglu #17, Beijing, 100083, China*
[2] *Department of Physics, Xiangtan Teacher's College, Xiangtan, 411201, China*
e-mail: jiaoqy@cau.edu.cn

Abstract

Reduce the damage of fruits is very important for harvesting, food processing and transportation engineering. Mechanical damage has been causing great lose each year. Most of the damage is because of the cell failure under external forces.

The primary goal of this paper is to study how the applied forces transfer and distribute inside a fruit tissue, in the cell level. According to the structural characteristic of the fruit tissue [1], a hexagon shaped mechanical model for a single fruit cell was proposed. Inside the cell, there is liquid. When a force was applied to the cell, the volume of the cell was kept un-change, although, it might have deformation. The cell pressure and stress and strain of the cell walls was studied for compressive and shear forces. It was found that the cell pressure changes with the applied force in the non-linear pattern. The cell pressure increases slowly at the beginning, then, rises very fast, when the compressive load increases. However, for the shear loading, the cell pressure changes with the applied force in a quit different manner. The cell pressure increases very fast with a small increasing of the force, while as the force increases continuous the cell pressure changes slowly.

Moreover, by integrating the hexagonal cells, a piece of fruit tissue was formed. To simulate the rheological property of the material, the cells were connected by spring and viscous damper elements. The cell pressure and stress of cell wall for each cell inside of the tissue under compression and shear was obtained using ANASYS software. The results indicate that two kinds of failure may happen for the fruit tissue. One is due to the high pressure inside the cell; the other one is because of the failure of the connection. When a compressive loading applied, the connection failure occurs at the direction of 45° with respect of the applied force. This is quit similar with the failure of dislocation for the solid material, under compression. The cell pressure distribution indicates that the largest pressure occurs in the area near the applied force. The simulated result agrees well with others work [2-3]. The results are much helpful for understanding the reason of the mechanical damage of fruits.

REFERENCES

[1] Q. Gao, R. E. Pitt, *Mechanics of parenchyma tissue based on cell orientation and micro-structrue*. Transactions of the ASAE, (1991), 232-238.
[2] R. E. Pitt, D. C. Davis, *Finite element analysis of fluid-filled cell response to external loading.* Transactions of the ASAE, (1984), 1976-1983.
[3] Zachary A. Henry, Haibing Zhang, D. O. Onks, *New model for lastic behaviour of cellular material*. Journal of Agricultural Engineering Research, 76, (2000), 399-408.

COMPUTATIONAL MECHANICS (Abstracts)
WCCM VI in conjunction with APCOM'04, *Sept. 5-10, 2004, Beijing, China*

Analysis of Seepage through Fracture Network in Rock Mass of Dam Foundation

Junrui Chai [1, 2*]

[1] *College of Civil and Hydroelectric Engineering, China Three Gorges University, Yichang, 443002, China*
[2] *College of Hydroelectric Engineering, Xi'an University of Technology, Xi'an, 710048, China*
e-mail: jrchai@xaut.edu.cn

Abstract

In this paper, the principle of seepage through an orthogonal network of widely (10m) spaced fractures in a discontinuous medium is used [1-4]. It is considered that the rock material matrix is impermeable so that the water flows only along the fractures [5-7]. Four different cases for a 100m high concrete dam were analyzed, namely: 1) without grout curtain or drain holes, 2) with grout curtain alone, 3) with drain holes alone and 4) with grout curtain and drain holes.
It was calculated that seepage discharge is reduced by 88% if only a grout curtain is provided, and by 86% if a grout curtain and drainage holes are installed. Hydrostatic pressure is reduced by 20% with only a grout curtain, by 48% with only drainage holes and by 52% with grout curtain and drainage holes. The distance between the action point of the resultant hydrostatic force and the dam heel is reduced by 8% with only a grout curtain, by 52% with only drainage holes and by 57% with grout curtain and drainage holes.
It has been concluded that 1) the correct method of seepage analysis according to the distribution of fractures in the rock mass of a dam foundation should be selected, and 2) that both a grout curtain and drainage holes are necessary to effectively reduce and control the seepage discharge and to reduce the hydrostatic pressure on the dam foundation.

REFERENCES

[1] R. W. Lewis, B. A. Schrefler, *The finite element method in the static and dynamic deformation and consolidation of porous media*, John Wiley & Sons Ltd., Chichester, England, (1998).
[2] J. R. Chai, *Analysis of coupled seepage and temperature fields in concrete dam*, Communications in Numerical Methods in Engineering, 18, (2002), 399-409.
[3] M. Oda, *An equivalent continuum model for coupled stress and fluid flow analysis in jointed rock masses*, Water Resources Research, 22, (1986), 1854-1865.
[4] D. F. Vittorio, *On non-Newtonian fluid flow in rough fractures*, Water Resources Research, 37, (2001), 2425-2430.
[5] M. Bai, *Numerical modeling of coupled flow and deformation in fractured rock specimens*, International Journal for Numerical and Analytical Methods in Geomechanics, 23, (1999), 141-160.
[6] J. R. Chai, S.Y. Li, Y. Q. Wu, *Multi-level fracture network model for coupled seepage and stress fields in rock mass*, Communications in Numerical Methods in Engineering, 20, (2004), 63-74.
[7] R. R. Huilgol, N. Phan-Thien, *Fluid Mechanics of Viscoelasticity*, Elsevier, New York, USA, (1997).

COMPUTATIONAL MECHANICS (Abstracts)
WCCM VI in conjunction with APCOM'04, *Sept. 5-10, 2004, Beijing, China*

Finite Element Analysis of the I-Lan Reinforced Soil Retaining Walls

S. J. Chao*

Department of Civil Engineering, National I-Lan University, No. 1, Shern-Nong Road, I-Lan, 26041, Taiwan, China
e-mail: chao@niu.edu.tw

Abstract

Reinforced Soil Retaining Structures are developed rapidly and constructed widely in civil engineering practices in Taiwan. However, this kind of structures is composed of backfill materials and reinforcements, which is relatively complicated in considering of the soil-structure interaction, and thus desires more study to understand the mechanical behavior. The backfill material is preferred to use sand or gravel for constructing reinforced soil retaining structures. Therefore, available design codes only allow this type of backfill materials. However, the soil deposits in the construction site can be any kind of materials. Following the principle of balancing the amount of cutting and filling to avoid construction pollution, accepting clayey soils as the backfill materials for the purposes of economical and ecological considerations is unavoidable today in Taiwan. The range of the allowable backfill materials are extended broadly from GW to CL by the existing specifications. In addition, to use clayey soil as the backfill material is also an important field of studying the mechanical behavior of reinforced soil retaining structures all over the world recently. In order to investigate the different mechanical behaviors of the reinforced soil retaining structures with granular and cohesive backfills, two 1.5m in height test reinforced soil retaining walls were constructed in laboratory site of the National I-Lan University, I-Lan, Taiwan, in the summer of 2002 (I-Lan Sand Wall and I-Lan Clay Wall). Strain gauges were attached along the length of the reinforcements for each layer to monitor the strain and stress distribution in the reinforcements. The measured results are presented in details firstly in this paper to illustrate the behaviors of the model structures. This paper then describes the increased measured data for adding surcharge on the walls to understand the development of stresses and strains along the reinforcements under loading. After that, the finite element method is employed to simulate the responses for the reinforcements and the backfill materials. The commercial finite element analysis program PLAXIS is used in this study. PLAXIS is specifically intended for the analysis of deformation and stability in geotechnical engineering projects. The two I-Lan Walls are simulated using PLAXIS to predict their stress distributions in the reinforcements, the deformation of soil mass, and the responses of the whole structures under different kinds of loading. The predicted results from FEM can be compared with the measured data of the two model structures to verify the proposed FEM model by this study. Since the interface behavior between backfills and reinforcements play an important role for numerical analysis of the reinforced soil retaining structures, the behavior of the soil-structure interaction is discussed and an appropriate interface element model for FEM is recommended to conclude this paper.

COMPUTATIONAL MECHANICS (Abstracts)
WCCM VI in conjunction with APCOM'04, Sept. 5-10, 2004, Beijing, China

Fluttering Instability in Geomaterials: Really or Fantasy?

K. T. Chau*, J. D. Zhao

Department of Civil and Structural Engineering, The Hong Kong Polytechnic University, Hung Hom, Kowloon, Hong Kong, China
e-mail: cektchau@polyu.edu.hk., Jidong.zhao@newcastle.edu.au

Abstract

The possibility of having a complex conjugate pair for acceleration speed of a discontinuous front of non-uniform deformation in solids was apparently first remarked by Rice [1] and it was coined as "fluttering instability" of solids as the mathematical form of the complex conjugate pair of the eigenvalues closely resemble those for the case of Hopf bifurcation of wind-induced vibration of slender-structures. Although this topic has subsequently been explored mathematically by some researchers, except for the liquefaction interpretation given by Bigoni and Willis [2] there has been no physical meaning attached to this mathematical possibility. Whether fluttering instability of solids actually occurs in nature? If yes, in what form? There are fundamental issues remain to be addressed. This paper presents the mathematical prediction of the possibility of fluttering instability for two pressure-sensitive dilatant solids: the Rudnicki-Rice [3] model and Chau-Rudnicki [4] model. Fluttering instability is searched within the physical range of the parameters, and potential implication on periodic pattern of non-uniform deformation in solids is proposed. Motivated by recent studies for compaction band, it was argued by Issen and Rudnicki [5] that both or either dilatant factor and pressure-sensitive coefficient can be negative if a two-yield surface model is used to interpret the failure yield surface of rocks. Therefore, we also allow for such possibility of parameter selection in the present study. Finally, it is speculated that fluttering instability can actually manifest itself in nature and can account for many regular patterns of deformation observed in rocks, soils and other materials.

REFERENCES

[1] J. Rice, *The localization of plastic deformation*, in: W. Koiter ed. *Theoretical and applied mechanics*, 14th IUTAM Congress, North-Holland, Amsterdam, (1976), pp. 207-220.
[2] D. Bigoni, J. R. Willis, *A dynamical interpretation of flutter instability*, in: R. Chambon, J. Desrues, I. Vardoulakis eds. *Localisation and bifurcation of rocks and soils*, A. A. Balkema Scientific Publishers, Rotterdam, (1994), pp.51-58.
[3] J.W. Rudnicki, J. R. Rice, *Conditions for the localization of deformation in pressure-sensitive dilatant materials*, J. Mech. Phys. Solids, 23, (1975), 371-394.
[4] K. T. Chau, J. W. Rudnicki, *Bifurcations of compressible pressure-sensitive materials in plane strain tension and compression*, J. Mech. Phys. Solids, 38, (1990), 875-898.
[5] K. A. Issen, J. W. Rudnicki, *Conditions for compaction bands in porous rock*, J. Geophys. Res., 105(B9), (2000), 21529-21536.

COMPUTATIONAL MECHANICS (Abstracts)
WCCM VI in conjunction with APCOM'04, *Sept. 5-10, 2004, Beijing, China*

A Decoupling FEM for Simulating Near-Field Wave Motions in Two-Phase Media

S. L. Chen [1*], Z. P. Liao [2], J. Chen [1]

[1] *The State Key Laboratory of Vibration, Shock & Noise, Shanghai Jiao Tong University, Shanghai, 200030, China*

[2] *Institute of Engineering Mechanics, China Seismological Bureau, Harbin, 150080, China*

e-mail: iemcsl@yahoo.com.cn

Abstract

A decoupling technique for simulating near -field wave motions in two -phase media is introduced in this paper, which includes lumped-mass FEM, explicit integration procedures and local artificial boundary condition. The decoupling feature of this method reduce s the implicit solution of a set of coupling algebraic equations to a successive explicit solution of the individual equations, and substantially reduces the computational cost, which renders the technique particularly suitable to cope with problems of nea r-field wave motion. First, An equivalent but direct weighted residual method is presented to solve boundary value problems in this paper, that is to apply the Green's theorem for integration by parts on the equivalent integral statement of the field contr ol equations, then introduce the Neumann conditions into directly. Using this method and considering the precision requirement in wave motion simulation, a lumped -mass FEM for two-phase media with clear physical concepts and convenience in employment is de rived. Then, considering the innate attenuation character of the wave in two -phase media, an attenuation parameter is introduced into MTF to simulating the attenuate outgoing wave in two -phase media. At last, two numerical experiments are presented, and the numerical results are compared with the analytical ones , which demonstrate that the lumped-mass FEM and the modified MTF introduced in this paper have good precision.

REFERENCES

[1] Z. P. Liao, H. L. Wong, *A transmitting boundary for the numerical simulation of elastic wave propagation*, Soil Dyn. Earthq. Eng., 3, (1984), 174-183.

[2] A. Gajo, A. Saetta, R. Vitaliani, *Silent boundary conditions for wave propagation in saturated porous media*, Int. J. Numer. Anal. Methods. Geomech., 20, (1996), 253-273.

[3] O. C. Zienkiewicz, T. Shiomi, *Dynamic behaviour of saturated porous media; the generalized Biot formulation and its numerical solution*, Int. J. Numer. Meth. Geomech., 8, (1984), 71-96.

[4] J. Chen, *Time domain fundamental solution to Biot's complete equations of dynamic poroelasticity, Part II: three-dimensional solution,* Int. J. Solids Structure, 31, (1994), 169-202.

COMPUTATIONAL MECHANICS (Abstracts)
WCCM VI in conjunction with APCOM'04, *Sept. 5-10, 2004, Beijing, China*

The Theory of Plasticity with Porosity and Porous Moisture Phase Balance for a Ground

Mikhail M. Dubina *, Yury A. Tchernyakov, Dmitry K. Teslenko

Tyumen`s State Architectural-Constructional Academy, Lounatcharsky Street, 2, Tyumen, 625001, Russia
e-mail: mmdubina@yandex.ru

Abstract

The mathematical model of thermoplastic media with phase equilibrium salted pore moisture is presented. The mathematical model include the following: the equations of distribution of heat in view of a spectrum of freezing temperature and dependence of a freezing beginning temperature on impurity concentration; the equations of a moisture phase equilibrium in pores, containing the dissolved impurity and gas; the stress-strain state equations of ponderable medium, containing a moisture in pores, with dependence of solid and liquid phases equilibrium according to thermodynamic equilibrium in pores. The temperature mode of a geo-massif calculations require to take into account, that the thermophysical parameters included in it depend not only on boundary conditions, but also on the process of redistribution of heat itself and on the stress-strain condition in the massif according to thermodynamic equilibrium of water in pores for which description we deduced the equations system. Such mathematical model attached to interconnected class of thermomechanical problems. The local thermal-mechanical balance equations in pores are connected to the general macro-equations of the porous media deformation mechanics. The pores filler and skeleton interaction is described within the framework of the mezomechanical theory. Examples of the non- frozen water content calculation in a rock massif around a circular excavation are presented.

REFERENCES

[1] M. M. Dubina, B. A. Krasovitsky, *Heat exchange and mechanics of interaction of pipelines and boreholes with grounds*, Novosibirsk: Nauka, (1987).
[2] V. N. Nikolaevsky et al., *Mechanics of the saturated porous medias*, Moscow: Nedra, (1979).
[3] A. I. Botkin, *About strength of non-cohesive and brittle materials*, News NIIG., 26, (1940), 64-69.
[4] M. V. Malishev, *Grounds strength and stability of the bases of structures*, Moscow: Stroyizdat, (1980).

COMPUTATIONAL MECHANICS (Abstracts)
WCCM VI in conjunction with APCOM'04, *Sept. 5-10, 2004, Beijing, China*

Flow of Granular Material in Tall Silos

Manfred Göttlicher*

Department of Civil Engineering, University of Applied Sciences, Altonaer Str. 25, Erfurt, D-99085, Germany
e-mail: m.goettlicher@t-online.de

Abstract

Solid behavior as well as liquid behavior characterizes the flow of granular material in silos. In the state of filling the theory of solid continua describes the resulting stress and displacement fields sufficiently. In the state flow the difference to solid behavior is negligible if the time of consideration is short and if the boundaries are far away from the point of consideration. A complete analysis requires reference to the aspects of liquid behavior also. The solid and liquid components together ensure the equilibrium of the overall system. The presented model is based on an appropriate interaction of a displacement field and a velocity field thus taking into account the two substantial aspects of flowing material.

The constitutive equations and the applied algorithm are developed from the exact solution for a standard system according to a bottom up approach. The standard system evolves from a very tall vertical plane strain silo containing material that flows at a constant speed. No horizontal displacements and velocities take place. No changes regarding the field values arise in the vertical direction and in time. Tension is not allowed at any point.

Coulomb friction represents the effects of the vertical walls. The interaction between the flowing material and the walls is covered by a forced boundary condition resulting in an additional matrix for the solid component as well as for the liquid component. The weight of the material is transferred completely to the associated part of the wall. The resulting integral equations are designed to be solved directly.

Three coefficients describe the properties of the granular material. They govern elastic solid behavior in combination with viscous liquid behavior. The presented paper focuses primarily on the solid liquid field interaction. A more accurate description of the solid and liquid properties of granular materials will improve the performance of the model and offer the opportunity for practical applications.

Representative single element systems covering all described aspects show the feasibility of developed algorithm. The chosen interpolation functions and the associated degrees of freedom enable the elements to accommodate to the important effects of steady state flow.

REFERENCES

[1] K. J. Bathe, *Finite-Elemente-Methoden*, Springer-Verlag, Berlin, (2002), (in German).

[2] O. C. Zienkiewicz, R. L. Taylor, *The finite element method, vol. 1: the basis, vol. 3: fluid dynamics*, Butterworth-Heinemann, Oxford, England, (2000).

COMPUTATIONAL MECHANICS (Abstracts)
WCCM VI in conjunction with APCOM'04, *Sept. 5-10, 2004, Beijing, China*

Dynamic Numerical Simulation of Deformation and Failure Process of Rock and Soil Mass Materials in the Low Ground Stress Regions

Xiasong Hu [1,2*], Yinghua Liu [1], Zhangzhi Cen [1]

[1] *Department of Engineering Mechanics, Tsinghua University, Beijing, 100084, China*
[2] *Department of Hydraulic and Hydropower Engineering, Qinghai University, Xining, 810016, China*

e-mail: hxs@mail.tsinghua.edu.cn, yhliu@tsinghua.edu.cn, demczz@tsinghua.edu.cn

Abstract

Low ground stress state plays a very important role in underground and slope engineering stabilities. In this paper, 2D discrete element method (DEM) is used to simulate the whole dynamic deformation and failure process of rock and soil mass materials in the low ground stress regions. One water conservancy underground excavation engineering project in Northwest China is taken as an example. The low ground stress states whose stress values are less than the value of gravity stress in both vertical and horizontal directions are mainly discussed here.

The established computational model consists of 5 kinds of stratum unite body. According to the stress distribution characteristic in the low ground stress regions, the boundary conditions of the computational model are determined. Basic discrete block is quadrilateral element, whose total number is 2190. The DEM software 2D-Block is used to conduct the numerical simulation. The adopted physical and mechanical material parameters are obtained by the test of experiments.

Following the iterative timing step, the dynamic process of rock and soil mass materials from initial deformation, failure to final stability has been simulated. The simulated results show that the deformation and failure would be formed at the roof and both sides of underground excavation, but the deformation and failure of rock and soil mass materials at the roof of underground excavation are especially more serious than those at both sides. These results agree well with the real cases. The achieved results will be of great significance for the further study of deformation and failure mechanism of rock and soil mass materials, meanwhile, they will also play a very important role in ensuring the progress in one underground engineering project.

REFERENCES

[1] X. P. Shen, Z. Z. Cen, B. Y. Xu, T. Q. Liu, *FEM computation on the overall design of unified reinforcement for opening-pit and ground-surface deformation,* Chinese Journal of Rock Mechanics and Engineering, 14, (1995), 201-210, (in Chinese).

[2] B. Y. Xu, X. P. Shen, Z. Z. Cen, *Rock-slope engineering and solid mechanics,* Advances in Mechanics, 24, (1994), 518-524, (in Chinese).

[3] Q. Yang, Y. G. Cheng, H. Zhang, *Simulation of cracking processes of rock materials by lattice model,* Engineering Mechanics, 20, (2003), 117-121, (in Chinese).

COMPUTATIONAL MECHANICS (Abstracts)
WCCM VI in conjunction with APCOM'04, *Sept. 5-10, 2004, Beijing, China*

A Bounding Surface Model for Granular Soils Subject to Cyclic Loading

N. Khalili *, M. Habte, S. Valliappan

School of Civil and Environmental Engineering, The University of New South Wales, Sydney, NSW, 2052, Australia
e-mail: n.khalili@unsw.edu.au

Abstract

The behaviour of granular soils is complex. It involves pressure and specific volume dependency of the stress-strain relationship and a strong non-linear response of the soil matrix, particularly under cyclic loading. Indeed, many of the characteristic features of granular soils cannot be captured using the classical plasticity models proposed for cohesive soils such as the Cam-Clay model. Another complicating factor, which is rarely taken into account in constitutive modelling of granular soils is the behaviour of granular materials, is the phenomenon of particle crushing, as observed in numerous experimental investigations. Particle crushing causes volumetric contraction in drained tests and pore pressure build in undrained tests. Such phenomena are of particular significance in cone penetration tests, pile foundations and dam engineering.
While numerous models have been presented for monotonic loading of granular soils, very few have successfully been applied to cyclic loading. Yu identified similarities between the state parameter concept for sands and the Cam-Clay model to develop a unified model for all soil types [1]. Crouch et al presented a model suited to a wider range of stresses and introduced highly complicated split function for the loading and bounding surfaces and radial mapping rule [2]. Pestana and Whittle introduced both shear and plastic volumetric strain as hardening parameters [3]. Cecconi et al used material parameters that were crushing dependent [4].
In this paper, an elasto-plastic model is presented for the cyclic analysis of sand. The model is formulated using the bounding surface plasticity, in a critical state framework. A new bounding surface along with a new mapping rule is introduced to obtain a more realistic response under cyclic loading. Other aspects taken into account are the non-associativity of the flow rule, and the crushing of the soil particles at high stresses. The model is validated against several cyclic experimental data from the literature.

REFERENCES

[1] H. S. Yu, *CASM: a unified state parameter model for clay and sand*, International Journal for Numerical and Analytical Methods in Geomechanics, 22, (1998), 621-653.

[2] R. S. Crouch, J. P. Wolf, Y. F. Dafalias, *Unified critical-state bounding-surface plasticity model for soil*, Journal of Engineering Mechanics ASCE, 120(11), (1994), 2251-2270.

[3] J. M. Pestana, A. J. Whittle, *Formulation of a unified constitutive model for clays and sands*, International Journal for Numerical and Analytical Methods in Geomechanics, 23, (1999), 1215-1234.

[4] M. Cecconi, A. DeSimone, C. Tamagnini, G. M. B. Viggiani, *A constitutive model for granular materials with grain crushing and its application to a pyroclastic soil*, International Journal for Numerical and Analytical Methods in Geomechanics, 26, (2002), 1531-1560.

COMPUTATIONAL MECHANICS (Abstracts)
WCCM VI in conjunction with APCOM'04, *Sept. 5-10, 2004, Beijing, China*

Analytical and Numerical Calculation of the Influence Zone inside the Subsoil

P. Kuklík [1*], M. Kopáčková [2], M. Šejnoha [1]

[1] *Department of structural mechanics, CTU in Prague, Faculty of Civil Engineering, Thákurova 7, 166 29, Czech Republic*
[2] *Department of mathematics, CTU in Prague, Faculty of Civil Engineering, Thákurova 7, 166 29, Czech Republic*
e-mail: kuklikpa@fsv.cvut.cz , marie.kopackova@fsv.cvut.cz , sejnom@fsv.cvut.cz

Abstract

For the calculation of foundation settlement it is recommended to take into account so called influence zone inside the subsoil bellow the carrying structure. Influence zone inside the subsoil is the region where the load has a substantial influence on the deformation of the soil skeleton. The soil skeleton is consolidated or over consolidated due to the original geostatic stress state. An excavation changes the original geostatic stress state and it gives the space for loads passed from upper structure. The consistent analytical solution is introduced in the analytical part of contribution [1], [2]. The theory of layered subsoil in Westergard manner is selected for the stress state calculation. The depth of influence zone is calculated from the condition, that the original vertical stress is equal the new vertical stress (excavation unloading plus the upper structure loads).
The analytical solution is compared with the numerical results using the critical state models [3]. The critical state models are forms of an isotropic strain-hardening (softening) law. It introduces a distinction between yielding and ultimate collapse by using the concept of a critical state line in conjunction with a strain-dependent yield surface. The yield surface is therefore strain dependent and expands or contracts as the soil hardens or softens. The initial size of the ellipse is governed by the maximum preconsolidation pressure to which the soil has previously been subjected during its past history [4].

REFERENCES

[1] Z. Bittnar, J. Šejnoha: *Numerical methods in structural mechanics*, ASCE Press, Thomas Telford, New York, USA, (1996).
[2] R. O. Davis, A. P. S. Selvadurai, *Elasticity and geomechanics*, Cambridge University Press, (1996), pp. 115-123.
[3] R. W. Lewis, B. A. Schrefler, *The finite element method in the static and dynamic deformation and consolidation of porous media*, John Wiley & Sons Ltd, Chichester, England, (1998), pp. 75-143.
[4] P. Kuklík, M. Šejnoha, J. Mareš, The structural strength of soil from the isotropic consolidation point of view, APCOM 99, Singapore, (1999), pp. 797-802.

COMPUTATIONAL MECHANICS (Abstracts)
WCCM VI in conjunction with APCOM'04, *Sept. 5-10, 2004, Beijing, China*

Numerical Study on the Stirring-up of Excess Pore Pressure in the Soil

Hsien Hua Lee*, Hong-Hua Sun

Dept. of Marine Environment & Engineering, National Sun Yat-sen University, Kaohsiung, Taiwan, China
e-mail: hhlee@mail.nsysu.edu.tw

Abstract

In this study a numerical model based on finite difference method was constructed to simulate the behavior of both the soil and the excess pore pressure during intensive dynamic loadings. The sudden increment of the excess pore pressure will reduce the effective bearing strength of the soil and then cause collapse of the foundation and structures. This has been a well-known behavior of liquefaction of the soil. Most studies on the liquefaction were focused on the initiation or trigger of the liquefaction at which the soil loses its bearing strength totally, while the study on the process of the liquefaction was limited. It is the purpose of this study to establish a model, which may accurately simulate the behavior of both the soil and the variation of the excess pore pressure. Not only the variation of the pore pressure and shear stress of the soil can be observed but also the interaction between them can also be realized during the loading process.

During the study firstly, a wave-propagation model is established and written into a numerical form, which is associated with radiation boundary conditions and then, the model of stirring-up of excess pore pressure is combined into the whole system. Usually, there are three stages for the stirring-up model of excess pore pressure namely, first stage of effective stress route, second stage of stable increase of the effective stress and the third stage of the reverse route of the effective stress. The propagation of the elastic wave will be affected by the media, the soil, where the wave has been traveling and therefore, the properties of the soil are important. However, the properties of the soil can also be influenced during the stirring-up of the elastic waves. This causes an interaction problem between the traveling waves and the soil that contains pore water.

In the numerical results it is found that the form of the exciting forces will have effect on the stirring-up of the excess pore pressure. When the soil is subjected to harmonic exciting forces with the same amplitude, the loading frequency can also affect the stirring-up process. It is also found that the loading direction may also affect the stirring-up of the excess pore pressure.

REFERENCES

[1] Kenneth L. Lee, Aurelio Albaisa, *Earthquake Induced Settlements in Saturated Sands*, Journal of the Geotechnical Engineering Division, ASCE 100, GT4, (1974), pp.387-406.

[2] Albert C. Reynolds, *Boundary Conditions for the numerical Solution of Wave Propagation Problems*, Geophysics, 43(6), (1978), 1099-1110.

[3] H. Bolton Seed, Philippe P. Martin, John Lysmer, *Pore-water pressure changes during soil liquefaction*, Journal of the Geotechnical Engineering Division, ASCE, 102(GT4), (1976), 323-346.

COMPUTATIONAL MECHANICS (Abstracts)
WCCM VI in conjunction with APCOM'04, *Sept. 5-10, 2004, Beijing, China*

A Study of the Impact of Desiccation Cracking of Soils on Performance of a Stiffened Slab Footing on Expansive Soils

Jie Li [1*], Stephen G. Fityus [1], David W. Smith [2], Wenxiong Huang [1]

[1] *Discipline of Civil, Surveying and Environmental Engineering, The University of Newcastle, Callaghan, NSW 2308, Australia*
[2] *Dept. of Civil and Environmental Engineering, University of Melbourne, Victoria 3010, Australia*
e-mail: jie.li@newcastle.edu.au

Abstract

Cracking is a fundamental and widespread phenomenon associated with the drying of most clay soils, especially expansive clay soils. Expansive or reactive clay soi l is defined as a soil that undergoes appreciable volume change in response to change in its moisture content. This volume change occurs as shrinkage upon drying, and swelling upon wetting. Volume changes in expansive clays manifest themselves as vertical and horizontal strains in the soils. The horizontal strains lead to shrinkage cracking, while vertical strains lead to distortions at ground surfaces, either upward or downward. Buildings constructed on expansive soils may be subjected to severe movement a rising from a change in subsoil moisture. The foundation distortion may be consequently manifested as cracking of walls and linings and possibly malfunction of doors and windows .

Cracking in expansive soils is a significant factor in ground movement predic tion, and in engineering design of foundation in expansive soils. First, cracks control the extent of lateral restraint within the soil mass, and this directly affects the magnitude of the vertical soil movement. Second, cracks can serve as pathways for th e transport of water, which can significantly increase the permeability of the soil. Third, cracking in expansive soils could lead to larger deflections of a slab and higher tensile stresses in the concrete footings. As an example, for a stiffened slab on a wetted foundation, the soil mass at the perimeter of the building will gradually lose water to the atmosphere by evaporation, while below the slab evaporation is prevented. Unequal horizontal soil pressures will develop on the either side of the edge bea m due to the fact that the soil dried out and shrunk away from the outside of the beam, whilst it stayed in close contact on the inside. The effect of these pressures on the overall behavior of a stiffened slab is ignored in all existing methods for analys is of expansive soil-footing interaction.

Despite recognition of its potential importance, very little research has been conducted to examine influence of lateral soil strains and desiccation cracking on the engineering design of foundation in expansive soils. In this research, a finite element model is developed for analysing a moisture flow-expansive soil-footing interaction problem. The environmental conditions assumed are typical of those prevailing in the semi -arid areas of southern Australia. Movemen t of the foundation and the footing are predicted in response to climatically driven suction changes applied at the boundary of a soil profile with a nominal initial soil suction distribution. The numerical model is used to investigate the effects of desic cation cracking of soil around the edge of a building on the behavi our of a stiffened slab footing. It is assumed that the footing was constructed at the end of the wet season and was then subjected to a six month drying period. The result has shown that c onsidering lateral shrinkage and desiccation cracking of the soil not only increases the differential deflection of the slab, but also alters the patterns of the slab movements.

COMPUTATIONAL MECHANICS (Abstracts)
WCCM VI in conjunction with APCOM'04, *Sept. 5-10, 2004, Beijing, China*

Finite Layer Analysis for Three-dimensional Biot Consolidation

G. X. Mei [1*], J. M. Zai [1], J. H. Yin [2]

[1] *College of Civil Engineering, Nanjing University of Technology, Nanjing, 210009, China*
[2] *Department of Civil and Structural Engineering, The Hong Kong Polytechnic University, Hong Kong, China*

e-mail: cemgx2003@yahoo.com.cn, Zai@jlonline.com

Abstract

Three-dimensional Biot consolidation is the most key problem in geotechnical engineering. That is due to: (1). Biot consolidation problem reflects the composition of soil; (2). Biot consolidation problem solves the three displacement variables and an excess pore water pressure variable; (c). Biot consolidation problem reflects that four variables above vary with time. One of the most popular numerical methods for the solution to the consolidation problem is the finite element method. However, the amount of computer storage and data preparation time required is usually large, especially for three-dimensional problems. Most analyses are restricted to the axisymmetric or p lane strain cases, which involve only two spatial dimensions. This disadvantage can be overcome by using the finite layer method for certain classes of problems.

The finite layer method makes use of the character of ground and its calculation process is li sted: (1). Split the ground into many layer elements along the depth direction of soil stratum due to the different properties and use polynomial approximation method to simulate, which is turned into realities by high efficient computer; (2). Choose an appropriate series function in response to the boundary condition to simulate the horizontal plane characters of soil stratum (layer element), which has the same soil parameter in all points in a layer element. Therefore, a three-dimensional problem can be became a one-dimensional problem; and the calculation work including pre/post processor, computing time and memory will be reduced a lot.

This paper presents a finite layer procedure for Biot consolidation analysis of layered soils using a cross-anisotropic elastic constitutive model. The program is first verified using published results. Then, using this program, the influences of cross-anisotropy on the consolidation behavior are investigated by changing one model parameter at one time. The results obtaine d using the cross-anisotropic elastic model are compared with results using an isotropic elastic model. It is found that the cross-anisotropy has very large influences on the consolidation behavior. Curves or tables of the average degree of consolidation are obtained and presented in the paper. These curves or tables can be easily used to estimate the consolidation behavior of a cross -anisotropic soil.

REFERENCES

[1] W. Q. Zhang, X. H. Zhao, J. M. Zai, *Finite layer analysis for layered and elastic half-space under arbitrary systems*, Proc. of Int. Conf. on FEM, Shanghai, (1982), vol.1, pp.460-464.

[2] Y. K. Cheung, L. G. Tham, *Numerical solutions for Biot's consolidation of layered soil*, Journal of Engineering Mechanics, 109(7), (1983), 669-679.

[3] G. X. Mei, *Consolidation finite layer theory and practice*, PhD thesis, Hohai University, (2002), (in Chinese).

COMPUTATIONAL MECHANICS (Abstracts)
WCCM VI in conjunction with APCOM'04, *Sept. 5-10, 2004, Beijing, China*

Evaluation of Multi-Physics Couplings Effects, in the Problem of Long Term Nuclear Wastes Storage

Alain Millard [1*], Amel Rejeb [2]

[1] *CEA/DM2S/LM2S, CEN Saclay 91191 Gif/Yvette France*
[2] *IRSN/DPRE/SERGD 60-68, av du Général Leclerc BP6 92265 Fontenay aux roses Cedex France*
e-mail: alain.millard@cea.fr, amel.rejeb@irsn.fr

Abstract

Among the various solutions considered for the treatment of high activity nuclear wastes, the storage in deep geological formations is the one favored by many countries. Different storage concepts have emerged, such as the one consisting in horizontal parallel galleries, where vertical boreholes are drilled for wastes emplacement. The wastes canisters are first placed in an overpack, and then installed in the boreholes which are then backfilled with an engineered barrier, generally consisting of bentonite. Finally, the galleries are also backfilled. After the excavation of the galleries and boreholes, the rock mass is in a certain state of stresses, temperature and fluid pressure. Then, after wastes emplacement, due to the heat released by the wastes, the thermo-hydro-mechanical state of the rock mass will evolve with time.

In order to assess the performance of such a repository, it is important to be able to analyse this coupled evolution. However, in view of the rock mass properties, some couplings may be disregarded, which simplifies the calculations. This has been investigated numerically on a specific repository configuration, in selected granite rock-mass [1]. From symmetry and periodicity considerations, a representative pattern has been identified. The different sequences are reproduced: excavation followed by a transient period until a new THM equilibrium is reached, then instantaneous emplacement of the canister and backfilling materials, and finally heating over a 100 years period. Material properties have been separately calibrated from available laboratory tests results. After a fully coupled THM analysis, partially uncoupled analysis has been performed: TH, TM, and HM. All results have been compared in terms of quantities which are relevant for performance assessment of the repository, such as maximum temperature, fluid pore pressure, time for resaturation of the engineered barrier, stress level with regards to a possible rock failure. In addition, the effect of the rock mass permeability on these couplings effects has been investigated. Concerning the results, the effect of couplings on the temperature is limited, mainly due to variations of the thermal conductivity of the porous medium. Conversely, the fluid pore pressure strongly depends on the couplings, because of the possible desaturation, and of the variations of the porosity. For stresses, the influence of the pore pressure is predominant on the thermally induced stresses. Of course, these results are greatly influenced by the rock-mass permeability, which controls the rate of resaturation of the engineered barrier. It turns out that for very low rock mass permeabilities, it is necessary to take into account all the couplings in order to have a realistic description of the phenomena.

REFERENCES

[1] T. S. Nguyen et al, *Implication of coupled thermo-hydro-mechanical processes on the safety of a hypothetical nuclear fuel waste repository,* Proc. GEOPROC Conf., Stockhlom, (2003).

COMPUTATIONAL MECHANICS (Abstracts)
WCCM VI in conjunction with APCOM'04, *Sept. 5-10, 2004, Beijing, China*

3-D Adaptive Wind Field Simulation Including Effects of Chimney Emissions

R. Montenegro *, G. Montero, J. M. Escobar, E. Rodríguez, J. M. González-Yuste

Institute of Intelligent Systems and Numerical Applications in Engineering, University of Las Palmas de Gran Canaria, Edificio Central del Parque Científico-Tecnológico, Campus Universitario de Tafira, Las Palmas de Gran Canaria, Canary Islands, 35017, Spain
e-mail: {rafa, gustavo, barrera}@dma.ulpgc.es, jescobar@dsc.ulpgc.es, jmgonzalezy@pas.ulpgc.es

Abstract

Air pollution models usually start from the computation of the velocity field of the fluid. In this paper, we present a model for computing such a field based on the contribution of the observed wind flow and the vertical buoyancy or momentum plume rise defined by a Gaussian plume model [1]. This initial velocity field is adjusted to verify incompressibility and impermeability conditions by using a mass consistent model [2]. In order to include effects of chimney emissions, we add to the interpolated wind field a vertical velocity along the trajectory of pollutant plume arising from chimney. In this environmental modelling that occur in a three-dimensional domain defined over complex terrain, a mesh generator capable of adapting itself to topographical data, to chimney geometry and to the numerical solution is essential [3]. Here, the unstructured tetrahedral meshes are generated by combining the use of a refinement/derefinement algorithm for two-dimensional domains and a tetrahedral mesh generator based on Delaunay triangulation. Occasionally in this process, low quality or even inverted elements may appear. For this reason, we have developed a simultaneous untangling and smoothing procedure to optimize the resulting meshes [4]. Once we have constructed the adapted mesh in accordance with the geometrical characteristics of our domain, we use an adaptive local refinement, based on the 8-subtetrahedron subdivision [5], in the plume trajectory and also for improving the numerical solution. Finally, this model is applied in several test problems.

REFERENCES

[1] R. W. Boubel, D. L. Fox, D. B. Turner, A. C. Stern, *Fundamentals of air pollution*, Academic Press, San Diego, USA (1994).

[2] G. Montero, R. Montenegro, J. M. Escobar, *A 3-D model for wind field adjustment*, J. Wind Eng. & Ind. Aer., 74-76, (1998), 249-261.

[3] R. Montenegro, G. Montero, J. M. Escobar, E. Rodríguez, J. M. González-Yuste, *Tetrahedral mesh generation for environmental problems over complex terrains*, Lecture Notes in Computer Science, 2329, (2002), 335-344.

[4] J. M. Escobar, E. Rodríguez, R. Montenegro, G. Montero, J. M. González-Yuste, *Simultaneous untangling and smoothing of tetrahedral meshes*, Comp. Meth. Appl. Mech. Eng., 192, (2003), 2775-2787.

[5] R. Löhner, J. D. Baum, *Adaptive h-refinement on 3D unstructured grids for transient problems*, Int. J. Num. Meth. Fluids, 14, (1992), 1407-1419.

COMPUTATIONAL MECHANICS (Abstracts)
WCCM VI in conjunction with APCOM'04, *Sept. 5-10, 2004, Beijing, China*

Unified DEM/FEM Approach to Geomechanics Problems

Jerzy Rojek [1, 2*], Eugenio Oñate [1]

[1] *International Center for Numerical Methods in Engineering, Universidad Politecnica de Catalunna, Gran Capitan s/n, 08034 Barcelona, Spain*
[2] *Institute of Fundamental Technological Research, Polish Academy of Sciences, Swietokrzyska 21, 00/049 Warsaw, Poland*
e-mail: jrojek@ippt.gov.pl, onate@cimne.upc.es

Abstract

The discrete element method is a well established method of modelling soil and rock materials and it is often a preferable analysis method of engineering problems in geomechanics. In many applications we deal with interactions between soil and structural components. Accurate modelling of such problems requires consideration of both soil and structure part of the model. In case of structure part finite element method seems to be the most suitable way of modelling. These are motivations of the unified DEM/FEM approach. Combination of both methods can be useful in many cases.
When using standard DEM codes [1] and FEM codes, combined FEM/DEM modelling would require combining two codes [2]. This could impose some limitations on the modelling possibilities. A better solution is combining discrete and finite elements in the same numerical program.
The paper presents the authors' own implementation of the discrete element method within the finite element explicit dynamic code [3]. The FEM presented in the paper allows large elasto-plastic deformations in solid regions. Such problems require the use of stabilized FEM to deal with the incompressibility constraint in order to eliminate the volumetric locking defect, especially when using triangular and tetrahedra elements with equal order interpolation for the displacement and the pressure variables. In the paper a stabilization based on the Finite Calculus (FIC) approach is used.
The paper presents some details of both formulations. Selected numerical results illustrate the possibilities and performance of discrete/finite element analysis in geomechanics problems. Numerical examples include simulation of rock cutting and soil/rock interaction for shallowly buried pipe.

REFERENCES

[1] Itasca Consulting Group, *PFC2D – particle flow code in 2 dimension*, Minneapolis, (1999).

[2] M. Seyferth, A. Henk, *Coupling of PFC2D and ANSYS – concepts to combine the best of two worlds for improved geodynamics models*, in H. Konietzky ed. *Numerical Modeling in Micromechanics via Particle Methods*, Proc. 1st Int. PFC Symposium, Gelsenkirchen, Germany, 6-8 November 2002, Balkema Publishers (2002), pp. 283-290.

[3] J. Rojek, E. Oñate, F. Zarate, and J. Miquel, *Modelling of rock, soil and granular materials using spherical elements*, in 2nd European Conference on Computational Mechanics ECCM-2001, Cracow, 26-29 June, (2001).

[4] E. Oñate, *A stabilized finite element method for incompressible flows using a finite increment calculus formulation*, Comput. Meth. Appl. Mech. Eng, 182, (2000), 355-370.

COMPUTATIONAL MECHANICS (Abstracts)
WCCM VI in conjunction with APCOM'04, *Sept. 5-10, 2004, Beijing, China*

Stress-Dilatancy Relation for Highly Structured Soils

A. M. Sarr [1*], A. Yashima [1], F. Zhang [2], H. Shigematsu [3], K. Sawada [1]

[1] *Department of Civil Engineering, University of Gifu, g3812102@guedu.cc.gifu-u.ac.jp,Gifu, 501-1193, Japan*
[2] *River Basin Research Center, University of Gifu, zhang-f@cive.gifu-u.ac.jp, Gifu, 501-1193, Japan*
[3] *Ishikawa National College of Technology, Kitacyujo, Tsubata, Ishikawa, 929-0392, Japan*
e-mail: g3812102@guedu.cc.gifu-u.ac.jp, *amadoumoctarsarr@hotmail.com*

Abstract

It is well known that, during shearing process, the structured geomaterials exhibit strain hardening and strain softening. Especially under the confining pressure near the consolidation yield stress, the structured geomaterials exhibit strain softening with negative dilatancy. In this paper, the stress-dilatancy relations for highly structured soils are analyzed carefully.
Since the mechanical behavior of structured soils exhibiting softening with large negative dilatancy are observed in many geotechnical problems, it is necessary to propose an appropriate constitutive model that can take into consideration the effect of the structure degradation. Based on more experimental results and model concepts considering the effects of structure, we have extended the constitutive model proposed by Adachi and Oka in order to describe the mechanical behavior by introducing new evolutional equation laws that take into consideration the effect of the structure degradation of geomaterials.
To evaluate the newly extended model's performances, a comparison is carried out with the original Cam-clay, the original elasto-plastic with strain hardening and softening model proposed by Adachi and Oka, its extended type in which the structure degradation effect is considered . All the above-mentioned are applied to experimental stress-dilatancy relations of structured geomaterials. It is found that the newly modified model predicts with sufficient accuracy the experimental stress-dilatancy relations of the structured geomaterials.

COMPUTATIONAL MECHANICS (Abstracts)
WCCM VI in conjunction with APCOM'04, *Sept. 5-10, 2004, Beijing, China*

Influence of Nonlinear Soil and Foundation on Behavior of Offshore Wind Energy Conversion Systems

T. Srisupattarawanit [1*], L. Lehmann [2], C. Corte [3], H. G. Matthies [1]

[1] *Inst .of Scientific Computing,,TU Braunschweig ,Hans -Sommer Str.65, Brunswick, 38106, Germany*
[2] *Inst. of Applied Mechanics, TU Braunschweig, Spielmann Str.11, Brunswick, 38106, Germany*
[3] *Inst. of Steel Structure, TU Braunschweig, Beethoven Str.51, Brunswick, 38106, Germany*
e-mail: {T.Srisupattarawanit, lutz-o.lehmann, c.corte, H.Matthies}@tu-bs.de

Abstract

Nowadays, the technology of wind energy is being developed over the world especially in the countries of the European Commission, in order to produce clean energy and for more stability in the future. The wind turbines are the tool that can convert the wind into electrical power. In history, the wind turbines are small and usually located on land, but presently they are built very large and more flexible and also one want to locate them offshore. So that not only the nonlinear influence of the dynamical system of structure becomes an important, but also the surrounding soil and foundation should be considered as well.

In practical terms, the engineers may ignore the influence of soil and foundation by designing an underground structure as a rigid-based foundation which is convenience but not corrected at all, in some conditions the soil surrounding quite influences the structural behavior (i.e. extreme wind and wave loading). Therefore the different modeling of soil and foundations were studied here in term of comparative results numerically. The different types of modeling include the following:

1) the rigid-based foundation modeling,
2) the elastic spring modeling,
3) the soil-structure interaction with linear soil modeling,
4) the soil-structure interaction with nonlinear soil modeling.

The dynamic behaviors of offshore wind energy conversion systems are observed and the comparative results are also presented in this paper in order to clarify the importance of nonlinear calculation of soil and foundation.

REFERENCES

[1] R. S. Crouch, T. Bennett, *Efficient EBE Treatment of the Dynamic Far-Field in non-linear FE Soil-Structure Interaction Analyses*, Europ. Congress on Comp. Methods in Appl. Sciences and Eng. ECCOMAS, (2000).

[2] Garrad Hassan & Partners et al, *Offshore Wind Energy Ready to Power a Sustainable Europe, Final Report: Concerted Action on Offshore Wind Energy in Europe, Duwind 2001.006,* Delft University of Technology, Netherlands (2001).

[3] L. Lehmann, *Schnelles Verfahren zur Berechnung der Baugrund-Bauwerk-Interaktion im Zeitbereich*, D-A-CH Mitteilungsblatt, Vol.22, No. 3 (2003), (in German).

COMPUTATIONAL MECHANICS (Abstracts)
WCCM VI in conjunction with APCOM'04, *Sept. 5-10, 2004, Beijing, China*

A New Numerical Approach for Coupled Rock-Coal Deformation and Gas Leak Flow

Peide Sun*

Department of Environmental Engineering, Hangzhou University of Commerce, 149 Jiao-gong Road, Hangzhou, 310035, China
e-mail: pdsun@hzic.edu.cn

Abstract

Underground coal seams are mostly distributed in adjacent multi-layers in most countries of the world, especially in China. Many problems in mining engineering and rock engineering remain unsolved such as prediction and control of methane gas emission from the coalface during mining, safety mining area prediction issues related to coal and gas outbursts in protective layer mining, prediction of methane gas drainage rate near to the mining seam and from original coal seam, as well as prediction of gas seepage rate in multi-coal-seam. These problems, in essence, are linked with rock mass deformation and gas leak flow. Under the action of a pore gas pressure gradient in the multi-coal-seam system, gas flow emission occurs near to the mining seam, which in turn leaks through the layer of intercalation with lower permeability, and flows into the mining coal faces or gas drainage bores. This is referred to as gas leak flow by Sun. It is a well-established fact that the description of the interaction between coal/rock mass and gas in the gas leak flow system is difficult. A new numerical approach for coupled coal/rock mass deformation and gas seepage in double coal seams are studied in this paper.

From the viewpoint of interaction mechanics for solid and gas, a coupled mathematical model is presented for solid coal/rock deformation and gas leak flow in parallel deformable coal seams. Numerical solutions using the SIP (Strong Implicit Procedure) method to the coupled mathematical model for double parallel coal seams are also developed in detail. Numerical simulations for the prediction of the safety range using protection layer mining are performed with experimental data from a mine with potential danger of coal/gas outbursts. Analyses show that the numerical simulation results are consistent with the measured data on the spot. The coupled model shows a positive future for applications in a wide range of gas-leak-flow-related problems in mining engineering, gas drainage engineering and mining safety engineering.

REFERENCES

[1] P. Sun, *Sun model and its applications*, Press of Zhejiang University, Hangzhou (2002), (in Chinese with English Abstract).

[2] P. Sun, X. Xian, *Coupled models for coal seam deformation – gas leakage and its applications*, J. China Coal Society, 24, (1999), 60-64, (in Chinese with English Abstract).

[3] P. Sun, *Numerical simulations for coupled rock deformation and gas leak flow in parallel coal seams*, Geotechnical and Geological Engineering, 22, (2004), 1-17.

COMPUTATIONAL MECHANICS (Abstracts)
WCCM VI in conjunction with APCOM'04, *Sept. 5-10, 2004, Beijing, China*

3D Modeling of Brittle Fracture in Heterogeneous Rocks

C. A.Tang [1*], Z. Z. Liang [1, 2], Y. B. Zhang [1, 2]

[1] *Center for Rock Instability and Seismicity Research, Northeastern University, Shenyang, 110004, China*

[2] *Research Center for Numerical Tests on Material Failure, Dalian University, Dalian, 116622, China*

e-mail: tca@mail.neu.edu.cn, crisr@ mail.neu. edu.cn, crisr@ mail.neu.edu.cn

Abstract

Owing to the vast developments in computer science and technologies, recent years have seen a renewal of interest in the computational modeling of material failure in meso-macro scale. The multi-scale capability of the method is recognized as a promising tool in attacking formidable problems in fracture mechanics for heterogeneous media, such as rock, concrete or ceramic, and, indeed, has been successfully applied to the various engineering problems for industrial materials. These models have established themselves as a powerful and realistic alternative to the nonlocal continuum models for softening damage and fracturing. However, the quantitative macro response (such as stress-strain) from most of the current meso-mechanical models is not close enough to the behaviour of real materials. One reason for this shortcoming is that most of these models are two-dimensional, and the three-dimensional effects are inevitably neglected.

In this paper, a 3-D material failure process analysis model, MFPA^{3D}, is proposed. The failure behavior of brittle materials can be simulated by this 3-D model. The model can realistically simulate the inelastic triaxial behavior, strength limits, and postpeak response for both tension and compression. The capabilities of the MFPA^{3D} model to generate a wide range of damage morphologies are examined in this paper.

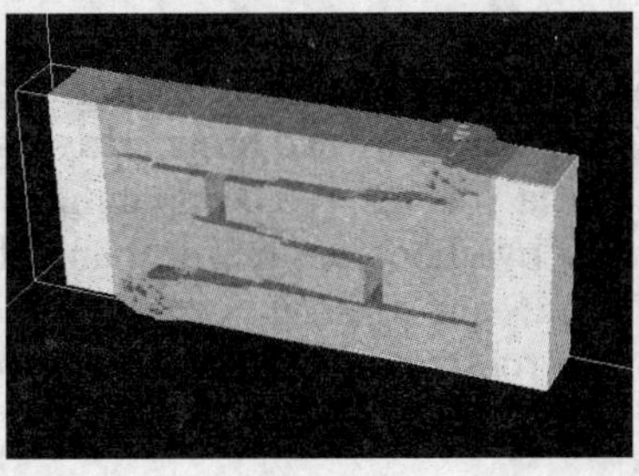
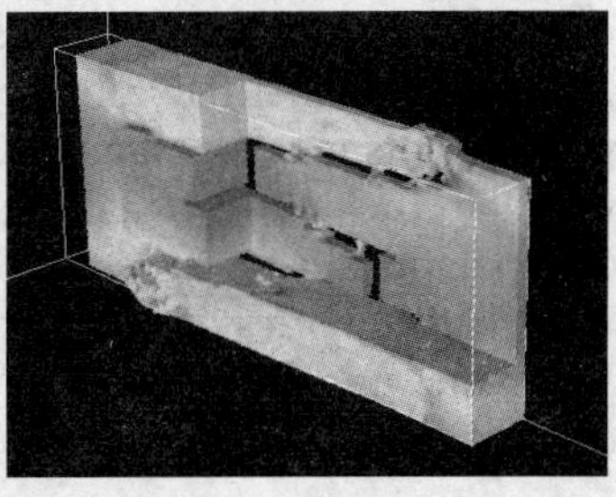

REFERENCES

[1] A. Arlsan, R. Ince, B. L. Karihaloo, *Improved lattice model for concrete fracture*, J. Eng. Mech., 128(1), (2002), 57-65.

[2] J. G. M. van Mier, M. R. A. van Vliet, T. K. Wang, *Fracture mechanisms in particle composites: Statistical aspects in lattice type analysis*, Mech. Mater., 34(11), (2002), 705-724.

[3] C. A. Tang, *Numerical simulation on progressive F.iilure lcading to collapse and associated seismicity*, Int. J. Rock Mech. Min. Sci., 34, (1997), 249-262.

COMPUTATIONAL MECHANICS (Abstracts)
WCCM VI in conjunction with APCOM'04, *Sept. 5-10, 2004, Beijing, China*

Another Way to Compute the Underground Room Support

K. Tor [*], L. Frgić , M. Hudec

Department of Mining and Geotechnics, University of Zagreb, Faculty of Mining, Geology and Petroleum Engineering, Pierottijeva 6, Zagreb, HR-10 000, Croatia
e-mail: ktor@rgn.hr, lfrgic@rgn.hr

Abstract

There is an obvious need in engineering practice for researching the complex problem of interaction between rock and primary support in making safe an underground opening driving.
As probably for no other construction, the static computation for the primary support is burdened with so many difficult definable factors, as the changes of geological conditions, unknown impacts of discontinuity system, insufficient knowledge of physical-mechanical characteristics, unknown relation between horizontal and vertical stresses, complexity of constitutive laws and failure criteria, development of shotcrete strengths in process of binding and hardening and other.
Intensity of the support loading with rock depends on realised deformations of the statically indefinite rock-support system. Rock deformations are increasing so long, until the stress condition in rock does not harmonise with reactive compression of the support and with equalizing of the deformations on the contact. Since it is difficult to determine the impact of the rock on the primary support, as well as the relation between support deformation (e.g. from shotcrete) and its bearing capacity, which would result from these deformations, because young, relaxed concrete is in question. Instead of computing the support condition resulting from an assumed deformation of the rock-support system, bearing capacity of the support is introduced into computation and the influence of the support on the rock is analysed. The rock loading on the opening boundary is therewith determined and it becomes the input information to analyse the stress condition and rock deformation. It is more reliable to assume the impact of boundary condition of the support on rock without regarding deformations. Beside limit bearing capacity – support reaction, it is also controlled if the rock satisfies the failure criteria conditions. The loading with support reactive compression and with anchors must be balanced. On the case of road tunnel driving, the method of boundary elements with the Hoek-Brown's failure criteria was used in computations.
Computations have to be performed by choosing the mechanical rock characteristics in accordance with the categories of rocks and the assumed limit bearing capacities of appropriate support types. When the support is well chosen, the rock near opening is strained under the failure limit, and such a computation proves the construction stability..

REFERENCES

[1] C. A. Brebbia, J. Domingues, *Boundary elements*, Mc. Graw Hill, London (1992), p. 313.
[2] E. Hoek, *Rock engineering - course notes*, Vancouver (1999), p. 303.
[3] M. Hudec, *Primjena principa graničnih stanja u tunelogradnji*, Zbornik referatov 5, Međunanarodno posvetovanje o gradnji predorov, Ljubljana (2000), II pp. 79–91, (in Croatian).
[4] M. Hudec, L. Frgić, *RUBNI – metoda rubnih elemenata*, Rudarsko-geološko-naftni fakultet Zagreb, Hrvatska, (1999), p. 39, (in Croatian).

COMPUTATIONAL MECHANICS (Abstracts)
WCCM VI in conjunction with APCOM'04, *Sept. 5-10, 2004, Beijing, China*

A New Format of Elasto-Plastic Constitutive Relation for Frozen Soil Porous Medium

Jianxing Wang [1, 2*], Jianguo Ning [2]

[1] *College of Architecture, North China University of Technology, Beijing 100041, China*
[2] *Key Lab for the Prevention and Control of Explosion Disaster, Beijing Institute of Technology, Beijing 100081, China*
e-mail: Wangjx6@ncut.edu.cn

Abstract

The constitutive equation about elasto-plastic problems was established by importing concerned concepts not only classical but also stop-press production, the equation could be used for 3-D porous medium. As a result of hypersensitive for temperature of frozen soil, when temperature altering, the constitutive equation would be temperature susceptive, so there was embodiment about the temperature susceptive coefficients. Through popularizing the research production, it could be found that temperature loading had an important influence on the constitutive expression, which embodimenting in the modality of performance of frozen soil porous medium.

Based our disquisition in theories and applications about elasto-plastic and other significative conclusions and calculating techniques, the control equations of stress-strain constitutive relationship was presented, and the temperature loading for elasto-plastic was considered. For the sake of efficacious theoretics, the research outcomes were significative. Weight relation for porous medium had been given when considering the Jaumann rate of Kirchhoff stress. Basic model and expresses in response to multifarious type of frozen soil porous medium, basic weight relation model could be depicted accordingly. Field equation of the elasto-plastic would represent in general modality, constitutive relation of elasto-plastic had been excogitated concern in the process of investigating material idiosyncrasy; this modality was adaptive for acquired for elasto-plastic idiosyncrasy. Finally, there are some original coefficients deservedly as the linear hypostasis for the flux and drives force with the elementary hypothesis of constitutive relation; the research established the ordinary constitutive relation law of frozen soil porous medium and inner thermal liquid, the temperature field distribution would variety in deferent circumstance. Several simulations for frozen soil porous medium have been excogitated; it indicated that the investigation would be significative.
This work was supported by the grant of the Knowledge Innovation Program of the Chinese Academy of Sciences, The Plans for Technology Development of Beijing Education Council and The Scientific Research Fund of North China University of Technology (NCUT).

REFERENCES

[1] J. X. Wang, J. G. Ning, *A new theory of constitutive relation for thermo-elasto-plastic problems in porous medium*, in Weizang Chien ed. *Proc. 4th Int. Conf. Non. Mech.*, Shanghai Univ. Press, (2002), pp.364-367.

[2] J. X. Wang, X. C. Wang, *Constitutive relation of thermo-elasto-plastic problems for frozen soil porous medium*, Mech. Prop. Adv. Eng. Materials, Tsinghua Univ. Press & Springer-Verlag, (2003), pp. 69-75.

COMPUTATIONAL MECHANICS (Abstracts)
WCCM VI in conjunction with APCOM'04, *Sept. 5-10, 2004, Beijing, China*

Study of the Soil-Structure Interaction of Suction Caissons with Advanced Finite-Element Methods

Shin-Tower Wang [1*], Lisa J. Novak [2], H. Theodore Lin [3]

[1] *Project Manager, Ensoft, Inc./Lymon C. Reese & Associates, P. O. Box 180348, Austin, Texas*
[2] *Project Engineer, Ensoft, Inc./Lymon C. Reese & Associates, P. O. Box 180348, Austin, Texas*
[3] *AMPS Technologies Company, 1138 Fox Chapel Road, Pittsburgh, PA 15238*
e-mail: stw@ensoftinc.com, lisa@ensoftinc.com, tedlin@ampstech.com

Abstract

The term "suction caisson" originated in the offshore industry where engineers conceived the idea of installing the bucket foundation for offshore structures using the self-weight and suction pressure inside the bucket. The suction caisson typically is made of steel and has a diameter of 8 to 12 meters with skirts around the periphery, rather like an upturned bucket. The installation is completed by reducing the water pressure inside the bucket, thus forcing the skirts into the seafloor. The in-service performance, particularly when combined vertical, lateral, and moment loads are applied, must be studied so that a safe design can be achieved.

Many proposals have been made for methods of analysis for the suction caisson. In general, there were three broad classes of methods of analysis: simplified limit-state methods; approximate computer-based semi-empirical methods; and more rigorous nonlinear finite-element methods (FEM). Two strong reasons dictate the use of the FEM: (1) the problem is so complex that simplified methods cannot model the soil-structure interaction correctly; and (2) codes for the FEM are available, powerful, and capable of being run on the personal computer. The three-dimensional FEM computer program is available for solving those complicated problems involving different soil behaviors inside and outside the bucket foundation.

One critical element in design of suction caissons in different environments is to analyze the penetration needed for the suction piles to sustain wave loads created by the maximum storm. With regard to modeling the problem, the vertical load coupled with the overturning moment on the foundation requires the soil to provide sufficient contact friction on the skirts. The excessive horizontal movement of the caisson pushed by the lateral load may leave a "gap" behind between the steel and the surrounding soil. All these special problems are beyond the capability of any simplified method. Thus, modeling of the suction caisson and supporting soil as a nonlinear/plasticity problem using FEM is a preferable method. Furthermore, the designer can easily investigate alternate geometries, can study the influence of the various parameters on a solution, and can produce results that are instructive and valuable.

REFERENCES

[1] D. Senpere, G. A. Auvergne, *Suction anchor piles – a proven alternative to driving or drilling,* Proc. 14th Offshore Technology Conference, Houston, (1982), pp. 483-493.

COMPUTATIONAL MECHANICS (Abstracts)
WCCM VI in conjunction with APCOM'04, *Sept. 5-10, 2004, Beijing, China*

Dynamic Soil-Sheet Pile Interaction Analysis Using a Cyclic Plasticity Model and Fully-Coupled Effective Stress Approach

Zhi-Liang Wang *, Larry Scheibel, John Egan, Chih-Cheng Chin

Geomatrix Consultants Inc., Oakland, California, USA
e-mail: {zlwang, lscheubel, jegan, ccchin}@geomatrix.com

Abstract

Sheet piling is often used in waterfront structures to retain a soil slope and maintain stability. In zones of high earthquake hazard (e.g., California), seismic stability is a significant factor in the design of such waterfront structures (e.g., port shipping berths). Conventional design methods are typically based on a pseudo-static approach that involves significant uncertainty and, in some cases, have resulted in overly-conservative designs.

In recent years, we have developed and implemented a fully-coupled, effective stress based, dynamic analytical method that uses a cyclic plasticity soil behavior model and a two-dimensional finite deference code capable of computing large deformations. A number of earth and rockfill dams, as well as waterfront embankments, have been analyzed using this method for design motions with peak ground accelerations as high as 1.0 g. The developed soil behavior model used is a key element of the computational tool because it allows generation of excess pore water pressure and liquefaction of saturated soils to be predicted and, in consequence, deformation of the soil-structure system to be computed. During recent projects, we have extended this approach to analysis of waterfront soil structures that include sheet piling.

In the state-of-practice approach, the sheet pile design involves the simulation of soil non-linearity through the p-y curve of the soil surrounding the sheet pile. For the state-of-art approach, it is currently acknowledged that the finite element (or finite deference) method can be used in studying the stresses and strains developed on the entire soil-sheet pile system and thus it can provide a better understanding of the behavior of the soil and the sheet piling. But, the key element is use of a comprehensive and practical constitutive model of the soil. The validated bounding surface hypo-plasticity model for sand [1] was utilized for the soil-sheet pile interaction analysis in this study. A numerical example, closely related to the design of a berth at the Port of Oakland, is presented that illustrates the computed performance of a strengthened soil slope supported by sheet piles at the waterfront under a set of design earthquake shaking conditions.

REFERENCES

[1] Z. L. Wang, Y. F. Dafalias, C. K. Shen, *Bounding surface hypoplasticity model for sand*, ASCE, Journal of Engineering Mechanics, 116(5), (1990), 983-1001.

COMPUTATIONAL MECHANICS (Abstracts)
WCCM VI in conjunction with APCOM'04, *Sept. 5-10, 2004, Beijing, China*

Three Dimensional Free Surface Seepage Analysis by Thermal Simulation

X. Yu [1*], S. Z. Peng [2]

[1] School *of Civil Engineering, Purdue University, 1284 Civil Engineering Building, West Lafayette, 47906, USA*
[2] *Department of Hydraulic Engineering, Tsinghua University, Beijing, 100084, China*
e-mail: xyu@purdue.edu

Abstract

Free surface seepage problem is a challenging topic in computational mechanics. There are two main approaches of FEM for this problem. The first approach only considers the saturated zone below free surface, whose location is generally uncertain. The process involves mesh iteration, where the location of free surface is changed iteratively and the computation mesh is rearranged until all the boundary conditions are satisfied [1]. The problem of mesh rearrangement is difficult for high dimensions. The second approach [2] computes the seepage field without mesh iterations. The computation is carried out on the whole model, which include both saturated zone and unsaturated zone. Much smaller coefficient of hydraulic permeability is assigned to elements in the unsaturated zone. Iterations are performed on the materials properties rather than mesh. A fixed mesh is used throughout the iteration process.

Seepage field, thermal field and static electrical field are all described by similar field equations. The similarity between static electrical field and seepage field is the basis for experimental study of seepage problem by water-electrical simulation. Commercial software (such as NASTRAN, SAP2000, ABAQUS) have the ability of conducting thermal analysis on sophisticated structures. However, they generally lack the capacity for free surface seepage analysis. This paper introduces an idea of studying free surface seepage problem combining the thermal analysis module and non-mesh-iteration approach. As an intermediate procedure, a simple computer program is developed to analyze the output file containing the results of thermal analysis by commercial software (NASTRAN [3] in this case). The program generate new input file with update material parameters (coefficient of thermal conductivity) for certain elements, which is subsequently analyzed by NASTRAN again. The process is iterated until results converge.

This approach was applied in a research involving study the three dimensional seepage problem of an arch dam. The results of seepage field distribution, seepage force and amount of seepage using thermal analogy are satisfactory. The approach can be implemented on any commercial software with thermal analysis module. It is advantageous especially for analysis of seepage problem with sophisticate geology structures.

REFERENCES

[1] W. D. L. Finn, *Finite element analysis of surface seepage through dams*, J. SMFD, ASCE, 93, SM6, (1967).

[2] K. J. Bathe, M. R. Khoshgoftan, *Finite element for surface seepage analysis without mesh iteration*, Int. J. Num. Ana. Methods in Geomechs, 3, (1979), 3-22.

[3] MSC Software, *MSC/NASTRAN handbook for thermal analysis.*

COMPUTATIONAL MECHANICS (Abstracts)
WCCM VI in conjunction with APCOM'04, *Sept. 5-10, 2004, Beijing, China*

Finite Element Implementation for Complex Critical State Models

Jidong Zhao *, Daichao Sheng

Department of Civil, Surveying & Environmental Engineering, University of Newcastle, Callaghan, NSW, 2308, Australia
E-mails: jidong.zhao@newcastle.edu.au, daichao.sheng@newcastle.edu.au

Abstract

This paper presents a general procedure for implementing complex critical state models in displacement finite element codes. These models are typically used for structured clays and sands, and are usually characterised by multiple yield surfaces, plastic yielding within yield surface, yield surface with vertices, and complex kinematic and isotropic hardening laws. The Backward Euler Return method, which leads to quadratic convergence when used with the consistent stiffness matrix in a Newton procedure, will have difficulties in integrating such models. The stress integration scheme used in this paper is based on the explicit Euler method with automatic substepping and error control [1-2]. This scheme employs the classic elastoplastic stiffness matrix and requires only the first derivatives of the yield function and plastic potential. A general purpose integrator can thus be designed for a wide range of models and is independent of the precise form of the stress-strain relations. Therefore, this scheme has obvious advantages when applied to complex constitutive laws A family of critical state models, including the Super/Subloading Yield Surfaces model (SSYS) [3], the Kinematic Hardening Bubble model (KHB) [4], as well as the classic critical state models[5], have been implemented into a finite element code. In both SSYS and KHB, apart from the normal Cam-Clay yield surface, two additional yield surfaces are introduced to account for the characteristics of overconsolidation and structures of soils and complex hardening laws are supposed. These models are then used to analyse various boundary value problems. It is demonstrated that the scheme used in this paper is generally robust and efficient. Different features of clays and sands are well captured.

REFERENCES

[1] S. W. Sloan, *Substepping schemes for the numerical integration of elastoplastic stress-strain relations*, International Journal of Numerical Method in Engineering, 24, (1987), 893-911.

[2] D. C. Sheng, S. W. Sloan, *Loading stepping schemes for critical state models*, International Journal of Numerical Method in Engineering, 50, (2001), 67-93.

[3] A. Asaoka, *Consolidation of clay and compaction of sand-an elasto-plastic description*, The 12th Asian Regional Conference on SMGE, Singapore, Aug., (2003).

[4] M. Rouainia, D. Muir Wood, *A kinematic hardening constitutive model for natural clays with loss of structure*, Géotechnique, 50(2), (2000), 153.

[5] A. Gens, D. M. Potts, *Critical state models in computational geomechanics*, Engineering Computation, 5, (1988), 178-197.

COMPUTATIONAL MECHANICS (Abstracts)
WCCM VI in conjunction with APCOM'04, *Sept. 5-10, 2004, Beijing, China*

A New Procedure for Searching Slip Surface of Geotechnical Slope with Genetic Algorithm

Yuanpai Zhou *

Department of Engineering Mechanics, Dalian University of Technology, Dalian, 116023 China
e-mail: zhouyuanpai1@sohu.com

Abstract

The search for critical failure surface of a general soil slope is difficult as the objective function of the factor of safety is non-convex and multiple minima exist in general. For this problem, b ased on genetic algorithm and Bishop Simplified Method, searching methods for the most dangerous slip surface of geotechnical slope are proposed and the according minimal safety factor is presented. A series of folding lines are applied for approximating to sliding surface, and the terminal endpoints of the folding lines describe the sliding surface. The most dangerous slip surface of geotechnical slope is investigated by searchin g the endpoints. The principles and the procedures of the computation are interpreted. Computational program for numerical computation and data process are developed using genetic algorithm and Bishop Simplified Method. Numerical example shows that the analysis method of slope stability based on the genetic algorithm is a global optimal method that can search the arbitrary-shaped slip surface of geotechnical slope and can overcome the drawbacks of local optimum widely existing in classic searching methods.

REFERENCES

[1] M. Paul, P. Wilkinson, *The use of the simple genetic algorithm in finding the critical factor of safety in slope stability analysis*, Computers and Geotechnics, 29(8), (2002), 699-714.

[2] L. M. Schmitt, *Theory of genetic algorithms*, Theoretical Computer Science, 259(1-2), (2001), 1-61.

[3] Y. M. Cheng, *Location of critical failure surface and some further studies on slope stability analysis*, Computers and Geotechnics, 2003;30(3):255-267.

COMPUTATIONAL MECHANICS (Abstracts)
WCCM VI in conjunction with APCOM'04, *Sept. 5-10, 2004, Beijing, China*

Modeling of Time-Independent and Time-Dependent Degradation in Rock Materials

Qizhi Zhu [1, 2*], Yan Zhang [1,2], Ziming Zhang [1]

[1] *Department of Engineering Mechanics, Hohai University, Xikang Road, Nanjing, 210098, China*
[2] *Laboratory of Mechanics of Lille, UMR 8107, 59655 Villeneuve d'Ascq, France*
e-mail: qizhi.zhu@eudil.fr

Abstract

Most geomaterials like rocks and concrete exhibit both instantaneous and time dependent irreversible deformations [1]. In this paper, a thermodynamically constitutive elastoplastic damage model is first present, in which a customary scalar internal damage variable is used to describe the instantaneous (time independent) degradation [2]. The deformation is supposed to be time dependent and is considered as a macroscopic consequence of progressive degradation of material structure at the microscopic scale. A second time dependant vvariable similar to damage variable, which is a function of plastic deformation, is introduced here to quantify the degradation of microstructure [3]. The constitutive model is then logically extended to characterize the long term behavior. This model is applied to predict the creep and relaxation tests. A good agreement between numerical simulations and experimental data is obtained. The proposed model is able to describe the main features observed in most cohesive frictional geomaterials (rocks and concrete), such as plastic deformation, damage, pressure sensitivity, rate dependency and creep. Incremental forms of stress-strain relations can be easily implemented into a numerical integration algorithm using the finite element method.

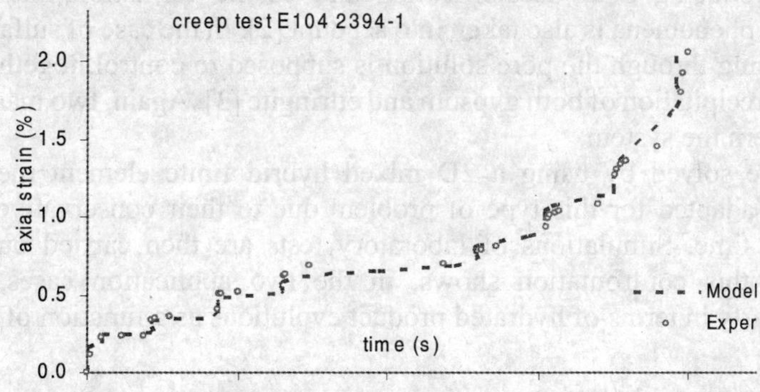

REFERENCES

[1] N. D. Cristescu, U. Hunsche, *Time effects in rock mechanics*, Wiley-Interscience-Europe; (1998).
[2] J. Lemaitre, *A course of damage mechanics*, 2nd ed. Berlin: Springer, (1992).
[3] S. Pietruszczak, D. Lydzba, J. F. Shao, *Modelling of inherent anisotropy in sedimentary rocks*, Int. J. Solids Struct., 39, (2002), 637-48.

COMPUTATIONAL MECHANICS (Abstracts)
WCCM VI in conjunction with APCOM'04, *Sept. 5-10, 2004, Beijing, China*

Simplified Chemo-Transfer Modeling of Carbonation and Sulfate Attack in Saturated Cement Pastes

B. Bary *, P. Le Bescop

LECBA, CEA Saclay, bât 158, 91191 Gif/Yvette, France
e-mail: bary@azurite.cea.fr, lebescop@azurite.cea.fr

Abstract

We present in this paper two simplified coupled chemo-transfer approaches which address the phenomena of carbonation and external sulfate attack in cement pastes under saturated conditions. These degradations appear typically when cementitious materials are exposed to the contact of water containing carbonate and sulfate ions, and they are of particular importance in the context of nuclear waste storage, where concrete could become an attractive option for the construction of disposal engineering barriers and underground structures exposed to ground water.

The basis of the approach proposed in this study relies on a simplified chemical description of the solid phase composing the cement paste [1]. The four main hydrated products are considered, and it is supposed that both portlandite and C-S-H phase dissolution/decalcification processes can be directly related to the calcium concentration in pore solution. In the case of carbonation, the variables retained for completely defining the state of the system are the carbonate concentration and calcium concentration in pore solution. The governing equations are then constituted by the mass balance equations for these two ionic species, which diffuse through the connected porosity from the core to the exterior of the material for the calcium, and conversely for the carbonate. Source terms are introduced for reproducing both decalcification and calcite formation; the porosity evolution resulting from these phenomena is also taken into account [2]. In the case of sulfate attack, the sulfate concentration diffusing through the pore solution is supposed to control, together with the calcium concentration, the precipitation of both gypsum and ettringite [3]. Again, two mass balance equations are assumed to govern the system.

These equations are solved by using a 2D mixed hybrid finite element method for the space discretization, well adapted for this type of problem due to their conservative properties, and an implicit scheme in time. Simulations of laboratory tests are then carried out and compared to experimental data; this confrontation shows, in the two application cases, a relatively good agreement, particularly in terms of hydrated product evolutions as a function of time.

REFERENCES

[1] F. Adenot, M. Buil, *Modelling of the corrosion of the cement paste by deionized water*, Cem. Concr. Res., 22, (1992), 489-495.

[2] B. Bary, A. Sellier, *Coupled moisture-carbon dioxide-calcium transfer model for carbonation of concrete*, Cem. Concr. Res., (2004), (to be published).

[3] J. Sercombe, D. Planel, P. Le Bescop, F. Adenot, A. Sellier, B. Capra, J.-M. Torrenti, *Calcium leaching of cement paste in sulfated water: relation between internal pressure kinetics and crack initiation*, Int. Conf. CONCREEP'6, Cambridge, USA, August 20-22, (2001).

COMPUTATIONAL MECHANICS (Abstracts)
WCCM VI in conjunction with APCOM'04, *Sept. 5-10, 2004, Beijing, China*

Seismic Response of the Confederation Bridge to Non-Uniform Support Excitation

Moe M. S. Cheung [*], Bahman Noruziaan, Chengyu Yang

Department of Civil Engineering, Hong Kong University of Science and Technology, Clear Water Bay, Kowloon, Hong Kong, China
e-mail: mscheung@ust.hk , bnoruzia@ust.hk , ceyangcy@ust.hk

Abstract

The Confederation Bridge has provided a fixed connection between the two Canadian provinces of New Brunswick and Prince Edward Island since 1997. The Bridge has a design life span of 100 years and with a total length of 12.9 km, it is the longest bridge built over ice-covered seawater. The Bridge is multi-span and the main part of it consists of 22 similar portal frames connected by simply supported drop-in girders. The portal frames have 95m long overhangs at their two sides and are 250m apart. The overhangs and drop-in members are of box girder type. The box girders rest on piers of 40 to 60m height above the water. They are 12 m wide with web spacing of 7m and depth of 4.5 m at mid-span to 14m over the piers. The bridge has been the subject of numerous studies. However, despite significant importance of non-uniform ground excitation in the seismic response of long-span bridges, so far, not enough attention has been paid to the effects of this phenomenon on the behavior of the Bridge. In this paper, the authors will study the effects of non-uniform ground motion on the behavior of the Confederation Bridge, especially the stress distribution in the frames and displacements at the bearings between the portal overhangs and the drop-in members. The ground tremor record will be modified in order to consider the effect of traveling seismic waves and before being applied to various piers of the Bridge.

REFERENCES

[1] G. Tadros, *The Confederation Bridge: an overview*, Canadian Journal of Civil Engineering, 24(6), (1997), 850-866.

[2] N. Naumoski et al, *Dynamic characteristics of the Confederation Bridge*, Canadian Journal of Civil Engineering, 29(3), (2002), 448-458.

[3] L. G. Jaeger et al, *Seismic design for the Confederation Bridge*, Canadian Journal of Civil Engineering, 24(6), (1997), 922-933.

[4] D. E. Becker et al, *Foundation design aspects of the Confederation Bridge*, Canadian Geotechnical Journal, 35(5), (1998), 750-768.

COMPUTATIONAL MECHANICS (Abstracts)
WCCM VI in conjunction with APCOM'04, *Sept. 5-10, 2004, Beijing, China*

Two-Dimensional Yee's Scheme for the Solution of Maxwell's Equation on Unstructured Meshes

O. Hassan [*], I. Sazonov, K. Morgan, N. P. Weatherill

School of Engineering University of Wales Swansea, Singleton Park, Swansea SA2 8PP, UK
e-mail: {o.hassan, i.sazonov, k.morgan, n.p.weatherill}@swansea.ac.uk

Abstract

Yee's scheme [1] is a very fast and accurate method for the computation of the scattering of electromagnetic waves. This algorithm is known for its low operation count as it requires few arithmetic operations to advance the solution in time in particularly when it is compared, for example, to the Time Domain Finite Element Method (TDFE). Initially proposed for structured grids, Yee's scheme can be generalized for unstructured mesh. This will enable its application to industrially complex geometries [2].

Yee's scheme is staggered algorithm both in space and time. It needs two orthogonal meshes for electric and magnetic fields. A dual Delaunay-Voronoi diagram forms such two orthogonal meshes. Since the time step limitation of Yee's algorithm is proportional to the shortest edge in the both dual meshes, it is critical that the Voronoi diagram is as well behaved as the Delaunay triangulation: to avoid very short Voronoi edges.

Using present meshing methods it is possible to build a smoothly varying Delaunay triangulation, with a high quality index, but a highly irregular Voronoi diagram with some very short Voronoi edges which will increase the total number of required time step per cycle and hence reducing the usability of the method.

The method proposed here will allow the generation of unstructured meshes, of which Delaunay and Voronoi edges are bounded from above and below. First, a high quality near-boundary triangulation is prepared. Structured mesh of equilateral triangles is used to fill the remainder of the domain. Then the near-boundary triangulation is 'stitched' to the ideal mesh so that the formed 'seam' has also high quality index for both the meshes.

Computations using meshes generated by this method for different geometries will be presented. The Radar Cross Section (RCS) will be compared to exact and other known solutions. The efficiency and accuracy of the methods will also be analysed and compared to the efficiency of a FE. Geometries which contain singularity will also be used to demonstrate the advantages of the proposed scheme.

REFERENCES

[1] K. S. Yee, *Numerical solution of initial boundary value problem involving Maxwell's equation in isotropic media*, IEEE Trans. Antennas and Propagation, 14, (1966), 302-307.

[2] K. Morgam, O. Hassan, J. Peraire, *A time domain unstructured grid approach to the simulation of electromagnetic scattering in piecewise homogeneous media*, Computer Methods in Applied Mechanics and Engineering, 134, (1996), 17-36.

COMPUTATIONAL MECHANICS (Abstracts)
WCCM VI in conjunction with APCOM'04, *Sept. 5-10, 2004, Beijing, China*

Applying CFD Techniques to Semiconductor Charge Transport

Shaoqiang Tang *

LTCS, Dept of Mechanics and Engineering Sciences, Peking University, Beijing, 100871, China
e-mail: maotang@pku.edu.cn

Abstract

Computer simulations of charge transport in semiconductor devices (like diodes and micro-chips) are used by the semiconductor industry as a tool for reducing the cost of developing new devices and new process technologies. At the scale of micron or sub-micron, the semiconductor Boltzmann equation is the most exact model. In order to alleviate computing load, macroscopic models have been derived, assuming that the state of the electron gas is described by certain averaged quantities. These models take similar forms as those in fluid mechanics, and we may apply CFD techniques to probe this promising field of academic importance and commercial value.

We shall present some of our recent results. First, by simulating a hydrodynamic model, we demonstrate the (direct) applicability of CFD techniques [1]. Secondly, the continuing trend of scaling-down and speed-up makes the modeling and computing of quantum effect and transient behavior among the top issues in semiconductor research. Interdisciplinary backgrounds are required in these explorations, and CFD community has experiences to share in combining physical, mathematical and computational techniques to tackle with complex systems. For instance, concerns of flow stability help constructing a PHD model for transient energy-transport model, whose convection part may be linearly unstable [2]. For another example, careful numerical tests helped identifying well-posedness problem in a quantum hydrodynamic (QHD) model. Viscous QHD model derived from a Wigner Fokker-Planck equation yields more reliable numerical results, and demonstrate interesting nonlinear phenomena, such as negative differential resistance and hysteresis [3]. Finally, semiconductor studies are rewarding, as it may provide new views and techniques to CFD, e.g. a positive entropic scheme for a quantum model applies to thin film [4].

REFERENCES

[1] A. Juengel, S. Tang, *A relaxation scheme for the hydrodynamic equations for semiconductors,* Appl. Numer. Math., 43, (2002), 229-252.

[2] S. Tang, *Pseudo-hydrodynamic approximation for energy-transport models,* (in print).

[3] A. Juengel, S. Tang, *Numerical approximation of the viscous quantum hydrodynamic model for semiconductors,* Appl. Numer. Math., (submitted).

[4] J. A. Carrillo, A. Juengel, S. Tang, *Positive entropic scheme for a fourth-order parabolic equation,* Discrete and Continuous Dynamical Systems, Series B, 3, (2003), 1-20.

COMPUTATIONAL MECHANICS (Abstracts)
WCCM VI in conjunction with APCOM'04, *Sept. 5-10, 2004, Beijing, China*

A Comparison of Two- and One-equation Finite Volume Computational Models for Simulating Low Temperature Drying

Chenhang Yang [1*], Fawang Liu [1, 2], Ian Turner [2]

[1] *School of Mathematical Sciences, Xiamen University, Xiamen, 361005, China*
[2] *School of Mathematical Sciences, Queensland University of Technology, GPO Box 2434, Brisbane, Qld. 4001, Australia*

e-mail: morning_sail@tom.com, fwliu@xmu.edu.cn, i.turner@qut.edu.au

Abstract

Drying of wood is an important industrial process that involves coupled heat and mass transfer in a complex biological porous medium [1]. In this work a comparison of the results offered by two finite volume computational models for simulating the low temperature drying of a radial-transverse cross-section of Australian Pinus Radiata is presented. The models considered here are simplifications of the complete drying model proposed in [2]. Under the assumptions of constant internal gaseous pressure within the porous medium being dried a two-equation model [3] can be derived, which uses as its dependent variables product temperature and moisture content. By further assuming that capillary pressure can be expressed as a function of moisture content alone, it becomes possible to express both vapour and liquid migration as diffusion mechanisms with so-called global "pseudo-diffusion" coefficients. Note however that through the difference of the vapour and liquid phase enthalpies the two-equation model is able to account for the latent heat of vaporization. Under the even more stringent assumption of constant product temperature during drying, the two-equation model can be reduced further to a one-equation model that governs only the moisture transport in wood. In this model, the product temperature is assumed to follow that of the drying chamber.
The results of both models will be compared against experimental data to examine their validity and to identify the weaknesses and strengths of the modeling assumptions for the low temperature drying conditions. An overview of the numerical strategies used in the implementation of the finite volume methodology will also be provided. In particular, an unstructured mesh finite volume method similar to the strategy given in [4], will be used for the spatial discretization and the DASSL solver is employed thereafter to resolve the system of ordinary differential equations.

REFERENCES

[1] P. Perré, M. Moserand, M. Martin, *Advances in transport phenomena during convective drying with superheated steam or moist air*, Int. J. Heat and Mass Trans., 36, (1993), 2725-2746.

[2] I. Turner, P. Perré, *A synopsis of the strategies and efficient resolution techniques used for modelling and numerically simulating the drying process*, in I. Turner, A. S. Mujumdar eds. *Mathematical modeling and numerical techniques in drying technology*, Marcel Dekker Inc., (1996), pp. 1-82.

[3] S. Ben Nasrallah, P. Perré, *Detailed study of a model of heat and mass transfer during convective drying of porous media*, Int. J. Heat and Mass Trans., 31(5), (1988), 957-967.

[4] F. Liu, I. Turner, V. V. Anh, *An unstructured mesh finite volume method for modelling saltwater intrusion into coastal aquifers*, Korean J. Comp. & Appl. Math., 9(2), (2002), 391-407.

COMPUTATIONAL MECHANICS (Abstracts)
WCCM VI in conjunction with APCOM'04, *Sept. 5-10, 2004, Beijing, China*

A Study on Steel Fiber and Nano Carbon Tube Improve the Toughness of HPC

Ping-Kun Chang *

Department of Civil Engineering, Vanung University, No.1 Vanung Rd., Chungli City, Taoyuan 32045, Taiwan, China
e-mail: pkchang@vnu.edu.tw

Abstract

In this paper, the nuclear magnetic resonance (NMR) and mercury intrusion porosity (MIP) technologies are adopted to analyze the pozzolanic reaction, and the tests of splitting strength and compressive strength are executed to find out the tensile-to-compressive strength ratio and the characteristic length and then to examine the brittleness of high performance concrete (HPC). The pozzolanic reaction of nano carbon tube can effectively transform $Ca(OH)_2$ to C-S-H gel to fill the pores and reduce their volume, so the densification of concrete is enhanced. The test results show that the splitting resistance of steel fiber can effectively strengthen the toughness of concrete, and then the brittleness of HPC is reduced as well.

COMPUTATIONAL MECHANICS (Abstracts)
WCCM VI in conjunction with APCOM'04, *Sept. 5-10, 2004, Beijing, China*

Molecular Dynamics Simulations of Boron-Nitride Nanotube Composites

M. Griebel, J. Hamaekers*

Department of Numerical Simulation, University of Bonn, Wegelerstraße 6, Bonn, D-53115, Germany
e-mail: griebel@ins.uni-bonn.de, hamaekers@ins.uni-bonn.de

Abstract

The elastic properties of boron-nitride nanotubes embedded in amorphous silicon-boron-nitride ceramics are examined by molecuar dynamics simulations. To derive stress-strain curves for various tensile and compressive load cases at given temperature and pressure, we use the Parrinello-Rahman approach to apply external stress to a periodic system. In addition to Young moduli and Possion ratios, we compare radial distribution functions, average coordination numbers, ring statistics and self-diffusion coefficients, in order to characterize the short-range, medium-range and long-range order of Si_3BN_5, $Si_3B_2N_6$ and $Si_3B_3N_7$ matrices. Here, our results show that $Si_3B_3N_7$ exhibits the highest Young modulus and the largest elastic range. Furthermore, we calculate stress-strain curves for capped BN-nanotubes and infinite BN-nanotubes embedded in Si-B-N matrices to compare the derived Young moduli with different macroscopic rule-of-mixtures predictions. Here, the influence of the nanotube/matrix ratio and of the temperature on the elastic modulus is examined. Our results show a good agreement with the rule-of-mixtures predictions. In particular, all systems are modeled with a reactive many-body bond order potential due to Tersoff.

REFERENCES

[1] M. Griebel, J. Hamaekers, *Molecular dynamics simulations of the elastic moduli of polymer-carbon nanotube composites,* Computer Methods in Applied Mechanics and Engineering, (in Press).

[2] S. J. Frankland, A. Caglar, D. W. Brenner, M. Griebel, *Molecular simulation of the influence of chemical cross-links and the shear strength of carbon nanotube-polymer interfaces,* J. Phys. Chem. B, 106, (2002), 3046-3048.

[3] M. Griebel, S. Knapek, G. Zumbusch, A. Caglar, *Numerische Simulation in der Moleküldynamik.* Numerik, Algorithmen, Parallelisierung, Anwendungen. Springer, Berlin, Heidelberg, Germany, (2003), (in German).

[4] M. Parrinello, R. Rahman, *Crystal structure and pair potentials: a molecular-dynamics study,* Phys. Rev. Lett., 45, (1980), 1196-1199.

[5] J. Tersoff, *Modeling solid-state chemistry: interatomic potentials for multicomponent systems,* Phys. Rev. B, 39, (1989), 5566-5568.

COMPUTATIONAL MECHANICS (Abstracts)
WCCM VI in conjunction with APCOM'04, *Sept. 5-10, 2004, Beijing, China*

Computational Cosserat Dynamics in MEMS Component Modelling

Dongsheng Liu, D. Q. Cao, Charles H. T. Wang *

Industrial Mathematics and Gravity Group, Lancaster University, Lancaster LA1 4YB, UK
e-mail: {c.wang, d.liu, d.cao}@lancaster.ac.uk

Abstract

This paper presents a new modelling for MEMS components based on Cosserat theory. The motion of a rod-like structure in three-dimensional space can be described by a reference curve and three perpendicular unit vectors in Cosserat theory [1-2]. The reference curve is called Cosserat curve and three unit vectors are called directors. The analysis of configuration of a rod in Cosserat theory has been studied by many authors such as [3-5] since the Cosserat approach can be used to model the components of many practical structures such as MEMS, flexible robotic arms even DNA strands.

The equations of motion for a rod are nonlinear partial differential equations about space variables and time. According to the Cosserat theory, suitable kinematical assumptions are imposed on the rod [5]. The basic kinematic quantities are the position of a point on the Cosserat curve and an orthogonal transformation that define the rotation of an orthogonal triad attached to the cross-section at each point of the Cosserat curve. This enables description of a rod using nonlinear ordinary differential equations in terms of end points of a rod.

A main feature of the Cosserat approach is benefit gained from its mathematical simplicity. Using the Cosserat theory that can accommodate to a good approximation the nonlinear dynamical behaviour of complex elastic structures, the kinetic and strain energies of a rod have been derived in terms of the shape functions obtained by solving the static model. This way the dynamics of the rod is reliably defined through an action principle that guarantees basic conservation laws. The derivative procedure can be implemented by using computer algebra softwares such as MAPLE.

The assumption that the cross-section of the rod remains perpendicular to the tangent to the Cosserat curve is employed for modelling a slender rod whose shear deformation can be negligible. Based on the Lagrange approach, the obtained kinetic and strain energies are used to derive the ordinary differential equations of motion with third order nonlinear terms for the Cosserat rod without considering the shear deformation. For illustration, the nonlinear dynamic model of a resonator in MEMS is investigated using the proposed approach. The corresponding nonlinear dynamical responses of the resonator have been presented through numerical simulations by Matlab software.

REFERENCES

[1] A. E. Green, P. M. Naghdi, M. L. Wenner, *On the theory of rods I: derivations from the three-dimensional equations,* Proc. Royal Soc. London A, 337, (1974), 451-483.

[2] A. E. Green, P. M. Naghdi, M. L. Wenner. *On the theory of rods I: developments by direct approach,* Proc. Royal Soc. London A 337, (1974), 485-507.

[3] S. S. Antman, *Nonlinear problems of elasticity*, New York, Springer, (1995).

[4] M. B. Rubin, *Numerical solution procedures for nonlinear elastic rods using the theory of a Cosserat point,* Int. J. Solids Struct., 38, (2001), 4395-4437.

COMPUTATIONAL MECHANICS (Abstracts)
WCCM VI in conjunction with APCOM'04, *Sept. 5-10, 2004, Beijing, China*

Structural Influence on the Mechanical Properties of Single-Walled Carbon Nanotubes under the Tension Deformation

Xianggui Ni *, Yu Wang, Xiuxi Wang

CAS Key Laboratory of Mechanical Behavior and Design of Materials, University of Science and Technology of China, Hefei, 230026, China
e-mail: xianggui@ustc.edu.cn, wyu@ustc.edu, xxwang@ ustc.edu.cn

Abstract

The study on mechanical properties of single-walled carbon nanotubes(SWCNTs) is important to investigate the properties of multiwalled carbon nanotubes and single-walled carbon nanotube bundles.

A molecular dynamics simulation program WMD was developed. The atomic interaction was represented by the second-generation multi-body Tersoff-Brenner potential. The tension deformation behavior of SWCNTs was numerically studied. Through tracing the evolution of the configuration of SWCNTs, the deformation process and the systematic energy change of SWCNTs were observed. It is found that the fracture behavior at nanoscale is different from that at macro scale. The simulation results show that an armchair nanotube (12,12) can be stretched by 36% with no sign of plasticity, or atomic rearrangement. This peculiar behavior was explained from the viewpoint of energy change. The elastic properties of armchair and zigzag SWCNTs under tension were further investigated. The range of the Young's modulus is about 750GPa-960Gpa, which agrees well with the experiment results of 1.28±0.59TPa reported by Treacy et al. The results also show that the Young's modulus decreases as the radii increase, which is consistent with the Popov's illustration. Finally, the effect of the configuration helicity on the mechanical behavior of SWCNTs was analyzed. The Young's modulus of the zigzag SWCNTs is higher than that of the armchair SWCNTs. This mechanism was explained from the viewpoint of structure mechanics. The results suggest that the micro structural characteristic of the configuration of a SWCNT has a great influence on its mechanical properties. These excellent properties will have great advantages to the applications of carbon nanotubes as nanotube-reinforced composite materials.

REFERENCES

[1] D. W. Brenner, O. A. Shenderova, J. A. Harrison, et al, *A second-generation reactive empirical bond order(REBO) potential energy expression for hydrocarbons*, J. Phys: Condens. Matter., 14, (2002), 783-802.

[2] M. M. J. Treacy, T. W. Ebbesen, J. M. Gibson, *Exceptionally high Young's modulus observed for individual carbon nanotubes*, Nature, 381, (1996), 678-680.

[3] V. N. Popov, V. E. Van Doren, M. Balkanski, *Elastic properties of crystals of single-walled carbon nanotubes*, Solid State Comm., 114, (2000), 395-399.

[4] C. Y. Li, T. W. Chou, *A structural mechanics approach for the analysis of carbon nanotubes*, Int. J. Solids Struct., 40, (2003), 2487-2499.

COMPUTATIONAL MECHANICS
WCCM VI in conjunction with APCOM'04, *Sept. 5-10, 2004, Beijing, China*

A Numerical Study of Fracture and Defect Nucleation in Carbon Nanotubes

Dongli Shi [1*], Xiqiao Feng [1], Hanqing Jiang [2], Yonggang Huang [2], Keh-Chih Hwang [1]

[1] *Department of Engineering Mechanics, Tsinghua University, Beijing, 100084, China*
[2] *Department of Mechanical and Industrial Engineering, University of Illinois at Urbana-Champaign, Urbana, IL 61801, USA*

e-mail: sdongli00@mails.tsinghua.edu.cn, fengxq@mail.tsinghua.edu.cn, hqjiang@uiuc.edu, huang9@uiuc.edu, huangkz@mail.tsinghua.edu.cn

Abstract

Owing to their superior mechanical and physical properties, carbon nanotubes (CNTs) have many potential applications such as in nanoelectronics and nanocomposites. These applications require the fundamental understanding of fracture behavior of CNTs. We present here a hybrid atomistic/continuum method to simulate fracture and defect nucleation in CNTs under tension or torsion. The continuum method based on the interatomic potential is used for atoms far from the potential defect, while the discrete, atomistic method is used for atoms around the defect. The combined atomistic/continuum method shows that, once the tensile strain reaches a critical value, defects may nucleate in the CNT in the form of the Stone-Wales transformation (90 degree of a carbon bond). The critical strain for defect nucleation and the subsequent fracture modes of the CNT are sensitive to the chiral angle. The critical strain of a zigzag CNT is about twice of that of an armchair CNT, and both armchair and zigzag CNTs fracture in a brittle manner at the low temperature. Our numerical results of the fracture strain and tensile stress-strain curves agree well with experimental results.

Studies on deformation and fracture of CNTs in composites are difficult both in experimental observation and in theoretical modeling. Using the above method, we can also analyze the fracture of CNTs embedded in a matrix. The deformation of CNTs is partly constrained by the surrounding matrix in this carbon nanotube-reinforced composite. The interaction among CNTs is accounted for via the Mori-Tanaka method. Our results suggest that, due to the CNT-matrix interaction, the fracture strains of both armchair and zigzag CNTs in a composite become smaller than those of stand-alone CNTs.

COMPUTATIONAL MECHANICS (Abstracts)
WCCM VI in conjunction with APCOM'04, *Sept. 5-10, 2004, Beijing, China*

Ultrafast Dynamics in Single-Walled Carbon Nanotube/Polymer Composites

David Styers-Barnett, John M. Papanikolas [*]

Department of Chemistry, University of North Carolina at Chapel Hill, Chapel Hill, NC 27599, USA
e-mail: john_papanikolas@unc.edu

Abstract

We have used femtosecond transient absorption spectroscopy to investigate the electron-hole recombination dynamics in composite materials consisting of single-walled carbon nanotubes (SWNT) dispersed in a colorless polyimide matrix at a 0.1% nanotube weight fraction. The polymer materials were obtained through collaboration with a group at the National Institute of Aerospace [1]. Steady-state absorption spectra in the near-IR and visible region of the spectrum exhibit several broad features that arise from interband transitions between the van Hove singularities. Closer inspection of the broad features reveals that these absorptions are inhomogeneously broadened by a number of smaller features that may correspond to transitions in individual nanotubes having different diameters and chiral wrappings. The structured spectrum is superimposed on top of a broad background that has been assigned to the red tail of an extended π-plasmon resonance. Thus, photon absorption can result in two different types of excitation: (1) the collective excitation of electrons via the plasmon resonance, and (2) the promotion of an electron from a π orbital of the valence band to a π* orbital in the conduction band to produce an electron-hole (*e-h*) pair.

In our experiments, the SWNT is excited with a femtosecond laser pulse at 550 nm and probed throughout the visible a transient absorption configuration. Electron-hole pairs give rise to sharp features in the transient spectra. These features decay in amplitude and exhibit rapid spectral shifts that are largely complete within 10 ps. The decay is attributed to electron-hole recombination, which occurs on both short (1.3 ps) and long (35 ps) time scales. Spectral shifts have not been seen before and could reflect either intra-band carrier relaxation or an electronic/structural relaxation process that arises from an electrostatic interaction between the (*e-h*) pair and the surrounding polymer. The transient spectra also exhibit a broad photobleach that is observed only at early times (<500 fs). The decay of this photobleach is attributed to the cooling of a hot electron gas that is created via excitation at the red edge of a broad π-plasmon band in the metallic SWNTs.

Examination of a range of SWNT/polymer composites shows excited state spectra and decay kinetics that depend upon the polymer environment. Because the (*e-h*) pairs in SWNT reside in extended surface states, their dynamical behavior may be influenced by their surroundings to a significant degree. These initial experiments represent the first step towards elucidating the role that the electrostatic environment plays in shaping the (*e-h*) pair dynamics.

REFERENCES

[1] K. E. Wise, C. Park, E. J. Siochi, J. S. Harrison, *Stable dispersion of single wall carbon nanotubes in polyimide: the role of noncovalent interactions*, Chem. Phys. Lett., (submitted).

COMPUTATIONAL MECHANICS (Abstracts)
WCCM VI in conjunction with APCOM'04, *Sept. 5-10, 2004, Beijing, China*

Reciprocal Transformation of Order and Disorder Behavior in Simulation of Carbon Nanotubes

Yujun Zhang*, Yuanzhong Hu, Hui Wang

Department of Precision Instrument and Mechanology, Tsinghua University, State Key Laboratory of Tribology, Beijing, 100084, China
e-mail: zhangyujun@tsinghua.org.cn, {huyz, wangh}@pim.tsinghua.edu.cn

Abstract

A new phenomenon has been observed in this study, which exhibits the instability of single-walled carbon nanotube (SWCNT) in equilibrate-state computation. The interactions among disordered vibrational atoms near their equilibration positions will result in a collective wavelike ordered behavior along the axis of carbon nanotube. Furthermore, this transformation is reversible. The amplitude of wavelike vibration is dependent on the temperature and diameters of carbon nanotubes, which are monotone increasing with the temperature increase and monotone decreasing with diameter increase. The diameters of SWCNTs used are less than 10 nanometers and Tersoff-brenner potential is applied in molecular dynamics simulation.
FFT (Fast Fourier Transform) Algorithm is applied to analyze the phonon propagation in carbon nanotube and as a tool to interpret this phenomenon. Phonon emissions while the thermal vibration of carbon nanotube and their interactions will be a key factor resulting in the orderd behavior of collective motion. The frequencies of collective wavelike vibration are in T Hz quantity level by spectrum analysis, which are comparable to the frequencies of sound wave in SWCNT.

REFERENCES

[1] D. W. Brenner, *Empirical potential for hydrocarbons for use in simulating the chemical vapor deposition of diamond films*, Physical Review B, 42(15), (1990), 9458-9471.
[2] D. W. Brenner, O. A. Shenderova, J. A. Harrison, et al, *A second-generation reactive empirical bond order (REBO) potential energy expression for hydrocarbons*, J. Phys: Condens. Matter., 14, (2002), 783-802.
[3] R. Saito, G. Dresselhaus, M. S. Dresselhaus, *Physical properties of carbon nanotube*, Imperical College Press, (1998).
[4] P. K. Schelling, S. R. Phillpot, P. Keblinski, *Phonon wave-packet dynamics at semiconductor interfaces by molecular-dynamics simulation*, Applied Physics Letters, 80(14), (2002), 2484-2486.
[5] P. K. Schelling, S. R. Phillpot, *Multiscale simulation of phonon transport in superlattices*, Journal of Applied Physics, 93(9), (2003), 5377.
[6] Y. Kogure, *Simulation of phonon propagation in fine particales*, Physica B, 263-264, (1999), 482-485.

COMPUTATIONAL MECHANICS (Abstracts)
WCCM VI in conjunction with APCOM'04, *Sept. 5-10, 2004, Beijing, China*

Revisiting the Dirichlet-to-Neumann Solver for Data Completion and Application to Some Inverse Problems

M. Azaïez [1*], A. Ben Abda [2], J. Ben Abdallah [2,3], F. Ben Belgacem [4]

[1] *MASTER, Ecole Nationale Sup'erieure de Chimie et de Physique de Bordeaux. 16 avenue Pey Berland, 33607 Pessac Cedex, France*

[2] *LAMSIN, Ecole Nationale d'Ing'enieurs de Tunis, Campus Universitaire, BP 37, 1002 Tunis-Belv'ed`ere, Tunisia*

[3] *Institut Sup'erieur des Sciences Appliqu'ees et de Technologie de Sousse, Cit'e affala -Ibn Khaldoun, 4003 Sousse, Tunisia*

[4] *MIP (UMR CNRS 5640), Universit'e Paul Sabatier, 118 route de Narbonne, 31 062 Toulouse Cedex 04, France*

e-mail: azaiez@enscpb.fr, {amel.benabda, jalel.benabdallah}@enit.rnu.tn, belgacem@mip.upstlse.fr

Abstract

We are concerned in this work with a data completion numerical procedure proposed by Koslov et al. in [1]. Computationnal examples are presented and discussed to test the capability of this algorithm and to show how extending lacking data allows the detection of cracks.

For some models, the Poisson problem or the elasticity equation, etc, the Dirichlet-Neumann solver is an affordable tool to achieve numerical data completion. The problem of reconstructing missing data on some portions of the boundary of the geometry where take place the physical phenomena we are concerned with, arises in many industrial situations. It can also be a part of a larger process such as cracks and defects detection where some first indications are needed for their localization. These mandatory information can be obtained by some fast inverse formulae such as the violation of the Maxwell-Betti Reciprocity principle (see [2]). In this case, a performing solver may be a complementary stencil to a full reconstruction of the cracks. As we have already mentioned above although the study has been carried out in 2D situations and in the case of the steady state heat conduction equation there is, conceptually, absolutely no obstacle to developing the same approach for other operators, in both 2D and 3D situations. What is practically needed is only a fast solver for the forward problem.

REFERENCES

[1] V. A. Koslov, V. G. Maz'ya, A. V. Fomin, *An iterative method for solving the Cauchy problem for elliptic equations*, Comput. Meth. Math. Phys., 31, (1991), 45-52.

[2] S. Andrieux, A. Ben Abda, *Identification of planar cracks by complete over-determined data: inversion formulae*, Inverse Problems, 12, (1996), 553-563.

COMPUTATIONAL MECHANICS (Abstracts)
WCCM VI in conjunction with APCOM'04, *Sept. 5-10, 2004, Beijing, China*

Topological Shape Optimization of Geometrically Nonlinear Structures Using Level Set Method

Seonho Cho*, Juho Kwak

Department of Naval Architecture and Ocean Engineering, Seoul National University, San 56-1 Sillim-Dong Kwanak-Gu, Seoul, 151-742, South Korea
e-mail: secho@snu.ac.kr

Abstract

Based on a level set method, we present a topology optimizati on method, for geometrically nonlinear structures, which meets a minimum compliance requirement under an allowable volume constraint. In this method, the whole structural domain is fixed and its boundary is implicitly represented by a moving boundary embedded in a level set function, which enables to handle complicated topological changes [1]. The "up-wind scheme" is employed to solve the "Hamilton–Jacobi" equation for the level sets [2]. The subsequent motion of boundaries Γ under a given velocity v can be tracked to determine the topological shape changes of structure in terms of a smooth function φ at zero level set. Total Lagrangian formulation is employed to describe the response of geometrically nonlinear structures. To solve the nonlinear problems, a Newton-Raphson method is employed. The convergence difficulty arising in the application of topology optimization method to geometrically nonlinear problems is actively discussed by many researchers. The difficulty in response analysis arises due to relatively sparse material distribution in isolated spots during the optimization process [3]. However, in the level set approach, the difficulty is greatly relieved by using the homogeneous material in the design domain. An adjoint variable method is employed for the efficient computation of the design sensitivity. The converged tangent stiffness, obtained from Newton-Raphson iterations at the final equilibrium configuration during the nonlinear response analysis, is directly used in solving the linear adjoint equation. The optimization process is to move the implicit moving boundary, according to the compliance sensitivity with respect to the boundary variations, while the allowable volume is kept constant. It is crucial to find an appropriate "velocity vector" such that it will iteratively lead the design boundary to the optimum shape [4]. The process would terminate if no further improvement in compliance is observed. Through several numerical examples, the applicability of the developed method and the necessity of nonlinear formulation for the topology optimization problems are demonstrated.

REFERENCES

[1] G. Allaire, F. Jouve, A. M. Toader, *A level-set method for shape optimization,* Numerical Analysis, 334, (2002), 1125-1130.

[2] S. Osher, R. Fedkiw, *Level set methods: An Overview and some recent results,* Journal of Computational Physics, 169, (2001), 475-502.

[3] S. Cho, H. Jung, *Design sensitivity analysis and topology optimization of displacement-loaded nonlinear structures*, Computer Methods in Applied Mechanics and Engineering, 192, (2003), 2539-2553.

[4] M. Y. Wang, X. Wang, D. Guo, *A level set method for structural topology optimization,* Computer Methods in Applied Mechanics and Engineering, 192, (2003), 227-246.

COMPUTATIONAL MECHANICS (Abstracts)
WCCM VI in conjunction with APCOM'04, *Sept. 5-10, 2004, Beijing, China*

Study on Optimization of the Cable-Net Deployed Satellite Antenna Structure

B. Y. Duan*, Y. Luo

School of Electromechanical Engineering, Xidian University, Xi 'an, China
e-mail: byduan@xiaidn.edu.cn

Abstract

According to the specific characteristics of large satellite antenna with flexible cable-net structures (shown in figure 1), a systematic study has been carried out. In the figure, there are two cable nets, the upper and bottom ones. The upper one is called the reflector-supported cable net, under which a refined and much flexible net called reflector net, which serves as the real antenna reflector, is mounted. The bottom cable- net is called the auxiliary towed cable net. Between the upper and bottom cable-nets there is a group of adjustable cables. By changing the length of adjustable cables, the reflector precision could be satisfied. To support the whole cable structure, a surrounded truss is utilized as shown in Figure 1.

To begin with, the characteristics of this kind of antenna are discussed and its difference from the other is shown by comparing it with ordinary earth antenna and non cable-net structure antenna in the orbit of space.

Secondly, the nonlinear cable structure analysis is experienced. A methodology of searching for its equilibrium state is developed.

Thirdly, the optimization with both deployed and folded states is developed. The optimization model with attempting to find minimum weight design whilst satisfying the reflector 's precision, eigenfrequency, material strength, side and the folded volume constraints, is given.

COMPUTATIONAL MECHANICS (Abstracts)
WCCM VI in conjunction with APCOM'04, *Sept. 5-10, 2004, Beijing, China*

Topological Shape Optimization of Heat Conduction Problems Using Level Set Approach

Seung-Hyun Ha *, Seonho Cho

Department of Naval Architecture and Ocean Engineering, Seoul National University, San 56-1 Sillim-Dong Kwanak-Gu, Seoul, 151-742, South Korea
e-mail: secho@snu.ac.kr

Abstract

We address the topology optimization method, for heat conduc tion problems using a level set method, which satisfies a minimum thermal compliance and an allowable volume constraint. In this method, the whole domain is kept fixed and its boundary is implicitly represented by a moving boundary embedded in the level set fun ction, which enables to handle complicated topological changes [1]. An up-wind scheme is employed to solve the "Hamilton –Jacobi" type equation for the level sets. The subsequent motion of boundaries Γ under a given velocity v can be tracked to determine the topological shape changes of the thermal domain in terms of a smooth function φ at the initial level set [2]. The response analysis to evaluate the o bjective function and constraint is carried out using the finite element method in the fixed domain with implicit moving boundary. Since the topology optimization necessarily involves many design variables, i.e. discrete level set function, gradient-based optimization methods are generally preferred. Therefore, the sensitivity of thermal compliance with respect to the boundary variations should be determined in a very efficient way. The efficient adjoint design sensitivity analysis (DSA) method is employe d for the heat conduction problems in steady state [3]. Using the developed DSA method, an automated topological shape optimization process is established. The objective function and constraint are the thermal compliance and allowable volume, respectively. The optimization process is to move the implicit moving boundary, according to the thermal compliance sensitivity with respect to the boundary variations, while the allowable volume is kept constant. It is crucial to find an appropriate "velocity vector" such that it will iteratively lead the design boundary to the optimum shape [4]. The process would terminate if no further improvement in the thermal compliance is obtained. Through several numerical examples, the topology design optimization is performed to demonstrate the applicability of the developed method. Also, the topology optimization yields physically meaningful results and could result in different topologies depending on the ratio of flux and internal heat generation rate.

REFERENCES

[1] G. Allaire, F. Jouve, A. M. Toader, *A level-set method for shape optimization,* Numerical Analysis, 334, (2002), 1125-1130.
[2] S. Osher, R. Fedkiw, *Level set methods: an Overview and some recent results,* Journal of Computational Physics, 169, (2001), 475-502.
[3] M. Y. Wang, X. Wang, D. Guo, *A level set method for structural topology optimization,* Computer Methods in Applied Mechanics and Engineering, 192, (2003), 227-246.
[4] A. Sluzalec and M. Kleiber, *Shape sensitivity analysis for nonlinear steady-state heat conduction problems,* International Journal of heat and mass transfer, 39, (1996), 2609-2613.

COMPUTATIONAL MECHANICS (Abstracts)
WCCM VI in conjunction with APCOM'04, *Sept. 5-10, 2004, Beijing, China*

An Inverse Analysis Algorithm of Material Parameters in Two-Dimensional Orthotropic Media

Lixin Huang, Xiushan Sun, Yinghua Liu, Zhangzhi Cen *

Department of Engineering Mechanics, Tsinghua University, Beijing, 100084, China
e-mail: {hlx01, sunxs99}@mails.tsinghua.edu.cn, {yhliu, demczz}@mail.tsinghua.edu.cn

Abstract

An inverse analysis algorithm is developed to identify the material parameters in two-dimensional orthotropic media. The inverse analysis problem of the material parameters is formulated as the problem of minimizing the cost function defined as a square sum of differences between the measured displacement and the computed displacement by a boundary element model. Levenberg-Marquardt method is used to solve the minimization problem. The sensitivities of displacements with respect to the material parameters in the minimization of the cost function are computed by the finite difference approximation method and the direct differentiation method respectively. The direct differentiation method is based on the direct differentiation of the discretized algebraic boundary element matrix equations with respect to the material parameters, in which the derivatives of fundamental solutions with respect to the material parameters are derived.
Numerical example is given to demonstrate the effectiveness of the proposed algorithm. Some computational issues, such as accuracy and computational cost, are discussed through the comparison between the numerical results obtained by the finite difference approximation method and the direct differentiation method. It can be found that both the finite difference approximation method and the direct differentiation method have high accuracy and stable convergence. Although the direct differentiation method is much more complicated to use than the finite difference approximation method due to the complicated expressions of the derivatives of fundamental solutions with respect to the material parameters and to the complicated implementation of the proposed algorithm, it has less computational cost, especially when the discretization with more boundary elements and nodes is used to analyze the direct problem. The direct differentiation method is a cheaper way of computing the sensitivities in the boundary element inverse analysis of material parameters for two-dimensional orthotropic media.

REFERENCES

[1] M. Tanaka, *Some advanced applications of the boundary element method to inverse problem in engineering*, European Conference on Computational Mechanics, Cracow, Poland, June 26 -29, (2001).

[2] L. X. Huang, X. S. Sun, Y. H. Liu, Z. Z. Cen, *Parameter identification for two-dimensional orthotropic material bodies by the boundary element method*, Engng. Anal. Bound. Elem., 28, (2004), 109-121.

[3] S. M. Chern, M. E. Tuttle, *On displacement fields in orthotropic laminates containing an elliptical hole*, J. Appl. Mech., 67, (2000), 527-539.

COMPUTATIONAL MECHANICS (Abstracts)
WCCM VI in conjunction with APCOM'04, *Sept. 5-10, 2004, Beijing, China*

A Computational Procedure for the Optimal Design of Large Structural Reliability Models

Hector A. Jensen *

Department of Civil Engineering, Federico Santa Maria University, Av. Los Placeres 401, Valparaiso, Casilla 100-V, Chile
e-mail: hector.jensen@usm.cl

Abstract

The reliability-based design of complex structural systems usually involves high-dimensional uncertain parameter spaces and complex failure regions. The numerical implementation of this type of problems is a challenging problem due to its complexity. At the same time the ability to perform reliability-based design of complex systems is quite compelling. In this work, the optimal design of structural system with reliability models of high dimension is considered. In particular, attention is directed toward the optimization of structures modeled as a series system of a large number of failure elements. The optimal design process is written as a two-level nonlinear optimization problem. The top-level includes the overall optimization in the design variables, whereas the sublevel problem corresponds to the failure probability estimates. The problem of calculating the system failure probability is first transformed to calculating the failure probability of the failure elements. The interaction of the failure elements is then considered by an efficient importance sampling technique that is integrated into the optimization process. The original implicit optimization problem is replaced by a sequence of approximate explicit sub-optimization problems. An explicit approximation of the system reliability in terms of the design variables is constructed in each sub-optimization problem. Each approximate problem is solved in an efficient manner due to the availability of the system reliability in explicit form. Several issues regarding the numerical implementation of the methodology are addressed. Finally, numerical examples are presented to demonstrate the efficiency of the proposed methodology.

REFERENCES

[1] I. Enevoldsen, J. D. Sorensen, *Reliability-based optimization in structural engineering*, Structural Safety, 15(2), (1994), 169-196.

[2] H. A. Jensen, A. E. Sepulveda, *Optimal design of uncertain systems under stochastic excitation*, AIAA Journal, 38(11), (2000), 2133-2141.

[3] S. K. Au, J. L. Beck, *First excursion probabilities for linear systems by very efficient importance sampling*, Probabilistic Engineering Mechanics, 16(3), (2001), 193-207.

COMPUTATIONAL MECHANICS (Abstracts)
WCCM VI in conjunction with APCOM'04, *Sept. 5-10, 2004, Beijing, China*

Sensitivity Analysis of Viscoplastic Deformation Process with Applications to Metal Forming Optimization

Zhan Kang [*], Yuanxian Gu

State Key Laboratory of Structural Analysis for Industrial Equipment, Dalian University of Technology, Dalian, 116024, China
e-mail: zhankang@dlut.edu.cn, guyx96@dlut.edu.cn

Abstract

Sensitivity analysis is crucial when gradient based optimization algorithms are used for solving structural and multi-disciplinary optimization problems. In this paper, the sensitivity analysis of rigid viscoplastic deformation process is addressed. For viscoplastic constitutive models, the deformation process is history-dependent in nature and thus the need for an incremental procedure for response sensitivity analysis is highlighted. The presented algorithm is derived on the basis of discrete finite element formulation of the direct analysis, where a time integration is used for the deformation history analysis and the contact conditions are treated with the penalty method. The equilibrium governing equations of the finite element analysis are directly differentiated with respect to the design variables, as well as the time integration scheme. The discrete form of the sensitivity equations are then solved with the similar procedures as in the direct analysis, where the tangent matrix decomposed in the direct analysis can be also utilized at each time instant. Thus the sensitivity of the deformation history is evaluated in a step-by-step procedure.
The presented algorithm is employed for optimization of the metal forming process. Numerical examples are given to demonstrate the applicability of the presented sensitivity analysis in this type of problems.
The computational efficiency of the presented algorithms is shown to be superior to the finite difference method and the accuracy of the numerical results obtained is compared with that of the finite difference method.

REFERENCES

[1] I. Doltsinis, *Inelastic deformation processes with random parameters – methods of analysis and design*, Comput. Methods Appl. Mech. Engrg., 192, (2003), 2405-2423.
[2] A. Srikanth, N. Zabaras, *An updated Lagrangian finite element sensitivity analysis of large deformations using quadrilateral elements*, Int. J. Num. Method. Eng., 52, (2001), 1131-1163.

COMPUTATIONAL MECHANICS (Abstracts)
WCCM VI in conjunction with APCOM'04, *Sept. 5-10, 2004, Beijing, China*

Structural Optimization of Biological Fiber-Reinforced Composites

Natalya N. Kizilova *

Department of Theoretical Mechanics, Kharkov National University, Svobody sq.,4, Kharkov, 61195, Ukraine
e-mail: nnk_@bk.ru

Abstract

Main part of biological tissues exhibit unique mechanical properties and can be considered as fiber-reinforced composite materials. Cartilage rings in trachea, trabecular system of compact bone, multilayer organization of articular cartilage are the most investigated structures of this kind. Plant leaves represent the excellent examples of the growing composite material with a network of fibers embedded in it. The fibers are viscoelastic thin tubes that supply the live cells of the leaf with a liquid and provide mechanical strength and stability of the leaf blade. Branching patterns of the network possess certain geometrical properties that correspond to a few optimal criteria [1]. Some leaf shapes and their vein structures have been already used in architecture (Victoria Regia). The principles and mechanisms of the optimal reinforced structure formation during the growth and development of the leaf are still unknown.

Here some results of the computer simulation of development of the growing biological material with a set of flexible growing fibers which are embedded in it and arranged in radial directions are presented. Two sorts of biological materials (the main tissue and the fibers) are considered as viscoelastic Maxwell liquids with different zero-stress state growths. No-slip conditions at the boundaries are specified. Variation principle for the growing material is formulated and the FEM method for numerical calculations is elaborated. The method is applied to the model of the leaf as a round plate with radial first-order veins (stringers). The cases of the unloaded perimeter of the plate and the zero-displacement conditions at the perimeter that correspond accordingly to the natural growth and to the experiments with perimeter-bounded leaf growth [2] are considered. The structure of the second-order small veins is modeled by the anisotropy of the main tissue of the plate. At the unloaded perimeter the zero-stress growth over the typical growth time t~1-2 days is achieved. Local discrepancies in the growth rates lead to appearance of small stresses that control the growth directions (i.e. keep the angles between the fibers within given ranges). At the perimeter-bounded growth the bulging of the plate is simulated that correspond to the experimental results. When fibers follow the principals of the stress tensor the optimal strength of the blade is achieved during its growth. Possible applications of the optimization principle to artificial composites are discussed.

REFERENCES

[1] N. N. Kizilova, *Transport system and growth of leaves*, in *Mechanics of growth and morphogenesis*, Moscow Univ. Press, Moscow, (2002), pp. 379-405.

[2] K. H. G. Ashbee, *Fundamental principles of fiber reinforced composites*, Technomic, Lancaster, (1993).

[3] N. N. Kizilova, *Analysis of stress distribution and leaf blade bending during bonded growth*, in *Summer Bioengineering Conference*, Royal Sonesta Resort, Key Biscayne, (2003).

COMPUTATIONAL MECHANICS (Abstracts)
WCCM VI in conjunction with APCOM'04, *Sept. 5-10, 2004, Beijing, China*

A Non-Gradient Approach to Topological Design Involving Impact

Qing Li *

School of Engineering, James Cook University, Townsville, QLD 4811, Australia
College of Mechanical & Automotive Engineering, Hunan University, Changsha, 410082, China
e-mail: Qing.Li@jcu.edu.au

Abstract

Over the past decade remarkable progress has been made in the development of software used in crash simulation and analysis. A number of commercial codes have been available with accepted accuracy. However, there has been little report in its inverse problem yet, i.e. topological and shape optimization for crashworthiness. The difficulty is primarily raised from two folders on 1) the nonlinear sensitivity analysis and computational cost [1], which become particularly challenging in various topological designs; 2) dynamic multi-modals and non-convex design space, which do not lend the crash problems themselves well to classical gradient techniques. The paper aims at developing an alternative approach to the crashworthiness design problems. The Evolutionary Structural Optimization (ESO) [2] will be applied to reduce the dependence on the continuum sensitivities.

In a crash simulation, it is frequently found that plastic deformation in some locations may be much higher than others. This implies that the material of the crashing elements may have different contributions to the crashworthiness goal. Ideally, the energy absorption levels in all location are near identical. To represent the relative performance of element's material, a dimensionless factor is formulated by dividing the crash energy absorption by each element to the highest one as,

$\alpha = U_e / U_{max}$

where $U_e = \int_{t_f} \int_{\Omega_e} \frac{1}{2} \sigma \varepsilon dx dt = \int_{t_f} \int_{\Omega_e} \frac{1}{2} E_{ijkl} \frac{\partial u_i}{\partial x_j} \frac{\partial u_k}{\partial x_l} dx dt$ computes the strain energy absorbed by the candidate element (*e*th) and t_f represents the final time of the crash process. It is worth pointing out that the total internal energy should contain the elastic and plastic components, i.e. $U = U^e + U^p$. Eq. (1) gives an indication to gradually shift material of the design domain from a lower energy absorption location to a higher one. As a consequence of this process, the resulting structural performance becomes more and more uniform in terms of the crashworthiness criterion.

In general, standard FE codes directly or indirectly provide the outputs for computing the internal energy of each element, which facilitates the use of any commercial package of crashworthiness analysis in industry. Therefore, the ESO method presented herein can readily be implemented in a variety of existing well-established analysis platforms. The examples of both topology and shape designs are also shown in this study to demonstrate the capabilities of the present method.

REFERENCES

[1] N. Zabaras, S. Ganapathysubramanian, Q. Li, *A continuum sensitivity method for the design of multi-stage metal forming processes*, Int. J. Mech. Sci., 45, (2003), 325-358.

[2] Y. M. Xie, G. P. Steven, *Evolutionary structural optimization*, Springer-Verlag, Berlin, (1997).

COMPUTATIONAL MECHANICS (Abstracts)
WCCM VI in conjunction with APCOM'04, *Sept. 5-10, 2004, Beijing, China*

Identification of Compressive Behavior of Aluminum Foams Using Genetic Algorithm

G. W. Ma*, C. Wang, C. K. Soh

School of Civil and Environmental Engineering, Nanyang Technological University, Nanyang Avenue, Singapore, 639798, Singapore
e-mail: cgwma@ntu.edu.sg, wangchaontu@pmail.ntu.edu.sg, csohck@ntu.edu.sg

Abstract

Split Hopkinson Pressure Bar (SHPB) has been recognized as one of the most effective methods for testing the stress-strain relation of a wide range of materials under high strain-rate condition. But the traditional SHPB technique fails in the study of the dynamic behavior of polymeric and metallic foams because the desired maximum strain of these materials is often up to 80% in the densification region.

When using modified SHPB setup with short pressure bars, the desired maximum strain for aluminum foams can be successfully obtained; however, the traditional data processing technique based on one-dimensional wave propagation theory becomes invalid. Genetic algorithm (GA), which is one of the more popular evolutionary algorithms, is superior in parameter identification and optimization due to its practicality and efficiency. For identifying the whole compressive behavior of aluminum foams, only some critical parameters are needed. In the present study, a genetic algorithm is developed for a modified SHPB test system to identify the compressive behavior of aluminum foams.

The illustrative example shows that the genetic algorithm is capable of identifying the whole compressive stress-strain curve of aluminum foams based on the modified SHPB. The outstanding advantages of this new method are that only one strain gauge is required and this gauge can be placed at any location on the pressure bars. In addition, this promising method can be easily extended to identify the dynamic behavior of other soft materials, for which the traditional SHPB test method is not effective.

REFERENCES

[1] R. M. Davies, *A critical study of Hopkinson pressure bar*, Philos. Trans. R. Soc. A, 240, (1948), 375-457.

[2] D. E. Goldberg, *Genetic algorithms in search, optimization, and machine learning*, Addison-Wesley, Reading, Massachusetts, (1989).

[3] H. Zhao, G. Gary, *A new method for the separation of waves. Application to the SHPB technique for an unlimited duration of measurement*, J. Mech. Phys. Solids, 45(7), (1997), 1185-1202.

[4] H. Zhao, *Material behaviour characterization using SHPB techniques, tests and simulations*, Computers & Structures, 81, (2003), 1301-1310.

COMPUTATIONAL MECHANICS (Abstracts)
WCCM VI in conjunction with APCOM'04, *Sept. 5-10, 2004, Beijing, China*

A Preliminary Investigation of the Application of Inverse Analysis to Ultrasonic Inspection of Structural Assemblies Containing Corrosion

W. Shah, T. Furukawa*, D. Kelly

Department of Mechanical Engineering, University of New South Wales, UNSW (Kensington), Sydney NSW 2052, Australia
e-mail: {wasim, t.furukawa, d.Kelly}@unsw.edu.au

Abstract

NDT techniques, such as ultra-sonic waveforms (UT) have been used successfully for the detection of cracks and damage assessment in hidden and buried areas of aircraft for a number of years including structural assemblies and fastener holes. A signal can be input into a component in an assembly, and propagates into the assembly before returning a reflection from a defect such as a crack. The ability to detect damage without disassembly is important to reduce the cost of inspection and maintenance.

The simultaneous use of this method to detect cracking & corrosion is more difficult as corrosion dissipates the signal and is detected by the loss of an image rather than a positive return. This can make the interpretation of signals and assessment of the reliability of the method particularly difficult.

This paper discusses the application of an inverse analysis using finite elements to enhance the assessment of ultra-sonic signals in inspection of these structural assemblies. The main challenges have been identified as: 1) the capture and analysis of the signal from real ultra-sonic output and 2) the use of inverse computation (with finite element analysis of propagation of ultra-sonic signals) to mathematically analyse the problem. The inverse problem is to define a flaw or area of corrosion that has led to the return of a real signal. This is achieved by comparing the output from a F.E.A with proposed damage to the real output, and modifying the damage in the finite element model to match the real signal.

The work here will focus on the interaction of reflected signals from cracks and the dissipated signals that warn of corrosion. Initial results from tests on a laboratory specimen will indicate the problem that is being addressed. The specimen represents a stiffener riveted onto the web of a helicopter engine support beam with a doubler plate covering the web from the other side [1]. The procedure for the inverse analysis presented by Furkarawa will then be reviewed and a strategy for application to the problem presented here will be developed. Procedures for finite element analysis of ultrasonic waveforms are discussed in [2]. The ability of the FE method to represent the signal from the inter-granular damage typical of corroded metals is the issue.

REFERENCES

[1] T. Chan, *An evaluation of NDT procedures for detecting cracking and corrosion within assembled structures*, Bachelor of Engineering Thesis, School of Mechanical and Manufacturing Engineering, University of New South Wales, (2003).

[2] M. Garton, *Refining automated ultrasonic inspections with simulation models*. The e-Journal of Nondestructive Testing, 2(10), (1997), http://www.ndt.net/article/garton/garton.htm. (26 August 2003).

[3] Shiro Kubo, *Computational inversion schemes for various categories of inverse problems*. International Symposium on Inverse Problems in Engineering Sciences, (IPES-94), July 27-30, (1994).

[4] P. H. Johnston, *Smart ultrasonic system for aircraft NDE (SUSAN)*, NESB. Internet, http://nesb.larc.nasa.gov/NESB/ndetasks/2000/susan.html. (19 February 2004).

COMPUTATIONAL MECHANICS (Abstracts)
WCCM VI in conjunction with APCOM'04, *Sept. 5-10, 2004, Beijing, China*

A Graph Theoretical Approach to Region Identification during Continuum Topology Optimization

Mohsen Shahroozi *

Department of Structural Engineering, International Institute of Earthquake Engineering and Seismology, No.26 Arghavan, North Dibajee, Farmanieh, Tehran, Tel.:+98 21 2830830 (Ext. 143) P.O.Box:19395/3913, I. R. Iran
E-mail: moshahruzi@noavar.com

Abstract

The process of well-known continuum topology optimization methods is based on defining an initial homogenized media covering the whole design domain and then altering its material distribution (subject to total mass constraint) in order to achieve the optimal layout. Consequently, the material density of some elements is forced to approach zero that is, they are eliminated in the final design. As a result, some discontinuous regions may arise in the design domain specially while interpreting mass-redistributed media to practical design with smooth boundary.

A review of recent interpretation methods is presented by Bulman et al [1]. In addition, the work of Hsu et al [2], which based on exploring density contours, these contours are extracted from the result of material density method for topology optimization. They also offered a procedure to distinguish region discontinuities based on computing the area enclosed by density contours.

It is a necessity to overcome such irregularities in order to achieve the correct optimal topology. This paper represents efficient algorithms based on graph theoretical approach regarding density priorities which can well recognize and handle stable continuous solid regions in pre-optimized media. However, other types of priorities can be treated in similar way depending on the optimization technique.

Compared with some other methods in literature, these graph theoretical algorithms can be simply extended to 3-dimensional space taking less computational effort. An illustrative example is treated to show the efficiency and application of the developed method.

REFERENCES

[1] S. Bulman, J. Sienz, E. Hinton, *Comparison between algorithms for structural topology optimization using a series of benchmark studies,* Comp. & Struct., 79, (2001), 1203-1218.

[2] Y. Hsu, M. Hsu, C. Chen, *Interpreting results from topology optimization using density contours,* Comp. & Struct., 79, (2001), 1049-1058.

COMPUTATIONAL MECHANICS (Abstracts)
WCCM VI in conjunction with APCOM'04, *Sept. 5-10, 2004, Beijing, China*

A Method to Determine Sub-Grade Reaction Coefficients of Piles

Zhitong Song [1, 2*], Ziming Zhang [1], Guofang Zheng [2]

[1] *Department of Engineering Mechanics, Hohai University, Nanjing, 210098, China*
[2] *Nanjing Hydraulic Research Institute, Nanjing, China, 210024, China*
e-mail: ziming58@163.com , yearover@163.com, gfzheng@a.njhri.edu.cn

Abstract

This paper assumes that the sub-grade reaction coefficients of laterally loaded piles are $K = c + my^n$, and internal forces and displacements of piles are calculated by use of Initial Parameter Method. The sum of the square of the difference of moments or displacements respectively between tested and theoretically computed results is taken as an objective function, and the minimum of the objective function is computed by Variable Polyhedron Method to determine foundational reaction forces. In the end, two examples computed by this method are compared with those computed by different methods. It proves that this method is accurate. No matter what lateral reaction coefficients are used, this method is also simple and effective for piles in multi-layer soils. The following conclusions are drawn: 1) the maximum relative error of displacements between tested and calculated results is about 2.7 percent, and maximum relative error of moments is about 2 percent. The fitting can satisfy engineering requirement, which shows that this method is effective and accurate, 2) Because of rapidity of the Initial Parameter Method and steadiness of this back analysis method, the sub-grade reaction coefficients for laterally loaded piles are calculated by using microcomputer. 3) Steadiness of The Variable Polyhedron Method is better. It may save time and do not need continuity of the objective function. Hence, it can be widely applied in engineering, 4) In the objective function f(c, m, n), magnitude of weight coefficient may be determined by specific requirement. If higher iterative accuracy of displacements is needed W_j is bigger; if higher accuracy of moments is needed, W_j is smaller.

REFERENCES

[1] G. F. Zhen. *Computing internal forces and displacements of pile by initial parameter method*, Ocean Engineering, (3). (1986), (in Chinese).

[2] J. Z. Chen et al, *Study on pile in sand under lateral loading by model testing*, NJHRI, (2), (1983), (in Chinese).

[3] R. G. Shen, *Analysis of finite element method under laterally loaded pile*, NJHRI, (2), (1983).

COMPUTATIONAL MECHANICS (Abstracts)
WCCM VI in conjunction with APCOM'04, *Sept. 5-10, 2004, Beijing, China*

Maximum Entropy Based Structure Two-Stage Optimum Design with Discrete Variables

Haiwen Teng [1, 2*], Da Huo [2]

[1] *Department of Civil Engineering, Tsinghua University, Beijing, 100084, China*
[2] *College of Architecture and Civil Engineering, Beijing University of Technology, Beijing, 100022 China*
e-mail: tenghaiwen@bjut.edu.cn, huoda@bjut.edu.cn

Abstract

With the transfer of optimization technique to practical engineering, research of structural optimum design with discrete variables catch people's attention [1]. Because many algorithms are difficult even unable to solve the problem of structural optimization design with discrete variables, based on the thinks of two-stage optimization and maximum entropy[2], a new method which can easily solve structure optimum design with discrete variables is present in this paper.
The step of the method is firstly, built mathematics model of structure optimum design according to load condition, structure material. Multiple loading cases and multiple constraints including stress constraints, displacement constraints and dimension constraints are all considered. Secondly, simplify the mathematics model. According to the function relationship of the variables, many design variables can be expressed by one design variable, which can simplify the constraint condition. Use Boolean variables express the design variable. Then simplify many inequality constraints with one inequality constraint by introduction of non-negative multipliers and surrogate constraints. Thirdly, uses Lagrange method to solve the simplified model and use maximum entropy principle optimize the multipliers.
Through introducing Boolean variables and additional constraint condition, discrete variables for structures optimization design problem is equivalently convert to continuous variables problem, which can be resolved by carrying out multiple-stage and structure-stage optimization process alternatively. Under this condition, the discrete variables structures optimization design is successfully designed and optimized. Examples show that this algorithm is concise and the iteration is converged rapidly. It can be applied to practical engineering efficiently, especially to the structure optimization of multiple loading cases and multiple constraints.

REFERENCES

[1] A. B. Templeman, X. S. Li, *A maximum entropy approach to constrained non-linear programming*, Engineering Optimization, 12, (1987), 36-46.

[2] X. S. Li, *A maximum entropy method for structural optimization*, Computational Structural Mechanics and Applications, 6, (1989), 191-205, (in Chinese).

COMPUTATIONAL MECHANICS (Abstracts)
WCCM VI in conjunction with APCOM'04, *Sept. 5-10, 2004, Beijing, China*

The Concurrent Design for the Optimum Heat Dissipation Efficiency of Cellular Materials

Bo Wang *, G. D. Cheng

State Key Laboratory of Structural Analysis for Industrial Equipments, Dalian University of Technology, Dalian, 116023, China
e-mail: wangbo@student.dlut.edu.cn, chenggd@dlut.edu.cn

Abstract

It is the mankind's desires that materials would be utilized as efficiently as possibly. Normally, the so-called efficiency is referred to as light-efficiency. Two groups of researchers, material and mechanical, attain the aim of the game in the two different pathways. In typical material design processes, a discretional local of structure is investigated based on the assumption that every part of the structure is the same like the local, thus, the macro-variation of the material distribution is put aside. Over the last decade, the researchers for topology optimization of two-dimension continuum structures have been mainly occupied with two different kinds of topology optimization formulations: the Material or Microstructure Technique and the Geometrical or Macrostructure Technique. The former is a material design because its design domain is a discretional cell just like the material researchers do, whereas the latter is a structural design, for the design domain is the structure, which would reduce the weight of structure obviously. It is well known that all the formulations of topology optimization technique are based on the assumption of the material models, such as Hole-in-cell microstructure model and SIMP model. Moreover, the interpretation of intermediate "grey" density of material is often questioned for such models, although Bendsøe and Sigmund can interpret black-white designs in many cases via micro mechanical model [1].After all, any type of present topology optimization technique does not apply physical material models, and too refined micro mechanical models are unpractical for manufacture. In this paper, a true cellular material model is applied in the formulation of optimum design for efficiency of heat dissipation. Since 1980, the cellular material, as a kind of lightweight materials, has been studied for the combinations of mechanical, thermal, acoustic and impact-absorb properties by more and more researchers. T. J. Lu investigated a sandwich panels with two-dimensional metal cores that be used to carry structural load as well as dissipate heat trough solid conduction and forced convection [2]. As the conclusion, regular hexagonal cores are found to provide the highest levels of heat dissipation. But it is impossible to obtain the optimal design by T. J. Lu's method because the method is a material design mentioned above. In this work we employ numerical techniques for heat dissipation optimization design for two-dimensional cellular structures, which is a concurrent design of materials and structures.

REFERENCES

[1] M. P. Bendsøe, O. Sigmund, *Material interpolation schemes in topology optimization*, Arch. Appl. Mech., 69, (1999), 635-645.

[2] S. Gu, T. J. Lu, A. G. Evans, *On the design of two-dimensional cellular metals for combined heat dissipation and structural load capacity*, Inter. J. of Heat and Mass Transfer, 44, (2001), 2163-2175.

COMPUTATIONAL MECHANICS (Abstracts)
WCCM VI in conjunction with APCOM'04, *Sept. 5-10, 2004, Beijing, China*

On the Estimation of Material Parameters for Internal Variable Thermoelastic-Viscoplastic Constitutive Models: Gradient-Based and GA Optimization Algorithms

A. Andrade-Campos[1*], S. Thuillier[2], P. Pilvin[2], F. Teixeira-Dias[1]

[1] *Departamento de Engenharia Mecânica, Universidade de Aveiro, Campus Universitário de Santiago, 3810-193 Aveiro, Portugal*

[2] *Laboratoire de Génie Mécanique et Matériaux (LG2M), Université de Bretagne-Sud, Rue de Saint Maudé, BP 92116-56321 Lorient cedex, France*

e-mail: gilac@mec.ua.pt, sandrine.thuillier@univ-ubs.fr, philippe.pilvin@univ-ubs.fr, ftd@mec.ua.pt

Abstract

Estimating material parameters for constitutive models, made in such a way as to describe with precision the behaviour of the material, is often a common need for the engineering community. The complexity of this process increases with the complexity of the material model itself. The work presented in this paper deals with the determination of material parameters used in internal variable constitutive models. In order to determine the best-suited material parameter set, in the less computationally expensive way, two optimization approaches are used: 1) a gradient-based method and 2) a continuous evolutionary genetic algorithm (GA) method. The first approach uses a combination of the steepest descent gradient and the Levenberg-Marquardt techniques. The performance of this method is known to be highly dependent on the starting set of parameters and its results are often inconsistent. The GA-based multiple-objective technique provides a better way to determine an optimized set of parameters (the overall minimum). The difficulty of choosing a starting set of parameters for this process does not exist. The main application, in this work, is with a 12 parameter thermoelastic-viscoplastic constitutive model. Experimental data was obtained from tensile and shear tests at different temperatures and used to compare numerical results and determine the correct set of material parameters. Both methods are used to determine the 12 material parameters needed for an AA1050-O aluminium alloy. Although the GA-based method achieved a slightly better result, it proved to be computationally more expensive than the gradient-based method.

REFERENCES

[1] A. Andrade-Campos, L. F. Menezes, F. Teixeira-Dias, *Large deformation processes on AA1050-O aluminium at elevated temperatures*, Advanced Materials Forum II, Key Engineering Materials, (2004), (in press).

[2] S. B. Brown, H. K. Kwon, L. Anand, *An internal variable constitutive mode for hot working of metals*, Int. J. Plast., 5, (1989), 95-130.

[3] G. Cailletaud, P. Pilvin, *Identification and inverse problems related to material behaviour*, International Seminar on Inverse Problems, Clamart, (1994), pp. 79-86.

[4] R. Leal, U. Ramos, *Algortimos genéticos na optimização de compósitos laminados*, in J. M. Goicolea, C. Mota Soares, M. Pastor, G. Bugeda eds. Proc. V Congreso de Métodos Numéricos en Ingenieria, SEMNI, Madrid, Spain, (2002), pp. 381.

COMPUTATIONAL MECHANICS (Abstracts)
WCCM VI in conjunction with APCOM'04, *Sept. 5-10, 2004, Beijing, China*

Variational Inverse Boundary-Value Problems of Aerodynamics for Subsonic Gas Flow

A. M. Elizarov, D. A. Fokin, A. N. Ihsanova *

Chebotarev Research Institute of Mathematic and Mechanics, Kazan State University, Universitetskaya St. 17, Kazan, 420008, Russia
e-mail: elizarov@ksu.ru, fokin@altair.dc, anisa@ksu.ru

Abstract

Variational inverse boundary-value problems of aerodynamics address the task of constructing wing profiles with optimal hydrodynamic or aerodynamic characteristics. The direct methods often applied to these problems involve the successive solution of problems of flow past a given airfoil by an incompressible fluid or compressible gas and modification of the airfoil's geometry to optimize its aerodynamic characteristics. A review of these results can be found in [1]. The fundamentals of the approach to the optimization of aerodynamic shapes developed here, based on the theory of inverse boundary-value problems, are given in [2]. Several problems of airfoil shape optimization of an airfoil shape in a subsonic gas flow are solved analytically or numerically with application of used Kuhn-Tucker approach [3].

Particularly, a problem of maximizing the airfoil lift coefficient C_y for a flow without separation and prescribed maximal velocity is solved analytically. The solution depends on free stream Mach number $M_\infty \in [0,1]$ and theoretical angle of attack $\beta \in [0,\pi/2]$. Analytical results are compared with results of numerical optimization. Examples of optimal airfoils are obtained and compared with results of numerical optimization for variety of the parameters $M_\infty \in [0,1]$ and $\beta \in [0,\pi/2]$.

This work was supported by Russian Foundation for Basic Researches, project 03-01-00015.

REFERENCES

[1] A. M. Elizarov, N. B. Il'inskiy, A. V. Potashev. *Mathematical methods of airfoil design: inverse boundary value problems of aerohydrodynamics*. – Berlin: Akad. Verlag, (1997).

[2] A. M. Elizarov, Yu. V. Fedorov *Optimization of aerodynamic forms by the method of inverse boundary-value problems,* Prikl. Mat. Mekh., 54 (4), (1990), 571-580.

[3] A. L. Peresini, *The mathematics of nonlinear programming*. – Springer-Verlag, (1988).

COMPUTATIONAL MECHANICS (Abstracts)
WCCM VI in conjunction with APCOM'04, *Sept. 5-10, 2004, Beijing, China*

Stochastic Models for Acoustic Scattering in Noisy Media

B. Faverjon[1*], R. Ghanem[2]

[1] *Department of Civil Engineering, University of Johns Hopkins, Latrobe Hall, 3400 N. Charles St., Baltimore, MD21218, United States of America*
[2] *Department of Civil Engineering, University of Johns Hopkins, Latrobe Hall, 3400 N. Charles St., Baltimore, MD21218, United States of America*
e-mail: beatrice_faverjon@yahoo.fr, ghanem@jhu.edu

Abstract

In complex areas such as an ocean, deterministic models of inverse problems in acoustic scattering are often inaccurate because of errors from models, numerical resolution, and l imited data. These errors can be represented and resolved using stochastic models. Models in acoustic scattering with a stochastic approach have already been presented, especially for cases of scattering by complex surfaces such as rough surfaces, inhomoge neities of the ambient medium (see e.g. [1 -3]). Especially, errors from experiments are due in part to noise levels carried through by experimental data. This work presents an investigation of the effect of noise levels on the inverse problem in acoustic scattering, for the domain of underwater acoustics which concerns the reconstruction of underwater objects and where noise levels can be very high. A stochastic approach is used for reconstructing, from the measured scattered acoustic field, with different levels of noise, the shape of a two-dimensional obstacle in an infinite medium. The measured scattered acoustic field is constructed from a deterministic field calculated from an analytical method on which uniform random noise level is added. The parameter ization of the shape used is random and is based on Polynomial Chaos expansion [4]. The inverse problem is solved in a statistical sense. For several noise levels, we present the measured scattered acoustic field and the optimally reconstructed random shap e as characterized by its mean, and its decomposition on the Polynomial Chaos basis.

REFERENCES

[1] H. Medwin, C. Clay, *Fundamentals of acoustical oceanography*, Academic Press, San Diego, United States (1998).
[2] J. A. Ogilvy, *Theory of wave scattering from random rough surfaces*, Institute of Physics Publishing, Bristol, Great Britain (1991).
[3] H. Ogura, H. Nakayama, *Scattering of waves from a random cylindrical surface*, J. Math. Phys., 29 (4), (1988).
[4] R.G., Ghanem, P.D. Spanos, *Stochastic finite elements: A spectral approach*, Springer-Verlag, New York, United States (1991).

COMPUTATIONAL MECHANICS (Abstracts)
WCCM VI in conjunction with APCOM'04, *Sept. 5-10, 2004, Beijing, China*

Solution of Inverse Nonlinear Heat Conduction Problem Using Regularization Technique

Shouju Li *, Yingxi Liu

State Key Laboratory of Structural Analysis of Industrial Equipment, Dalian University of Technology, Dalian 116023, China
e-mail: lishouju@sina.com

Abstract

The accurate knowledge of the heat transfer coefficients is of importance in many engineering applications, including the cooling of continuously cast slabs and of electronic chips. In order to determine the heat transfer coefficients of materials, some identification methods have been developed for solving the problem. Direct heat transfer problems can be mathematically classified as well-posed. The solution of a well-posed problem is required to satisfy the conditions of existence, uniqueness, and stability with respect to the input data. The inver se heat transfer problem is usually ill-posed. An ill-posed problem is characterized by the non-uniqueness and instability of solution. The regularization technique has been employed to overcome the ill-posedness of inverse heat transfer problems. Several such techniques have been introduced in the literature. The proposed parameter identification scheme is based on the minimization of the least-squared errors between measured and calculated temperatures at observation points. To stabilize the optimization process and to overcome numerical difficulties caused by the non-uniqueness of the solution, a regularization function is added to the objective function. The regularization factor is adjusted by the reduction factor during the optimization process to main tain a suitable ratio of the regularization function to the error function. It is shown that the proposed method will converge very fast and can identify temperature-dependent material properties and the boundary conditions. The numerically computational results prove the insensitiveness of the proposed scheme to measurement errors.

REFERENCES

[1] H. S. LEE, *A new spatial regularization scheme for the identification of the geometric shape of an inclusion in a finite body*, Int. J. for Numerical Methods in Engineering, 46, (1999), 973-992.

[2] A A Tseng, *Direct sensitivity coefficient method for solving two-dimensional inverse heat conduction problems by finite-element scheme*, Numerical Heat Transfer, 27, (1995), 291-307.

[3] J. Marcelo, *Comparison of different versions of the conjugate gradient method of function estimation*, Numerical Heat Transfer, 36, (1999), 229-249.

[4] Tervola P. *A method to determine the thermal conductivity from measured temperature profiles*, Int. J. Heat Mass Transfer, 32, (1989), 1425-1431.

COMPUTATIONAL MECHANICS (Abstracts)
WCCM VI in conjunction with APCOM'04, *Sept. 5-10, 2004, Beijing, China*

Shape Optimization for Contact Problems with Multiple Load Cases

Wei Li [1,2*], Qing Li [2,3], Grant P. Steven [4]

[1] *School of Aerospace, Mechanical & Mechatronic Engineering, University of Sydney, NSW 2006, Australia*
[2] *School of Engineering, James Cook University, Townsville, QLD 4811, Australia*
[3] *College of Mechanical & Automotive Engineering, Hunan University, Changsha, 410082, China*
[4] *School of Engineering, University of Durham, Durham DH1, 3LE, United Kingdom*
e-mail: wei@aeromech.usyd.edu.au, Qing.Li@jcu.edu.au, grant.steven@durham.ac.uk

Abstract

Most structures in the real life are subject to multiple load cases. This paper aims at extending the evolutionary structural optimization algorithm [1] to shape design involving unilateral elastic contact [2] with the multiple load cases. To evaluate the reference stresses of each contact node in a finite element framework, an extreme stress criterion (the worst case design) and a weighted average criterion (Pareto design) are presented. In the extreme stress method, the highest nodal contact stress under all load cases is adopted as the reference level. In the weighted average method, the weighted sum of nodal contact stresses over all the load cases is regarded as the reference. It is found that these two criteria can produce different results. In this paper, the several examples are presented to demonstrate some new features of contact shape optimization in the presence of the multiple load cases. Shown in Fig.1 is the initial FE contact model of one example, where the nodal gaps are treated as design variables. Fig. 2 depicts the evolutionary process of design optimization based on the weighted average criterion. As a result of this, the Pareto frontier is identified for those two load cases.

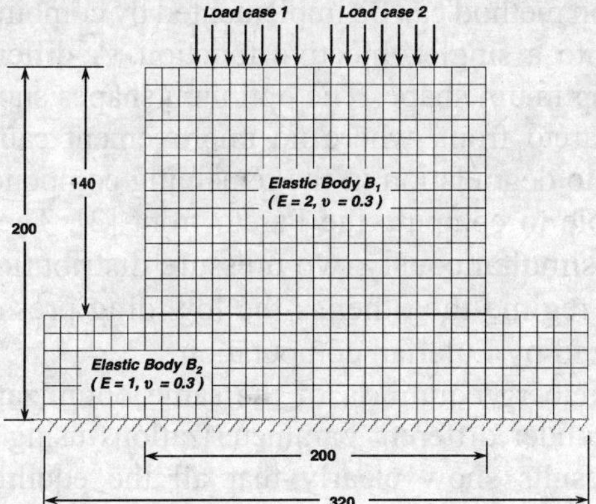

Fig. 1 FE modeling of contact bodies with multiple load cases

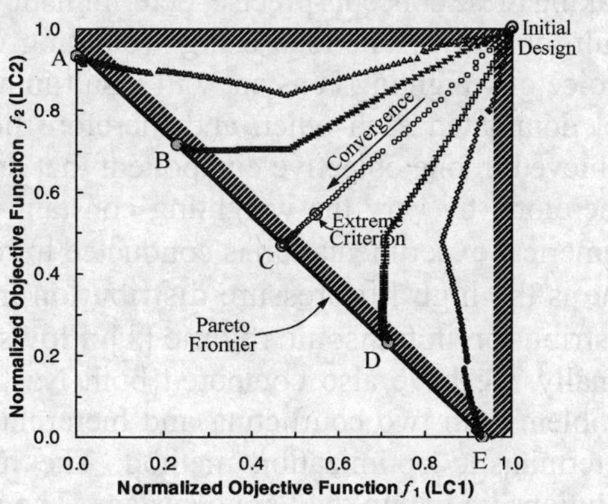

Fig. 2 Optimization processes and the Pareto frontier for the multiple load cases

REFERENCES

[1] Y. M. Xie, G. P. Steven, *Evolutionary Structural Optimization*, Springer-Verlag, Berlin (1997).
[2] W. Li, Q. Li, G. P. Steven, Y. M. Xie, *An evolutionary shape optimization procedure for contact problem in mechanics designs*, IMechE. J. Mech. Eng. Sci., 217, (2003), 435-446.

COMPUTATIONAL MECHANICS (Abstracts)
WCCM VI in conjunction with APCOM'04, *Sept. 5-10, 2004, Beijing, China*

Multi-Objective Inverse Problem in Aerodynamics Using Adjoint Method and Game Theory

Jacques Periaux [1], Jean-Antoine Desideri [2], Zhili Tang [3*]

[1] *Pole Scientifique Dassault Aviation/UPMC, 78 Quai Marcel Dassault 99214 Saint-Cloud, France*
[2] *Opale Project INRIA, 2004 route des Lucioles –BP 93 – 06902 Sophia Antipolis, France*
[3] *Aerodynamics Department, Nanjing University of Aeronautics & Astronautics, 29 Yudao Street Nanjing 210016, China*

e-mail: jacques.periaux@dassault-aviation.fr, Jean-Antoine.Desideri@inria.fr, tangzhili@hotmail.com

Abstract

This paper approaches the question of multi-objective inverse problem for optimum shape design in aerodynamics using deterministic method and Game Theory. The employed basic optimizer is a gradient-based control theory, more precisely an adjoint method. This method requires the solution of adjoint problem, in addition to the analysis problem, to evaluate the gradient of cost function. The complexity of the adjoint problem is equivalent to that of the analysis problem, therefore the computational cost of this method is approximately twice the evaluation of aerodynamic function, regardless of the number of design parameters, and is well suited to shape design problem.

In a multi-objective optimization problem, there is non unique optimal solution but a whole set of potential solutions since in general no solution is optimal with respect to all criteria simultaneously, instead, one identifies a set of non-dominated solution referred to as the Pareto optimal front. After making these concepts precise, deterministic optimization method can be implemented by combining multi-objective functions using weighting constants into a single objective function. A different choice of weighting constant will result in a different optimum shape. The optimum shapes should not dominated each other, and therefore lie on the Pareto front, where no improvement can be achieved in one objective component that doesn't lead to degradation in the remaining component. Therefore, by vary the weighting constant, it is possible to compute the Pareto front [2]. Then a numerical experimentation is conducted to reconstruct simultaneously two pressure distributions, one is the high lift pressure distribution in subsonic regime the other is the low drag pressure distribution in transonic regime [1]. Flows are analyzed by Eulerian computation.

Finally, we have also computed both Nash and Stackelberg equilibria of the same optimization problem with two conflicting and hierarchical targets under different parameterizations using the deterministic optimization method. The numerical results show clearly that all the equilibria solutions lie within the Pareto surface. The non-dominated Pareto Front are obtained, this is due to the factor that the best non-dominated solution by cooperative strategy however the CPU cost to capture a set of solutions makes the Pareto Front an expensive to the designer engineer.

REFERENCES

[1] N. M. Marco, J. A. Desideri, S. Lanteri, *Multi-objective optimization in CFD by genetic algorithms*, INRIA Report RR-3686, April, (1999).

[2] K. Leoviriyakit, S. Kim, A. Jameson, *Viscous aerodynamic shape optimization of wings including planform variables*, AIAA paper 2003-3498, June, (2003).

COMPUTATIONAL MECHANICS (Abstracts)
WCCM VI in conjunction with APCOM'04, *Sept. 5-10, 2004, Beijing, China*

Energy Criteria for Optimizing Material Distribution

Z. G. Sun *

Institute of Road and Bridge Engineering, Dalian Maritime University, Dalian, 116026, China
e-mail: zgsun@newmail.dlmu.edu.cn

Abstract

Material optimal distribution, or shape optimization for continuous structure is an important domain in structural optimization. Under certain constraint conditions, the optimal structure is realized through the adjustment of geometric shape. Because the geometry of the structure is always changing during computation, it is a problem to correlate the changing geometry to the finite element model and algorithms related [1]. Many researchers have paid much attention to the problem, and obtained many fruits in the area in the past decades [2-4]. Energy Criteria is an important aspect of optimization theories. This paper propounds an energy method of structural shape optimization. Under the constraint conditions of displacement and material volume, it assumes that the maximum absorbability of structure strain energy is equivalent to its maximum load-bearing capacity in determining the material distribution. According to this assumption, a Lagrange function in term of energy is provided. By maximizing the function using variational principle, the optimal shape of the structures is determined. The same result can be obtained if the term of the displacement constraint condition in the Lagrange function is replaced by potential energy of the external force.

Taking a cantilever beam as an example, an analytic solution of optimal material distribution is derived by the method. The numerical method based on the theory is demonstrated. In numerical solution, the entire region D can be divided into n subregions, or known as elements d_i. The assumption of material distribution can be made on element d_i. Some unknown constants are involved in the function of material distribution assumed. Based on the assumption of the material distribution, the Lagrange function can be constructed. The unknown constants involved in the function are determined by maximizing the Lagrange function, and then the material distribution of the structure is determined. The numerical solution converges to the analytic solution.

REFERENCES

[1] Bei Wang, Runkang Sui, *Shape optimization design of three-dimensional continuity*, Chinese Journal of Applied Mechanics, 16(2), (1999), 127-132, (in Chinese).

[2] J. A. Tarrago, J. Canalest, *CODISYS, an integrated system for optimal structural design*, Comput. Struct., 52(6), (1994).

[3] M. H. Imam. *Three-dimension shape optimization*, Int. J. Num. Meth. Eng., 18, (1982), 661-673.

[4] Yuanxian Gu, Gengdong Cheng, *Study and application of numerical method for structural shape optimization*, Computational structural mechanics and its application, 10(3), (1993), 321-335, (in Chinese).

COMPUTATIONAL MECHANICS (Abstracts)
WCCM VI in conjunction with APCOM'04, *Sept. 5-10, 2004, Beijing, China*

A Computational Statistics Approach to Stochastic Inverse Problems and Uncertainty Quantification in Heat Transfer

J. Wang *, N. Zabaras

Materials Process Design and Control Laboratory, Sibley School of Mechanical and Aerospace Engineering, Cornell University, 188 Frank H.T. Rhodes Hall, Ithaca, NY, 14853 -3801, USA
e-mail: jw332@cornell.edu, zabaras@cornell.edu

Abstract

As most engineering systems and processes operate in an uncertain environment, it becomes increasingly important to address their analysis and inverse design in a stochastic manner using statistical data-driven prior and concurrent information on the system response. Recent advances in computational Bayesian and spatial statistics enable complete and efficient solution procedures to such problems. Herein, a novel framework based on Bayesian inference is presented for the solution of multi-dimension and multi-mode inverse heat transfer problems. The posterior probability density function (PPDF) of unknowns (modeled as random variables or stochastic processes), such as the boundary heat flux or heat sources, is computed given finite set of thermocouple temperature measurements. Markov Random Field (MRF) was adopted for prior distribution modeling. Markov Chain Monte Carlo (MCMC) algorithms are exploited to obtain estimates of statistics of random unknowns. One and two-dimensional linear boundary heat flux reconstruction problems are first discussed to demonstrate stochastic regularization through prior modeling and MCMC posterior state space exploration. A one-dimensional inverse heat conduction problem (IHCP) with consideration of uncertainty in thermo-physical properties and sensor location is also presented for complete stochastic description of the IHCP. Finally, a problem of reconstructing heat sources in three-dimensional radiation heat transfer is presented. This example will demonstrate the need for model reduction in the likelihood calculation when applying MCMC algorithms to complex continuum problems. Simulation results demonstrate the great potential of applying a Bayesian approach to stochastic estimation and design problems. Although discussed in the context of thermal systems, the methodology presented is general and applicable to design and estimation problems in diverse areas in engineering.

REFERENCES

[1] J. Wang, N. Zabaras, *A Bayesian inference approach to the stochastic inverse heat conduction problem*, Int. J. Heat Mass Trans., in press.
[2] J. Wang, N. Zabaras, *Using Bayesian statistics in the estimation of heat sources in radiation*, Int. J. Heat Mass Trans., submitted for publication.
[3] C. Andrieu, N. Freitas, A. Doucet, M. Gordan, *An introduction to MCMC for machine learning*, Mach Learning, 50, (2003), 5-43.
[4] J. Besag, P. Green, D. Higdon, K. Mengersen, *Bayesian computation and stochastic systems*, Statistical Science, 10, (1995), 3-41.

COMPUTATIONAL MECHANICS (Abstracts)
WCCM VI in conjunction with APCOM'04, *Sept. 5-10, 2004, Beijing, China*

Optimum Reinforcement in Concrete Using Genetic Algorithms

Rena C. Yu*, Gonzalo Ruiz, Jacinto R. Carmona

School of Civil Engineering, University of Castilla-La Mancha, Avda. Camilo Jose Cela s/n, Ciudad Real, 13071, Spain
e-mail: rena@uclm.es, Gonzalo.Ruiz@uclm.es, jacintoruiz@yahoo.es

Abstract

We tackle the problem of minimum reinforcement in concre te using Genetic Algorithms (GAs), the searching procedures based on the fundamental evolutionary principle: better individuals survive and reproduce more often than other individuals [1]. GAs work very well on mixed (continuous and discrete) combinatorial problems; they are less susceptible to getting "stuck" at local optima than other gradient search methods. But they nevertheless tend to be computationally expensive, since a "pool" of samples with variations is needed to start and evolve the searching pr ocess. If the fitness value has to be fed by a finite element method calculation, thousands or millions of calculations are needed before a near-optimum solution can be finally found; and this would render the usage unrealistic. In particular, in dealing w ith optimization in lightly-reinforced concrete beams, the peak and near post-peak in the load-displacement and its neighborhood are known to b e controlled by three factors: 1) the steel ratio; 2) the bond-slip properties; and 3) the steel cover [2]. But how to adjust these three factors to obtain a global ductile response of the structure is only by trial-and-error both numerically and experimentally. In this paper, we adopt a cohesive crack model developed by Ruiz [3], which deals with the propagation of a cohesive crack crossing a reinforcement layer in a beam made out of a cohesive matrix and lightly reinforced with steel bars. This model accounts for the aforementioned three factors, but above all, it gives a fast numerical solution, which makes the use of GAs in this particular problem feasible. Specifically, the case of a reinforced prismatic beam loaded at three points is modeled. GAs are adopted to study carefully the influence of those three parameters, optimization process is carried out toward an objective function which is the ductile response of the beam structure, i.e. a linear elastic perfect -plastic response in the dimensionless load-displacement curve.

REFERENCES

[1] D. E. Goldberg, *Genetic algorithms: in search, optimization and machine learni ng*, Addison Wesley Publishing Longman, Inc., Reading, Massachusetts, USA , (1989).
[2] Z. P. Bažant, J. Planas, *Fracture and size effect in concrete and other quasi-brittle materials*, CRC Press, (1998).
[3] G. Ruiz, *Propagation of a cohesive crack crossing a reinforcement layer*, Int. J. Fract., 111, (2001), 265-282.

COMPUTATIONAL MECHANICS (Abstracts)
WCCM VI in conjunction with APCOM'04, *Sept. 5-10, 2004, Beijing, China*

An Integrated Structural Design Procedure Using Genetic Algorithm

R. Balamurugan *, C. V. Ramakrishnan

Department of Applied Mechanics, Indian Institute of Technology, New Delhi, 110 016, India
e-mail: rbala_udt@yahoo.com, cvrama@am.iitd.ernet.in

Abstract

The problem of optimum structural design can be solved using an iterative two stage process [1] , [3] as an integrated topology and shape design problem using Genetic Algorithm (GA) as the optimization tool. In the integrated approach, the topology optimization problem, which is ill -posed, is regularized using constraints for the exclusion of small h oles and perimeter and solved using GA. Since, the algorithm is not dependent on design derivatives there is a smooth convergence of the results. The topology optimization problem has been studied for various benchmark problems and found to be adapting wel l with GA technique and constitutes the first phase. The number of variables in the topology optimization module is high and hence the module has to be used effectively to save the computing time.

The shape optimization problem is analyzed using GA starti ng from the tentative topology design of the previous phase and constitutes the second phase of the integrated problem. We attempt to find the optimum design of the benchmark cantilever plate problem with single load acting in lateral and axial directions separately using the integrated design approach.

The topology optimization problem is solved first and then the internal and external boundaries given by this module is fed into the shape optimization module. The boundaries of this tentative topology are defined through five straight lines and three spline curves which permit good control over the shape with less number of design variables. A Design window concept is used to identify the variables and their range of variation in the design space. These desi gn variables are then manipulated by genetic operators within the specified range to explore the design space for the optimum. The entire domain is divided into number of patches and is automatically meshed using a mapped mesh generation procedure and eigh t noded isoparametric elements have been used. Thus using finite element analysis for simulation and GA for optimization the integrated topology and shape design problem is solved.

REFERENCES

[1] N. Swaminathan, C. V. Ramakrishnan, R. Balamurugan, *Optimum topology design of products using genetic algorithm,* Int. Conference on Numerical Methods in Industrial Forming Processes, Columbus, USA, (2004).

[2] R. Balamurugan, C. V. Ramakrishnan, S. Gajbhiye, *Optimum shape design using genetic algorithm,* Proc. 48th Congress of The Indian Society of Theoretical and Applied Mechanics, Ranchi, India, (2003), pp. 204-212.

[3] K. Maut, E. Ramm, *Adaptive topology optimization of shell structures,* AIAA J., 35, (1997), 1767-1773.

COMPUTATIONAL MECHANICS (Abstracts)
WCCM VI in conjunction with APCOM'04, *Sept. 5-10, 2004, Beijing, China*

Optimization of Piezoelectric Fiber Distribution in Composite Smart Materials

Niels Bohn *, Sreedhar Kari, Harald Berger, Ulrich Gabbert

Institute of Mechanics, Chair of Computational Mechanics, Otto-von-Guericke University, Universitaetsplatz 2, 39106 Magdeburg, Germany
e-mail: {bohn, kari, berger, gabbert}@ mb.uni-magdeburg.de

Abstract

The increasing demand for low cost and effective use of materials in the industry can be satisfied by the use of intelligent (smart) materials. In this context intelligence is defined as the ability of the structure or material to detect changes in its environment and to react in such a way that certain system conditions are maintained or avoided. Typical applications are the use of adaptive elements in mechanical structures which, for example, minimize acoustic noise radiation, damp vibrations or compensate deformations due to thermal loads [1-4]. The efficiency of such intelligent systems can be improved by the use of smart composite materials, consisting of piezoelectric fibers embedded in a matrix material. A new approach to obtain some desired composite properties is given by the optimization of the piezoelectric fiber distribution for a given pair of materials. For this purpose an optimization tool was developed by the first author during the last year. The optimization is carried out by Evolution Strategies, using an interface to the finite-element-software ANSYS. Two applications are presented: the first example covers the optimization of the effective properties of a composite material via a homogenization algorithm, and the second one deals with the optimization of a structure, consisting of a smart composite material.

REFERENCES

[1] J. K. Dürr, U. Herold-Schmidt, H. W. Zaglauer, D. Scheulen, R. Sipple, *Smart composites for adaptive satellite mirrors*, Proceedings of the Adaptronic Congress, Potsdam, Germany, (2000), pp. 96-107.
[2] U. Gabbert, *Research activities in smart materials and structures and expectations to future developments*. J. of Theoretical and Applied Mechanics, 40(3), (2002), 549-574.
[3] U. Gabbert, H. Köppe, T. Nestorović Trajkov, Controller design for engineering smart structures based on finite element models, in V. S. Rao ed. *Smart structures and materials 2002: modeling, signal processing, and control*, SPIE Proceedings Series, 4693, (2002), 430-439.
[4] F. Seeger, U. Gabbert, *Optimal placement of distributed actuators for a controlled smart elastic plate*, PAMM, 2(1), (2003), 262-263.

COMPUTATIONAL MECHANICS (Abstracts)
WCCM VI in conjunction with APCOM'04, *Sept. 5-10, 2004, Beijing, China*

Evolutionary Optimization of the 3-D Structures for Various Criteria

T. Burczyński [1, 2], A. Poteralski [1*]

[1] *Department for Strength of Materials and Computational Mechanics, Silesian University of Technology, Gliwice, 44-100, Poland*
[2] *Institute of Computer Modeling, Cracow University of Technology, Cracow, 31-155 Poland*
e-mail: Tadeusz.Burczynski@polsl.pl, Arkadiusz.Poteralski@polsl.pl

Abstract

The paper is devoted to an application of the evolutionary algorithms and the finite element method to genetic generation of 3-D structures. The shape, topology and material of the structure are generated for various optimization criteria: the minimum of volume, minimum of the stress functional, the minimum displacement functional, the maximum of the first eigenfrequency and the maximum of difference between first and second eigenfrequency.
Evolutionary methods have found various applications in mechanics, especially in structural optimization [1]. The main aim of these methods is the simulation of biological processes based on heredity (genetics) and on the natural selection (the theory of evolution) to create the optimal individuals (solutions) presented by single chromosomes. Evolutionary models of computation can be performed by using genetic algorithms.
During the evolutionary process the domain, its boundary, and the field of Young's modulus can change for each generation. The distribution of values of Young's modulus and as a consequence the shape and topology of the structure is described by hyper surface which is controlled by genes. Genes create a chromosome which is a member of a population of solutions.
Numerical example: Optimization of 3-D structure resembling stairs – minimum of the volume. Some results are shown in Fig. 1.

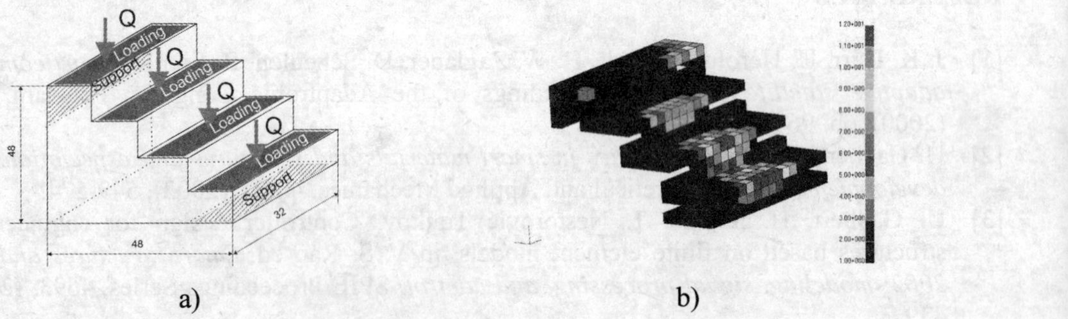

Fig.1 Optimization of 3-D structure resembling stairs, the best structure after 122nd generations
a) The scheme of loading; b) Distribution of Young's moduli

REFERENCES

[1] T. Burczyński, A. Osyczka eds. *Evolutionary methods in mechanics*, Proc. IUTAM Symposium, Kluwer, Dordrecht, (in press).

COMPUTATIONAL MECHANICS (Abstracts)
WCCM VI in conjunction with APCOM'04, *Sept. 5-10, 2004, Beijing, China*

New Bionics Algorithm-Immuned Evolutionary Programming

W. Gao *

Department of Civil Engineering, Wuhan Polytechnic University, Wuhan, 430023, P. R. China
e-mail: wgaowh@hotmail.com

Abstract

The inverse problem in geotechnics is a very complicated optimization problem. To solve this problem, the traditional local optimization method has many shortcomings, such as relying on initial values, et al. So, some global optimization methods, such as genetic algorithm, neural network et al, are introduced into this problem. But the efficiency of those methods is low. In order to overcome this shortcoming, a new bionics algorithm-immunized evolutionary programming is introduced here.

Evolutionary programming is another evolutionary algorithm to solve global optimization problem. But the traditional evolutionary programming also has some shortcomings. Here, from analysis the mechanism of the traditional algorithm, the immune algorithm is introduced to improve it, and a new bionics algorithm-immunized evolutionary programming is proposed here. In this new algorithm, the two main aspects-the mutation operation and selection operation of traditional evolutionary programming are improved. Firstly, the traditional mutation operation is Gauss mutation, but the previous study has indicated that the Cauchi mutation is better, for which can search in a wider space. But the traditional Cauchi mutation is a static search process and cannot be changed with evolutionary extent. Here a new adaptive mutation operation is proposed. This adaptive mutation can make the disturbing extent of mutation adaptively changeable through the evolutionary extent, so the search performance of the whole algorithm can be improved. Secondly, the traditional selection operation is a kind of stochastic tournament modal that is called random competition with parameter q. But this operation has the shortcomings of losing diversity of population and "premature". In order to overcome those shortcomings, the principles of artificial immune system are introduced into selection operation, and a new selection operation based on thickness adjustment is proposed here. This new operation can adjust the balance between thickness of individual and its fitness, so the evolution process is smooth and the "premature" phenomenon can be avoided.

At last, this new algorithm is verified by simulation experiment of a typical optimization function. The results of experiment show that, the proposed bionics algorithm can improve not only the convergent speed of original algorithm but also the computation effect of original algorithm, and is a very good optimization method.

REFERENCES

[1] D. B. Fogel, L. J. Fogel, *An introduction to evolutionary programming*, in J. M. Alliot, E. Lutton, et al eds. European Conf. on AE'95, Berlin: Springer, (1996), pp. 21-33.

[2] K. Kumar, J. Neidhoefer, *Immunized artificial systems-concepts and applications*, in D. Quagliarella, J. Periaux, et al eds. *GA and ES in engineering and computer science*, London: John Wiley & Sons Ltd., (1997), pp. 85-104.

COMPUTATIONAL MECHANICS (Abstracts)
WCCM VI in conjunction with APCOM'04, *Sept. 5-10, 2004, Beijing, China*

Optimization of Stiffened Plates by the Coupled BEM/FEM and the Evolutionary Algorithm

Radosław Górski *, Piotr Fedeliński

Department for Strength of Materials and Computational Mechanics, Silesian University of Technology, Konarskiego, 18a, Gliwice, 44-100, Poland
e-mail: Radoslaw.Gorski@polsl.pl, Piotr.Fedelinski@polsl.pl

Abstract

In recent years a great interest in composite materials having new properties is observed. The composite materials, which consist of two or more different materials, have usually better mechanical properties than the traditional homogenous materials. The optimization of these materials plays an important and practical role.
Nowadays the boundary element method (BEM) and the finite element method (FEM) are the most popular computer techniques of continuum solid mechanics. Both methods have advantages and disadvantages. There are certain applications where one technique is more suitable than the other. Combining both techniques in the same computer program seems to be the most efficient way in problems which require both boundary element (BE) and finite element (FE) features.
The aim of this work is to present the results of the analysis and optimization of stiffened structures using a coupled BEM/FEM. An example of a linear-elastic body subjected to static or dynamic loads is shown in Fig.1.

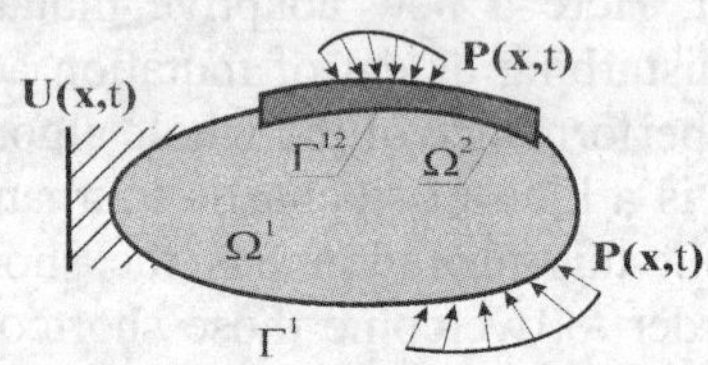

Fig. 1 A stiffened structure

The plates, which are modeled by the BEM, are stiffened by frames (or beams) modeled by the FEM. The aim of this combination of the two kinds of regions having different geometrical and material properties (plate and frame) is to model the composite materials.
There are several ways of combining of BEs and FEs. In the present approach, the subregion technique is used, in which the direct coupling of the BE and FE matrices, by transforming the FE matrices into an equivalent BE-type, is performed. In this approach the FE nodal forces are transformed into the BE tractions by using a special transformation matrix. The equations for the composite structure are obtained by assembling the set of equations for each subregion using compatibility of displacements and equilibrium of tractions between the common interfaces.
The problem of optimization is solved using an evolutionary algorithm. The composite plates are optimized using stiffness or strength criteria. The design variables of the problem are geometrical or material properties of the considered regions.
In order to present the accuracy and efficiency of the proposed method, several numerical examples are given. The results of analysis are compared with the solutions computed by the FEM. The results of optimization are compared with the solutions obtained using the systematic search method.

COMPUTATIONAL MECHANICS (Abstracts)
WCCM VI in conjunction with APCOM'04, *Sept. 5-10, 2004, Beijing, China*

Parameter Identification of Ground Elastic Modulus at Excavation Site of Tunnel

Atsushi Hikawa [1*], Mutsuto Kawahara [1], Noriyoshi kaneko [2]

[1] *Department of civil engineering, Chuo University, Kasuga 1-13-27, Bunkyo-ku, Tokyo 112-8551, Japan*
[2] *Department of SATO KOGYO CO., LTD, 12-20, Nihonbashi-honcho 4-Chome, Chuo-ku, Tokyo 103-8639, Japan*
e-mail: atsu1005@kc.chuo-u.ac.jp, kawa@civil.chuo-u.ac.jp, n.kaneko@satokogyo.co.jp

Abstract

This paper presents a numerical technique to predict a soft stratum ahead of an excavating face of a tunnel. The natural ground consists of many faults. The strata intricately exist in the longitudinal and lateral directions to the section of a tunnel face. If a tunnel face is confronted with a soft stratum, the ground on the surface of the tunnel sometimes turns to be dangerous. It is necessary to evaluate the geological behav ior ahead of a tunnel face at the early excavating stage.
An easy and accurate method to predict a soft stratum ahead of a tunnel face is expected. To predict the stratum ahead of a tunnel face, many methods, such as TSP(Tunnel Seismic Prediction) and bore hole, have been used until now. These predictions delay the construction and need the particular equipment. The prediction of the stratum ahead of a tunnel face without troublesome procedure is expected.
As a new solution, the prediction of the stratum ahead of a tunnel face using the numerical model can be introduced. If the Young's modulus which expresses the strength of ground can be determined, the stratum ahead of a tunnel face can be predicted by digital evaluation. The functional technique is a parameter identification of the Young's modulus which is one of the inverse analysis.[1] This technique replaces the estimation with the minimization problem of the performance function. The conjugate gradient method is applied to the minimization technique.
In this paper, a parameter identification of the Young's modulus considering excavation using 3-D model has been presented.[2] The 3-D model is expected to verify more exactly the properties at the cross section of the tunnel face. As a numerical model to be identified, the ground ahead of a tunnel face is divided into some strata. In this numerical study, the effectiveness of this prediction is initially verified using the ideal model. For the calibration, the calculated displacements by the forward analysis are used for the measured data. The prediction of a soft stratum ahead of a tunnel face using the measured data at the actual site is examined. The 3-D model considering the full section of tunnel can be applied to the actual problem.

REFERENCES

[1] M. Sugimoto, S. Sakurazawa, S. Kageyama, *Three dimensional reverse analysis on ground properties with time-dependence*, 9th International Congress on Rock Mechanics, Vol.2, (1999).

[2] Y. M. Hsieh, A. J. Whittle, *A computational strategy for solving three-dimensional tunnel excavation problems*, Second M.I.T. Conference on Computational Fluid and Solid Mechanics, (2003).

COMPUTATIONAL MECHANICS (Abstracts)
WCCM VI in conjunction with APCOM'04, *Sept. 5-10, 2004, Beijing, China*

GA Technique by Special Feature Extracting of Gene and the Application

J. Oda [1*], R. Kanai [2]

[1] *Department of Human and Systems Mechanical Engineering, University of Kanazawa, 2-40-20, Kodatsuno, Kanazawa, 920-0867, Japan*
[2] *Graduate School of Natural Science and Technology, Kanazawa University*
e-mail: oda18@t.kanazawa-u.ac.jp, kanai@nihonkai.kanazawa-u.ac.jp

Abstract

The Genetic Algorithm (GA), which is optimum method based the natural evolution theorem, is applicable on the wide scale problems. By using the GA, various evolutionary optimum methods are studied. But, in these methods, the meanings of gen etic-code are not considered at all.
In this study, the Special Feature Extracting GA (SFE -GA) is proposed. The SFE-GA is based on the idea that the genes adapting to the environment extend to all over the chromosome, and is the GA with an operation of ste pping up the function. A superior group with the high fitness value includes the schemata of building blocks of good solutions. The schemata are extracted from the superior group, and the extracting schemata overwrite the gene locus in the group with the l ow fitness value.
In order to extract the schemata theoretically, t -test is selected. In this method, c hromosomes are divided into two groups according to the value of gene locus. The two groups are tested. If the result of t-test is regarded as difference from fitness value of two groups dependent on gene locus, value of gene locus in higher fitness value is extracted as special feature from comparing sample mean of two groups.
This method is applied to the problem that searches the maximum value in a si mple function. The behavior in the search is clarified and it is confirmed to extract the effective schemata sequentially. Additionally, the influence of new parameters to the searching behaviors is checked. The technique is applied to the layout design pr oblem of truss structure with 9 nodes at first and the useful range of parameter is checked. Moreover, the layout design problem of truss with 25 nodes as the wide scale problem is analyzed. From the numerical results, the validity of SFE -GA is checked.

REFERENCES

[1] M. Sano, T. Hiroyasu, M. Miki, *Distributed probabilistic model-building genetic algorithm*, JSME OPTOS2002, (02-03), (2003), pp. 65-70, (in Japanese).
[2] H. Sato, I. Ono, S. Kobayashi, *A new generation alternation model of genetic algorithms and its assessment*, Trans. JSAI, 12(5), (1997), pp. 734-744, (in Japanese).
[3] M. Pelikan, D. E. Goldberg, F. Lobo, *A survey of optimization by building and using probabilistic model*, IlliGAL Report 99018, Sep, (1999).
[4] J. Oda, R. Kanai, Y. Kawato, *The study of new GA technique by special feature extracting of gene*, JSME Hokuriku-Shinetsu Meeting, No. 037-1, (2003), pp.133-134, (in Japanese).
[5] Y. Sakaki, *Human Genome*, Iwanami Shoten, (2001), pp. 83-94, (in Japanese).

COMPUTATIONAL MECHANICS (Abstracts)
WCCM VI in conjunction with APCOM'04, *Sept. 5-10, 2004, Beijing, China*

Sensitivity Analysis of the SPR Error Estimator and Its Application to an Adaptive Remeshing Strategy

G. Bugeda [*]

International Center for Numerical Methods in Engineering (CIMNE), Universitat Politècnica de Catalunya (UPC), Escola Universitària d'Enginyers Tècnics Industrials (EUETIB), C/ Comte d'Urgell, 187 - 08036 Barcelona, Spain
e-mail: bugeda@cimne.upc.es

Abstract

Traditionally, adaptive remeshing strategies for linear elastic problems are based on the control and distribution of a specific error norm over the entire analysis domain. The most commonly used error norm is the energetic one and, typically, it is equally distributed between all the elements of a mesh. Alternative mesh optimality criteria based on the equal distribution of the density of the energetic norm of the error have also been proposed [1]. In addition, the equal distribution of alternative error measurements like the point wise error in stresses, the point wise error in the von Misses stress and the point wise error in the Tresca stress have also been proposed in [2].
All the mentioned strategies provide control over the error over the whole domain. Nevertheless, in most occasions the main objective of a linear elastic analysis is to determine the highest value of the stresses over the domain and, typically, the point where this high value occur is perfectly known. This fact indicates that the derivation of adaptive remeshing strategies specially suited for providing control over the point wise error in the stresses at that point, without taking excessive care over the rest, could produce much more economical meshes than traditional strategies.
This work proposes a new adaptive remeshing strategy based on the sensitivities of the point wise error in stresses. These sensitivities are computed through the derivation of the point wise error in stresses e with respect to the mesh nodal coordinates. This error can be computed using the SPR stress recovery procedure (see [3]). For the derivation of the error it is assumed to depend of the nodal coordinates of the mesh $\mathbf{x}$ and the nodal displacements $\mathbf{d}$ ($e = e(\mathbf{x}, \mathbf{d})$) and the corresponding sensitivities are obtained using the adjoint-state method (see [4]).

REFERENCES

[1] E. Oñate, G. Bugeda, *A study of mesh optimality criteria in adaptive finite element analysis*, Eng. Comp., 10, (1993), 307-321.

[2] G. Bugeda, *A comparison between new adaptive remeshing strategies based on point wise stress error estimation and energy norm error estimation*, Commun. Numer. Meth. Engng, 18, (2002), 469-482.

[3] O. C. Zienkiewicz, J. Z. Zhu, *Superconvergence derivative recovery techniques and a posteriori error estimation in the finite element method. Part I and part II*, Int. J. Num. Meth. Engng, 33, (1992), 1331-1382.

[4] G. A. Galarza, J. Carrera, A. Medina, *Computational techniques for optimization of problems involving non-linear transient simulations*, Int. J. Num. Meth. Engng, 45, (1999), 319-334.

COMPUTATIONAL MECHANICS (Abstracts)
WCCM VI in conjunction with APCOM'04, *Sept. 5-10, 2004, Beijing, China*

hp Adaptive Refinement Strategy in the PZ Environment

Philippe R. B. Devloo *, Edimar C. Rylo, Cedric M. A. A. Bravo

Department of Structure Faculty of Civil Engineering, University of Campinas, Av. Albert Einstein 951, Campinas - SP, CP 6021, CEP: 13083-852, Brazil
e-mail: phil@fec.unicamp.br, ecrylo@fec.unicamp.br, cedric@fec.unicamp.br

Abstract

hp refinement applied to the finite element method should lead to faster convergence than *h* or *p* refinement [3,4]. On the other hand, *h* and *p* refinements are easier to implement. The complexity of an *hp* refinement is increased when the process is implemented for any type of mesh [1] In the PZ environment [2], the *hp* adaptive method was designed to work with heterogeneous mesh.
In this paper the implementation of *hp* adaptivity in the PZ environment will be explained. The focus will be on the implementation. The complexity of the code is managed by precise mathematical definition of topological concepts and by their translation in the Object Oriented language C++. *hp* Adaptivity is implemented in two steps: the geometric element division, where element neighboring information is discussed together with the requirement of new nodes creation and the second step where the *p*-order interpolation space enrichment including restriction requirements is discussed. The main result of this research is that a single concept is applicable to all known types of elements. As an extension of this study, a new geometric element division based on a refinement pattern is presented. A refinement pattern is defined based on an example mesh. Directional refinements are used to illustrate the concept of refinement patterns.
The research in *hp*-adaptivity has evolved from a set of classes which each individually implemented *h*-refinement to a single templated class where the elements are solely defined based on their geometry and interpolation functions.

REFERENCES

[1] L. Demkowicz, D. Pardo, W. Rachowicz, *3D hp-adaptive finite element package (3dhp90) version 2.0*, TICAM—Report 02-24, TICAM, The University of Texas at Austin, Austin, TX 78712, USA, (2002).

[2] P. R. B. Devloo, *PZ: An object oriented environment for scientific programming*, Computer Methods in Applied Mechanics and Engineering, 150, (1997), 133-153.

[3] John Tinsley Oden, F. Carey, ?. Graham, E. B. Becker, *Finite elements - an introdution*, Vol.1, Prentice Hall Inc., New Jersey, USA, (1981).

[4] B. Szabo, I. Babuska, *Finite element analysis,* A Wiley-interscience publication, (1991).

COMPUTATIONAL MECHANICS (Abstracts)
WCCM VI in conjunction with APCOM'04, *Sept. 5-10, 2004, Beijing, China*

FER/View— an Interactive Finite Element Post-Processor

Z. Q. Feng *, Z. G. Feng

Laboratoire de Mécanique d'Evry, Université d'Evry, 40, rue du Pelvoux, 91020 Evry, France
e-mail: feng@iup.univ-evry.fr

Abstract

Development of user-friendly and flexible scientific programs is a key to their usage, extension and maintenance. This paper presents the deve lopment process, the architecture and the main features of FER/View, an interactive finite element post -processor. Based on the OOP (Object -Oriented Programming) approach and the graphic standard OpenGL [1, 2], this software has been specifically designed to fulfill most of the common requirements of engineers and numericians, in the context of finite element simulations. FER/View has a mouse oriented graphical user interface, which makes the program intuitively easy to use. Therefore, the user does not rea lly need any specific learning stage prior to be able to use it. Several examples with graphical representations illustrate the efficiency and some functionalities of FER/View: to visualize and manipulate mesh data structures via the mouse, to deal with is ovalues associated with a mesh and to deal with more advanced features: interactive animations, cutting planes or texture transparency, etc. (Fig.1, Fig.2).

Fig. 1 Isosurface plot

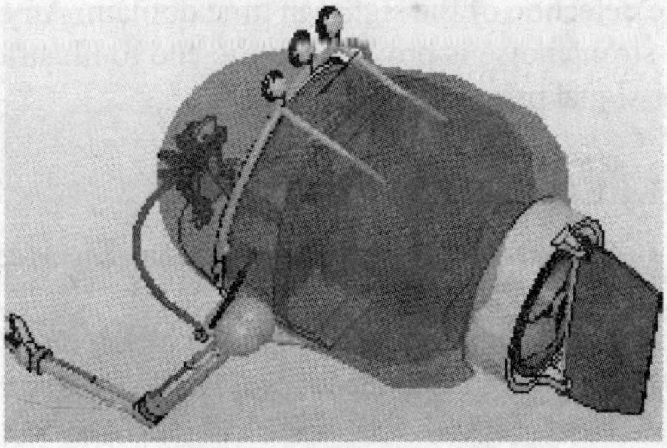

Fig.2 Transparent view

REFERENCES

[1] R. S. Jr Wright, M. Sweet, *OpenGL Superbible: the Complete Guide to OpenGL Programming for Windows NT and Windows 95,* Waite Group Press, (1996).
[2] http://gmfe16.cemif.univ-evry.fr:8080/~feng/FerSystem.html

COMPUTATIONAL MECHANICS (Abstracts)
WCCM VI in conjunction with APCOM'04, *Sept. 5-10, 2004, Beijing, China*

Information Detection Based on Bistable Systems Connected in Series

Y. Guo [1*], S. Y. Zhang [1], Y. G. Leng [2], B. Guo [1], M. W. An [3]

[1] *Management School, Tianjin University, Beiyang Kexue Building No. 105, Tianjin, 300072, China*
[2] *Mechanical Engineering School, Tianjin University, Tianjin, 300072, China*
[3] *Institute of Applied Mechanics, Taiyuan University of Technology, Taiyuan, 030024, China*
e-mail: zeyangang@yt-public.sd.cninfo.net, zeyangang @tdme.tju.edu.cn

Abstract

Stochastic resonance (SR) is a phenomenon wherein signal, noise and nonlinear system cooperate to enhance the signal-to-noise ratio(SNR) of the output signal. The SR characteristics of two bistable systems connected in series have been investigated in the present work. The viewpoint of the cascaded bistable system being considered as a low pass filter has been proposed, and hence the cascaded system can get rid of high frequency of the output of the former system but keep its time wave shape. With the analysis of the law of the signal spectral peak varying with noise intensity for the cascaded system, it is deduced that within a limited low frequency area, the cascaded system is able to promote the height of the signal spectral spike by a certain amount of noise. If the former of the two bistable systems does not stay in SR, the latter could not optimize the former output to become a SR state. For large parameters, because of too high spectral peaks of low frequency noise accumulated twice by the cascaded system, it is hard for the signal peak to be the highest amplitude in the whole spectral graph. Therefore, the bistable systems connected in series is generally benefit to the detection of the signal in time domain. An example of detecting weak aperiodic signal embedded in strong noise is provided in the end to illustrate that the cascaded bistable system is a powerful too for signal processing.

REFERENCES

[1] R. Benzi, A. Sutera, A. Vulpiana, *The Mechanism of Stochastic Resonance*, Phys. A, 14, (1981), L453-457.
[2] L. Gammaitoni, P. Hänggi, et al, *Stochastic resonance*, Rev. Mod. Phys., 70, (1997), 223-246.
[3] F.Chapean-Blondeau, *Stochastic Resonance at Phase Noise in Signal Transmission*, Phys. Rev. E, 61, (2000), 940-943.
[4] Y. G. Leng, T. Y. Wang, *Numerical research of twice sampling stochastic resonance for the detection of a weak signal submerged in a heavy noise*, Acta Phys. Sin., 52, (2003), 2432-2437.
[5] Y. G. Leng, T. Y. Wang, et al, *Power Spectrum Research of Twice Sampling Stochastic Resonance Response in A Bistable System*, Acta Phys. Sin., 53, (2004), in press.

COMPUTATIONAL MECHANICS (Abstracts)
WCCM VI in conjunction with APCOM'04, *Sept. 5-10, 2004, Beijing, China*

A Novel Blocking Algorithm for Large-Scale Sparse Matrix: the Auto-Optimized Blocking and Equivalent Block Method

Y. Q. Li [1*], Z. Y. Shen [1], Y. Tamura [2]

[1] *Department of Structural Engineering, Tongji University, Shanghai, 200092, China*
[2] *Wind Engineering Research Center, Tokyo Polytechnic University, Atsugi 243-0297, Japan*
e-mail: liyq@mail.tongji.edu.cn, zyshen@mail.tongji.edu.cn, yukio@arch.t-kougei.ac.jp

Abstract

Blocking algorithm is often used in many kinds of large-scale structural analysis. Many structural problems always finally become problems with linear equation group, which can be shown by solving a sparse matrix in most cases. Since it usually means a huge computational work in structural analysis, a more suitable and effective solution algorithm for this problem should be searched [1-3]. In this paper, a novel method for blocking algorithm, named "Auto-optimized Blocking & Equivalent Block method", was presented for effectively solving the large-scale sparse matrix equation group. In this method, an equivalent block $A^e_{i,i}$, which can simply represent all the influence of the influencing blocks to current active block, is defined by the Choleski decomposition results, $L_{i,i}$, D_i, of the previous sub-block as follows:

$$A^e_{i,i} = L_{i,i} D_i L^T_{i,i}$$

In addition, "Auto-optimized Blocking" means that blocking of current stiffness matrix of structures could be conducted after the band of sparse matrix was auto-optimized so that the equivalent block has the smallest dimension. The optimum controlling aims are shown as follows:

$$\left.\begin{array}{l} DOF_i \le DOF_{\max} \\ DIM(A^e_{i,i}) \Rightarrow \min \end{array}\right\}$$

By using the equivalent block, address finding of the influencing rows can be avoided, and the exchange between internal and external memories in that block is no longer necessary. Combined with auto-optimized blocking, the computational efficiency can be greatly improved. Finally, a simple example was given, which proved that this method was efficient and suitable to solve large-scale sparse matrix equation group. At the same time, this algorithm can save a lot of storage locations of computer, which is also important in a FEM analysis of large-scale structural problems.

REFERENCES

[1] Z. Y. He, X. D. Lin, *The numerical methods in nonlinear problems of structural engineering,* Beijing, National Defense Industry, (1994), (in Chinese).

[2] K. J. Bathe, *Finite element procedures in engineering analysis,* Prentice Hall, New Jersey, USA, (1982).

[3] Y. Q. Li, *Wind-resistant theoretical research on large-span reticulated shell structures and analysis of some problems in practical projects,* Report of Postdoctoral Research Work, Hangzhou, Zhejiang University, (2000), (in Chinese).

COMPUTATIONAL MECHANICS (Abstracts)
WCCM VI in conjunction with APCOM'04, *Sept. 5-10, 2004, Beijing, China*

Effective Implementation of a Time Discontinuous Galerkin method for Non-Linear Structural Dynamics

Massimo Mancuso [1], Francesco Ubertini [2*]

[1] *DIMC, University of Modena and Reggio Emilia, Via Vignolese 905/B, Modena, 41100, Italy*
[2] *DISTART, University of Bologna, Viale Risorgimento 2, Bologna, 40136, Italy*
e-mail: mancuso.massimo@unimo.it, francesco.ubertini@mail.ing.unibo.it

Abstract

In the last years much research has been devoted to developing accurate and robust time integration methods for non-linear structural dynamics. To this end, filtering out the unwanted high-frequency response has been recognized as a desirable feature and various attempts have been made to introduce such a dissipative feature (see [1] and the references therein, amongst other). An attractive approach is offered by the time discontinuous Galerkin (TDG) method, which has been successfully applied to linear and non-linear structural dynamics, owing to its good stability and accuracy properties (see for example [2-3]). Although it properly combines higher order accuracy and high-frequency dissipation, standard implementations turn out to be much more expensive than classical algorithms, since a larger system of equations should be solved at each time step. Thus, developing effective implementations plays a key role in the practical applicability of the method.
The purpose of this work is to establish an efficient iterative procedure to implement the TDG method based on piecewise linear time interpolation for non-linear dynamics. This is achieved by generalizing the algorithm proposed by the authors in [3] for linear problems. The solution strategy is rather different from previous efforts, which are essentially based on conventional iterative schemes (such as Gauss-Jacobi or Gauss-Seidel scheme). In particular, the present procedure is based on an implicit predictor-two-corrector scheme, designed to search for the TDG solution while preserving the TDG stability and dissipative properties at each iteration. Improving accuracy is demanded to iterative corrections. The proposed procedure is simple and offers several advantages. Remarkable computational savings are achieved, since iterative corrections are performed using an effective stiffness matrix, as it occurs in standard algorithms. In linear regime, no more than two iterations are needed to obtain the same overall accuracy of the TDG method. Moreover, the algorithm can be easily implemented into existing finite element codes. Numerical tests confirm the attractiveness of the present procedure, which appears to be competitive with commonly used algorithms and available TDG implementations.

REFERENCES

[1] D. Kuhl, M. A. Crisfield, *Energy conserving and decaying algorithms in non-linear structural dynamics,* Int. J. Numer. Meth. Engng, 45, (1999), 569-600.

[2] N. E. Wiberg, X. D. Li, *Adaptive finite element procedures for linear and non-linear dynamics,* Int. J. Numer. Meth. Engng, 46, (1999), 1781-1802.

[3] M. Mancuso, F. Ubertini, *An efficient integration procedure for linear dynamics based on a time discontinuous Galerkin formulation,* Comput. Mech., 32, (2003), 154-168.

COMPUTATIONAL MECHANICS (Abstracts)
WCCM VI in conjunction with APCOM'04, *Sept. 5-10, 2004, Beijing, China*

Evaluating the Homogeneity of Asphalt Mixture Using Digital Image Processing Techniques

Yong Peng *, Lijun Sun

The Key Laboratory of Road and Traffic Engineering, Tongji University, Shanghai, 200092, China
e-mail: y_peng22@hotmail.com ,ljsun@mail.tongji.edu.cn

Abstract

Homogeneity of asphalt mixture is dependent on the distributions of its constituents. Hereinto, constituents should be interpreted to mean asphalt cement, aggregates, additives, and air voids, and among which the distribution of aggregates is especially important and homogeneity of asphalt mixture is highly dependent on it, because the distributions of the other three constituents in asphalt mixture are generally determined by it. Using digital image processing techniques, the purpose of this study is to assess homogeneity of asphalt mixture by studying directly the aggregate distribution in asphalt mixture. The present methods for evaluating homogeneity of asphalt mixture, or segregation in asphalt mixture were reviewed. The digital image processing techniques were compendiously introduced. The method for evaluating homogeneity of a section in asphalt mixture was described in detail. The method for evaluating homogeneity of the whole was also presented. In the end, the validity of the method for evaluating homogeneity of a section in asphalt mixture was tested. The test results indicate that if the four aggregate distribution parameters in asphalt mixture, i.e. aggregate distributed situation, aggregate distributed number, area ratio coefficient which is defined as in a section the ratio of the area of aggregates retained on the same sieve size to the area of whole aggregates, and a coefficient concerning gradation are considered comprehensively, it is feasible to evaluate quantitatively homogeneity of asphalt mixture.

REFERENCES

[1] Qinglin Sha, *Freeway early distresses and prevention on Asphalt concrete pavement*, China Communications Press, Beijing, (2001), pp323-364, (in Chinese).

[2] Kenneth R. Castleman, *Digital image processing*, Prentice Hall, Inc. (1996), pp. 2-5.

[3] Jingna Zhang, *Study of microstructure analysis methodology for asphalt concrete based on digital image processing*, PhD Thesis, Tongji University, China, (2000), pp. 55-67, (in Chinese).

COMPUTATIONAL MECHANICS (Abstracts)
WCCM VI in conjunction with APCOM'04, *Sept. 5-10, 2004, Beijing, China*

Parallel Cell Based Computing System Using Cellular Automaton

Yoshitaka Wada[1], Takayoshi Matsumoto[2*], Masanori Kikuchi[3]

[1] *Department of Mechanics and Systems Design, Tokyo University of Science, Suwa, Tyohira 5000-1, Chino, Nagano, 391-0292, Japan*
[2] *School of Mechanical Engineering, Tokyo University of Science, Yamazaki 2641, Noda, Chiba, 278-8510, Japan*
[3] *Department of Mechanical Engineering, Tokyo University of Science, Yamazaki 2641, Noda, Chiba, 278-8510, Japan*
e-mail: wada@rs.suwa.tus.ac.jp, takayosh@me.noda.tus.ac.jp, kik@me.noda.tus.ac.jp

Abstract

Many numerical simulation techniques have been developed and utilized in engineering. But model generation is still one of the major problems in numerical simulation. On the other hand, development of recent high performance computing can realize detailed and realistic analysis. If we effectively use recent computer hardware and software, we should develop a new software to solve problems of model generation and use of high performance. Therefore we propose a cell based parallel computing [1] system using a cellular automaton [2] model to satisfy two requirements.
The proposed system is designed by object-oriented technique [3] to realize extensibility for parallel cellular automaton model. Because the system needs only information of model as cells which are represented by pixel and voxel mesh, model generation is very easy from any data formats. We implement several applications, which are 1-D life game, 2-D life game, 2-D diffusion, 2-D parallel diffusion, 2-D wave propagation [4], 3-D wave propagation and 3-D parallel wave propagation on the system. Status of cells is updated by adjacent status of cells. In other words, behavior of system is completely specified in terms of a local relation. Cell can hold many properties to represent its status Cellular automata system can calculate physical phenomena and discrete problem like a life game simultaneously. We present some results of a physical analysis and discrete analysis, and discuss effectiveness and parallel efficiency of proposed system.

REFERENCES

[1] P. S. Pacheco, *Parallel programming with MPI*, Morgan Kaufman Publisher, (1997).
[2] S. Wolfram, *Cellular automata and complexity-collected papers*, Addison-Wesley, (1994).
[3] E. Gamma, R. Helm, R. Johnson, J. Vlissides, *Design patterns elements of reusable object-oriented software*, Addison Wesley, (1995).
[4] T. Komatsuzaki, H. Sato, Y. Iwata, S. Morishita, *Simulation of wave propagation using cellular automata*, Transactions of JSCES, Paper No.19990017, (1999).

COMPUTATIONAL MECHANICS (Abstracts)
WCCM VI in conjunction with APCOM'04, *Sept. 5-10, 2004, Beijing, China*

Development of Web Based CAE System: CASOW

Yoshitaka Wada [1*], Shunsuke Rikitake [2], Masanori Kikuchi [2]

[1] *Department of Mechanics and Systems Design, Tokyo University of Science, Suwa, Tyohira 5000-1, Chino, Nagano, 391-0292, Japan*
[2] *Department of Mechanical Engineering, Tokyo University of Science, Yamazaki 2641, Noda, Chiba, 278-8510, Japan*
e-mail: wada@rs.suwa.tus.ac.jp, riki@me.noda.tus.ac.jp, kik@me.noda.tus.ac.jp

Abstract

CAE software is developed year after year and many engineers use it and number of the users is increased. CAE software becomes complicated as CAE software's advances. Indeed topologically same model are analyzed in engineering work. In short, in such a case we don't need generalized software of CAE, and we just need simple specialized CAE software. On the other hand, internet environment becomes better and faster and then ubiquitous computing [1] is being realized. Many network based CAE systems are studied and developed [2]. WebCS [3] is an integrated CAE system to join researchers each others. WebCS has many functions which are source code analyzer, FEM analysis platform, result recorder (database system) and so on. WebCS are implemented using XML, XSL, and XSLT [4], Java and Servlet and other internet technologies. As a result the WebCS possesses extensibility of system, percolation of firewall and standard user interface using HTML, but the system becomes complex and it is difficult to add a new CAE software. In this presentation we propose web based CAE system: CASOW (Computer Aided Simulation On Web). CASOW has following characters: 1) Anyone uses the CASOW at any place through web; 2) Simple implementation to develop the system without database and other complicated technology; 3) Various modeling for analysis GeoFEM [5] is used as an FEM application which is a parallel FEM platform software. We demonstrate that advantages of the interface and easy operation and discuss effectiveness of the proposed system.

REFERENCES

[1] Ubiquitous computing resource, http://www.ubiq.com/hypertext/weiser/UbiHome.html
[2] M. Matsumoto, K. Nishimori, S. Kawata, *WWW application system for a numerical simulation server development environment*, Trans. of JSCES, Paper No.199990005 (1999).
[3] K. Minami, T. Ohtani, N. Jimbo, K. Sakane, *Software development for the researcher community formation that used GeoFEM as an example*, Trans. of 15th Comp. Mech. Conf., (2002), pp.89-90.
[4] XML and XSL resource, http://www.w3c.org/XML/
[5] GeoFEM project web page, http://geofem.tokyo.rist.or.jp/

COMPUTATIONAL MECHANICS (Abstracts)
WCCM VI in conjunction with APCOM'04, *Sept. 5-10, 2004, Beijing, China*

Multigrid Technique for Hexahedral Mesh Using Unstructured Local Mesh Refinement

Yoshitaka Wada [1], Jun'ichi Shinbori [2*], Masanori Kikuchi [2]

[1] *Department of Mechanics and Systems Design, Tokyo University of Science, Suwa, Tyohira 5000-1, Chino, Nagano, 391-0292, Japan*

[2] *Department of Mechanical Engineering, Tokyo University of Science, Yamazaki 2641, Noda, Chiba, 278-8510, Japan*

e-mail: wada@rs.suwa.tus.ac.jp, riki@me.noda.tus.ac.jp, kik@me.noda.tus.ac.jp

Abstract

3-D finite element method (FEM) is widely used as an effective numerical simulation technique. In the simulation technique, accuracy is one o f the most significant issues. In case of FEM, the accuracy is affected by number of freedom and shape of each element. On the other hand, iterative solvers based on conjugate gradient method are widely used, but its convergence depends on meshes and a kind of analyses.

Multigrid method[1] is one of the solutions of such a convergence problem and reducing calculation time. Generally multigrid method requires two or more meshes in order to propagate and convergence residual of error norm. Cost of calculation and convergence depend on multigrid cycles. From the view point of preconditioning, we can change degree of precondition and calculation cost. Hexahedral mesh generation and its adaptation are not enough to use for practical applications, because its mesh generation is very difficult and still labor intensive work by hand. We have developed hexahedral local refinement[2] technique controlled by posterior error estimation using Zienkiewicz and Zhu method[3]. A multi-refined mesh can be obtained by just appl ying refinement some or several times. In local refinement process a hexahedral element is divided into 27 elements and transition mesh patterns are generated around refined elements. Transition mesh patterns are connected between refined and coarse elemen ts and contain several distorted elements. Indeed high error norm occurs in such a distorted element, but total error norm are certainly decreased. A multi-refined mesh is an optimal mesh as concerned number of nodes and accuracy. Therefore we utilize multi-refined meshes as a multigrid.

We propose multigrid method for hexahedral mesh using adaptive unstructured local mesh refinement. We evaluate effect of transition mesh patterns and compare between preconditioned CG and multigrid technique.

REFERENCES

[1] W. Hacksbusch, *Multi-grid methods and applications*, Springer Verlag, (1985).

[2] Y. Wada, H. Okuda, *Effective adaptation technique for hexahedral mesh*, ACES Special Issue of Concurrency and Computation: Practice and Experience, 14, (2002), 451-463.

[3] O. C. Zienkiwicz, Z. Zhu, *A simple error estimator and adaptive procedure for practical engineering analysis*, Int. J. Numer. Meth. Eng., 24, (1987), 337-357.

COMPUTATIONAL MECHANICS (Abstracts)
WCCM VI in conjunction with APCOM'04, *Sept. 5-10, 2004, Beijing, China*

Study of Large Parameter Stochastic Resonance in Signal Processing

T. Y. Wang [1], Y. G. Leng [1*], X. D. Qin [1], Y. Guo [2], S. B. Fan [1], W. J. Wang [1]

[1] *School of Mechanical Engineering, Tianjin University, Beiyang Science Building No. 105, Tianjin, 300072, China*
[2] *School of Management, Tianjin University, Tianjin, 300072, China*
e-mail: tywang@tdme.tju.edu.cn, zeyangang@tdme.tju.edu.cn

Abstract

In the context of signal processing, stochastic resonance (SR) is generally suitable for small parameters (the frequency and amplitude of a periodic signal and noise intensity are all smaller than one) such as the amplification of a small signal, but for large parameters (frequency and/or amplitude and/or noise intensity can be much larger than one) such as the detection of a weak, large-frequency signal submerged in strong noise for short data records. However, a novel way of frequency transformation stochastic resonance (FTSR) proposed here is able to not only realize the large parameter SR-like response, but also successfully extract the weak signal from heavy noise. With the large parameter FTSR technique, the behavior study of the response power spectrum of a bistable system demonstrates that, to obtain an explicit maximal SR-like spectral spike at the periodic signal frequency, the ratio of sampling frequency and the signal frequency should be larger than fifty and the compressed sampling frequency in the FTSR calculation can be caught by the estimation of noise intensity. In addition, the analysis of the output SN ratio and the input SN ratio of the bistable system under the condition of heavy noise indicates that the former is always larger than the latter, which interprets why the signal spectral peak could be viewed in the response power spectrum of the bistable system but in the input spectrum. Therefore, the approach of large parameter FTSR is of promising future in practical engineering signal processing.

REFERENCES

[1] L. Gammaitoni, P. Hänggi, P. Jung, F. Marchesoni, *Stochastic resonance*, Rev. Mod. Phys., 70, (1998), 223-246.
[2] J. J. Collins, C. C. Chow, T. T. Imhoff, *Stochastic resonance without tuning*, Nature, 376, (1995), 236-238.
[3] L. Andrey, Pankratov, M. Salerno, *Resonant activation in overdamped systems with noise subjected to strong periodic driving*, Phys. Lett. A, 273, (2000), 162-166.
[4] Y. G. Leng, T. Y. Wang, *Numerical research of twice sampling stochastic resonance for the detection of a weak signal submerged in heavy noise*, Acta Phys. Sin., 52, (2003), 2432-2438.
[5] F. Chapeau-Blondeau, *Stochastic resonance at phase noise in signal transmission*, Phys. Rev. E, 61, (2000), 940-943.
[6] Z. Gingl, R. Vajtai, L. B. Kiss, *Signal-to-noise ratio gain by stochastic resonance in a bistable system*, Chaos, Solitons & Fractals, 11, (2000), 1929-1932.

COMPUTATIONAL MECHANICS (Abstracts)
WCCM VI in conjunction with APCOM'04, *Sept. 5-10, 2004, Beijing, China*

Seismic Response of MDOF Systems by Wavelet Transform

Meiling Xiao [1*], Liaoyuan Ye [2]

[1] *Institute of Yunnan Province for Earthquake Engineering Research, College of Civil Engineering, Kunming University of Science and Technology, Kunming, 650051, China*
[2] *Yunnan University, Kunming, 650051, China*
e-mail: *xmlgct@public.km.yn.cn, lyye@ynu.edu.cn*

Abstract

This paper proposes a wavelet transform method based on mode shape decomposition for linear structure system of Multi-degree-of-freedom (MDOF) seismic response. On basis of Duhamel integral that structure responses can be expressed as earthquake ground motion's wavelet transform. Due to analyzing linear structure seismic response and the solution includes the impulse function. From the point of wavelet, if the impulse function has scale factor, the seismic solution of structure based on duhamal integral is just a wavelet transform. That is, the structure's displacement response, velocity response and absolute acceleration response can be expressed by earthquake ground motion acceleration process's wavelet transform. These special wavelets are called odd exponent wavelets, their local characteristic of time-frequency are associated with the structure dynamic characteristic such as natural frequency and damp ratio. Since wavelet analysis techniques have made it possible to represent the temporal variations of frequency content and amplitude in a signal elegantly, this proposed method is not only very simple but also considers two factors of earthquake ground motion: amplitude and spectrum. Wavelet coefficients reflect amplitude; wavelet transform by scale factor variation reflects spectrum of earthquake ground motion. This approach has been formulated and also has been validated through shake-table test. Experiment for the two-storey frame model that the scale for model to prototype is 1:2.7 shake-table test has been proved this method is practicable. In this test, the base acceleration which input the structure is EL-CENTRO (N-S) and the maximum peak is 0.1g. The dynamic parameters (mode shape, natural frequency and damp ratio) of test structure were tested by dynamic character test. The odd exponent wavelets' time domain curves and frequency domain curves for different natural frequency in this test structure were given. And also explain their local characteristic of time-frequency. Then seismic solutions of the structure for the proposed approach were calculated. By comparing wavelet method with the experiment, the results of each storey were in good agreement.

REFERENCES

[1] Biswajit Basu, Vinay K. Gupta, *Seismic response of SDOF systems by wavelet modeling of non-stationary processes*, Journal of Engineering Mechanics, 124, (1998), 1142-1150.

[2] Sushovan Mukherjee, Vinay K. Gupta, *Wavelet-based characterization of design ground motions*, Earthquake Engng Struct. Dyn., 31, (2002), 1173-1190.

COMPUTATIONAL MECHANICS (Abstracts)
WCCM VI in conjunction with APCOM'04, *Sept. 5-10, 2004, Beijing, China*

The Design of the Framework for Object-oriented Nonlinear Finite Element Program

Xin Zong *, Jun Mo

Institute of Structure Mechanic, CAEP, Mianyang, Sichuan, 621900, China
e-mail: songofthelife@sina.com

Abstract

The paper show an approach to nonlinear finite element analysis based on object -oriented programming techniques, and aimed at the general design of the application framework of the finite element analysis core.

By comparing with the conventional procedural finite element programming, the object -oriented programming method is introduced to the fini te element programming, and a series of data object classes such as FEObject, Node, Element, Load and etc., used frequently in the process of finite element analysis are constructed, based on which the concept FEFC (Finite Element Foundation Class Library) is presented. FEFC contains classes which work together by associating with each other, and the relationship of the se classes is comparative determined. Furthermore, FEFC can make the development of finite element programs more effective and apt to the id ea of OOP (Object-Oriented Programming). Based on FEFC, an object-oriented finite element application framework related with analysis core is designed, and the framework can provide templates of common finite element programs. Under this framework, nonline ar finite element programming can be convenient, and linear finite element programming can also be contained. Instantiating objects of classes in FEFC or classes derived from it can drive a message routing when a nonlinear finite element program executing, and then the work can be done. The main idea of this paper can be show as the figure, and the encapsulation, inheritance and polymorphism in object -oriented methodology are embodied.

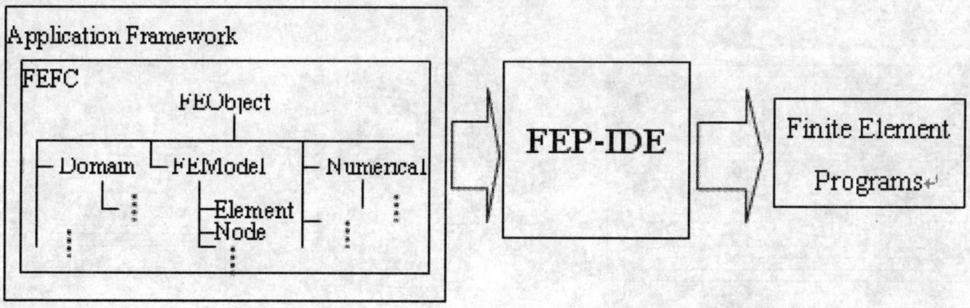

REFERENCES

[1] Bruce W. R. Forde, Ricardo O. Foschi, Siegfried F. Stiemer, *Objecet-oriented finite element analysis*, Computers & Structures, 1990, 34(3), 355-374.

[2] Thomas Zimmermann, Yves Dubois-Pelerin, Patricia Bomme, *Objecte-oriented finite element programming: (I) Governing principles*, Computer Methods in Applied Mechanics and Engineering, 98, (1992), 291-303.

[3] Ph. Menetrey, Th. Zimmermann, *Object-oriented non-linear finite element analysis: application to J2 plasticity*, Computers & Structures, 49(5), (1993), 767-777.

COMPUTATIONAL MECHANICS (Abstracts)
WCCM VI in conjunction with APCOM'04, *Sept. 5-10, 2004, Beijing, China*

Adaptive Finite Element Mesh Generation Oriented to Geometrical and Mechanical Features

Qungui Du [1*], Xiaodong Huang [1], Bangyan Ye [2]

[1] *College of Traffic & Communication, South China University of Technology, Wushan, Guangzhou, 510641, China*
[2] *College of Mechanical Engineering, South China University of Technology, Wushan, Guangzhou, 510641, China*

e-mail: ctqgdu@scut.edu.cn, ctdhuang@ctang.com, hgycby@cnuninct.com

Abstract

Four Properties are found in our research, which are relative with initial Delaunay triangulation of a point set that is generated by dis cretizing the boundary of a 2-dimension domain with a constant finite element size. The first property describes the relation between initial Delaunay triangulation and the medial axis of the domain. The second property points out the maximal circum-circle diameter of the triangle element whose center is in the domain is a constant value. The third property describes the relation between initial Delaunay triangulation and the curvature of the boundary edge. The fourth property describes that the circum-circles of triangles whose center is in the domain are tend to be enveloped by the boundary. These four properties build relation between initial De launay triangulation and the geometrical features of the domain.

According to these four properties, methods to recognize geometrical and mechanical features automatically are provided and adaptive mechanism for mesh generation is built. A new method to generate adaptive mesh is developed, which is based on Delaunay triangulation. Grade adaptive mesh with high qual ity can be automatically generated oriented to geometrical and mechanical features and represent a better approximation of the exact geometrical model. Examples show that i t is advantageous to improve efficiency of modeling process and calculating accuracy in finite element analysis.

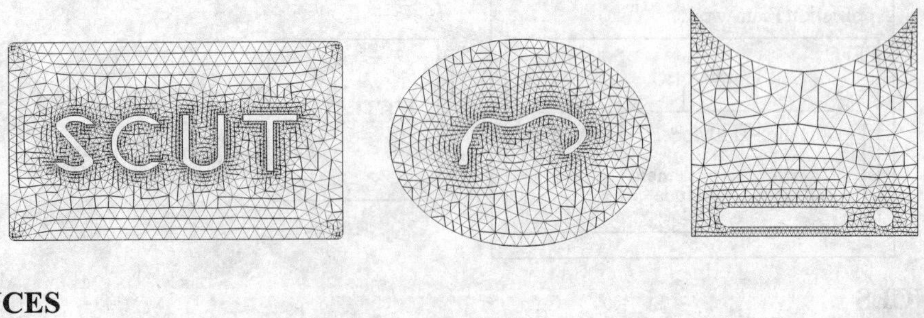

REFERENCES

[1] Weixin Wang, Dahua Deng, *An adaptive mesh generation algorithm based on geometric features and mechanical properties*, Journal of Computer Aided Design & Computer Graphics, 13(1), (2001), 52-55.

[2] S. H. Lo, C. K. Lee, *Selective regional refinement procedure for adaptive finite element analysis*, Computers &Structures, 68, (1998), 325- 341.

COMPUTATIONAL MECHANICS (Abstracts)
WCCM VI in conjunction with APCOM'04, *Sept. 5-10, 2004, Beijing, China*

Triangulation with Internal Points Generated by Crossed Circles

Y. Ezawa *

Department of Computational Science & Engineering, Faculty of Engineering, Toyo University, 2100 Kujirai, Kawagoe, Saitama, 350-8585, Japan
e-mail: ezawa@eng.toyo.ac.jp

Abstract

Delaunay method [1] is a powerful method for triangulations. The quality of mesh generated by this method, however, depends on the allocation of nodes. For getting suitable allocations, Bubble method [2] was developed. In Bubble method all nodes should be defined in advance. Especially it is important how to determine internal nodes. In this study a meth od for generating internal points is presented. This method is suitable to Delaunay method and is called Crossed circle method because crossed circles are used for generating internal points. In this method the following process is inserted after swapping circumcircles in the process of Delaunay method.
1) Calculate the positions of the centers of the swapped circumcircles that form crossed circles. 2) Check if both centers are included in both circumcircles. 3) Calculate the position of the middle point of these centers. 4) Check if any nodes exist in the neighborhood of this middle point. 5) Add this middle point.
The figure 1 shows the example that is the initial nodes for testing the Crossed circle method. The figure 2 is the mesh by Crossed circle me thod. This result shows that the Crossed circle method can make a high-quality mesh where each triangle is not depressed. And this new method does not require troublesome addition of nodes by hand.

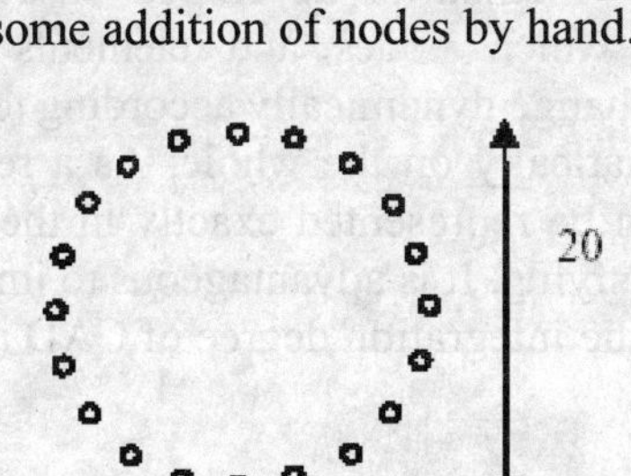

Fig. 1 Initial nodes

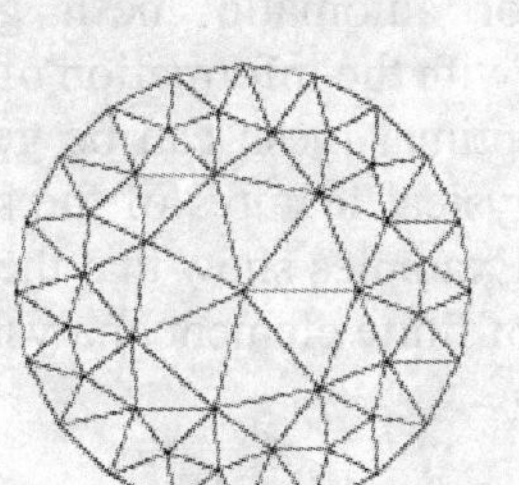

Fig. 2 Crossed circle method

REFERENCES

[1] P. L. George, *Automatic mesh generation – application to finite element methods* –, Wiley, (1991).

[2] K. Shimada, D. C. Gossard, *Bubble mesh generation: automated 2D/3D triangulation via physically-based node placement and modified Delaunay tessellation*, WCCM III, Vol. I, Tokyo, August, (1994), pp.900-902.

COMPUTATIONAL MECHANICS (Abstracts)
WCCM VI in conjunction with APCOM'04, *Sept. 5-10, 2004, Beijing, China*

3-D Adaptive Finite Element Mesh Generation Based on Projecting Method

Xiaodong Huang [1*], Qungui Du [1], Bangyan Ye [2]

[1] *College of Traffic & Communication, South China University of Technology, Guangzhou, 510641, China*
[2] *College of Mechanical Engineering, South China University of Technology, Guangzhou, 510641, China*
e-mail: ctdhuang@etang.com, ctqgdu@scut.edu.cn, hgyeby@cnuninet.com

Abstract

With the development of 3-Dimension finite element mesh automatic generation, more attention has been paid to full 3d adaptive finite element analysis, which is able to adjust the mesh according to the physical features of the object to be analyzed and guarantee geometrical features of the solid been represented exactly in mesh. For this purpose, refined mesh should be generated in the special areas of the solid in an optimal manner. A new method is developed in this paper for tetrahedral adaptive mesh automatic generation.
According to the geometrical information provided by the solid model and the physical features of the object estimated by analyzer, a field of the element size information on the whole is built in a projecting manner after appropriate modes are selected to act on the special areas for refined mesh generation. Typical modes are provided for refined mesh generation in the special areas. They are center-mode, curve-mode, surface-mode, etc. New method is developed based on Delaunay triangulation for automatic mesh generation, in which nodes and elements are generated simultaneously with the information of element size change dynamically according to the position of element. The optimal mesh can be generated automatically on the whole. As a result, geometric features and physical features of the solid object can be represented exactly in the adaptive finite element mesh. Examples show that the method is satisfying. It is advantageous to improve accuracy and efficiency of finite element analysis and elevate the integration degree of CAD and CAE.

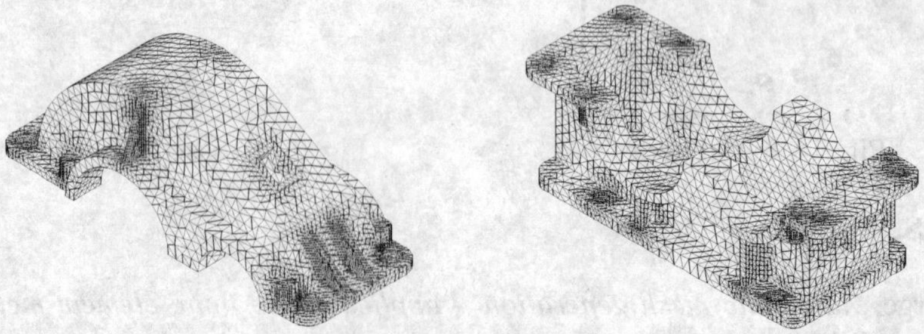

REFERENCES

[1] M. L. Staten, N. L. Jones, *Local Refinement of Three-Dimensional Finite Element Meshes*, Engineering with Computers, 13, (1997), 165-174.
[2] Weixin Wang, Dahua Deng, *3D Automatic Mesh Generation Method for Solid Object Simulation Modeling*, Chinese J. Computers, 25, (2002), 44- 48, (in Chinese).

COMPUTATIONAL MECHANICS (Abstracts)
WCCM VI in conjunction with APCOM'04, *Sept. 5-10, 2004, Beijing, China*

Design, Optimization and Simulation of the Focal Structural of LAMOST

J. F. Jin [1*], H. B. Chen [1], P. F. Shao [2], H. Z. Hu [2], C. Zhai [2]

[1] *CAS Key Laboratory of Mechanical Behavior and Design of Materials, Department of Modern Mechanics, University of Science and Technology of China, Hefei, Anhui230026, China*
[2] *Department of Precision Machinery and Precision Instrumentation, University of Science and Technology of China, Hefei, Anhui230026, China*
e-mail: jjf@ustc.edu.cn, hbchen@ustc.edu.cn

Abstract

The Large Sky Area Multi-object Fiber spectroscopic Telescope (LAMOST) project [1] is a large scale national scientific project in China at the turn of the century. LAMOST is a meridian reflecting Schmidt telescope with a clear aperture of 4-meter, a focal length of 20-meter and a wide field of view of 5-degree and can be used to observe about 4,000 celestial b odies simultaneously. The focal structural of LAMOST [2] is an available large focal plane of 1.75 meter in diameter which may accommodate up more than 4,000 fiber positioning mechanisms, by which the collected light of distant and faint celestial objects down to 20.5 magnitude is fed into the spectrographs. The spectroscopic survey carried out by LAMOST of tens of millions of galaxies and others will make substantial contribution to the study of extra-galactic astrophysics and cosmology, such as galaxies, quasars and the large-scale structure of the universe [3].

In this paper, the gravity and thermal deformation of the focal structure using the finite element package ANSYS is simulated since the large deformations of focal structural will affect celestial body image. Because of the complicate of the LAMOST structural and so much fiber positioning mechanisms even more than 4000, a large FEA model more than 1.6 millions nodes and more than 200,000 quadratic pentahedral elements are obtained in preprocessing and we find an economically way to complete so large-scale engineering project on a personal microcomputer. Based on the simulation data of whole LAMOST structural, the optimization size of support radius and focal thickness is proposed. Finally we make some suggestions for proper material and the design size of focal structural.

REFERENCES

[1] Yaoquan Chu, Xiangqun Cui, *The LAMOST project*, in N. Kaifu ed. Proc of 3rd East-Asian Meeting on Astronomy, (1995), p.327.

[2] Xiangquan Cui, Dehua Yang, *Support structure of LAMOST Schmidt plate,* Ma, SPIE, 3352, (1998), 378.

[3] Xiangquan Cui, Dingqiang Su, Shouguan Wang, *A new configuration for a 10-metre optical/infrared telescope*, Chinese Astronomy and Astrophysics, 25, (2001), 253-258.

COMPUTATIONAL MECHANICS (Abstracts)
WCCM VI in conjunction with APCOM'04, *Sept. 5-10, 2004, Beijing, China*

Network-Based FEA and Engineering Collaborative Design System

Haijiang Li [1*], Gengdong Cheng [2]

[1] *Institute of Road and Bridge Engineering, Dalian Maritime University, Linghai Road 1, Dalian, 116026, China*

[2] *Department of Engineering Mechanics, Dalian University of Technology, Linggong Road 2, Dalian, 116024, China*

e-mail: haijiang@newmail.dlmu.edu.cn, chenggd@dlut.edu.cn

Abstract

This paper represents a new network-based engineering software system called NetFEF (Network-based Finite Element Analysis and Engineering Collaborative Design Framework).Combined with some new computer technologies, like high performance computing cluster, network collaborative design, parallel and distributed computing, and OO method, NetFEF integrates three important parts: OOParaFEA (Object-Oriented Parallel Finite Element Analysis kernel), WebFEA (Web-based distributed Finite Element Analysis) and NetGWCAD (Network-based Gravity Wharf CAD integrated system, which has passed the appraisal of China Ministry of Transportation with a comment which says "holds the first-class level in China").

First, the whole structure of NetFEF is described generally. Next, the process for implementing the system is given, which consists of several steps as follows, using newly created cluster class hierarchy to build the PCs computing cluster as the hardware platform of NetFEF; building the object-oriented parallel FEA kernel which realizes linear/nonlinear, serial/parallel analysis, sub-structure method and C++ socket-based communication; building WebFEA which integrates HttpServer to build internet project web site with ComputingServer to deal with different computing tasks, and combines Web-based distributed FEA with cluster parallel FEA to form a seamless framework; building NetGWCAD which uses Java applet as client interface, web server extended by Java servlet as a control center, PCs cluster with parallel FEA ability as a comput ing center, self-developed project data transfer module as the communication layer, successful testing in our university and other institutes of construction has shown that NetGWCAD can increase the wharf analysis and design efficiency by several times. Then some examples are introduced to show the applications of NetFEF, and a concept named "charging only for service" is put forward to make the NetFEF practical in business on the Internet. Finally, some conclusions are given in the paper.

REFERENCES

[1] Mark Baker, *Cluster computing white paper*, University of Portsmouth, UK, (2000).

[2] H. Adeli, G. Yu, *An integrated computing environment for solution of complex engineering problems using object-oriented programming paradigm and a blackboard architecture*, Computers & Structures, 54(2), (1995), 255-265.

[3] J. Peng, et al, *An open source model for collaborative development of finite element program*, International Conference on Computing in Civil and Building Engineering, Palo Alto, CA, August, (2000), pp. 1309-1316.

COMPUTATIONAL MECHANICS (Abstracts)
WCCM VI in conjunction with APCOM'04, *Sept. 5-10, 2004, Beijing, China*

A Common Geometry Interface for Mesh Generation

J. L. Shan*, Z. Q. Guang, Y. X. Gu

State Key Laboratory of Structural Analysis for Industrial Equipment, Dalian University of Technology, Dalian 116024, China
e-mail: sjulin@student.dlut.edu.cn

Abstract

How to access to the CAD geometry in an accurate and efficient manner is one of the major issues of mesh generation in recent years. Commonly, The techniques for CAD geometry accessing include: Translation & Healing, Discrete Representations, Direct Geometry Access, but these techniques have many limitations. On the other way, high-end CAD systems provide access to their geometry through APIs. However, differences between these CAD systems' APIs are quietly distinct, both at the syntax level and in the underlying topological models. This paper proposed a common geometry interface for mesh generation (CGIM) to solve this problem.
CGIM is implemented using the C++ object-oriented language. The implementation of the algorithm using object-oriented methods including popular Design Patterns allows flexibility in programming and increases the computational efficiency of mesh generation. Three main Design Patterns called "Bridge", "Factory Method" and "Singleton" to achieve one of the main traits: crossing the CAD system, are also presented in this paper.
CGIM is a wrapper of CAD systems' geometry APIs, which translate function calls such that client access geometry in CAD systems' native format. However, CGIM also provides its own topology entities corresponding to CAD geometry, such as vertex, edge, loop, shell and so on. Some tool functions, like generation of boundary representation tree (Brep Tree), discretion of CAD geometry module, and persist file object, for mesh generation are also contained. CGIM is designed to be extensible, thus, it is easy to build a new driver component to access to a new CAD geometry-modeling engine. Currently, the access component to Autodesk® Mechanical Desktop® has been established. The access components to other CAD systems such as SolidWorks® and SolidEdge® are underway.

REFERENCES

[1] Mark W. Beall, Joe Walsh, Mark S. Shephard, *Accessing CAD geometry for mesh generation*, Proc. 12th International Meshing Roundtable, Sandia National Laboratories, Sept, (2003), pp. 33-42.

[2] Tim Tautges, *The common geometry module: a solid model-based geometry library*, http://legacy.ep.wisc.edu/~tjtautg/CGM-seminar/CGM-seminar_frame.htm.

COMPUTATIONAL MECHANICS (Abstracts)
WCCM VI in conjunction with APCOM'04, *Sept. 5-10, 2004, Beijing, China*

A Mesh Adaptation Schema for Unstructured Mesh of 3D Solids

Chao Song*, Zhenqun Guan, Yuanxian Gu

State Key Laboratory of Structural Analysis for Industrial Equipment, Department of Engineering Mechanics, Dalian University of Technology, Dalian, 116024, China
e-mail: songchao@student.dlut.edu.cn, guanzhq@dlut.edu.cn, guyx96@dlut.edu.cn

Abstract

The accuracy of finite element analysis (FEA) strongly depends on the quality of the mesh generated in the domain. In complicated cases, the desired size and shape of elements vary over the global domain to meet the requirement on computational accuracy. The mesh adaptation can provide FEA program with elements holding such property. In this paper, a mesh adaptation schema of the fully automatic unstructured mesh generation for 3D solid models is proposed to feed FEA program. Local remeshing technique is used to perform the mesh adaptation and FEA software called JIFEX [1] is adopted to perform finite element analysis in this schema. Mesh generation algorithm, element size control method, as well as error indicator are three key issues in local mesh regeneration, and they will be addressed with following approaches.

The advancing front technique (AFT) [2] is employed to generate tetrahedral elements automatically for its advantages of local element size control and boundary integrity. Particularly, linear programming is applied to solve the final hull problem commonly existing in 3D AFT to guarantee the convergence of the algorithm. The element size control method, which specifies element size distribution over the whole domain, is another important issue. It determines the transition of element sizes, and influences the robustness of mesh generation method. Background control space is utilized as size control method to govern the element size distribution over the whole domain. The mesh of former analysis stage is taken as the background mesh for the mesh adaptation of current stage. A structured grid technique that helps to manage the background mesh elements is introduced to accelerate the size computational speed. Error indicator [3], which gives the feedback to mesh generation algorithm, is an essential part for the mesh adaptation process.

This mesh adaptation schema is implemented successfully for unstructured mesh of 3D solid models as the preprocessor of JIFEX. An example is given in the paper to illustrate that the schema can provide adaptive finite element analysis the high quality mesh with desired element sizes and gradient in the iterative adaptation process es.

REFERENCES

[1] Y. X. Gu, H. W. Zhang and et al. *New generation software of structural analysis and design optimization-JIFEX.* Int. J. Struct. Eng. Mech., 7, (1999), 589-599.

[2] R. Lohner, *Progress in grid generation via the advancing front technique,* Engineering with computers, 12, (1996), 186-210.

[3] J. Z. Zhu, O. C. Zienkiewicz, *A posteriori error estimation and three-dimensional automatic mesh generation, Finite Elements in Analysis and Design*, 25, (1997), 167-184.

COMPUTATIONAL MECHANICS (Abstracts)
WCCM VI in conjunction with APCOM'04, *Sept. 5-10, 2004, Beijing, China*

Visual Structure Expert System VSES03

Jixin Yang [1, 2*], Yan Zeng [1], Yifeng Huang [1]

[1] *College of Transportation, Wuhan University of Technology, Wuhan, 430063, China*
[2] *College of Civil Engineering and Mechanics, Huazhong University of Science & Technology, Wuhan, 430074, China*

e-mail: whutvses@163.com, hnqyyjx@sohu.com

Abstract

A structure analysis system—Visual Structure Expert System (VSES) is introduced. VSES is a visual software for mechanics analysis which can be widely used in engineering. Its first version is formed in 2000 and the newest version VSES03 is formed in 2003. It is developed on the base of building structure vibration calculation procedure SUNV and the visual bridge computation system VMAS. In the aspects of solving structures, it has the similar function with the Super SAP, ANSYS and so on. The system is running on windows'98 or the higher edition. It does not need any other software to support except operation system. The procedure is written with the Visual Basic, roughly 50,000 sentences.

In this paper something about VSES are introduced, including its function, structure, designing principle and some key techniques such as the triangle mesh generation, the hexahedron mesh generating technique, the hidden method for 3-D models, the approach of forming stress images, etc. VSES is a integrated software which can build structure models, do mechanics analysis and show the results in several ways. VSES has many advantages such as exact calculation, high speed, large computing scale, high degree visualization and so on. It can be widely used to analyze the structures in mechanics engineering, civil engineering and transportation engineering, etc.

REFERENCES

[1] Jixin Yang, *Analysis on vertical vibration of multi-story buildings and the program researched*, Master's Thesis of Tianjin University, (1998), (in Chinese).

[2] Jixin Yang, Dingfang Chen, *Visual bridge structure computation system*, Transportation and Computers, 18(3), (2000), 49-52.

[3] Jixin Yang, Dingfang Chen, *Design and realization of visual finite element system*, Journal of Wuhan Transportation University, 4(2), (2000), 142-146, (in Chinese).

[4] Jixin Yang, Dingfang Chen, *Mesh generation method for triangle elements in VSES*, Journal of Wuhan University of Technology (Transportation Science & Engineering), 26(1), (2002), 28-31, (in Chinese).

[5] Jixin Yang, Dingfang Chen, *A practical method of generating hexahedron mesh for finite element method*, Journal of Wuhan University of Technology, 23(11), (2001), 42-45, (in Chinese).

[6] Jixin Yang, Jie Liu, Dingfang Chen, *A efficient and applied hidden method for FEM models*, Journal of Wuhan University of Technology, 24(8), (2002), 60-62, (in Chinese)

[7] Jixin Yang, Dingfang Chen, *A method for forming stress images*, Computer Engineering and Applications, 38(16), (2002), 113-115.

COMPUTATIONAL MECHANICS (Abstracts)
WCCM VI in conjunction with APCOM'04, *Sept. 5-10, 2004, Beijing, China*

Parallel Numerical Simulation of Reactive Flow using Finite Volume Element Method

A. Banaeizadeh*, M. Darbandi

Department of Aerospace Engineering, Sharif University of Technology, Azadi Ave., PO *Box 11365-8639*, Tehran, Iran
e-mail: banaeizadeh@hotmail.com, darbandi@sharif.edu

Abstract

A highly accurate and efficient method is developed to solve the laminar reactive flows using parallel computers. Based on control-volume-based finite-element methods, the formulation retains the geometrical flexibility of the pure finite element methods while derives the governing discrete algebraic equations using the conservation balance applied to control volumes. The physical influence upwinding scheme is used to approximate the convection terms at cell faces. This treatment respects the physics of flow and provides the necessary coupling of velocity and pressure fields. The parallel processing is achieved using MPI (Massage Passing Interface) library. The solution of laminar diffusive flame is obtained in three main steps: 1. the mesh generation and domain decomposition, which is required for parallel computations, is performed in a manner that causes minimum interfaces between sub-domains, and hence, decreases the communication time between the processes. 2. A fully implicit solution of the governing fluid flow equations is obtained using the finite volume element method in each decomposed domain. 3. The equations for chemical species are solved for mass fractions of species in a semi-coupled manner. From the parallel computation point of view, the primitive variables of the fluid flow and the mass fraction of the chemical species are updated in the ghost points at the end of individual iterations. This data exchange is compulsory in the current fully implicit procedure. The results are excellent comparing with the experimental results and benchmark solutions. Additionally, the algorithm provides very good parallel processing performances.

REFERENCES

[1] D. Elkaim, M. Reggio, R. Camarero, Numerical solution of reactive laminar flow by a control volume based finite element method and the vorticity-streamfunctions formulation, Numerical Heat Transfer, Part B, 22, (1991), 223-240.

[2] J. R. Edwards, Advanced implicit algorithms for finite-rate hydrogen-air combustion calculation, 32nd AIAA/ASME/ASEE Join Propulsion Conference, July, (1996).

COMPUTATIONAL MECHANICS (Abstracts)
WCCM VI in conjunction with APCOM'04, *Sept. 5-10, 2004, Beijing, China*

Parallel Computation Applied to 3D FE Simulation of Polymer Injection Molding

H. Digonnet *, L.Silva, T. Coupez

Ecole des Mines de Paris, Centre de Mise en Forme des Matériaux, 1 rue Claude Daunesse, 06904 Sophia Antipolis Cedex, France
e-mail: Hugues.Digonnet@ensmp.fr, Luisa.Silva@ensmp.fr, Thierry.Coupez@ensmp.fr

Abstract

In this paper, we consider 3D modelling of the polymer injection moulding process, which involves the study of the fluid thermoviscoelastic behaviour with free surface evolution. The platform used in this work is Rem3D®, a software developed at the CEMEF and used to model polymer an d metal forming processes. It is based on an Eulerian formulation, by solving conservation equations in the whole computational domain to determine stress, temperature, filling, and velocity, functions of time and part location. To compute velocity and pre ssure, we extend a mixed finite element method for incompressible viscous flow to account for viscoelasticity and compressibility (stabilised mixed velocity-pressure formulation). To compute the extra -stress field, we chose the PomPom model. Evolution equations solvers for molecular orientation and chain stretch were built using the Space-Time Discontinuous Galerkin method [1]. This same method is used to determine the free surface position as well as the temperature field [2], this later important in shrin kage and non-relaxed stresses computation, that will produce part deformation.

To capture the different phenomena and to consider all the effects simultaneously, fine domain discretisations are required, involving a large number of degrees of freedom. Thus, parallelisation of Rem3D® became essential. However, adaptation of simulation software to allow computation over several processors generally finishes with different sequential and parallel versions of the same code. To avoid this (or recruiting parallel experts to develop physical applications), interface classes were implemented, which allow the development of parallel applications by the common developer, interfaces that "hide" parallel instructions and that are easy to understand and use [3]. Rem3D® has therefore a global "master-slave" parallel architecture, with local SPMD modules, structure described in this paper. Validations tests are shown, using classical benchmark examples, with results concerning also speed-up and execution times. Examples in 3D complex industrial injection moulded parts underline the robustness and the efficiency of our model and implementations.

REFERENCES

[1] L. Silva, R. Valette, T. Coupez, *3D MFE for compressible viscoelasticity with moving free surfaces*, J. Non-Newt. Fluid Mech., (submitted).

[2] S. Batkam, J. Bruchon, T. Coupez, *A space-time discontinuous Galerkin method for convection and diffusion in injection molding*, Int. J. Form. Proc., (accepted).

[3] H. Digonnet, T. Coupez, *Object-Oriented Programming for "fast-and-easy" development of parallel application in forming processes simulation*. Proc. 2nd MIT Conf. on Comp. Fluid and Solid Mech., Elsevier (2003), pp. 1922-1924.

COMPUTATIONAL MECHANICS (Abstracts)
WCCM VI in conjunction with APCOM'04, *Sept. 5-10, 2004, Beijing, China*

Advances in High-Performance Domain Decomposition Methods

Yannis Fragakis [*], Manolis Papadrakakis

Institute of Structural Analysis & Seismic Research, National Technical University of Athens, 9, Iroon Polytechniou st., Zografou Campus, Athens, 15780, Greece
e-mail: fragayan@central.ntua.gr, mpapadra@central.ntua.gr

Abstract

Over the last decades, Domain Decomposition Methods (DDM) have emerged as powerful solvers for the analysis of large-scale problems of Computational Mechanics in both serial and parallel computational environments [1]. Today, the field of DDM is spanned by many different methods and computational techniques. A unified framework for formulating DDM, independently of several choices in the methods' formulation was recently introduced in [2] and extended in [3]. In particular, this framework encompasses several different choices, regarding the cases of homogeneous or heterogeneous problems and the different definitions of the Lagrange multipliers that are often used in DDM.
Furthermore, a new family of DDM, under the name "Primal class of FETI methods", or simply "P-FETI methods", was proposed in [2]. The theoretical relations of the new category of methods to the previously existing methods were studied in static structural analysis [2] and in implicit dynamic analysis [3] problems. In particular, the theory of P-FETI methods was found to be closely related to the theory of the respective previously known dual methods. In addition, a comparison of the computational performance of the methods [2-3] showed that the P-FETI methods are faster and more robust than previous methods, when applied to heterogeneous problems and/or plate and shell problems.
In the present contribution, the authors extend the theory and benchmark tests presented in [2] and [3]. Issues related to the consistency of the unified theory are revisited, while the computational performance of the methods is compared in challenging large-scale problems. The tests reveal novel aspects of the differences of the methods in their CPU time and robustness.

REFERENCES

[1] C. Farhat, F.-X. Roux, *Implicit parallel processing in structural mechanics*, Comput. Mech. Adv., 2, (1994), 1-124.

[2] Y. Fragakis, M. Papadrakakis, *The mosaic of high performance domain decomposition methods for structural mechanics: formulation, interrelation and numerical efficiency of primal and dual methods*, Comput. Methods Appl. Mech. Engrg., 192, (2003), 3799 -3830.

[3] Y. Fragakis, M. Papadrakakis, *The mosaic of high performance domain decomposition methods for structural mechanics – part II: formulation enhancements, multiple right-hand sides and implicit dynamics*, Comput. Methods Appl. Mech. Engrg., (To appear) .

COMPUTATIONAL MECHANICS (Abstracts)
WCCM VI in conjunction with APCOM'04, *Sept. 5-10, 2004, Beijing, China*

Development of Parallel Multi-Scale Finite Element Method Based on Crystalline Homogenization Considering Dynamic Load Allocation

Hiroyuki Kuramae [1*], Eiji Nakamachi [2]

[1] *Department of Industrial Management, Osaka Institute of Technology, 5-16-1, Omiya, Asahi-ku, Osaka, 535-8585, Japan*
[2] *Department of Mechanical Engineering, Osaka Institute of Technology, 5-16-1, Omiya, Asahi-ku, Osaka, 535-8585, Japan*
e-mail: kuramae@dim.oit.ac.jp, nakamati@mmsun1.med.oit.ac.jp

Abstract

A parallel elastic/crystalline viscoplastic finite element analysis (FEA) code based on a crystallographic homogenization theory for the multi-scale analysis [1-2] is developed on PC cluster system using message passing interface (MPI) library. The homogenization method is introduced to calculate macro-continuum plastic deformation and material properties by considering a micro-structural in-homogeneity, which is characterized by the crystal texture. The microstructure is defined as the unit cell, which satisfies the periodicity condition, where both compatibility and equilibrium are satisfied.

This multi-scale analysis is performed using the dynamic explicit finite element method by employing the updated Lagrangian formulation and the rate type elastic/viscoplastic constitutive equation. Since the dynamic explicit method is able to calculate the time integration without solving simultaneous equations, parallel processing technique is applied to macro-structure. Whole region of macro-structure is fictitiously divided into sub-domains and allocated to processor nodes. During the time integration process of the dynamic explicit method, computation of the homogenized stress using the micro-structure is performed for each sub-domain in parallel. After collecting the homogenized stress from each sub-domain, updating the deformation can be performed sequentially. The dynamic workload balancing technique, in which sub-domain allocation is performed automatically and dynamically during calculation, is introduced to this parallel multi-scale analysis to improve the parallel performance. It is well demonstrated through numerical examples that the parallel elastic/crystalline viscoplastic finite element analysis code shows very high parallel efficiency on 30 PCs (Intel Pentium IV 2.4 GHz CPU, 512 MB memory) connected to the Giga-bit Ethernet (1000Base-TX, 1 Gbps).

REFERENCES

[1] N. Nakamachi, C. L. Xie, M. Harimoto, *Drawability assessment of BCC steel sheet by using elastic/crystalline viscoplastic finite element analyses*, Int. J. Mech. Sci., 43, (2001), 631-652.
[2] N. Nakamachi, C. L. Xie, M. Harimoto, K. Morita, N. Yokoyama, *Formability assessment of FCC aluminum alloy sheet by using elastic/crystalline viscoplastic finite element analysis*, Int. J. Plasticity, 18, (2002), 671-632.

COMPUTATIONAL MECHANICS (Abstracts)
WCCM VI in conjunction with APCOM'04, *Sept. 5-10, 2004, Beijing, China*

Expert System for Assessing the Fatigue Resistance of Thin-Wall Profile Welds

M. Ruzicka [1*], Jan Papuga [1], R. Penicka [1], M. Hejman [2]

[1] *Czech Technical University in Prague, Technicka 4, Praha 6, 166 07, Czech Republic*
[2] *SKODA VYZKUM s.r.o., Tylova 57, Plzen, 316 00 Czech Republic*
e-mail: milan.ruzicka@fs.cvut.cz, marek.hejman@skoda.cz

Abstract

The basis of many transportation machines and building constructions are welded thin -wall frame structures. An operational reliability of welded constructions was discussed in [1]. For designing such structures and analysing stresses in them, finite element metho ds (FEM) are commonly used, whether only single-dimensioned bar elements, plate elements or, less frequently, 3D elements. During their lifetime, the structures are mostly subjected to dynamic forces that result in failure resulting from fatigue, often in the welded joints. An analysis and assessment of the resistance of welded joints against fatigue is the main topic of the presented paper. The objective is to develop an expert system as a set of computer aid programs. The paper describes the project's str ucture, based on a four-tier computer-aided analysis. The first level of the assessment is to classify the welded joint into a particular type category. Categories are defined by the material used and the weld node's shape arrangement. The second tier comp rises operations performed on top of a database of limit fatigue states. Using selected and computed flags, Smith's diagrams are chosen to assess the structure for an unlimited fatigue life. If the analysis is to be conducted for a limited fatigue life, appropriate stress fatigue curves are assigned to the node. Their selection is performed in accordance with European standards (Eurocode 3), although curves for individual assessment can also be selected. The third tier of the assessment system is based on d etermining weld node stress. The approach of analysing nominal and shape stresses in the weld can be applied here, or an approach that determines local stresses. In the first case, stresses are determined by classic methods of elasticity. For either approa ch, it is possible to determine stresses by applying 3D FEM analysis. The fourth and last level of the assessment system entails decomposing the actual operation stress curves or applying design load spectra to determine a failure due to fatigue and an ass essment of the limiting state of fatigue strength and life of the weld node. In the assessment, criteria based on equivalent stresses according to a strength hypothesis (uni-axial approach), as well as fatigue criteria from the point of view of multi-axial loading (multi-axial fatigue criteria), can be used. The paper describes in detail certain numeric peculiarities as well as the procedures used to determine the weld's lifespan by applying these methods. This project No. GA 101/02/0141 has been supported by the Grant Agency of the Czech Republic.

REFERENCES

[1] M. Ruzicka, R. Penicka, M. Hejman, M. Kepka, *Fatigue strength of thin-walled welded tubes*, in GESA Symposium 2003, VDI Berichte 1757, Düsseldorf, VDI Verlag, (2003), pp. 517-524.

COMPUTATIONAL MECHANICS (Abstracts)
WCCM VI in conjunction with APCOM'04, *Sept. 5-10, 2004, Beijing, China*

Comparison of Geometric Partitioning Algorithms for Parallel Analysis of Finite-Discrete System

Fujun Wang [1]*, Yaojun Li [1], Y.T. Feng [2], D. R. J. Owen [2]

[1] *College of Water Conservancy & Civil Eng., China Agricultural University, Beijing, 100083, China*
[2] *Civil & Computational Engineering Centre, University of Wales Swansea, Swansea, SA2 8PP, U.K.*
e-mail: wangfj@cau.edu.cn, cau_lee@163.com, y.feng@swansea.ac.uk, d.r.j.owen@swansea.ac.uk

Abstract

The efficient execution of parallel finite-discrete element analysis requires a mapping of the computational objects onto processors such that each processor get s roughly an equal amount of object weight and that the amount of interprocessor communication is minimi zed. Such a mapping is commonly found by partitioning the entire domain. This paper reviews the major classes of partitioning algorithms and discusses the suitability of three typical geometric partitioning algorithms (i.e. RCB, UCB and Octree) on finite-discrete element problems. The main inclusions are as follows: (1) Although topological algorithms can produce better partitions than geometric algorithms, but are more expensive. Geometric methods are more suited to finite-discrete element simulation than topological methods. (2) The quality of partitions generated by RCB is slightly poorer than those produced by UCB and Octree. However, its simplicity an d speed are appealing. Good load balance can be obtained by using a modified RCB algorithm proposed by the authors [1]. (3) Jones and Plassmann reported that URB algorithm leads to a modest reduction in communication cost [2]. However, this result was not observed in the current study. The fact is that the idle seconds for synchronization increase explicitly with the contact developing. Although the communication amount is kept small, but the time used for communication and synchronization is large. It tend s to work well when only no contact occurs. The overall communication performance of URM is much worse than those of RCB. (4) The partitions produced by Octree algorithm are not confined in simple boxes. Octree algorithm is fast and is easy to make data mi gration [3]. However, the long boundary size increases the buffer between two adjacent subdomains. Thus, it results more communications. The programming complicacy of Octree algorithm is larger than any of other methods. If we consider the overall performances of all partitioning algorithms, RCB is the first choice for the parallel finite-discrete element analysis on distributed-memory environments.

REFERENCES

[1] F. Wang, Y. T. Feng, D. R. J. Owen, *Parallelisation for finite-discrete element analysis on distributed- memory environment*, Int. J. Comp. Eng., 5, (2004), (to be published).

[2] M. T. Jones, P. E. Plassmann, *Computational results for parallel unstructured mesh computations,* Computing Systems in Engineering, 5, (1994), 297-309.

[3] B. Hendrickson, K. Devine, *Dynamic load balancing in computational mechanics*, Comp. Meth. Appl. Mech. Engng., 184, (2000), 485-500.

COMPUTATIONAL MECHANICS (Abstracts)
WCCM VI in conjunction with APCOM'04, *Sept. 5-10, 2004, Beijing, China*

Mechanistic Analysis of Expandable Solid Tubulars

Jan A. Aasen*, Bernt S. Aadnoy

Department of Petroleum Engineering, Stavanger University, P.O.Box 8002, 4068 Stavanger, Norway
e-mail: jan.aasen@tn.his.no, bernt.aadnoy@tn.his.no

Abstract

Considerable development has taken place within oil well technology the past decade. At present there is significant focus on advanced technology to save money and to make projects technically feasible. One of the new areas with considerable potential is expandable solid tubulars [1].
Conventional oil wells step down significantly in size as they go deeper, thus restr icting the size of the final production tubing. Expandable solid tubular technology does not decrease diameter as much, enabling wells to be started with small diameter holes and target depths to be reached using unrestricted production tubing. Expandable solid tubulars are used in conjunction with conventional tubulars, either planned from the outset or as contingency measures should unexpected geological conditions be encountered in the well. The expandable tubular is lowered into the hole and is then permanently expanded outwards by forcing a hardened metal mandrel through it in either an upwards or downwards direction.
The expanded tubular is subjected to three separate events, namely yielding throughout the pipe wall, followed by plastic expansion and finally elastic recovery. Residual stresses remain in the pipe after completion of these events. In the paper we consider analytical autofrettage solutions [2] for zero outside pressure and zero axial stress. These results are then used to develop a model applicable to pressure and load situations encountered in the petroleum industry. The theory is based on the maximum distortion energy yield criterion and zero strain -hardening. Model outputs include incipient yield, ultimate yield, plastic growth, residual stresses, burst and collapse. Burst resistance is enhanced while the collapse pressure is reduced after expansion.
The paper presents theory and a dimensionless graphical solution applicable to expandable solid tubulars. The complete process is defined as paths on the dimensionless graph. The elastic recovery and the residual stresses are analyzed; including burst and collapse properties of the pipe after expansion.

REFERENCES

[1] A. Filippov, R. Mack, L. Cook, P. York, L. Ring, T. McCoy, *Expandable tubular solutions,* paper SPE 56500, Society of Petroleum Engineers Annual Technical Conference and Exhibition, Houston, Texas, (1999).

[2] H. E. Good, *Effect of choice of the failure criteria on the residual stress distribution of thick-walled cylinders,* unpublished thesis, University of Colorado, Boulder, Colorado, (1971).

COMPUTATIONAL MECHANICS (Abstracts)
WCCM VI in conjunction with APCOM'04, *Sept. 5-10, 2004, Beijing, China*

Applications Program for the Design of Systems with the Distributed Parameters

Elena N. Artamonova [*], Vyacheslav Y. Artamonov

Saratov State Technical University, Russia
e-mail: derevyankina@yahoo.com

Abstract

In a number of technological processes the problem of the temperature control of the current inert gas in the heating chamber of the cylindrical form of the limited length is posed. The following basic problems are solved with the design of data of technological installations:
1) Selection of optimum design parameters of heating chamber on the assigne d requirements for the accuracy of the stabilization of temperature field.
2) Synthesis of the regulator of control system.
The projected object of control is described by system of equations parabolic type. Control of object is accomplished through the b oundary conditions. The developed batch of programs was used for the calculation of design parameters and system of control of cylindrical heating chamber.

COMPUTATIONAL MECHANICS (Abstracts)
WCCM VI in conjunction with APCOM'04, *Sept. 5-10, 2004, Beijing, China*

Powder Filling Simulation with the Particle Finite Element Method

J. C. Cante [1*], J. Oliver [2], C. Gonzalez [2]

[1] *Technical University of Catalonia, C/. Colom 11,Edf TR45, Terrassa, 08222, Spain*
[2] *Technical University of Catalonia, Jordi Girona 1-3,Campus Nord, mod C1, BCN, 08034, Spain*
e-mail: juan.cante@upc.es, xavier.oliver@upc.es, ferrari@cimne.upc.es

Abstract

A new approach for numerical modelling of the powder filling stage, in powder metallurgy manufacturing process, is explored. The most innovative aspect of the work is *the full exploitation of the particle method technology to trace the motion of a representative set of powder particles* during the filling process. The final goal is to obtain relevant and representative information about the density evolution of the powder particles in the die, which can be used in modelling subsequent stages of the manufacturing process.

Typical limitations of classical finite element methods, in terms of transferring information during remeshing, are automatically circumvented by the recently explored meshless methods. On the other hand, unlike other "particle-following" methods like the discrete element methods, the approach keeps the full physical significance of the problem and consistence with its mathematical approach. Implementation of the new method does not start from zero, since the finite element technology developed in previous works can be applied directly to solve the incremental problem. The implementation effort can be even smaller if the point of departure is a code provided with an automatic remeshing. Characterization of the material parameters follows the usual strategy outlined by the continuum approach. The contact algorithm is based on the generation of a contact interface defined through a family of mixed elements composed by nodes of the powder and nodes of the rigid solid, and a constitutive model provided for them. This novel contact interface traces more carefully the changes in curvature around convex and concave zones of the rigid solid. These properties increase robustness of the contact problem especially in the region of sharp corners. The performance of the approach is assessed via numerical simulation of actual industrial problems

REFERENCES

[1] S. Idelsohn, E. Oñate, F. Del Pin, *A Lagrangian Meshless Finite Element Method applied to Fluid Structure Interaction Problems*, Computer and Structures, 81, (2003), 655-671.
[2] J. C. Cante, J. Oliver, C. Gonzalez, On the numerical simulation of powder compaction processes: powder transfer modelling and characterization, *Powder Metallurgy*, (accepted).
[3] J. Oliver, S. Oller, J. C. Cante: Int. J. Solids and Struct, 33, (1996), 3161-3178.

COMPUTATIONAL MECHANICS (Abstracts)
WCCM VI in conjunction with APCOM'04, *Sept. 5-10, 2004, Beijing, China*

Finite Element Analysis and Comparison of Two USX Inlet Models

Zengliang Gao *, Lihua Liang, Weiming Sun, Jinsong Zheng, Weiya Jin, Kangda Zhang

Institute of Process Equipment and Control Engineering, Zhejiang University of Technology, Hangzhou 310032, China
e-mail: zlgao@zjut.edu.cn

Abstract

USX inlet header is the key part of quench boiler of cracked gas for ethene plant. It is loaded by internal pressure, pipe load and moment, thermal loads. The quite uneven temperature field in the USX can result in high thermal stress. Proper design of the USX is very important for safety and effectiveness of the USX. Two USX inlet header models are analyzed and compared based on the finite element results.

Temperature field in the USX and heat conduct in still steam layer between two tubes of the USX are calculated by FEA. Plane 77 element and solid 90 element are used for the still steam layer and channel cover structure of the USX respectively. A 3D FE model is established for the USX inlet header to take into consideration of non-symmetry of the loads and coherence of element and mesh of FE for temperature and stress calculation. The elements of the still steam in the models for temperature field calculation are deleted, and then the solid 90 element of the structure is changed to solid 95 element for calculating the stress field. Elastic FEA and Elastic-plastic FEA with two kinds of simplified stress-strain relationships of the material are carried out. Thermal loads and mechanical loads are taken into consideration in FEA calculation.

The results of the FEA calculations show that the stresses in the new USX inlet header are much lower than those in the original one. The stresses and strains in both models meet the requirements of ASME T-1310 and T-1332. But the accumulated creep and fatigue damage over the service life of the original USX, including hold time and strain rate effects, should be considered, if there are temperature and load cycles in the USX. The new USX inlet header has been safely used in a ethene plant for two years.

REFERENCES

[1] ASME Boiler & Pressure Vessel Code Section III, Division I -Subsection NH "Rules For Construction of Nuclear Facility Components", (2001).

COMPUTATIONAL MECHANICS (Abstracts)
WCCM VI in conjunction with APCOM'04, *Sept. 5-10, 2004, Beijing, China*

Aerodynamics and Flight Dynamics Aspects of Controlled Missile

Abdulhamid A. Ghmmam*, Saad M. Issa

Department of Aeronautical Engineering, Al-Fateh University, Tripoli-Libya

Abstract

The performance requirements of today's missile are increasing by imposing more sever demands. The high maneuverability, controllability, and good tracking of target, are among those requirements. However, meeting those requirements is not an easy task, since the behavior of the missile while flying in air, is very difficult to predict, because of the interaction between the aerodynamics and the flight dynamics.

This paper puts the emphases on some aerodynamic and flight dynamics aspects, especially those related to large angles of attack, and the resulting flight dynamics problems.

At high angles of attack significant flow phenomena take place, among these phenomena; flow separation, unsteadiness of the flow, formation of fore-body and wing vortices, time-dependent and hysteresis effects, and non-linear behavior of the aerodynamic forces and moments.

Since the dynamic problems at high angles of attack are the logical consequence of the interaction between the highly complex flow field over the missile, and the dynamic characteristics dictated by the geometry of the configuration, the flight dynamics phenomena associated with flying at high angles of attack, need to be considered when designing a control law to control the path of the missile.

To asses the significance of some selected stability derivatives, and their effect on the predicted motion, a numerical simulation of the six-degrees-of freedom equations of motion is performed. The numerical integration of the full non-linear equations is done using the fourth order Rung-Kutta method.

Even though reliable data for missiles is not readily available in the open literature, some data which are believed to be representative of high performance configuration (such as a missile) are used in the present study.

Results indicate the importance of non-linear cross-coupling stability derivatives; hence the inclusion of such effects in the control law is very significant in order to predict the motion of the missile to an acceptable degree of accuracy.

COMPUTATIONAL MECHANICS (Abstracts)
WCCM VI in conjunction with APCOM'04, *Sept. 5-10, 2004, Beijing, China*

Application of Pseudo-Elastic Model to Compressible Materials

Z. Guo *, L. J. Sluys

Faculty of Civil Engineering and Geosciences, Delft University of Technology, Stevinweg 1, 2600GA Delft, the Netherlands
e-mail: z.guo@citg.tudelft.nl, l.j.sluys@citg.tudelft.nl

Abstract

It is well known that rubber-like materials exhibit a strongly non-linear behaviour characterized by large strains and a non-linear stress-strain response. Moreover, when a rubber specimen is subjected to cyclic loading, not only non-linear behaviour but also *stress-softening* phenomena (the Mullins effect) are observed. Under cyclic loading conditions, the load required to produce a given stretch during the second loading cycle is smaller than that required to produce the same stretch during the primary loading cycle. Ogden and Roxburgh [1] have recently developed a pseudo-elastic model for incompressible materials to capture the Mullins effect, treating it as a quasi-static phenomenon. This energy-based damage model is distinct from the strain-based models since the extent of damage sustained by the material is controlled by the maximum energy state. Therefore, this model is readily applicable to three-dimensional states of deformation, especially, for computational purposes.

In this paper, we extend Ogden and Roxburgh's pseudo-elastic model to compressible materials, for which the strain energy function is described by the distortional part and the hydrostatic part in a strictly separated fashion. As suggested by Simo [2] the damage-caused softening effect is assumed to affect the isochoric response only. So, the Ogden & Roxburgh model is extended with the hydrostatic part. Furthermore, the damage variable is modified in order to adopt the model to very large strain cycles. This modification greatly improves the simulation of the three-dimensional inhomogeneous problem, in which inhomogeneous deformations of different ranges may cause convergence problems.

The model is also combined with a strain energy density function proposed by Gao [3], based on the idea that a material element should possess stiffness to resist both shape change and volume change separately. A modified pseudo-elastic model was implemented together with the Gao model in the finite element code DIANA. Necessary assessments and comparison demonstrate that this modified pseudo-elastic model is reasonable to describe the Mullins effect. The analysis of a three-dimensional inhomogeneous problem illustrates the capability of the modified damage model. Moreover, the calculations for the inhomogeneous problem converge very fast with the modified damage formulation.

REFERENCES

[1] R. W. Ogden, D. G. Roxburgh, *A pseudo-elastic model for the Mullins effect in filled rubber*, *Proc. R. Soc. London*, A 455, (1999), 2861-2877.

[2] J. C. Simo, *On a fully three dimensional finite strain viscoelastic damage model: Formalation and computation aspects*, *Comput. Mech. Appl. Mech, Engng.*, 60, (1987), 153-173.

[3] Z. Guo, P. Stroeven, L. J. Sluys, Y. Cheng, *New constitutive models in DIANA for rubber-like materials, Finite elements in civil engineering applications;* Proceedings of the third DIANA world conference, Tokyo, 9-11 Oct., Rotterdam: Balkema, (2002), pp. 361-368.

COMPUTATIONAL MECHANICS (Abstracts)
WCCM VI in conjunction with APCOM'04, *Sept. 5-10, 2004, Beijing, China*

Nose-Deck Interaction in ILM Bridge Proceeding with Tapered Sectional Launching Nose

Hwan Woo Lee*, Tae Wook Ahn, Kwang Yang Kim

Department of Civil Engineering, Pukyong National University, San 100 Yongdang-dong Namgu, Busan 608-739, South Korea
e-mail: hwanwoo@pknu.ac.kr

Abstract

ILM(incremental launching method) bridge, which is one of PS (prestressed) concrete bridge construction methods is widely adopted owing to its effectiveness for the quality control. The structural behavior of superstructure in the bridge strongly depends on the process of launching nose during construction. In this study, the nose-deck interaction is analyzed for the bridge according to the parameters of the ratios of length, weight, and flexural stiffness of launching nose to those of superstructure. The section of nose is assumed as the tapered shape to consider the real project. Design formulas are also suggested to pursue the optimum values for relative length, weight, and flexural stiffness between launching nose and superstructure.

REFERENCES

[1] W. Zellner, H. Svensson, *Incremental launching of structures,* Journal of Structural Engineering, ASCE, 109(2), (1983), 520-537.
[2] M. Rosignoli, *Launched bridge*, ASCE Press, (1998).
[3] M. Rosignoli, *Nose-deck interaction in launched prestressed concrete bridges,* Journal of Bridge Engineering, ASCE, 3(1), (1998), 21-27.

COMPUTATIONAL MECHANICS (Abstracts)
WCCM VI in conjunction with APCOM'04, *Sept. 5-10, 2004, Beijing, China*

The Analysis of Compression Forces on Casing and Cement during Production

Yukun Li*, Xinghua Tong

Department of Mechanics Engineering, University of Petroleum (East China), DongYing, Shandong, 257061, China
e-mail: mliyk@hdpu.edu.cn

Abstract

The effect of borehole compression forces on casing and cement is a major concern to engineers completing well. According to API standards of Casing Design, casing is compressed by hydrostatic pressure and continuous changing reservoir pressure that occurs around the oil well and water well during production. Evaluation of effects of changing loads on casing and cement is involved in this paper.

First, the compression forces on casing and cement during production under condition just following completion are discussed. It is investigated that the loads on casing and cement may increase gradually with sediment accumulation and pore pressure drop after long production in Stratigraphic oil-gas reservoirs. Second, by applying the theoretical model of immiscible saturated two-phase (oil & water) flow in nonlinear porous media, the finite element codes for computation are programmed. Finally, with the assumption that the casing and cement consolidate well with porous media in reservoir, a 3D model is established, in which the parameters of the oil well are adopted from Gudong Production Plant, Shengli Oilfield, China. It is shown that stresses in casing and cement will increase after high-intensity development, which might result in the casing deformation.

REFERENCES

[1] R. W. Lewis, *Finite element modeling of two-phase heat and fluid flow in deforming porous media,* Trans. Porous Media, 4, (1989), 319-334.

[2] A. V. Hadad, J. Bensabat, H. Rubin, *Simulation of immiscible multiphase flow in porous media: a focus on the capillary fringe of oil-contaminated aquifers,* Trans. Porous Media, 12, (1996), 245-269.

[3] S. F. Xue, X. H. Tong, B. Q. Yue, H. Z. Song, *Progress of seepage2rock mass coupling theory and its application.* Journal of the University of Petroleum, 24(2), (2000), 109-114, (in Chinese).

COMPUTATIONAL MECHANICS (Abstracts)
WCCM VI in conjunction with APCOM'04, *Sept. 5-10, 2004, Beijing, China*

Application of Computational Fracture Mechanics in Fatigue Life Prediction and Materials/Processes Selection for Truck and Automotive Components

Hong Lin [1*], Robert R. Binoniemi [2], Gregory A. Fett [3]

[1] *Dana Corporation, Technology Development, 8000 Yankee Road, Ottawa Lake, MI 49267, USA*
[2] *Dana Corporation, Heavy Vehicle Technologies Center, P O Box 4097, Kalamazoo, MI 49003, USA*
[3] *Dana Corporation, Automotive Systems Technology Center, 3939 Technology Drive, Maumee, OH 43537, USA*

e-mail: hong.lin@dana.com, bob.binoniemi@dana.com, greg.fett@dana.com.

Abstract

This paper summarizes our work in using computational fracture mechanics for fatigue life simulations, and for comparing the fatigue performance of different steels and processes of truck and automotive components during the last decade. Computational fracture mechanics tools have enabled us reduce product design cycle time, select high performance and cost effective materials and quantify the effect of different peening technologies on fatigue life, and above all improve the durability and increase the power density of our products.

Fatigue life (S-N, stress vs. cycles, curves) simulation results by using the computational fracture mechanics CRACKS software will be presented. Based on LEFM, the program has various K solutions and can perform both constant amplitude and variable amplitude load crack growth analysis. It can also simulate the effect of non-uniform residual stress distributions on fatigue life. First, fatigue life simulation results by using CRACKS were verified by comparing with gear teeth bending fatigue test data and three point bending fatigue test data. Then the CRACKS program is used to evaluate the effect of various factors on fatigue life S-N curves computationally, such as steels with different alloy contents, residual stresses from standard shot peen, high intensity shot peen, ultrasonic impact process, and laser shock peen, initial defect size, steel properties such as fracture toughness and crack growth threshold values. The computational mechanics tool was effective in reducing prototype testing and reducing design cycle time.

In addition, recent research results in computational fatigue life simulations will also be presented. These include the investigations and simulations on sample volume effect on fatigue endurance limit using both CRACKS and a stochastic fracture mechanics based analysis software. Finally contact fatigue life analysis results will be presented including contact fatigue test data, 3D rough surface elasto-hydrodynamic lubrication analysis, and contact fatigue life analysis. Contact fatigue analysis results are compared with test results.

COMPUTATIONAL MECHANICS (Abstracts)
WCCM VI in conjunction with APCOM'04, *Sept. 5-10, 2004, Beijing, China*

Large Strain Deformation and Fracture of Quasi-brittle Materials with Application to Process Comminution Equipment

J. G. Loughran [1*], S. I. Anderson [2]

[1] *School of Engineering, James Cook University, Queensland 4811, Australia*
[2] *Rockfield Technologies Australia Pty. Ltd., Queensland 4817, Australia*
e-mail: jeffrey.loughran@jcu.edu.au, scott.Anderson@rocktech.com.au

Abstract

Many industrial processes involve the impact of foreign debris against quasi-brittle materials. The subsequent damage can be erosion or in extreme cases catastrophic failure. This paper reports a finite element investigation of the impact of steel debris against sugar mill shredder hammer tips. During cane preparation, billets of cane are comminuted against tungsten carbide (WC) tips with impact velocities approaching one-third the speed of sound. Foreign debris (often hardened steel pins) in the input stream, result in a high probability of fracture to the WC tiles. During fabrication the WC tiles are brazed to a mild steel or white iron backing-block using a tri-ply laminate. The laminate on average is several orders of magnitude less stiff than the WC and besides acting as a fastener, serves to minimize the build-up of residual stress in the WC due to the marked difference in coefficient of thermal expansion between adjacent materials. The laminate also influences the impedance as compression waves are reflected and refracted.

The impact process when steel debris comes into contact with the WC is modelled using explicit finite element dynamics under plane strain conditions. Following traditional large strain finite element procedures the deformation gradient is split in a multiplicative sense into elastic and plastic components. Incremental procedures are required since the problem is invariably path dependent. This requires a numerical approximation to the constitutive law for update of local stress and other internal variables during each increment. The finite plastic evolution equations are discretised by an exponential map integrator while the elastic components are discretised by the standard backward-Euler scheme. Key issues addressed are: 1) appropriate modelling of deformation mechanisms at an elemental level to enable prediction of deformation, strain, stress and contact laws; 2) inclusion of finite strain elasto-plasticity; 3) development of fracture criteria; 4) detection procedures for monitoring contact between deformable discrete elements (fragments) and continuum regions; 5) Representation of interaction laws for contacting elements. A Rankine rotating crack model is used to characterize the WC and the projectile is modelled as elasto-plastic. The evolution of cracking which occurs due to impact is studied. The effect of the WC geometry on resistance to crack initiation is also noted and explored numerically. Agreement with experiment is excellent.

REFERENCES

[1] E. A. De Souza Neto et al, *Continuum modeling and numerical simulation of material damage at finite strains*, Arch. Comput. Meth. Engng, 5, (1988), 311-384.

[2] D. R. J. Owen et al, *A hybrid continuous/discrete representation of fracturing solids*, Proc. of IMECE'03, (2003), pp. 1-25.

COMPUTATIONAL MECHANICS (Abstracts)
WCCM VI in conjunction with APCOM'04, *Sept. 5-10, 2004, Beijing, China*

Analysis of Excessive Deformation of the Mahabad Dam

K. Soleimani *, M. H. Baziar, H. Salehzadeh

Geotechnical Engineering, Iran University of Since and Technology, Tehran, Iran
E-mail: k_soleimany@yahoo.com, Baziar@iust.ac.ir, H-salehzadeh@yahoo.com

Abstract

Mahabad dam is an earth-rock fill dam one with a thin clay core. From the very beginning of its construction unusual settlement and lateral displacement occurred in the total body of the dam. The amount of settlement of the foundation from the beginning to the end of the construction was 130 cm and during the utilization period is has been 60 cm. The maximum settlement and the horizontal displacement from time of construction were 150 cm and 100 cm respectively. The behavior of the dam has been back analyzed using software called PLAXIS during construction and for 30 years after construction. The cross-section of the body of the dam and its foundation has been meshed by a triangular 15-node element. For the bedrock part of the foundation and the diaphragm, a linear elastic model has been used. Material from the foundation and the body of the dam containing fine grains and with the long-term possibility of settlement has been modeled by soft-soil-creep model. In parts lacking a long-term possibility of settlement Moher Colomb model has been used. Accordingly, considering the result of the back-analysis and the bank of precise instrument of the dam, its condition including the reasons for deformation, the condition of the foundation and existing of crack being studied.

REFERENCES

[1] D. T. Bregado, C. Teerawattanasuk, S. Youwai, P., Voottipruex, *Finite element modeling of hexagonal wire reinforced embankment on soft clay*, Canadian Journal of Geotech, 37, (2000), 1209-1226.

[2] D. J. Naylor, J. R. Maranha, E. Maranha das Neves, A. A. Veiga Pinto, *A back-analysis of Beliche Dam*. Journal of Geotechnique, 47(2), (1997), 221-233; Discussion, Geotechnique, 51(4), (2001), 377-381.

[3] D. J. Naylor, E. Maranha Das Nevas, J. R. D. Mattar, A. A. Veiga Pinto, *Prediction of construction performance of Beliche Dam*. Geotechnique, 36(3), (1986), 359-376.

[4] James L. Sc.D. Sherard, Richard J. M. S. Woodward, Stanley F. B. S. Gizienger, *Earth-rock dams*, John Wiley & Sons, (1963).

[5] A.Desideri Pagano, F. Vinale, *Interpeting settlement profiles of erth dams*, ASCE Journal of Geotechnic and Geoenvironmental Eng, 124(10), (1998), 923-932.

COMPUTATIONAL MECHANICS (Abstracts)
WCCM VI in conjunction with APCOM'04, *Sept. 5-10, 2004, Beijing, China*

Numerical Partition of Energy Absorption of Foam-filled Top-Hat and Double-Hat Sections

Hongwei Song [1*], Gang Yu [1], Zijie Fan [2], Qingchun Wang [2]

[1] *Institute of Mechanics, Chinese Academy of Sciences, Beijing, 100080, China*
[2] *State Key Laboratory of Automotive Safety and Energy, Department of Automotive Engineering, Tsinghua University, Beijing, 100084, China*
e-mail: songhw@tsinghua.edu.cn, gyu@imech.ac.cn

Abstract

Recently, much attention is given to the foam-filled thin-walled structures[1-6]. The studies showed that the interaction between foam fillers and the supporting structures produce excellent crushing behaviors and energy absorption characters. However, up to now little work has been done to quantitatively determine this effect, i.e. the contribution of each component of the compound structure to the total energy absorption.

In the present study, the interaction effect, i.e., the contribution of each component to the total energy absorption of an axially crushed foam-filled hat section was investigated quantitatively via numerical simulation. The FE results were first verified by experimental work of aluminum foam-filled top-hat and double-hat sections, then the contribution of foam-fillers and that of hat sections to the overall energy absorption were quantitatively obtained, respectively. When foam-filled, increase in energy absorption was found both in hat section component and foam-filler component, whereas the latter contributes predominantly to the interaction effect.

If the collapse modes in Fig.8 and Fig.10 stand for the mean crushing force or energy absorption of corresponding components and individuals, there exist

$(a)=(b)+(c)$; $(b)>(d)$; $(c)>(e)$; and $(a)>(d)+(e)$

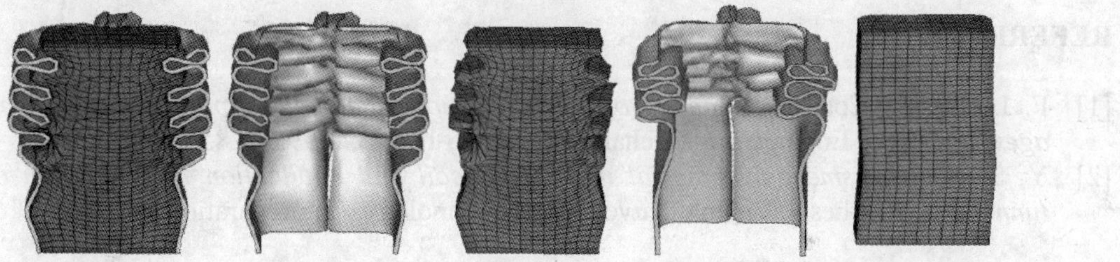

(*a*) (*b*) (*c*) (*d*) (*e*)

Fig. 1 Interaction effect expressed in the form of simulated collapse mode (*a*) filled double-hat; (*b*) double-hat component; (*c*) foam-filler component; (*d*) empty double-hat; (*e*) unbounded foam column

REFERENCES

[1] A. G. Hanssen, M. Langseth, O. S. Hopperstad, *Static and dynamic crushing of square aluminium extrusions with aluminium foam filler*, International Journal of Impact Engineering, 24, (2000), 347-383.

[2] S. Santosa, T. Wierzbicki, *Crash behavior of box columns filled with aluminum honeycomb or foam*, Computers and Structures, 68, (1998), 343-367.

COMPUTATIONAL MECHANICS (Abstracts)
WCCM VI in conjunction with APCOM'04, *Sept. 5-10, 2004, Beijing, China*

Numerical Assessment of the Interaction between Creeping Ground and Hydrating Shotcrete Shell in Tunnel Excavations

Y. Spira, R. Lackner, H. A. Mang *

Institute for Strength of Materials, Vienna University of Technology, Karlsplatz13/202, A -1040 Vienna, Vienna, Austria
e-mail: yvonne.spira@tuwien.ac.at, roman.lackner@tuwien.ac.at, herbert.mang@tuwien.ac.at

Abstract

In tunnels constructed according to the New Austrian Tunneling Method (NATM), the ground surrounding the tunnel opening together with the shotcrete lining provide the main support of the tunnel. Hereby, the relatively large compliance of young shotcrete permits deformations, resulting in the activation of the load-carrying capacity of the ground.
In order to highlight the interaction between the creeping ground and the aging shotcrete shell in different geological conditions, material models for both granular and cohesive soil as well as for shotcrete are reported in this work [Spira, 2004]. Both material models for soil are characterized by non-associative viscoplasticity including hardening/softening behavior. For the numerical simulation of shotcrete, a chemomechanical material model is employed [Ulm and Coussy, 1995].
As an example problem, the effect of tunnel excavation and the subsequent application of the shotcrete shell on deformation and stress states in granular and cohesive soil is investigated. Based on the obtained numerical results, the complex structural interaction between the creeping ground and the hydrating shotcrete shell characterized by the time-dependent increase in strength and stiffness is analyzed. Consideration of two types of soil and different viscous properties for the soil provides insight into the deformation and loading of the shotcrete shell and, hence, the performance of the aging shotcrete shell with respect to the different geological conditions.

REFERENCES

[1] F. J. Ulm, O. Coussy, *Modeling of thermochemomechanical couplings of concrete at early ages*, Journal of Engineering Mechanics (ASCE), 121(7), (1995), 785 -794.
[2] Y. Spira, *Assessment of material models for soil and application to simulation of NATM tunneling*, PhD thesis, Vienna University of Technology, (in preparation).

COMPUTATIONAL MECHANICS (Abstracts)
WCCM VI in conjunction with APCOM'04, *Sept. 5-10, 2004, Beijing, China*

Simulation of an optical interferometry experiments in EHL study of Cylindrical Roller

Haoyang Sun *, Xiaoyang Chen

Research Institute of Bearings, Shanghai University, Shanghai, China
e-mail: hysun@mail.shu.edu.cn, xychen@mail.shu.edu.cn

Abstract

In this paper, a complete numerical solution of the finite line contacts EHL problems has been presented. With the multilevel and multi-integration method, some dedicated numerical technique are employed to adapt the particularity of this problem, as a result, convergence could be obtained even when the contact pressure exceed 1.0 GPa and the length/width ratio of roller contact area over 100. The famous experiments carried out by Wymer and Cameron [1] has been simulated quantitatively, the rollers axially end profiles were fitted to a function for calculations based on the experimental data. As show in Fig. 1, it has been found that the calculated film contours and the experimental interferograms agree well between each other, and the numerical film thickness has been showed some agreements with the experimental results. In the middle of the rollers, their differences are less than 10%; while on the roller edge, the numerical minimum film thickness are only about 40% to 60% of the experimental ones. The discrepancy seems to be mainly attributed to two factors: the fitting error of the roller end profiles and the thin film lubrication effects. It could be concluded that the calculated results can provide satisfactory predictions of the finite line contact EHL performance, therefore, some further study could be carried out on the crowning design based on current work. It has also been found that on the roller edge the thin film effect might play an important role in film conformations.

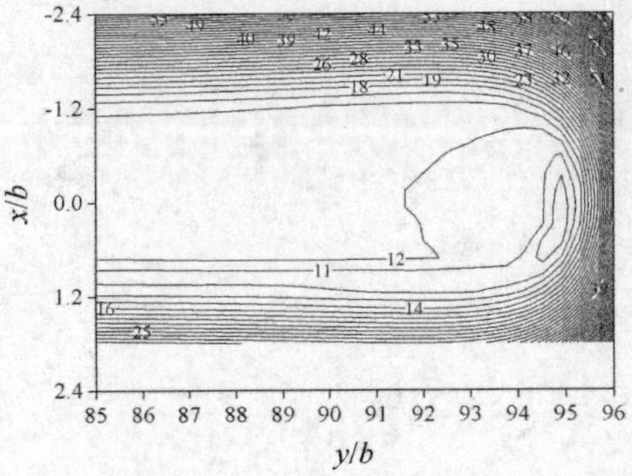

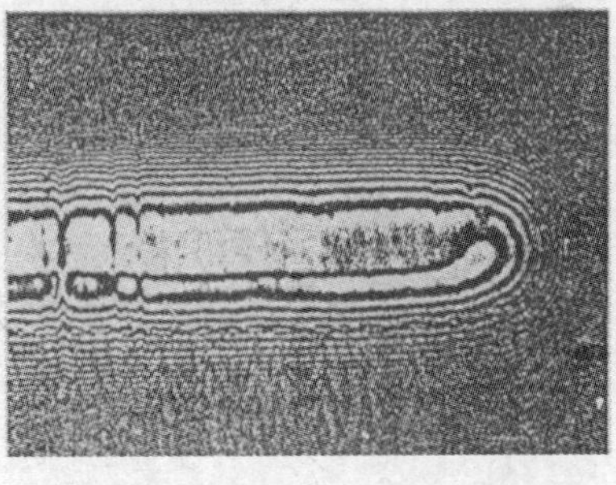

Fig.1 Comparison of current solutions and interferograms obtained in [1]
($U^* = 25 \times 10{-}11$, $W = 1.1458{\times}10{-}4$ ($p_H = 0.4997$ GPa), $E' = 117$ GPa, $G = 5616$, $S = 0$)

REFERENCES

[1] D. G. Wymer, A. Cameron, *Elastohydrodynamic lubrication of a line contact*. Proc. Inst. Mech. Engrg, 188, (1973), 221-238.

COMPUTATIONAL MECHANICS (Abstracts)
WCCM VI in conjunction with APCOM'04, *Sept. 5-10, 2004, Beijing, China*

Multi-Axle Vehicle Load Identification

Ling Yu [1*], T. H. T. Chan [2]

[1] *Blasting and Vibration Research Department, The Yangtze River Scientific Research Institute, Wuhan, Hubei, 430010, China*
[2] *Department of Civil and Structural Engineering, The Hong Kong Polytechnic University, Kowloon, Hong Kong, China*
e-mail: L_yu_1997@yahoo.com.cn

Abstract

Both the time domain method (TDM) and frequency-time domain method (FTDM) are introduced and modified for the indirect identification of multi-axle vehicle loads from the bending moment responses of bridges. Two solutions to the over-determined set of equation involved in the identification methods are adopted. One is direct calculation of the pseudo inverse and another calculation of the pseudo inverse via the singular value decomposition technique for the ill-conditioned problems encountered. A few multi-axle vehicles were designed and constructed in the laboratory for validation purposes based on the ASSHTO standard specifications of highway bridges. Two kinds of frames between the tractor and trailer of trucks and, three types of vehicle suspension systems were simulated. Different multi-axle vehicle loads were identified from the measured bending moment responses of bridges under different operation conditions in the laboratory. The effects of various vehicle and bridges parameters were evaluated. Comparative studies show that both TDM and FTDM methods involved in the moving force identification system (MFIS) are good identification methods and could efficiently identify the multi-axle vehicle loads on bridges.

COMPUTATIONAL MECHANICS (Abstracts)
WCCM VI in conjunction with APCOM'04, *Sept. 5-10, 2004, Beijing, China*

Computational Strategies for Element Volume Change Control in the Simulation of Metal Forming Using Rigid-Plastic FEM

Song Yu [*], Guoqun Zhao, Guangchun Wang

Mold & Die Engineering Technology Research Center, Shandong University, Ji'nan, Shandong, 250061, China
e-mail: yusong@sdu.edu.cn

Abstract

In this paper, several reasons for element volume change in FEM analysis of bulk metal forming are discussed in detail. In the method of rigid-plastic FEM with a penalty function, the accuracy of solver depends on the value of penalty constant $\boldsymbol{K}$ effectively. Too small $\boldsymbol{K}$ results in unacceptably large volumetric strain, while too large may cause ill-conditioned matrix and difficulties in convergence. Element volume loss appears in every time-increment step although introducing the incompressibility to penalty function. If reducing the time increment, the volume loss will be decreased, but the simulation time will be increased. Density of mesh has the same effect as the time-increment. Meshing or remeshing is main reason of volume loss. If the same remeshing criteria taken by different deformation problems, the complex shape and large deformations will have more remeshing times than the simple shape and small deformations. Contrarily if the different remeshing criteria are taken by forming process, times of remeshing will be less, but the interference volume will be large and the volume loss is more. For the touch-die nodes, the nodes position must be adjusted in every step in simulation. It brings the touch-die element volume increased and affects the workpiece volume. The result of analysis shows that the volume loss caused by time-increment and remeshing is inevitable, the penalty constant and touch-die nodes adjustment are considered to compensate the volume change. The magnitude of penalty constant synchronizes with flow stress, and it is greater in the case of two-dimension than in three-dimension. A new strategy, original length method, is presented to solve the touch-die nodes adjustment on "the bland area of normal vector" of die surface. It can keep the touch-die element volume unchanged and be easily realized in three-dimensional programming. An example of the block upsetting process is used to demonstrate the effectiveness of the proposed techniques in the paper. The result shows that the techniques can improve the accuracy and trustiness of FEM simulation effectively.

REFERENCES

[1] S. Kobayashi, S. Oh, T. Altan, *Metal Forming and the Finite-Element Method*, Oxford University Press, Oxford, New York, (1989).

[2] Guoqun Zhao, Guangchun Wang, Xinwu Ma, *Control method of volume loss in simulation using rigid-plastic FEM*, Metal Forming Technology, 17(6), (1999), 42-45, (in Chinese).

[3] Yanran Zhan et al., *Determination of penalty coefficient during rigid plastic FEM simulation*, Forging & Stamping Technology, 1, (2000), 7-9, (in Chinese).

[4] Guoqun Zhao, *Forward and backward numerical simulation of forging processes*, Ph.D. Thesis, Shanghai Jiaotong Univ., (1992), (in Chinese).

[5] Mei Zhan et al., *A new method of modifying the node touching on the die during 3-D FEM simulation of the metal plastic forming*, Journal of plasticity engineering, 7(2), (2000), 62-65.

COMPUTATIONAL MECHANICS (Abstracts)
WCCM VI in conjunction with APCOM'04, *Sept. 5-10, 2004, Beijing, China*

A Study on Wheel-Rail Frictional Contact Problem by FE Parametric Quadratic Programming Method

Jun Zhang [1], Changhua Wu [2*], Zheng Zhong [1]

[1] *Key Laboratory of Solid Mechanics of Ministry of Education, Tongji University, Shanghai 200092, China*
[2] *Department of Engineering Mechanics, Dalian University of Technology, Dalian 116024, China*
E-mail: zhangjun611@hotmail.com, chwu57735@sohu.com, zhongk@online.sh.cn

Abstract

Since the conventional models used for computing wheel-rail creep forces in the world have not broken the limit of elastic semi-space assumption, it is impossible to conduct a research on three-D complicated wheel-rail contact problems, such as elastic-plastic contact problem, two-point contact problem and conformal contact problem.

In this paper, the elastic-plastic wheel-rail contact problem with rolling friction is solved by using FE parametric quadratic programming (PQP) method. Thus, we can cope with the complex and practical elastic-plastic contact problem, avoiding Hertz's hypothesis and elastic semi-space assumption, and gain higher computational accuracy and efficiency. This makes it possible to precisely analyze the wheel-rail contact problem.

The three-D computational models featuring "one-point" contact of wheel tread with rail top are constructed, respectively, according to the true dimensions of the locomotive wheel sets with the cone tread and the worn tread and the rail widely used in China, and elastic-plastic contact computation is carried out for different axle loads, tractive efforts, and friction factors between wheel and rail. The changing laws of wheel-rail contact forces and contact states (i.e. distribution of adhesion and creep regions in contact spot) versus various parameters are gained.

Based on the wheel-rail "one-point" contact calculation, the computational models of "two-point" contact (for cone tread) and conformal contact (for worn tread) produced as wheel flange is close to rail are firstly established, and the computation is completed. In the case of "two-point" contact, the changing laws of wheel-rail contact states and forces in various load cases are carefully analyzed, and the main reason both of wheel flange wear and rail side wear are found out. Based on computational models constructed according to different contact positions between wheel and rail, the conformal contact computation is made and the changing law of wheel-rail contact forces in the process of closing the wheel flange to the rail is obtained.

Having analyzed and compared the influence of the two kinds of treads on wheel-rail wear, this paper points out the reason why the wheel with the worn tread has stronger wear resistance, and meanwhile draws some important conclusions, such as: with transverse forces applied between wheel and rail increased, adhesive power of locomotive will be decreased.

COMPUTATIONAL MECHANICS (Abstracts)
WCCM VI in conjunction with APCOM'04, *Sept. 5-10, 2004, Beijing, China*

Nonlinear Analysis of Wet-Sprayed Fiber Reinforced Concrete Lining Construction Process for Railway Tunnel within Hard Rock

Taiquan Zhou [1*], Yuan Hua [1], Junying Lian [1], Renhe Wang [2]

[1] *Department of Civil Engineering, Southern Yangtze University, Wuxi, 214063, China*
[2] *Department of Civil Engineering, Anhui University of Science and Technology, Huainan, 232001, China*

e-mail: zhoutaiquan@sytu.edu.cn, rhwang@aust.edu.cn

Abstract

The New Austrian Tunnel Method is widely accepted in the tunnel construction field. The relatively flexible anchor sprayed lining structure is used instead of the common steel and wood lining, which has the good virtue of timely lining and easy construction. The fiber reinforced concrete has higher tensile, flexural strength and lighter weight compared with that of the ordinary concrete. The wet-sprayed fiber reinforced concrete is first introduced in the tunnel lining construction in China. It is difficult to design such lining structure according to the existing tunnel design codes. In order to investigate the nonlinear deformation and the internal force distribution of such lining structure to assure the structure safety, the nonlinear finite element method is carried out to investigate the wet-sprayed fiber reinforced concrete lining construction process. The nonlinear materials of lining structure, the interaction between the lining structure and the rock and the whole construction process modeling are discussed in detail. As a case for study, the nonlinear analysis for practical railway tunnel engineering within hard rock is studied using the nonlinear finite element software ADINA. The numerical analysis result for the wet-sprayed concrete stress and the top tunnel displacement agrees with the measured fairly well. The numerical analysis result shows that the stress field for the lining structure remains elastic during the large section excavation and the lining construction process and the new lining structure contributes to the tunnel section deformation decrease.

REFERENCES

[1] Y. Hua, H. B. Oi, Z. Q. Jiang, S. Z. Huang, S. B. Zhang, *Study on the bending fatigue damage of the carbon and the polypropylene hybrid fiber reinforced concrete*, Key Engineering Materials, 183, (2000), 571-576.

[2] K. J. Bathe, *Finite Element Procedures in Engineering Analysis*, Prentice-Hall, Englewood Cliffs (1982).

[3] O. C. Zienkiewicz, A. H. C. Chan, M. Pastor, B. A. Shrefler, *Computational Geomechanics*, John Wiley & Sons Ltd., Chichester, England (1999).

COMPUTATIONAL MECHANICS (Abstracts)
WCCM VI in conjunction with APCOM'04, *Sept. 5-10, 2004, Beijing, China*

Study of Structural Assemblies in the Presence of Uncertainties

P. A. Boucard [1*], C. Blanzé [1], L. Champaney [2]

[1] *LMT Cachan – ENS Cachan, 61 avenue du Président Wilson, Cachan Cedex, 94235, France*
[2] *LEMA – Univ. Versailles/StQuentin, 45 avenue des Etats-Unis, Versailles Cedex, 78035, France*
e-mail: boucard@lmt.ens-cachan.fr, blanze@lmt.ens-cachan.fr, champaney@meca.uvsq.fr

Abstract

The objective of the work presented is to develop a suitable strategy for the sensitivity analysis of assemblies including contact with friction, gaps, elastically bonded joints… The approach consists in developing an application of the LArge Time INcrement (LATIN) method [1] for the parametric analysis of static problems with multiple joints. For structural engineers, the incorporation of parametric uncertainties of a system into a mechanical model represents a challenge. In order to take such uncertainty into account, it is necessary to calculate the response of the structure for all possible sets of values of the design parameters or to use a probabilistic approach.

Our approach is based on a decomposition of the assembly into substructures and interfaces. The interfaces play the vital role of enabling local nonlinearities, such as contact and friction, to be modeled easily and accurately. The problem is solved in each substructure by the finite element method, and an iterative scheme based on the LATIN method is used for the global resolution.

The first approach proposed consists in calculating response surfaces such that each point of a surface is associated with a set of values of all the variable parameters (friction coefficients, prestress). Here, as an alternative to carrying out these full calculations, we propose to take advantage of the capability of the LATIN method to reuse the solution of a given problem (for one set of parameters) to solve similar problems (for the other sets of parameters) [2]. The numerical examples presented in the paper concern 3D assemblies. The comparison of the computation costs with those of classical industrial codes shows the algorithm is very efficient.

The second technique proposed addresses the random response of assemblies whose interfaces have uncertain characteristics. Thus, the randomness of the response comes from the random behavior of the connections. In this context, the random response is developed using a Polynomial Chaos Expansion (PCE) [3] coupled with the LATIN approach. A dedicated computational strategy is presented [4]. The numerical example concerns an assembly of structures connected by adhesively bonded joints with random stiffness characteristics. The results are compared with those of the Monte Carlo method and the stochastic finite element method.

REFERENCES

[1] P. Ladevèze, *Nonlinear Computational Structural Mechanics, New Approaches and Non-Incremental Methods of Calculation*, Springer Verlag, Berlin, Germany (1999).

[2] P. A. Boucard, L. Champaney, A suitable computational strategy for the parametric analysis of problems with multiple contact, Int. J. Num. Meth. Engng, 57(10), (2003), 1259–1281.

[3] R.G. Ghanem, P.D. Spanos, *Stochastic Finite Elements: A Spectral Approach*, Springer Verlag, Berlin, Germany (1991).

[4] C. Blanzé, L. Champaney, *Computational strategy for the analysis of assemblies containing interface uncertainties*, in K.J Bathe editor, Computational Fluid and Solid Mechanics 2003, Proc. 2nd Int. MIT Conference, Cambridge, MA, USA, (2003) pp. 1863–1866.

COMPUTATIONAL MECHANICS (Abstracts)
WCCM VI in conjunction with APCOM'04, *Sept. 5-10, 2004, Beijing, China*

An Explanation for Abrasive Critical Size Effect in Two-Body Abrasion by 3D FEM and Monte-Carlo Simulation

Liang Fang *, Qihong Cen, Kun Sun

The State Key Laboratory for Mechanical Behavior of Materials, Xian Jiaotong University, Xian, 710049, China
e-mail: fangl@xjtu.edu.cn, qhcen@163.com, sunkun@mail.xjtu.edu.cn

Abstract

The size of abrasives has a critical effect on the wear loss of materials in two -body abrasion. An abrasive paper in the tests was produced by Monte -Carlo stochastic simulation in this study. The indented depth of abrasive was estimated for different loads from Johnson 's contact theory. Meanwhile, a three dimensional finite element method (FEM) was used to simulate the grooving process of SiC abrasives against ANSI 1080 steel. The indented depths varied from 10 μm to 100μm and the abrasive particle radius was limited in 150 μm in FEM. The material removal factor f_{ab} from groove was calculated in across sectional images by FEM. The result showed that particle area density in the abrasive paper is reduced with the increase of abrasive sizes. As a result, the indented depth of abrasives is increased at the same normal load. The factor f_{ab} can be taken as a monitor of indented depth. The factor f_{ab} increases with a small indented depth. After indented depth increases further to some extents, the factor f_{ab} shows a level-off characteristic. It is considered that the characteristic of the factor f_{ab} can be applied to explain the critical size effect of abrasives in abrasive wear. That should be one of explanations that we ar rate also behaves a critical size effect with increasing the particle sizes.

REFERENCES

[1] K. Kato, *Micro-mechanisms of wear— wear modes*, Wear, 153, (1992), 277-295.

[2] G. Subhash, W. Zhang, *Investigation of the overall friction coefficient in single-pass scratch test*, Wear, 252, (2002), 123-134.

[3] J. L. Bucaille, E. Felder, G. Hochstetter, *Mechanical analysis of the scratch test on elastic and perfectly plastic materials with the three-dimensional finite element modeling*, Wear, 249, (2001), 422-432.

[4] V. Jardret, H. Zahouani, J. L. Loubet et al., *Understanding and quantification of elastic and plastic deformation during a scratch test*, Wear, 218, (1998), 8-14.

COMPUTATIONAL MECHANICS (Abstracts)
WCCM VI in conjunction with APCOM'04, *Sept. 5-10, 2004, Beijing, China*

Damage Estimation Parameters Based on Field Data of Steam Turbine

Yoshiro Imai [1*], Nagatoshi Okabe [2], Hiroshi Ishikawa [1]

[1] *Department of Reliability-based Information Systems Engineering, Kagawa University, 2216-20, Hayashi-cho, Takamatsu City, 761-0396, Japan*
[2] *Department of Mechanical Engineering, Ehime University, 10-13, Dogo-Himata, Matsuyama City, 790-85773, Japan*
e-mail: imai@eng.kagawa-u.ac.jp, okabe@en1.ehime-u.ac.jp, ishikawa@eng.kagawa-u.ac.jp

Abstract

Structural components of steam turbine are prone to crack initiation and growth due to combined effects of creep and cyclic loading. In such a case, damage analysis of structural components can be divided into two approaches: creep rupture damage analysis and thermal fatigue damage one. Especially, thermal fatigue of structural components results from thermal stress accumulation based on start-up/shut-down fatigue cycles (i.e. iteration of stop and restart). Manson-Coffin and Basquin's results [1] teach us a relation between a number of cycle to rupture generation N, total strain range $\Delta\varepsilon$, and so on. During service of steam turbines, start-up/shut-down cycles are iterated continuously. Almost turbines consist of following operation patterns: restarting at cold temperature, restarting at warm one and restarting at hot one. From Manson-Coffin and Basquin's Equation, N_c, N_w and N_h by can be obtained as estimated numbers of cycles to rupture generation at according three operation patterns. n_c, n_w and n_h by denote the according numbers of cumulative start-up/shut-down cycles according above mentioned three patterns (Cold-start, Warm-start, and Hot-start), respectively. With these numbers, thermal fatigue damage ϕ_f may be expressed as:

$$\phi_f = n_c / N_c + n_w / N_w + n_h / N_h$$

In order to treat thermal fatigue damage uniformly among different operating conditions, we have introduced a thermal fatigue damage estimation parameter and evaluated thermal fatigue damage with this parameter in the unified criteria. Our estimation parameter q is expressed as,

$$q = n(\sigma_c \frac{n_c}{n} + \sigma_w \frac{n_w}{n} + \sigma_h \frac{n_h}{n})^{\mu}$$

where $n = n_c + n_w + n_h$, and μ is a constant to be determined according to the target structure component. q can give us an optimal approximation of ϕ_f for each structural component of turbine. With our parameter, fatigue damage may be expressed in terms of constant K as,

$$\phi_f = Kq$$

Statistical analysis has to be performed for each structural component of steam turbine to evaluate variability of field data quantitatively. We have employed Weibull statistical analysis approach .

REFERENCES

[1] S. Suresh, *Fatigue of materials—2nd edition*, Cambridge University Press, UK, (1998)

COMPUTATIONAL MECHANICS (Abstracts)
WCCM VI in conjunction with APCOM'04, *Sept. 5-10, 2004, Beijing, China*

Time Domain Reliability Analysis of Ships with Correlated Corrosion and Fatigue

Xiaoli Jiang[1*], Rob E. Melchers[2]

[1] *School of Transportation, Wuhan University of Technology, Wuhan, 430063, China*
[2] *School of Engineering, University of Newcastle, Callaghan, 2308, NSW, Australia*
e-mail: xiaoli.Jiang@newcastle.edu.au, rob.melchers@newcastle.edu.au

Abstract

There are a great number of papers on the reliability assessment of various kinds of aging ships. In recent two decades, corrosion and fatigue crack are taken into account either individually or simultaneously. Although it is well known that the existence of corrosion will enhance the development of fatigue crack and vice verse, the influences of corrosion and fatigue are always considered respectively or a simple parameter, e.g. a corrosion-enhanced factor is used to represent the inter-relationship of corrosion and fatigue. In this paper, a short review on corrosion and fatigue research is presented. The most recent time-dependant corrosion wastage model is applied to simulate the corrosive degradation process, which thoroughly reflects the influence of location and application (such as oil tank or water tank, etc.) on the corrosive wastage growth; the fracture mechanics method is utilized to calculate the fatigue cracks. According to the comparison results of 5 classification societies, although hot spot stress is more suitable for the calculation of fatigue crack, the uncertainty of load treatment and the application of different FEM elements result in the considerable scatter data. Therefore, the nominal stress is still used to determine the maximum stress range; Above all, corrosion and fatigue are treated as correlated variables not independent ones as they used to be regarded. The transformation of correlated coefficient before and after normalization of non-normal variables is presented by means of Nataf transformation. Herein, progressive collapse analysis, named Smith's method, is used to carry out ultimate hull girder strength calculation. Compared with FEM and ISUM, high efficiency and short time-consumption of Smith' method is considerably apparent. The loads considered consist of two components, still water bending moment and vertical wave-induced bending moment, both of which are simulated by means of extreme type I distribution. The modified FOSM and surface response approach are utilized to carry out time-variant ultimate hull girder strength reliability analysis. An example application foe a VLCC "Energy Concentration" is presented. A series of positive correlated coefficients ranging from 0.2, 0.5 to 0.8 are taken to analyze their influence on reliability index. A comparison is made between the present method and current traditional methods, which illustrates the reliability index is sensitive to the values of correlated coefficients and lack of sensitivity to the corrosion-enhanced factors. Moreover, the traditional methods to some extent underestimate the influence of combined corrosion and fatigue on ultimate hull girder strength, the present method, by contrast, is reasonable and acceptable. It is also proved that the correction of corrosion database has significant impact on the time-variant reliability analysis through alternation of coefficients of variance of corrosion rates. Therefore, more precise and sea state or routes-classified corrosion databases are necessary for ship hull strength and reliability analysis.

COMPUTATIONAL MECHANICS (Abstracts)
WCCM VI in conjunction with APCOM'04, *Sept. 5-10, 2004, Beijing, China*

Reliability of Uncertain Linear Systems Subjected to Stochastic Dynamic Loads

Lambros S. Katafygiotis [1*], Sai Hung Cheung [2]

[1] *Dept. of Civil Eng., HKUST, Clearwater Bay, Kowloon, Hong Kong, China*
[2] *Graduate Student, Caltech, Pasadena, CA, USA*
e-mail: lambros@ust.hk, sai@caltech.edu

Abstract

A number of very efficient methodologies for calculating the reliability of linear systems subjected to stochastic dynamic loads have been recently proposed. Such methods, e.g. Domain Decomposition Method [1], assume that the linear system is deterministic and the only random variables for consideration are those used to model the random input. The resulting reliability problem in such a case can be shown to have a corresponding limit state function (LSF) that comprises of a union of linear failure domains. The remarkable efficiency of these algorithms is due to their full exploitation of this particular structure of the failure domain.

Many reliability problems of interest involve additional uncertainties that violate the aforementioned structure of the failure domain. For example, consideration of structural model uncertainties and/or certain type of excitation model uncertainties, result in a nonlinear reliability problem. It is well known that the treatment of general nonlinear reliability problems is a very challenging computational task. Herein, we propose a very efficient methodology for treating this particular nonlinear problem. The proposed method is based on treating separately the parameters $\mathbf{Z}$ that lead to a linear problem, such as the random variables modelling the Gaussian excitation, from the remaining uncertain parameters θ. It is worth noting that the dimension of $\mathbf{Z}$ is much larger than that of θ. The probability of failure P_F, originally expressed as an integral over the $(\mathbf{Z}, \theta)$ domain, can be rewritten as an integral over the relatively low-dimensional θ -domain, involving the conditional probability of failure given θ. The latter can be calculated very efficiently using algorithms such as DDM. The integral involving the conditional probability given θ is rewritten as a reliability integral by introducing an auxiliary random variable. The resulting nonlinear reliability integral is then calculated by using a very efficient methodology developed by the authors, referred to as Auxiliary Domain Method. Numerical examples demonstrate the accuracy and efficiency of the proposed method.

REFERENCES

[1] L. S. Katafygiotis, S. H. Cheung, *Domain decomposition method for calculating the failure probability of linear dynamic systems subjected to Gaussian stochastic loads*, J. Engng Mech., ASCE, (accepted).

COMPUTATIONAL MECHANICS (Abstracts)
WCCM VI in conjunction with APCOM'04, *Sept. 5-10, 2004, Beijing, China*

Mechanical Reliability of Anisotropic Conductive Adhesive Films under Moisture/Reflow Sensitivity Test

W. K. Kim*, T. Ikeda, N. Miyazaki

Department of Chemical Engineering, Kyushu University, 6-10-1 Hakozaki, Higasi-ku, Fukuoka, 812-8581, Japan
e-mail: {wkkim, ikeda, miyazaki}@chem-eng.kyushu-u.ac.jp

Abstract

Anisotropic conductive adhesive film (ACF) is a thermoset adhesive containing conductive particles that supply electrical conduction. The ACF has been used for assemblies such as the connection between a liquid crystal display (LCD) panel and a flexible print circuit board (FPC). The ACF is expected to be one of the key technologies for flip chip packaging, system in packaging and chip size packaging. The goal of our work is to provide an optimum design scheme to achieve the best combination of electrical performance and mechanical reliability for electronic packages using the ACF. The stress intensity factors of an interface crack between jointed dissimilar materials are utilized for evaluation in a flip chip connected by ACF under moisture/reflow tests, together with the vapor pressure calculated from the moisture absorption analysis using three-dimensional finite element method. The predicted results are compared with the results of moisture/reflow sensitivity tests of a flip chip connected by ACF. It is concluded that the durability of a flip chip using the ACF connection can be predicted by using the delamination toughness of the interface crack between the ACF and adherend.

REFERENCES

[1] T. Ikeda, I. Arase, Y. Ueno, N. Miyazaki, *Strength evaluation of electronic plastic packaging using stress intensity factors of V-notch*, Computer Modeling and Simulation in Engineering, 1, (2000), 91-97.

[2] T. Ikeda, I. Arase, Y. Ueno, N. Miyazaki, N. Ito, M. Nagatake, M. Sato, *Strength evaluation of plastic package during solder reflow process using stress intensity factors of V-notch*, J. Electronic Packaging, 25, (2003), 31-33.

[3] W. K. Kim, T. Ikeda, N. Miyazaki, *Reliability evaluation of flip chip using stress intensity factors of an interface crack*, Proc. IPACK03, International Electronic Packaging Technical Conference and Exhibition, Paper IPACK2003-35146, (2003).

COMPUTATIONAL MECHANICS (Abstracts)
WCCM VI in conjunction with APCOM'04, *Sept. 5-10, 2004, Beijing, China*

On Applications of Asymptotic Coefficient of Variation of Structural Response for Reliability Analyses

P. Ngamcharoen *, W. Ouypornprasert

Department of Civil Engineering, College of Engineering, Rangsit University, 52/347 Paholyothin Rd., Pathumthani, 12000 Thailand
e-mail: Phadet_ng@Thai.com, winai@rangsit.rsu.ac.th

Abstract

The objective of this technical paper is to apply concepts of asymptotic coefficient of variation of structural response (Ω_R) for reliability analyses. Firstly basic concepts of structural reliability are reviewed [1]. Reliability analyses in terms of load effects and structural resistance imply that, for a pre-specified value of target failure probability, there must be a corresponding asymptotic Ω_R [2], in this case i.e. the upper limit of Ω_R, which is defined as the ratio of its standard deviation to its mean. For the case of the normal structural response this asymptotic Ω_R is inversely proportional to, independent of variation of load effects, the safety index (β), which could be interpreted as the shortest distance of the limit state to the origin of the standard normal space. Below this upper limit value of Ω_R structures could resist safely with the expected acceptable level of risk. For instances the asymptotic Ω_R of 0.131 and 0.210 would correspond to the target values of failure probability of 10^{-14} and 10^{-6}, respectively. However based on existing engineering data it is not uncommon that structural responses, in some cases, have to be represented by non-normal distributions. Therefore it is the aim of this technical paper to generalize the concept of asymptotic to non-normal structural responses. The applicability of the proposed concept could be demonstrated in numerical examples. It should be emphasized that the relationship among basic variables in the structural resistance could be either linear or non-linear or even so complicated that the structural response can only be obtained from complete finite element analyses. Then the fitted distribution of the structural response could be specified by Goodness-of-Fitted Tests based on random experiments. If a lognormal distribution is fitted, the asymptotic Ω_R would be 0.074 for $p_f = 10^{-14}$. It is interesting to note that this generalized concept of asymptotic Ω_R could also be extended for any kind of output for civil engineering systems i.e. pile, concrete frame and pre-stressed concrete beam or even might be applied for other engineering systems e.g. electrical circuits and piping, etc.

REFERENCES

[1] G. I. Schuëller, C. G. Bucher, U. Bougund, W. Ouypornprasert, *On efficient computational schemes to calculate structural failure probabilities*, In Y. K. Lin, G. I. Schuëler eds. Lecture Notes in Engineering (31): Stochastic Structural Mechanics, Springer-Verlang, New York.

[2] P. Ngamcharoen, W. Ouypornprasert, *Asymptotic coefficient of variation of structural response against target reliability*, International Journal of Materials & Structural Reliability, 2(1), (2004), 59-74.

COMPUTATIONAL MECHANICS (Abstracts)
WCCM VI in conjunction with APCOM'04, *Sept. 5-10, 2004, Beijing, China*

Chaos Polynomial Representation of Parametric Uncertainties in Aeroelasticity

Fabrice Poirion*

ONERA/DDSS, 29 avenue de la Division Leclerc, B.P. 72, Châtillon CEDEX 92322, France
e-mail: fabrice.poirion@onera.fr

Abstract

Stability of the coupled system aircraft/aerodynamics is a key issue in civil airplane manufacturing. During the design stage of an airplane, many structural parameters are neither clearly fixed nor known, but nevertheless the final project must comply to the various international certification regulations. One popular approach to deal with this problem is to model the parametric uncertainties through random variables. Then one has to study the stability of a random parameter dynamical system. Various steps, based on Monte Carlo simulation, have been developed at ONERA in order to construct an effective numerical procedure which can be utilized together with standard structural and aerodynamic codes by manufacturers. Basically two procedures were used: a direct simulation method which randomly draws the values of the uncertain parameters directly in the finite element model of the aircraft, and performs a complete flutter calculation at each step. It yields exact statistics in the sense that no reduction is performed on the random parameters. But, although this method is straightforward, it is not applicable at the present time for the large d.o.f models which are currently used (hundreds of thousands of d.o.f.).

A second approach has been developed based on a projected simulation method. It necessitates the construction of an unique basis of projection. Results have shown that the mean modal basis, which seems a priori to be a good candidate, is not rich enough to restitute the exact coupling behavior between the structure and the fluid. Moreover, this method is generally used in conjunction with a perturbation technique in order to represent the random parameter dependency of the generalized mass and stiffness matrices, valid solely for small uncertainties. We present an approach for extending the mean modal basis to a richer one. Moreover we show that polynomial chaos can be introduced in order to replace the perturbation technique with a representation capable of taking into account important uncertainties.

In order to illustrate this last point, we present a thorough study based on a beam elementary stiffness matrix where the length is considered as a random parameter. The corresponding random variables can be projected on a Gaussian polynomial chaos using exactly two Gaussian random variables. The polynomial expansion coefficients are estimated by a Monte Carlo method. Comparisons between a first order perturbation method, the chaos decomposition model and the analytical results are given for the stiffness matrix mean value, for its eigen values and also its eigen vectors. Convergence results are illustrated. Finally we conclude by giving the first results obtained on a complete aircraft model.

COMPUTATIONAL MECHANICS (Abstracts)
WCCM VI in conjunction with APCOM'04, *Sept. 5-10, 2004, Beijing, China*

Collocation Scheme for Computing the Coefficients of the Polynomial Chaos Expansion in Stochastic Finite Element Analysis

Bruno Sudret [1*], Marc Berveiller [1, 2], Maurice Lemaire [2]

[1] *Electricité de France R&D Division, Site des Renardières, F-77818 Moret-sur-Loing, France*
[2] *Institut Français de Mécanique Avancée, LaRAMA, Campus des Cézeaux, BP265, F-63175 Aubière, France*

e-mail: bruno.sudret@edf.fr, marc.berveiller@edf.fr, maurice.lemaire@ifma.fr

Abstract

The Stochastic Finite Element Method developed by Ghanem *et al* [1] allows to solve stochastic boundary value problems involving spatially randomly varying materials usually described as Gaussian or lognormal random fields. The method is based on the discretization of the input random fields and the expansion of the mechanical response onto the so-called *polynomial chaos*. A similar procedure allowing to model random material properties and loading by means of any number of random variables of any type has been recently proposed [2-3].

In both cases, the coefficients of the response expansion are computed using a Galerkin procedure, which leads to a linear system whose size is equal to the number of degrees of freedom of the system multiplied by the number of coefficients retained in the response expansion. This approach presents three main drawbacks, namely 1) the programming of *ad-hoc* software to assemble and solve this system, 2) the total computational cost, 3) the difficulty to address non linear problems (although an attempt can be found in [4]).

In this paper, a *collocation scheme* is proposed to compute the response coefficients [5]. This calculation reduces to solving selected *deterministic* finite element problems. This task may be carried out with any finite element code at hand without additional implementation. The three drawbacks mentioned above are solved *ab initio*. The approach is illustrated by the analysis of a crack in a pipe weld in the context of fracture mechanics.

REFERENCES

[1] R. G. Ghanem, P. D. Spanos, *Stochastic finite elements – a spectral approach*, Springer Verlag, (1991).

[2] B. Sudret, M. Berveiller, M. Lemaire, *Application of a stochastic finite element procedure to reliability analysis*, Proc. 11th IFIP WG 7.5 Working Conference on Reliability and Optimization of Structural Systems, Banff, Canada, (2003).

[3] B. Sudret, M. Berveiller, M. Lemaire, *A stochastic finite element method in linear mechanics*, Comptes Rendus Mécanique, (2004), (accepted).

[4] A. Keese, H. Matthies, *Efficient solvers for nonlinear stochastic problems*, Proc. 5th World Congress on Computational Mechanics, Vienna, Austria, (2002).

[5] S. S. Isukapalli, *Uncertainty analysis of transport-transformation models*, PhD Thesis, The State University of New Jersey, (1999).

COMPUTATIONAL MECHANICS (Abstracts)
WCCM VI in conjunction with APCOM'04, *Sept. 5-10, 2004, Beijing, China*

Application of Element Free Galerkin Method on Metal Forming Process

Weidong Wang [1*], Guoqun Zhao [2], Yiguo Luan [2]

[1] *Department of Engineering Mechanics, Shandong University, Jinan, 250061, China*
[2] *ERC for Mould & Die, Shandong University, Jinan, 250061, China*
e-mail: wangwd@sdu.edu.cn

Abstract

A new meshless method for simulating the metal forming process is given by combining the element free Galerkin method with the flow theory of rigid plastic/viscoplastic mechanics. To alleviate the difficulty of the implementation of essential boundary conditions in element free Galerkin method, the full transformation method, which proposed in the reproducing kernel particle method, is utilized to implement the velocity boundary conditions directly and exactly at nodes by modifying MLS approximation. The modified moving least square appropriation is used to construct velocity field function. The matrix forms of some variables, such as velocity field function, strain rate, effective strain rate, and volume strain rate, are established. The penalty method is proposed to satisfy the volume incompressibility condition in rigid plastic/viscoplastic mechanics. The discrete control equations of element free Galerkin method is developed for the problems of two dimensional metal forming by utilizing the incomplete generalized variational principle of rigid plastic/viscoplastic. The methods, such as the treatment of contact frictional boundaries, the criteria and control of convergence, are given to improve the method proposed in the present paper. Numerical simulation of an axisymmetric upsetting process is given to show the effectiveness and feasibility of the present method.

REFERENCES

[1] T. Belytschko, Y. Y. Lu, L. Gu, *Element-free Galerkin methods*, Int. J. Num. Meth. Eng., 37, (1994), 229-256.

[2] Y. Y. Lu, T. Belytschko, L. Gu, *A new implementation of the element-free Galerkin methods*. Int. J. Num. Meth. Eng., 37, (1994), 397-414.

[3] Y. Krongauz, T. Belytschko, *Enforcement of essential boundary conditions in meshless approximations using finite elements*, Comput. Meth. Appl. Mech. Engrg, 131, (1996), 133-145.

[4] L. Gavete, J. J. Benito, S. Falcon, A Ruiz, *Implementation of essential boundary conditions in meshless method*, Communications in numerical methods in engineering, 16, (2000), 409-421.

[5] J. S. Chen, C. Pan, C. T. Wu, W. K. Liu, *Reproducing kernel particle methods for large deformation analysis of non-linear structures*. Comput. Meth. Appl. Mech. Engrg., 139, (1996), 195-227.

[6] S. Kobayashi, S. I. Oh, T. Altan, *Metal forming and the finite element method*, Oxford University Press, New York, (1989).

COMPUTATIONAL MECHANICS (Abstracts)
WCCM VI in conjunction with APCOM'04, *Sept. 5-10, 2004, Beijing, China*

Simulation Approach in Stiffness Reliability Analysis

Qi Zhang*, Xingsi Li

State Key Lab. of Structural Analysis for Industrial Equipment, Dalian University of Technology, Dalian, 116023, China
e-mail: zhangqi@student.dlut.edu.cn, lixs@dlut.edu.cn

Abstract

Structural reliability analysis under the condition of limit state function unknown is most important for large and complex structure. Generally, Response Surface Method (RSM) is used to evaluate the structural limit state function, and then the Moment Method is used to serve. On account of the assumption of polynomial form, the RSM can not solve the problem with high accuracy.
A semi-parameter interpolating technique, Kriging Simulation (KS), which could avoid the restriction and the influence of different limit state function formulation, could evaluate the structure response with high accuracy[1]. There is no explicit limit state function (LSM) in KS, which is the main difference between KS and RSM[2]. Compared with other approximated methods, KS is a more 'statistically sound' method for approximating deterministic computer experiment. KS differs from the traditional interpolation techniques in two aspects: dynamic configuration of points; local and global approximation are considered. Such a feature enables the technique to take into account possible trends[3]. KS does not require a large amount of data to be tuned and a reference baseline to define the unknown model.
Combined the KS with the optimization method, the reliability index could be solved. An efficient computational procedure include: selction of sampling points; structure response computation of sampling points; construction of KS model; computation of raliability index or failure probability. Random sampling (two level sampling) or Latin hypercube sampling are used in the selction of sampling points.Door shaped frame and three bay twelve-storey frame are investigated in the numerical examples. Compared the results of KS with RSM and Monte Carlo Method (MCM), the failure probability of KS is more accurate than RSM and similar to MCM, the times of structure analysis is similar to RSM. KS is shown to be a new efficient scheme with respect to both computational effort and accuracy for structural reliability analysis. Especially in two level simulations, the global and local simulation can be done effectively. The course of constructing the KS model is a learning phase, so it can be performed as an evaluation system for structure reliability analysis.

REFERENCES

[1] G. Matherton, *Principles of geo-statistics*, Economic Geology, 58, (1963), 1246-1266.
[2] T. W. Simpson, *Comparison of response surface and kriging models in the multi- disciplinary design of an aerospike nozzle*, NASA/CR-1998-206935, ICASE Report No. 98-16, (1998).
[3] A. A. Giunta, L. T. Watson, *A comparison of approximation modeling techniques: polynomial vs. Interpolating models*, AIAA-98-4758, (1998).

COMPUTATIONAL MECHANICS (Abstracts)
WCCM VI in conjunction with APCOM'04, *Sept. 5-10, 2004, Beijing, China*

Numerical Tools for Defect Detection in Predictive Maintenance Using Vibration Analysis

K. Debray*, P. Estocq, F. Bogard, Y. Q. Guo

Group of Mechanics, Materials & structures, Laboratory GMMS, EA2617, University of Reims Champagne Ardenne, Moulin de la Housse, BP 1039, 51687, Reims cedex 2, France
e-mail: {karl.debray, philippe.estocq, fabien.bogard, yq.guo}@univ-reims.fr

Abstract

The follow-up, by vibration analysis, of the damage of the sensitive components in a revolving machine is a classical technique used in predictive maintenance [1]. A traditional experimental approach consists in following the evolution of the vibration spectrum at various measurement points on the machine in order to supervise the appearance of the characteristics relative to the frequencies of defects. Generally, the measurement points are situated near by the considered component, but it will be better to numerically determine optimum measurement areas, particularly when the detection relates to several bodies. This work proposes a numerical methodology for the defect detection by vibration analysis in the predictive maintenance. This study is composed of two phases: Firstly, the finite element method is used for the vibration analysis of the entire structure (shaft, bearing, casing ...), a particular care is taken for the modelling of the sensitive and essential bodies such as the ball bearings [2]. The static nonlinear analysis allows obtaining the tangent stiffness matrices of the bearing under a given static load. These tangent matrices are then used for the linear dynamic analysis by supposing that the dynamic effects of defects are much smaller than the static ones. The exploitation of the dynamic behaviour excited by the defects in the bearing allows determining the degree of the damage of the bearing [3]. The most adapted measurement areas are those with the strongest responses due to the defect excitation. Secondly, the measurement areas obtained by the numerical analysis are used for experiments. Several distinct approaches are planned: 1) vibration measurements are carried out directly using several accelerometers, 2) measurements are taken by using a mechanical resonator to amplify the defects on the inner or/and outer rings of the ball bearing. 3) the machine's frame is modified by adding a wave guide in order to transfer the vibrations due to a defect towards an accessible measurement area. The numerical results highlight that the optimum measurement zones are not necessarily close to the defects. These results made it possible to install a mechanical resonator for the defect detection. The application of the resonator is simple, economic and efficient for the detection of the defects. This study shows that the defect detection by vibration analysis can be notably improved without using expensive and sophisticated measuring apparatus.

REFERENCES

[1] A.F.N.O.R.: Surveillance des machines par analyse des vibrations. ISBN: 2123090123 (1995).

[2] A. Bourdon, J. F. Rigal, D. Play, *Static rolling bearing models in a CAD Environment for the study of complex mechanisms: part I – rolling bearing model*, J. of Tribology, 121, (1999), 205-214.

[3] F. Bogard, K. Debray, Y. Q. Guo, *Determination of optimum sensor positions for defect detection on revolving machines by a finite element vibration analysis*, Int. J. of Solids and Structures, 39(12), (2002), 3159-3173.

COMPUTATIONAL MECHANICS (Abstracts)
WCCM VI in conjunction with APCOM'04, *Sept. 5-10, 2004, Beijing, China*

Numerical Simulation with Finite Element Method for the Infrared Image of Cracked CFRC

Li Huang *, Zhuoqiu Li, Xianhui Song

Department of Engineering Structures and Mechanics, Wuhan University of Technology, Wuhan, 430070, China
e-mail: hLwhut@yahoo.com.cn, wgdhL@163.com, science@mail.whut.edu.cn

Abstract

Carbon fiber reinforced concrete (CFRC) is a composite material with the advantages of lightweight, corrosion resistance and electrical conductivity and so on. Based on the infrared thermography crack inspection of CFRC by making use of the electro-thermal effect of it, a theoretical model has been built up in this paper for analyzing the innovative application.
Differing from the conventional infrared inspection for concrete usually with the aid of an external heat source [1], this inspection involves a heat transfer problem with internal heat sources. In the case that the analytical solution is difficult to be found, we adopt the numerical simulation with finite element method for the infrared image, in order to discuss the special detection mechanism and its quantitative limit under various conditions. In fact, a few investigations only for conventional infrared inspection have been addressed this aspect besides Inagaki Terumi in [2] and J. K. C. Shih in [3].
The simulation process indicates that when a cracked CFRC specimen is applied a low voltage, the electric current distribution in specimen will be heterogeneous due to the crack existence. Correspondingly, the internal heat sources are heterogeneous. Further, the temperature distribution is heterogeneous especially near the tip of the crack. Meanwhile, heat transfer also lies in both convection and radiation between the specimen surface and the circumstance. Actually, this is an unsteady heat transfer problem with internal heat sources.
As a result, the numerical distribution of the surface temperature is given by programming computation. The differences between the numerical result and the infrared inspection result are analyzed, and the reasons causing these differences are analyzed as well.

REFERENCES

[1] Büyüköztürk Oral, *Imaging of concrete structures*, NDT & E International, 31(4), (1998), 233-243.
[2] Inagaki Terumi, Ishii Toshimitsu, Iwamoto Toshikatsu, *On the NDT and E for the diagnosis of defects using infrared thermography*, NDT & E International, 32, (1999), 247-257.
[3] J. K. C. Shih, R. Delpak, C. W. Hu, et al, *Application of infrared thermography for damage detection in structural concrete — a preliminary report*, NDT.net – February, 5 (2), (2000).

COMPUTATIONAL MECHANICS (Abstracts)
WCCM VI in conjunction with APCOM'04, *Sept. 5-10, 2004, Beijing, China*

Improved Damage Indicator Function Using Localization-Based Mode Combination Criteria

Guo Sun*, Yuanxian Gu

State Key Laboratory of Structural Analysis for Industrial Equipment, Dept. of Engineering Mechanics, Dalian University of Technology, Dalian 116024, China
e-mail: sunguo@dlut.edu.cn, guyx96@dlut.edu.cn

Abstract

The damage detection and the health monitoring of large-scale structures are important challenges to structural engineering and research of computational mechanics. One common approach of the damage detection is to employ the vibration characteristics of a structure to predict the damage locations and to estimate the damage degrees. The damage indicators have noticeable influence to the efficiency and reliability of such kind approach.
An improved damage indicator function, $\{\eta\}$ using localization-based mode combination criteria is proposed in this paper. By combining the natural frequencies and the mode shape data at a few selected freedoms and the mode weight coefficients, the damage indicator function $\{\eta\}$ is obtained with better localization property to the damage factor distribution. For the beam-like and plate-like structures, this method with proposed new damage indicator function needs only the easily measured vertical DOF and gets an advantage over the conventional damage indicator methods such as flexibility matrix method and mode shapes method. The sensitivity matrices of the indicator functions are studied, and the strategies to analyze the localization property of indicator function quantitatively are given. Some numerical simulations are performed and the results using different indicator functions are compared. The numerical results show that the proposed method with new indicator function can detect and approximately quantify the damages of beam-like and plate-like structures effectively.

REFERENCES

[1] S. W. Doebling, et al, *A review of damage identification methods that examine changes in dynamic properties*, Shock and Vibration Digest, 30(2), (1998), 91-105.
[2] A. K. Pandey, M. Biswas, M. M. Samman, *Damage detection from changes in curvature mode shapes*, J. Sound and Vibration, 145(2), 1991, 321-332.
[3] A. K. Pandey, M. Biswas, *Damage detection in structures using changes in flexibility*, J. Sound and Vibration, 169(1), (1994), 3-17.
[4] G. Sun, Y. X. Gu, *Improved flexibility matrix method of damage identification for multi-span beams*, Engineering Mechanics, 20(4), (2002), 50-54, (in Chinese).
[5] G. Sun, Y. X. Gu, *Damage identification of continuum structure using blanketing effect*, Proc. CJKOSM-2, Busan, Korea, (2002), pp. 507-514.

COMPUTATIONAL MECHANICS (Abstracts)
WCCM VI in conjunction with APCOM'04, *Sept. 5-10, 2004, Beijing, China*

A Study on the Optimal Design of the Flare Nut for Automobiles Using Finite Element Analysis

K. T. Han [1*], S. D. Kim [2]

[1] *School of Mechanical Engineering, Pukyong National University, Busan, 608-739, Korea*
[2] *Graduate School of Mechanical Engineering, Pukyong National University, Busan, 608-739, Korea*
e-mail: kthan@pknu.ac.kr, air-kim6755@hanmail.net

Abstract

Flare nut is a small and important part used to joint a brake tube-end in automobiles. The flare nut has been made by machining during the past 20 years but in recent years, there has been a growing interest in the forming by forging technology. According to the report, the quality of the flare nut produced by machining is not uniform and also the production cost is expensive. However, the production using forging processes could not be tried due to the lack of technology. Recently, to produce the flare nut economically, the forging processes have been recommended.
The purpose of the present paper is to investigate the optimal shape of the flare nut for automobiles using the DEFORM^{TM}-3D, commercially available finite element code. That is, in this paper, we studied optimal forging processes for the flare nut. Finite element analysis has been carried out to predict an optimal shape of the flare nut and its simulation results were reflected to the forging processes design for the flare nut. The shape of the flare nut is in agreement with the finite element simulation and the test results. Through this research, we could get the optimal shape of the flare nut.

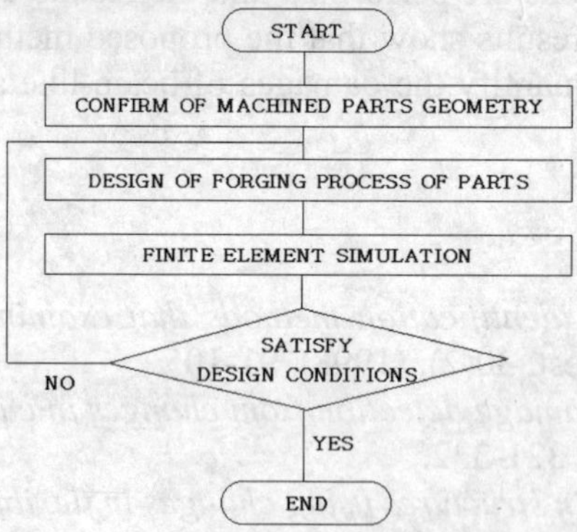

Fig. 1 Flow chart of process planning

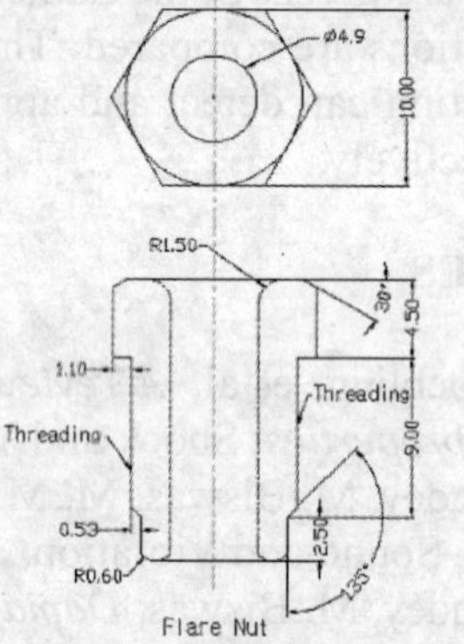

Fig. 2 Geometry of flare nut

REFERENCES

[1] C. H. Lee, S. Kobayashi, *New solutions to rigid-plastic deformation problems using a matrix method*, J. of Eng. for Ind., Trans. ASME, 95, (1973), 865-873.
[2] O. C. Zienkiewicz, P. N. Godbole, *Flow of plastic and visco-plastic solids with special reference to extrusion and forming processes*, Int. J. Num. Meth. Eng., 8, (1974), 3-16.
[3] G. B. Yu, T. A. Dean, *A practical computer aided design approach to design for axisymmetric forging die cavities*, Int. J. Mach. Tool, Des. Res., 25(1), (1985), 1-13.

COMPUTATIONAL MECHANICS (Abstracts)
WCCM VI in conjunction with APCOM'04, *Sept. 5-10, 2004, Beijing, China*

Optimization of High Temperature Furnace Design Using FEM Model and DFSS Analysis

Yi-Fei Zhang [1*], H. P. Wang [2]

[1] *FEM Inc., 772 Red Oak Dr, Niskayuna, NY 12309, USA*
[2] *GE Global Research Center, Niskayuna, NY 12309, USA*
e-mail: zhangyf@crd.ge.com, wanghp@crd.ge.com

Abstract

The basic DFSS (Design for Six Sigma) methodology will be briefly introduced. In order to establish the relationship between the design/processing parameters and the product (tube) dimensions, or the transfer functions, a significant number of experiments, either physical or numerical, are required. It was found that the most cost-effective way to develop the DFSS transfer functions for this particular tube drawing process is through a semi-empirical approach, integrating an advanced numerical model with experimental database.

Finite element model is applied to develop the transfer functions for the furnace design and processing parameters of a high temperature glass tubing process. Some of the uncertain properties as well as boundary conditions are difficult to quantify or measure under such high temperature environment. For example, the correlation between the furnace coil power and the die exit temperature and the equivalent total emissivity of glass can be calibrated against some existing data. The numerical DOE (Design of Experiments) is applied in the furnace design, which helps in choosing the right fraction of the many possible experiments to get the maximum amount of information for a minimum cost. In addition, the designed numerical experiments can detect and quantify the effect of the parameters on the process.

The goal of developing transfer functions between the inputs Xs (design and processing parameters) and the outputs Ys (CTQ, i.e. Critical-to-Quality characteristics) is to identify both the process window and the design space. Operating conditions should be away from the edge of the process window to ensure a stabled output while the design parameters should be in low gradient regions as a function of Ys to follow the robust design principle. The critical furnace design parameters, such as the die, core, and recess dimensions and the glass processing parameters, such as draw speed and die temperature, are finally optimized to meet the stringent requirements for the glass tube dimensions such as ID, OD, and wall thickness.

REFERENCES

[1] H. P. Wang et al., *A semi-empirical approach for developing a profile extrusion die design methodology*, FED, Vol, 231, Book No. H01028-1995, (1995), pp 173-188.

[2] H. P. Wang, H. S. Lee, *Numerical techniques for free and moving boundary problems*, in C. L. Tucker ed., *Fundamentals of Computer Modeling for Polymer Processing*, Hanser Publishers, New York, (1989), pp.369-401.

[3] H. P. Wang, *Semi-Empirical Manufacturing Transfer Functions: A Core Element in Design for Six Sigma*, Technical Report, GE Research & Development Center, (2001), pp. 1-4.

[4] O. C. Zienkiewicz, *The finite element method, 3rd edition*, McGraw-Hill, (1977), pp. 113-115.

COMPUTATIONAL MECHANICS (Abstracts)
WCCM VI in conjunction with APCOM'04, *Sept. 5-10, 2004, Beijing, China*

Numerical Simulation of Heat Transfer and Kinetics in the Bulk Growth of Silicon Carbide

Zibing Zhang, Jing Lu, Qisheng Chen*

Institute of Mechanics, Chinese Academy of Sciences, 15 Beisihuanxi Road, Beijing 100080, China
e-mail: qschen@imech.ac.cn

Abstract

Wide-bandgap silicon carbide (SiC) is a promising semiconductor material for electronic and opto-electronic devices involving high power, high temperature, high frequency and intense radiation. The bulk growth of SiC by physical vapor transport (PVT), or modified Lely method involves many important physical phenomena, such as electromagnetic field, radio frequency induction heating, conduction and radiation heat transfer, sublimation and condensation, mass transfer, and so on. Since direct measurement of temperature, flow velocity, species concentration, and growth rate is extremely difficult, physics-based modeling is an important research tool by which one can develop a better understanding of the transport phenomena in a SiC growth system. A comprehensive process model for SiC bulk growth has been developed that incorporates the calculations of radio frequency (RF) induction heating, heat and mass transfer and growth kinetics. The electromagnetic field is calculated by using a magnetic vector potential theory; temperature distribution is computed by using a coupled radiation and conduction model. A kinetics model for silicon carbide growth is used that is based on Hertz-Knudsen equation and relates supersaturation to the growth rate. The vapor species concentration profiles in the growth chamber are obtained by using a one-dimension advective mass transfer model. The transport equations for electromagnetic field, heat transfer, and species transport are solved using a numerical scheme called MASTRAPP (Multizone Adaptive Scheme for Transport and Phase Change Process), which is based on the finite volume method [1]. MASTRAPP employs high-resolution schemes for grid refinement and clustering for phase change and moving boundary problems, and has been extensively used for accurate and efficient simulation of various industrial processes involving different heat and mass transfer mechanism that occur on irregularly shaped domains and regions with moving and free boundaries, especially for crystal growth modeling. The scheme employs the multizone adaptive grid generation technique (MAGG) for the discretization of physical domains of arbitrary shape, and the curvilinear finite-volume approach for the discretization of the governing equations. Electromagnetic field and temperature field distribution has been obtained, and growth rate has been predicted and its dependence on temperature, temperature gradient and pressure has been studied in details. The predicted data have been compared with the experimental ones.

REFERENCES

[1] Q.-S. Chen, H. Zhang, V. Prasad, C. M. Balkas, N. K. Yushin, *Modeling of heat transfer and kinetics of physical vapor transport growth of silicon carbide crystals*, J. Heat Transfer, 123, (2001), 1098-1109.

COMPUTATIONAL MECHANICS (Abstracts)
WCCM VI in conjunction with APCOM'04, *Sept. 5-10, 2004, Beijing, China*

Mode Enrichment Techniques for Dispersive Modeling within a Multiscale Mixed Finite Element Framework

Mahmoud I. Hussein, Gregory M. Hulbert*

Department of Mechanical Engineering, University of Michigan, Ann Arbor, MI 48109 -2125, USA
e-mail: husseinm@umich.edu, hulbert@umich.edu

Abstract

The propagation of elastic waves in periodic media is increasingly attracting the attention of researchers because of the unique dispersive characteristics that such media exhibit. An essential tool for analyzing the dynamical characteristics of a periodic medium is the means for computing the frequency spectrum. A diagram of the frequency band structure provides insights into wave dispersion, and its frequency dependence, associated with the particular configuration of the medium. This information could be used for the analysis, as well as the design, of realistic structures composed of a periodic material with the same configuration.
Recently, the finite element method [1] was directly applied to solving the eigenvalue problem governing the dispersion relation of a periodic medium. Although practical and accurate, this approach leads to prohibitively high computational costs for large models. To overcome this problem, a multiscale reduced order modeling methodology has been developed [2] for computing frequency spectra of elastic periodic composite materials. The methodology is based on an assumed strain, mixed finite element formulation in which the solution field is projected onto a sub-space spanned by a set of vectors that sufficiently characterize the local dynamics of the medium. Originally, the methodology employed, as basis functions, high order characteristic functions generated by a hierarchical homogenization analysis scheme. In the present study, the projection space is enhanced [3], or alternatively replaced [4], with periodic eigenvectors selected in both the temporal and spatial frequency domains. Using a model of a layered periodic structure, the new enrichment techniques are shown to provide good accuracies across wide frequency ranges. Following an "order of operation count" analysis, substantial savings in computational costs are achieved for large problems compared to direct application of the finite element method.

REFERENCES

[1] D. C. Dobson, *An efficient method for band structure calculations in 2d photonic crystals*, Journal of Computational Physics, 149, (1999), 363-376.

[2] T. McDevitt, G. M. Hulbert, N. Kikuchi, *An assumed strain method for the dispersive global-local modeling of periodic structures*, Computer Methods in Applied Mechanics and Engineering, 190, (2001), 6425-6440.

[3] M. I. Hussein, G. M. Hulbert, *A fast multiscale method for computing dispersion curves of periodic materials*, Proc. 7th USNCCM, Albuquerque, New Mexico, (2003).

[4] M. I. Hussein, G. M. Hulbert, *A multiscale reduced order model for computing frequency spectra of periodic materials*, Proc. 40th Annual Meeting of the Society of Engineering Science, Ann Arbor, Michigan, (2003).

COMPUTATIONAL MECHANICS (Abstracts)
WCCM VI in conjunction with APCOM'04, *Sept. 5-10, 2004, Beijing, China*

Effect of Local Structure on Wave Propagation in Sandwich Structures

Ying Liu [1*], Kaishin Liu [1], Tongxi Yu [2]

[1] *Department of Mechanics and Engineering Science, Peking University, Beijing, 100871, China*
[2] *Department of Mechanical Engineering, Hong Kong University of Science& Technology, Clear Water Bay, Kowloon, Hong Kong, China*

e-mail: y_liu@mail.china.com, kliu@mech.pku.edu.cn, metxyu@ust.hk

Abstract

This is an attempt to study the local effects of inner topology on wave propagation in sandwich structures. The usual method used to treat sandwich panel is to get them equivalent properties through homogenization method, et al. The effects of the inner cellular geometry on the properties of the sandwich structures are mainly reflected through the relative density of cellular cores. This is perfectly true in the static or quasi-static situation. However, under impact loading, some dynamic characteristics, such as local dynamic stress concentration, wave propagation, wave reflection/ refraction, et al., which are highly shape sensitivity, will play an important role in the dynamic response of the sandwich structures, and to some extent, affect the damage modes.
In order to reflect the influence of the complex topology of cores on the dynamic responses of the sandwich structures, instead of using homogenization method, wave propagation in the true structure of the sandwich panels is simulated by adopting explicit dynamic finite element method. The wave propagation in the sandwich panels under impact loading is studied. The distribution of the stress under different loading conditions is presented. The effects of the cell wall thickness and the materials of the core and skins are analyzed in detail. The results show that due to the influence of the inner structure, under symmetric impact loading, the wave propagation exhibits anisotropy in the sandwich panels with regular hexagon inner structure, which have the uniform cell wall thickness and given isotropic in-plane properties by a homogenization treatment. This reveals the importance of the effects of the inner geometry on the dynamic characterization of the sandwich panels with cellular cores.

REFERENCES

[1] L. J. Gibson, M. F. Ashby, *Cellular solids: structure and properties*, Cambridge: Cambridge University Press, (1997).

[2] A. Hönig, W. J. Stronge, *In-plane dynamic crushing of honeycomb, part I: crush band initiation and wave trapping*, International Journal of Mechanical Sciences, (1997), 1665-1696.

[3] A. Hönig, W. J. Stronge, *In-plane dynamic crushing of honeycomb, part II: application to impact*, International Journal of Mechanical Sciences, (2002), 1697-1714.

[4] D. Ruan, G. Lu, B. Wang, T. X. Yu, *In-plane dynamic crushing of honeycombs—a finite element study*, International Journal of Impact Engineering, (2003), 161-182.

COMPUTATIONAL MECHANICS (Abstracts)
WCCM VI in conjunction with APCOM'04, *Sept. 5-10, 2004, Beijing, China*

2-D Numerical Simulation of Explosion and Impact Problems

Jianguo Ning *, Tianbao Ma

State Key Lab of Prevention and Control of Explosion Disasters, Beijing Institute of Technology, Beijing, 100081, China
e-mail: jgning@bit.edu.cn, madabal@bit.edu.cn

Abstract

In this paper, based on Eulerian algorithm of multi-material in cell, the 2-D elastic-plastic hydrodynamics finite difference calculation code using C++ is compiled, and the preprocessing and the visualization software VISC2D [1] are developed. In addition, some typical explosion and impact problems are simulated by this code.
The physical model can be described by 2-D unsteady elastic-plastic hydrodynamics equations. In computation, the "and" splitting scheme is used, and the forward difference scheme and center difference scheme are adopted for time and space respectively. For each direction, the computation of equations can be divided into two steps: Lagrange step and Transport step (Euler step). In Lagrange step, considering the pressure gradient's and partial stress's effect, the Lagrange velocity and internal energy can be got; In Transport step, calculating the transportation of mass, momentum, energy carried in or out through the grid boundaries, the final values of these physical parameters of the mediums in grid are gained. In this code, many solid and fluid material models such as explosive, metal, soil, concrete, air, water and so on are available. In addition, the constitutive equations, state equations and failure criterions corresponding to these materials are provided. The code can accurately treats the interface problems of mixed grids that include at most three materials. For the mixed grids with two materials, Young's interface technology [2] is used, and for the mixed grids with three materials, SLIC (simple line interface calculation) is adopted.
The preprocessing code provides a friendly interface so that the users can easily modeling and meshing cells. The visualization software VISC2D can display the variation of every physical parameter in animated-cartoon mode.
The finite difference code is all-purpose for simulating all kinds of explosion and impact problems such as shaped charge jet, explosion in air and water, penetration and so on. In this paper, the numerical simulation results of shaped charge jet and the explosive shock wave with or without protective wall are provided and compared with the theoretic and experimental results. The simulation study indicates that the model and algorithm presented in the code are reasonable and the simulation method can be used for project design.

REFERENCES

[1] Wenyao Zhang, Jianguo Ning, Zuguo Zheng, *Visualization of explosive field,* Journal of Beijing Institute of Technology, 8(4), (1999), 424-429.
[2] D. L. Youngs, *Time-dependent multi-material flow with large fluid distortion,* in K. W. Morton, M. J. Baines eds. *Numerical Method for Fluid Dynamics,* Academic Press, New York, (1982).

COMPUTATIONAL MECHANICS (Abstracts)
WCCM VI in conjunction with APCOM'04, *Sept. 5-10, 2004, Beijing, China*

An Efficient Homotopy Continuation Procedure for Obtaining the Fields in Electromagnetic Problems when using the MEF with Curvilinear Elements

Luis Nuño [1*], Juan V. Balbastre [1], Sebastián Rodríguez-Mattalia [1], Lucas Jódar [2]

[1] *E.T.S.I. Telecomunicación, Universidad Politécnica de Valencia, Camino de Vera, s/n, Valencia, 46022,Spain*

[2] *Multidisciplinary Mathematics Institute, Universidad Politécnica de Valencia, Camino de Vera, s/n, Valencia, 46022,Spain*

e-mail: lnuno@upv.es, jbalbast@itaca.upv.es, srodriguczmattalia@hotmail.com, ljodar@mat.upv.es

Abstract

Curvilinear elements have been reported to be one of the more advantageous features of the finite element method (FEM), since they allow fitting arbitrarily shaped geometries without staircase approximations, thus leading to more accurate results. Nevertheless, the graphical representation of the computed fields involves quite complex computations, due to the fact that systems of polynomial equations arise.

A classical method for solving polynomial systems is the Newton's method, but this method is not suitable for this problem due to its local convergence. On the other hand, homotopy based methods have been reported as a valid tool for solving systems of polynomial equations [1]. These methods have a global convergence, i.e. all the solutions can be found from the solutions of a suitable starting system by using the appropriate path track procedure. Since the whole set of solutions is found, there will be always one among them lying within the standard element.

Independently, a general FEM scheme for analyzing lossy inhomogeneous and anisotropic waveguides was developed [2]. In this paper, an efficient homotopy continuation procedure has been developed and optimized for using along with the FEM program. Propagation constants and field distributions have been obtained for several waveguides with curvilinear boundaries. The results were much more satisfactory and reliable than those obtained by means of the Newton's method.

REFERENCES

[1] Van Verschelde, Homotopy *continuation methods for solving polynomial systems,* PhD Dissertation, Katholieke Universiteit Leuven, May, (1996).

[2] L. Nuño, J. V. Balbastre, H. Castañé, *Analysis of general lossy inhomogeneous and anisotropic waveguides by the finite-element method (FEM) using edge elements*, IEEE Trans. Microwave Theory Tech., MTT-44, (1997), 446-449.

COMPUTATIONAL MECHANICS (Abstracts)
WCCM VI in conjunction with APCOM'04, *Sept. 5-10, 2004, Beijing, China*

An Approximation Method for Multiple Scattering of Elastic Wave

Kwannate Tharmmapornphilas *, Sohichi Hirose

Department of Mechanical and Environmental Informatics, Tokyo Institute of Technology, O-okayama, Meguro-ku, Tokyo, 152-8552, Japan
e-mail: kwannate@qnde.mei.titech.ac.jp, shirose@cv.titech.ac.jp

Abstract

Noises usually emerged in ultrasonic testing system are electronic circuit noise, structure noise, ringing noise and digital noise. Among these noises, structure noise is considered the most serious noise in the system [1]. In materials with large noises, the flaw echo is often embedded in noise echo. As a result, flaws and defects cannot be detected and may lead to unreliable inspection. Therefore, scattering characteristic of waves in materials must be studied to understand the nature of structure noise. Unfortunately, with the restriction of computer capacity, the memory size of computational model is limited.

In this paper, an approximation method is presented for a multiple scattering problem of elastic waves in frequency domain to reduce the computation time and memory consumption as compared with a conventional boundary element method (BEM). The numerical model is a number of cylindrical inclusions, which are randomly distributed in an infinite matrix subjected to an inplane incident wave. The multiple scattering displacement at a specific point in the matrix is approximated as a summation of displacements of all possible wave scattering paths. Considering a combination of L and T-waves, the scattering displacement for each path is obtained by multiplying the displacement of the incident wave by the single scattering coefficients for inclusions on wave path in sequence. The single scattering coefficient is a scattering displacement of a single inclusion subjected to a unit L or T-wave, calculated by using a BEM with the far field approximation [2].

The numerical analysis is performed using several orders of approximation. In the case of the 4th order multiple scattering approximation of 50 inclusions randomly distributed, the computation time and memory consumption are reduced to 25% and 4% of a conventional BEM with 30 elements per inclusion, respectively. The scattering displacements obtained by the multiple scattering approximations are comparable with the traditional BEM solutions.

REFERENCES

[1] J. Chen, Y. Shi, S. Shi, *Noise analysis of digital ultrasonic nondestructive evaluation system*, Int. J. Pressure Vessels and Piping, 76, (1999), 619-630.

[2] J. D. Achenbach, A. K. Gautesen, H. McMaken, *Ray methods for waves in elastic solids*, Pitman, (1982).

COMPUTATIONAL MECHANICS (Abstracts)
WCCM VI in conjunction with APCOM'04, *Sept. 5-10, 2004, Beijing, China*

Adaptive Simulation of Ultrasonic Guided Waves in Railroad Tracks Using CHARMS

A. Trivedi [1*], P. Krysl [1], J. McNamara [2], F. Lanza di Scalea [1]

[1] *Department of Structural Engineering, University of California, San Diego, 9500 Gilman Drive, La Jolla, CA 92092, USA*

[2] *Department of Civil and Geological Engineering, New Mexico State University, P.O. Box 30001 MSC 3CE, Las Cruces NM, 88003-8001, USA*

E-mail: atrivedi@ucsd.edu, pkrysl@ucsd.edu, mcnamara@nmsu.edu, flanza@ucsd.edu

Abstract

Simulation of ultrasonic guided waves in large finite domains presents an interesting challenge. If the size of the discreet domain is relatively large in comparison to the minimum wavelength of the collection of waves traveling within the domain with different frequencies, the size of the resultant problem increases dramatically due to the requirement of a finer resolution. The solutions of similar problems sometimes become very cumbersome or almost impossible even after using the most advanced computational resources.

This document describes the solution of general guided wave problems by tracking the active energy fronts using a naturally conforming hierarchical refinement method (CHARMS). The problem described in this document was validated with respect to test results on rail road tracks at the UCSD Non destructive evaluation lab for San Diego Trolley.

Our research addresses complexities of simulations of general guided waves in a finite non-symmetric domain with close to optimal computational effort by tracking active energy fronts with mesh refinement and un-refinement. Our method also demonstrates that the conforming CHARMS can be used effectively with time dependent solvers.

REFERENCES

[1] Abhishek Trivedi, Petr Krysl, *A framework of adaptive mesh refinement for time dependent solvers*. Unpublished, (2003).

[2] Petr Krysl, Abhishek Trivedi, Baozhi Zhu, *Object-oriented hierarchical mesh refinement with CHARMS*, International Journal for Numerical Methods in Engineering, (2003), (to appear).

[3] John D. McNamara, *Health monitoring of rail road tracks by elastic waves based non destructive testing*, Ph.D. Theses, University of California, San Diego, (2003).

[4] Francisco Lanza di Scalea, J. McNamara, *Ultrasonic NDE of railroad tracks: Air-coupled cross-sectional inspection and long range inspection*, INSIGHT – Non-Destructive Testing and Condition Monitoring, EURO issue on NDT in the rail industry, 45(6), (2003), 1-8.

[5] Friedrich Moser, Laurence J. Jacobs, Jianmin Qu, *Modelling elastic wave propagation in waveguides with the finite element method*, NDT&E International, 32, (1999), 225-234.

COMPUTATIONAL MECHANICS (Abstracts)
WCCM VI in conjunction with APCOM'04, *Sept. 5-10, 2004, Beijing, China*

Numerical Simulation of the Effect of Waving Bed on the Surface Wave

Songling Wang, Youliang Cheng, Zhengren Wu *

School of Energy & Power Engineering, North China Electric Power University, Baoding, Hebei, Province 071003, China
e-mail: zhengren_wu@yahoo.com.cn, ylcheng001@163.com, hdwangsl@163.com

Abstract

The effect of waving bed on the surface wave has been researched. The wave equation was reduced from the potential flow theory with the little parameter perturbation technique, and then was solved by using the pseudo-spectral method. Finally, the waterfall of the surface wave was simulated with Matlab software. The results are shown in Fig. 1.

The figure shows the surface wave for the waving bed is obviously different from the waveform of the non-waving bed. An additional harmonic wave seems to appear on the surface together with the solitary, which exists for the non-waving bed. Two kinds of waves do not interfere with each other. With the development of time, the wave form for the waving bed keeps invariable, just the amplitude is reduced gradually; broken phenomenon of the wave for the non-waving bed does not appear.

The project is supported by the National Natural Science Foundation of China under Grant No.10272044.

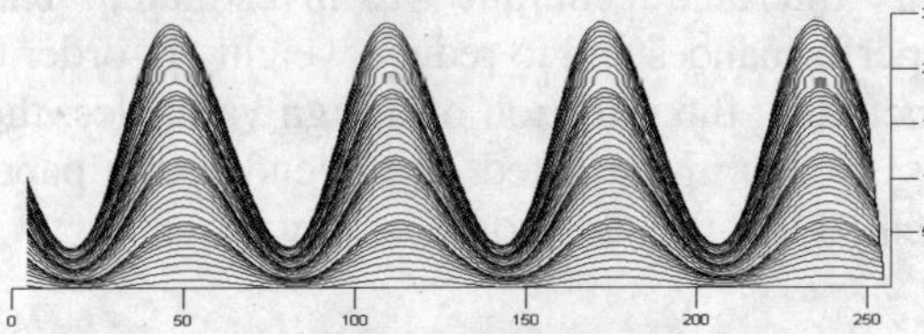

(a) for the waving bed

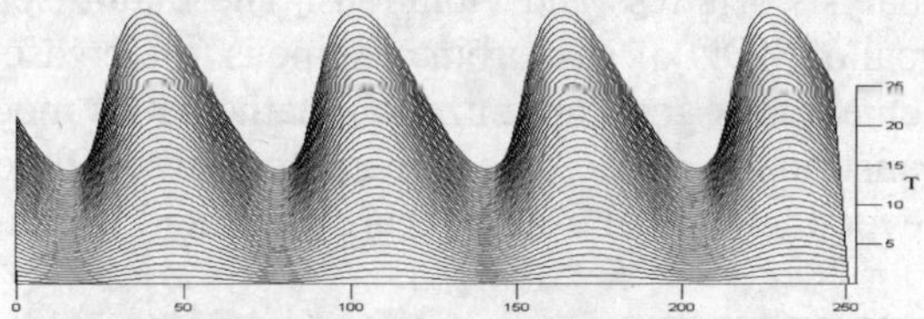

(b) for the non-waving bed

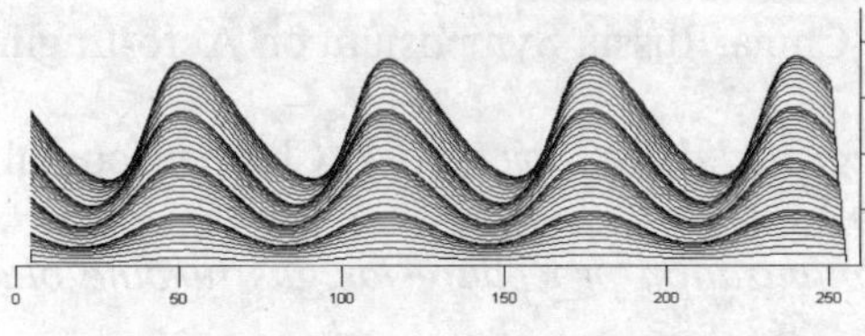

(c) for the waving bed

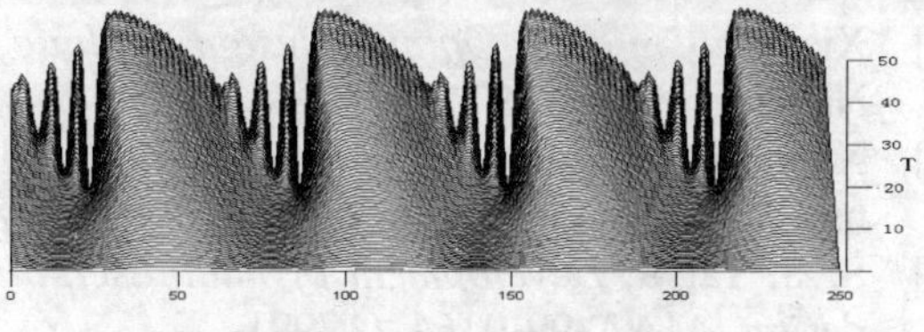

(d) for the non-waving bed

Fig. 1 the numerical results for surface wave

REFERENCES

[1] Mian Lin, Jiachun Li, *Effects of surface waves and marine soil parameters on seabed stability*, Applied Mathematics and Mechanics, 22, (2001), 806-816.

[2] Hongnong Cui, Xueque Yang, Xiaofeng Pang, et al, *A study on the behavior of non-propagating solitary wave*, Journal of Hydrodynamics, 6, (1991), 18-25.

COMPUTATIONAL MECHANICS (Abstracts)
WCCM VI in conjunction with APCOM'04, *Sept. 5-10, 2004, Beijing, China*

MDO of Gas Turbine Engines

Xianxin Cai [*], Zeyong Yin, Xiangming Jin, Mi Dong, Liqiang Wu

Dep.9, Chinese Aviation Power Plant Research Institute, Zhuzhou, 412002, China
e-mail: newvilacai@yahoo.com

Abstract

A gas turbine engine which is an extremely complicated system mainly consists of a compressor, a turbine, and a combustion chamber. The design of it involves many disciplines such as computational fluid dynamics (CFD), heat conduction, structural design and computational solid mechanics (CSM) and some of these disciplines interact from each other strongly. The high performance of the advanced gas turbine engine requires that these disciplinary design should be integrated so multidisciplinary design optimization (MDO) is vitally important for the next generation gas turbine engines. The procedure of multidisciplinary design optimization (MDO) for gas turbine engine was investigated in the present paper. In CDF, the Navier-Stokes equations are solved to get the flow field. For the design of a blade, quasi-three dimensional calculation which was composed of the iteration between the stream function calculations on a mean S_2 stream surface and several S_1 stream surface of revolution for transonic flow was employed and in the structural design of a blade, the controlled diffusion airfoils (CDA) technique was adopted to get the airfoil of a blade. In the heat conduction analysis, FEM was used to solve the Poisson's equation. In CSM, FEM was used to carry out the analysis of stress and vibrations. The multidisciplinary interface technique was investigated. The goal of MDO of gas turbine engine is to increase the performances and to reduce weight. In order to achieve this goal, optimization mathematics model including the selection of design valuables, the construction of the objective and constraint functions is also investigated. In the end of the paper several numerical examples are presented which indicated that this procedure is very efficient.

REFERENCES

[1] Xianxin Cai et al, *Optimal design of blade*, 1997 China-Russia Symposium on Aero-Engines, Nanjing, July, (1997).
[2] Xianxin Cai et al, *The conformal and optimal integrated design for a blisk*, Chinese Journal of mechanical engineering, 35(3), (1999), 81-84.
[3] S. S. Talya, *Development of multidisciplinary optimization procedure for gas turbine blade design*, AIAA-99-0364, (1999).

COMPUTATIONAL MECHANICS (Abstracts)
WCCM VI in conjunction with APCOM'04, *Sept. 5-10, 2004, Beijing, China*

2-D Simulation of Air Intake and Compression in an I.C. Engine

R. Y. Hong [1*], H. Z. Li [2], J. Ding [3]

[1] *Dept. of Chemistry & Chemical Engineering, Soochow University, Suzhou, 215006, China*
[2] *Agere Systems, 555 Union Boulevard, Allentown, PA 18109, USA*
e-mail: rhong@suda.edu.cn, dingjianmin@agere.com

Abstract

The emission control of engines is becoming increasingly tighter in the world. Efficient fuel utilization is required due to the decreasing oil reservation. Appropriate approaches have to be made to obtain optimal flow structure inside the cylinder and to understand the mixing process of fuel and air, the propagation of flame, and the influences of various operation factors on combustion. The temperature and the pressure of the combustion chamber are very high, while the cyclic time is very short. Only in the past few years, did there appear some experimental methods. The progresses in experimentation provided possibilities for further numerical modeling. The combustion flow processes in the engine cylinders obey the conservation of mass, momentum and energy, which are described by the so-called Navier-Stokes (N-S) type equations with combustion reaction kinetics and proper boundary conditions. These governing partial differential equations can be solved by adequate numerical method on a computer. Detailed information of the flow field, temperature and concentration fields, effects of geometric parameters on flow field and combustion can be obtained. It will be easier to compare different design schemes thereafter. Therefore, experiment and design of the I.C. engines can be modeled and guided cost-effectively by computer simulations.

In this paper, two-dimensional (2-D) modeling and simulation were conducted to simulate air intake and compression in an I.C. engine. A 2-D general orthogonal curve-linear coordinate [1-4] was employed to account for the complex geometry of the I.C. engine. The governing equations were discretized and solved using the so-called control volume approach [5] on staged grids. The SIMPLEC algorithm was used to treat the pressure-velocity coupling. Four turbulence models, the standard *K-ε* model, the Realizable *K-ε* model, the Reynolds stress model, and the *subgrid-scale* model were used, and the predicted results were analyzed. The computed flow fields were compared with the experimental measurements. Some interesting flow patterns in the engine were predicted using the present model.

REFERENCES

[1] G. Ryskin, L. G. Leal, *Orthogonal mapping*, J. Comp. Phys., 50, (1983), 71-100.

[2] L. Eca, *2D Orthogonal grid generation with boundary point distribution control*, J. Comput. Phys., 125, (1996), 440-453.

[3] R. Y. Hong, G. L. Zheng, H. Z. Li, *Numerical simulation of a radio-frequency plasma reactor using 2-D orthogonal coordinates*, in C. K. Wu ed. *Proc. of 13th Int. Conf. on Plasma Chem.*, Beijing, China, (1997), pp. 1690-1695.

[4] R. Y. Hong, J. Y. Zhang, M. Y. Cheng, W. C. Fan, *Hydrodynamic modeling of gas-solids fluidization*, in A. A. Balkena ed. *Proc. of 2nd Asia-Pacific Conf. on Comp. Mech. (APCOM-2)*, Sydney, Australia, (1993), pp. 647-652.

COMPUTATIONAL MECHANICS (Abstracts)
WCCM VI in conjunction with APCOM'04, *Sept. 5-10, 2004, Beijing, China*

Mathematical Simulation of Hydraulic Transient and Load Rejection Analysis in Francis Turbine Power Plant

A. Nourbakhsh [1], A. Riasi [1*], R. Mirian[2]

[1] *Depart*ment *o*f Mechanic, University *o*f Tehran, Iran
[2] *F*arab Co ., Tehran, Iran
e-mail: anour@ut.ac.ir, riasi@me.ut.ac.ir

Abstract

Hydraulic transients in a multiunit hydro power plant which is recently installed in Iran has been studied in this paper. For this purpose a computer program has been developed. The continuity and momentum equations are formed a pair of quasi-linear hyperbolic partial differential equations in terms of two depended variables, velocity and head, and two independent variables, distance along the waterways and time. The equations are solved using complex finite difference and characteristics method. The equations are described along closed conduits from upstream reservoir to the downstream reservoir considering the turbine in the midway. No data have been reported in the literature for turbine characteristics during transient state conditions. Therefore the steady state model test results (Hill Chart) are used to plot the expected prototype turbine characteristics and these are assumed to be valid during transient state as well. The upstream conduits (penstock) convey water from upstream source, such as, a reservoir or lake to the turbine. Outflow from the turbine is discharged through a draft tube to the downstream water passage (tailrace). An electric generator is mechanically coupled to the turbine and the transmission line carries the electric output to the load centers. A governor is provided to correct any changes in the system frequencies by opening or closing the wicket gate of the turbine. In this paper, a mathematical model is developed for analyzing hydraulic transient caused by various turbine operations such as, start up, load acceptance or load rejection. Finally, the load rejection results including speed rise of turbine, maximum penstock pressure and minimum draft tube pressure, are compared with measured values of this Francis power plant. These results have good agreement with measured data.

COMPUTATIONAL MECHANICS (Abstracts)
WCCM VI in conjunction with APCOM'04, *Sept. 5-10, 2004, Beijing, China*

A Cracked Rotor Bending Model with Effect of Shear Forces

P. Verrier *, C. Varé

EDF R&D, Acoustics and Mechanical Analysis Department, 1, Avenue du Général de Gaulle, 92141 Clamart, France
e-mail: pierre.verrier@edf.fr, christophe.vare@edf.fr

Abstract

A propagating crack affecting a rotor opens and closes in time due to the rotation and the vibration amplitude. In some cases such as Jeffcot rotor or horizontal axis power generation turbines, the weight and flexion dominant load enable to assume that the local flexibility introduced by the crack depends on angular position of the rotor. This leads to a linear by part system. Assuming switching crack (bi-level direct stiffnesses), equations become linear when the crack is completely open or closed [1]. However, the system to be solved is generally non linear [2] especially in slightly loaded machine (vertical axis rotor).

This paper faces with a lumped three-dimensional cracked beam model. Formulation of the potential energy –positively homogeneous of degree two – of the system is conducive to reduce the number of parameter [3]: the closure of the crack is only determined by loading direction which induces a displacement discontinuity with respect to a non cracked rotor. Convex analysis is used to properly account for displacement discontinuities constrained by contact condition on the lips of the crack whereas any end-loads can be prescribed.

The fact is emphasized that the behaviour is not only influenced by the bending moment but also by the shear forces, which for example may not be neglected in overhung cracked rotor. The strain energy, which is related to compliance, can thus be computed for any loading direction rather than assuming sine or cosine variation between open and closed crack flexibility [4]. Nonlinear lumped cracked beam element is derived as a "current" stiffness matrix so as to be incorpor ated in lineshaft FE analysis.

REFERENCES

[1] C. W. Lee, *Vibrations of a rotor under combined effects,* in: *vibration analysis of rotors*, Kluwer Academics Publishers, Dordrecht, The Netherlands, (1993), pp. 277 -308.

[2] A. D. Dimarogonas, *Vibration of cracked structures: a state of the art review*, Engng Fracture Mech., 55(5), (1996), 831-857.

[3] S. Andrieux, C. Varé, *A 3D cracked beam model with unilateral contact, application to rotors*, Eur. J. Mech. A, 21, (2002), 793-810.

[4] P. N. Saavedra, L. A. Cuitiño, *Vibration analysis of rotor for crack identification*, J. Vib. Control, 8, (2002), 51-67.

COMPUTATIONAL MECHANICS (Abstracts)
WCCM VI in conjunction with APCOM'04, *Sept. 5-10, 2004, Beijing, China*

CFD Analysis of a Low Head Axial Flow Pump

Haisong Wang [1*], Fujun Wang [1], Yaojun Li [1], Yanli Wang [1], Shuhe Bu [2]

[1] *College of Water Conservancy and Civil Engineering, China Agricultural University, Beijing, 100083, China*
[2] *Water Resources and Hydropower Planning and Design General Institute, Beijing, 100011, China*
e-mail: hans_cn@163.com, wangfj@cau.edu.cn, cau_lee@163.com, lelei21c@163.com

Abstract

The construction of South-to-North Water Transfer project in the east of China demands a number of low head axial flow pumps with high performances. However, the characteristic curve of an axial pump can not be evaluated in the design phase. It is obtained just after the practical testing on the laboratory platform. Therefore, it will take long time and is too expensive to develop a good pump model.

In this study, a CFD approach is introduced to the performance evaluation of axial flow pumps. Hydraulic design of axial flow pumps could be improved by combining this method with pump design process. The details of CFD analysis of a low head axial flow pump is given in this paper. A 3D solid model of the pump flow zone is generated and meshed on a base of its hydraulic design. Because the flow field is greatly impacted by the interaction of transferring fluid and the rotating blades of the impeller, the region near blades is meshed subtly. The standard $k-\varepsilon$ turbulent model and SIMPLEC algorithm are applied for solving the momentum equations (Navier-Stokes) coupled with the continuity equation. The flows at several conditions, including design condition and off-design condition, are simulated. The total and static pressure, the velocity contours and vectors at blade surface are presented.

The CFD performance estimation of the pump shows that the back flow is evident when mass flow rate is very low. The calculated pump head, hydraulic efficiency and axial force are compared with experiment data. The difference between predicted performance and experimental result is within 9%. The design approach [3] and prediction method presented herein can be used as a unified hydraulic design process of axial flow pumps. This can contribute to an efficient and time-saving design procedure.

REFERENCES

[1] Akira Goto, Nohmi Motohiko, Sakurai Takaki, Sogawa Yoshiyasu, *Hydrodynamic design system for pumps based on 3-D CAD: CFD and inverse design method*, Proceedings of the ASME Fluids Engineering Division Summer Meeting, ASME, Vol. 2, (2003), pp. 215-223.

[2] Plutecki Janusz, Skrzypacz Janusz, *CFD simulations of 3D flow in a pump stator with a spherical surface*, World Pumps, 443, (2003), 28-31.

[3] Haisong Wang, Fujun Wang, Zhimin Zhang, *Automatic feature-based modeling of axial flow impeller*, Journal of Hydrulic Engineering, 20, (2004), 285-291, (In Chinese).

COMPUTATIONAL MECHANICS (Abstracts)
WCCM VI in conjunction with APCOM'04, *Sept. 5-10, 2004, Beijing, China*

Modelling and Experimental Validation of an Industrial Sheet Metal Forming Process

Diego J. Celentano[1*], Claudio M. García[1], Fernando G. Flores[2], Jean-Philippe Ponthot[3]

[1] *Department of Mechanical Engineering, University of Santiago de Chile, Av. B. O'Higgins 3363, Santiago, Chile*

[2] *Department of Structural Engineering, National University of Córdoba, Casilla de Correo 916, 5000 Córdoba, Argentina*

[3] *LTAS-Continuous medium and Thermomechanics, University of Liège, 1 Chemin des Chevreuils, B-4000 Liège-1, Belgium*

e-mail: dcelenta@usach.cl, claudio.garcia@usach.cl, fflores@efn.uncor.edu, JP.Ponthot@ulg.ac.be

Abstract

Relevant advances in research and development of sheet metal forming processes have been proposed and discussed during the last decades (see e.g. [1]). Although the finite element analysis is successful in simulating complex industrial forming operations, a full experimental validation of the results provided by this technique is still a hard task due to the inherent difficulties associated with the measurements of the different variables involved in this kind of problems. Nevertheless, this task is a necessary step for a further control and/or optimization of such processes.

The aim of the present work is the modelling and corresponding experimental validation of an industrial sheet metal forming process consisting in the deep drawing of a EK4 steel component of a commercial washing machine. To this end, the simulation of the material response during the deformation process of the sheet is carried out through a hyperelastic elastoplastic constitutive law written in terms of Hencky stress and strain measures [2]. Strain hardening and normal anisotropy (along the sheet thickness) effects are both considered via the Hill 48 associate plasticity model. The elastic and plastic material parameters necessary for this model have been experimentally derived by means of an exhaustive material characterization of its mechanical behaviour [3]. Moreover, the mechanical interactions between the sheet and the different tools (punch, blank holder and die) are taken into account through a Coulomb-type contact-friction law. The resulting equilibrium equation is solved in the context of the finite element method by using a large strain shell formulation based on thin triangles with only translational degrees of freedom [2].

The punch force, thickness distributions and principal deformations predicted by the model are all compared with the respective experimental measurements. An overall good agreement is achieved.

REFERENCES

[1] Dong-Yol Yang, Soo Ik Oh, Hoon Huh, Yong Hwan Kim eds. *NUMISHEET 2002*, Proceedings of the 5th International Conference and Workshop on Numerical Simulation of 3D Sheet Forming Processes, Jeju Island, Korea, October 21-25, (2002).

[2] F. G. Flores, E. Oñate, *A basic thin shell triangle with only translational DOFs for large strain plasticity*, Int. J. Num. Meths. In Engng, 51, (2001), 57-83.

[3] C. García, D. Celentano, J. P. Ponthot, *Characterization of the mechanical behaviour of deep drawing steels*, Proceedings of Jornadas SAM-CONAMET-2003, Bariloche, Argentina, November (2003), (in Spanish).

COMPUTATIONAL MECHANICS (Abstracts)
WCCM VI in conjunction with APCOM'04, *Sept. 5-10, 2004, Beijing, China*

Computational Modeling of Forming Processes Using an Arbitrary Lagrangian-Eulerian Method

Amir R. Khoei *, Masood Anahid

Department of Civil Engineering, Sharif University of Technology, 11365-9313, Tehran, Iran
e-mail: arkhoei@sharif.edu

Abstract

In numerical simulation of forming processes, there are two approachcs in modeling of problems, Lagrangian and Eulerian methods. In Lagrangian approach, the mesh nodes are implemented to material points, therefore this formulation is well suited for simulation of problems containing materials with path dependent behavior, and free surface motion. However, the method suffers from numerical difficulties in problems with very large deformation, because the mesh may undergo severe distortion, due to this fact that elements are deformed with material deformation [1-2]. In Eulerian approach, the mesh is fixed in space and material flows through it, which is very suitable for problems with very large material deformation such as fluid dynamics. But the very important disadvantageous of this method is that it is less suited for history dependent material and problems with free boundary motion. It can be concluded that neither Lagrangian nor Eulerian formulation alone is well suited for simulation of processes involving large deformation and free boundary motion.

In this paper, an ALE technique is developed in simulation of forming processes. The basic idea of ALE technique is to combine advantageous of both Lagrangian and Eulerian approaches. In this method, the reference configuration is used for describing the motion, instead of material configuration in Lagrangian, and spatial configuration in Eulerian formulation. This formulation introduces some convective terms in the finite element equations and consists of two phases [3]. In Lagrangian phase, the mesh and material movements are identical. In Eulerian phase, it is allowed the mesh to have an arbitrary motion, independent of material motion, keeping the mesh regular. In this study, each time step is analyzed according to Lagrangian phase until required convergence is attained. Then, the Eulerian phase is applied to keep mesh configuration regular. Because of relative displacement between mesh and material, all dependent variables, such as: stress and strain are converted through the Eulerian phase. Finally, several numerical forming processes are provided to illustrate that the ALE method can be effectively used to solve problems which will encounter some difficulties in Lagrangian and Eulerian formulation.

REFERENCES

[1] R. W. Lewis, A. R. Khoei, *Numerical modeling of large deformation in metal powder forming*, Comput. Meth. Appl. Mech. Eng., 159, (1998), 291-328.

[2] A. R. Khoei, R.W. Lewis, *Adaptive finite element remeshing in a large deformation analysis of metal powder forming*, Int. J. Num. Meth. Eng., 45, (1999), 801-820.

[3] A. R. Khoei, M. Anahid, M. Mofid, *An application of arbitrary Lagrangian-Eulerian method in numerical simulation of forming processes using cap plasticity model*, 12th AMME conference, Poland, (2003).

COMPUTATIONAL MECHANICS (Abstracts)
WCCM VI in conjunction with APCOM'04, *Sept. 5-10, 2004, Beijing, China*

A New Nonlinear Shell Element Taking the Normal Stress in Thickness Direction into Account for Simulation of Sheet Metal Forming

Hong Lu [1*], Koichi Ito [1], Koichi Kazama [2], Y. Nagai [2], Hua Zhang [3]

[1] *Graduate School of Information Science, Tohoku University, 04 Aramaki-Aoba, Aoba-ku, Sendai, 980-8589, Japan*
[2] *Press Kogyo Co., Ltd, Engineering Departmen t, 2003-1 Endo, Fujisawa,Kanagawa, Japan*
[3] *Department of Engineering Mechanics, College of Resource and Environment Science, ChongQing University, Chongqing, 400044, China*
e-mail: honglu512@hotmail.com, ito@cmm.is.tohoku.ac.jp, {k-kazama,y-nagai}@presskogyo.co.jp, cquzhh@email.cta.cq.cn

Abstract

In simulation of sheet metal forming, the prediction of springback , which is a major defect, is not accurate. One reason for this is that the distributions of internal stresses are not predicted accurately. The conventional degenerated shell elements based on Mindlin-Reissner's theory of plates are usually used to simulate sheet metal forming process. The degenerated sh ell element uses the assumption that the thickness of the element remains constant and the normal stress in thickness direction is zero. These lead to problems in simulation of sheet metal forming, especially in prediction of springback. In this paper, we present a new nonlinear shell element that takes the normal stress in thickness direction into account which allows for large deformation. Two pseudo nodes, which have only one degree of freedom, are introduced on the top surface and bottom surface respectively. As a result, the normal strain in thickness direction can be calculated and the normal stress in thickness direction can be calculated using constitutive relation directly. It's possible to treat the boundary condition that pressure acts on top surfaces and bottom surface is simultaneously constrained. The element is tested in several n umerical examples. The simulation results show that stress prediction is improved. The normal stress in thickness direction exhibit agreement with that calculated by surface loads and bending moment decreases as normal stress in thickness direction increases, which indicates that the element can simulate the effect of bottoming, i.e. sheet metal is stamped with a great vertical load after the punch reaches dead point in the process of V-type bending to reduce springback.

REFERENCES

[1] O. C. Zienkiewicz, R. L. Taylor, *The finite element method, fifth edition, vol. 2: solid mechanics*, Butterworth-Heinemann, (2000).

[2] Boštjan Brank, Jože Korelc, Adnan Ibrahimbegović, *Nonlinear shell problem formulation accounting through-the-thickness stretching and its finite element implementation*, Computers & Structures, (2002), 699-717.

[3] J. C. Simo, M. S. Rifai, D. D. Fox, *On a stress resultant geometrically exact shell model, part 4: variable thickness shells with through-the-thickness stretching*, Computer Methods in Applied Mechanics and Engineering, (1990), 91-126.

COMPUTATIONAL MECHANICS (Abstracts)
WCCM VI in conjunction with APCOM'04, *Sept. 5-10, 2004, Beijing, China*

Numerical Modeling of a Two-Step Forming Process for Springback Reduction

Cedric Xia *, Feng Ren

Scientific Research Laboratories, Ford Motor Company, Dearborn, MI 48121, USA
e-mail: zxia@ford.com; fren@ford.com

Abstract

Numerical modeling is conducted to analyze a two-step forming process for springback analysis. In this process a set of lockbeads in the binder activates just before the end of punch stroke, locking the remaining blank in the binder. The continuation of the punch stroke then creates a final increment of pure stretch. It is most effective for deep drawings of U-channel type of structures such as rails.

This technique is also known as "post-stretch" or "shape set" in the automotive industry and in the literature. Finite element simulations for a straight channel are conducted in order to understand the springback reduction mechanism of the process and to determine its effectiveness. After an examination of deformation profile after drawing and wall curl as a result of springback, various magnitude of post-stretch amount is modeled and their deformation history is analyzed. It is found that a post-stretch strain around 2% almost completely eliminates wall curl. CAE investigations demonstrate that the technique is equally applicable to more complex 3D channel, where a step channel is examined. The effectiveness of this concept is demonstrated by laboratory experiment on the forming of a U-channel. Various implementation techniques for the process in an industrial environment are also suggested, together with a discussion on the associated benefits and costs for production use.

COMPUTATIONAL MECHANICS (Abstracts)
WCCM VI in conjunction with APCOM'04, *Sept. 5-10, 2004, Beijing, China*

Quality Prediction and Die Design Optimization for Sheet Metal Stamping Using Monte Carlo Simulation and Inverse Finite Element Modeling

Maolin Yu*, Yiping Chen, R. Du

Department of automation and compute-aided engineering, The Chinese University of Hong Kong, Shatin N. T., Hong Kong, China
e-mail: myu@acae.cuhk.edu.hk

Abstract

Sheet metal stamping is one of the most commonly used manufacturing processes. However, although much research has been carried out (e.g., [1 -2]), few can predict the product quality without expensive experiment tryout, not to mention the design optimization. This paper introduces a new method for quality evaluation and die design optimization using a combination of Monte Carlo simulation and inverse FEM. The use of Monte Carlo simulation for manufacturing quality prediction has been discussed in a number of papers, such as [3]. However, because of the excessive computation load in classical incremental FEM, it has not been used for stamping application. Recently, the inverse method [4], also called the one -step method, is developed, which is very fast and its accuracy is acceptable. Based on a combination of Monte Carlo simulation and the inverse method, we studied the quality variation of drawing a square cup. In the study, the workpiece loading/unloading variation and blank holding force variation are modeled as the variation of blank holding force with Gaussian probability distribution. The effects of the lubrication (i.e., the friction) and workpiece material (including the thickness and the plastic deformation properties) are also included in the simulation. Based on the analysis of the simulation results, the critical design parameters are identified. Accordingly, die design optimization can be carried out to minimize the variations.

REFERENCES

[1] A. Tezuka, S. H. Kim, H. Huh, *Process parameter design in sheet stamping processes with rigid-plastic finite element analysis*, Trans. of JSCES, Paper No. 20000011.

[2] W. Sosnowski, I. Marczewska, A. Marczewski, *Sensitivity based optimization of sheet metal forming tools*, J. of Materials Processing Technology, 124(1-3), (2002), 319-328.

[3] W. Chen, C. Yuan, *A probabilisitic-based design model for achieving flexibility in design*, Transaction of the ASME Journal, Journal of Mechanical Design, 121(1), (1999), 77-83.

[4] Y. Q. Guo, J. L. Batoz, H. Naceur, S. Bouabdallah, F. Mercier, O. Barlet, *Recent developments on the analysis and optimum design of sheet metal forming parts using a simplified inverse approach*, Computers and Structures, 78, (2000), 133-148.

COMPUTATIONAL MECHANICS (Abstracts)
WCCM VI in conjunction with APCOM'04, *Sept. 5-10, 2004, Beijing, China*

Finite-Element Simulation of Coating in Drawing Process

Liqun Zhou [1,2*], Yichun Zhou [1], Yong Pan [1]

[1] *Institute of Fundamental Mechanics and Material Engineering, Xiangtan University, Hunan, 411105, China*
[2] *Mechanical Engineering School of Xiangtan University, Hunan, 411105, China*
e-mail: johnzlq@163.com

Abstract

An electrodeposited nickel coating which thickness 3~10μm was prepared on the substrate of low carbon steel sheet. After special heat treated, it was found that the nickel coating and substrate sheet had high interface strength, excellent toughness. It could be used as the materials of forming advanced battery shell. This process that uses the already electrodeposited nickel sheet to drawing battery shell do not need later electroplating after formed, and can realize continuous production. The elastic-plastic finite element method of dynamic explicit algorithm was used to simulate the drawing process. The Belytschko-Tsay four-node shell element was used to mesh the materials of nickel coating and substrate sheet, the tools and the blank holder were adopted rigid shell elements, and the nickel coating and substrate sheet were adopted ideal elastic-plastic model by experimental test. The tied with failure contact was employed for the interface combination of the coating and substrate. Rate type elastic-plastic constitute law was modeled to handle the large deformation in the drawing process, and the central difference method was utilized to solve the finite element equations for dynamical analysis explicitly.

By simulation, the stress and strain fields were worked out, the extent of the coating being thinned after drawn was clear, the interface stress between the coating and substrate was illustrated, and the computation showed that the nickel coating and substrate sheet would not delaminate in its combined interface during the forming process, and the materials were suitable for manufacturing advanced structure. The study also showed that the dynamical explicit finite element method can be used for analyzing complicated mechanical problems in coating and substrate, and in that the Belytschko-Tsay shell element is valid for forming process.

REFERENCES

[1] G. Pasquinelli, C. N. R. Pisa, *Simulation of metal-forming processes by the finite element method*, Int. J. Plasticity, 11(5), (1995), 623-651.

[2] H. Y. Kim, B. C. Hwang, and W. B. Bae, *An experimental study on forming characteristics of pre-coated sheet metals*, J. Mater. Process. Tech., 120, 290-295, (2002).

[3] R. Parisot, S. Forest, A. F. Gourgues, et al, *Modeling the mechanical behavior of a multicrystalline zinc coating on a hot-dip galvanized steel sheet*, Comp. Mater. Sci., 19, (2000), 189-204.

[4] K. Tunvisu, N. P. O'Dowd, E. P. Busso, *Use of scaling functions to determine mechanical properties of thin coating from microindentation tests*, Int. J. Solids Struct., 38, (2001), 335-351.

COMPUTATIONAL MECHANICS (Abstracts)
WCCM VI in conjunction with APCOM'04, *Sept. 5-10, 2004, Beijing, China*

Role of Logarithmic Strain in Finite Deformation Plasticity

Zhe Zhou *, Mingrui Li

Institute of Applied Mechanics, China Agricultural University (East Campus), Beijing, 100083, China
e-mail: cmrli@public.fhnet.cn.net

Abstract

In the small deformation plastic theory, the concept of effective stress and strain is adopted widely. It is well known that metal in its plastic state keeps its volume unchanged. This fact is equivalent to the vanishing of the first invariant of the strain tensor, i.e. $I_1 = 0$. This concept faces great challenge in the field of finite deformation theory. There are various definitions of strain measure [1] in the finite deformation theory. Though the chosen of strain measure is arbitrarily in the elastic problem, it is strongly restricted in the plastic problem. The dilatation property of material undergoing finite deformation is characterized by the determinant J of the deformation gradient $\boldsymbol{F}$ and can be expressed as $J = \det(\boldsymbol{F}) = \det(\partial \boldsymbol{x}/\partial \boldsymbol{X}) = 1$, where $\boldsymbol{X}$ is the coordinate of a material point in the initial configuration and $\boldsymbol{x}$ is that in the current configuration [2]. The matrix $\boldsymbol{F} = \partial \boldsymbol{x}/\partial \boldsymbol{X}$ is called the deformation gradient and its determinant J is called the Jacobian. When and only when the logarithmic strain measure is used we can have $I_1 = \varepsilon_1 + \varepsilon_2 + \varepsilon_3 = \ln(J)$. Thus we arrive at the conclusion that the logarithmic strain is the only suitable strain measure for the multi crystal metal material. In this way the traditional small deformation plasticity can be transplanted to the finite deformation plasticity. Obviously, the flow rule [2, 3] and the useful return mapping algorithm [4] should also be modified accordingly. We propose a new algorithm which has been applied to solve several metal forgi ng problems by using the total Lagrangian (TL) method. Which means the only reference configuration is the initial configuration and the mesh regeneration technique is not invoked at all. The given examples show not only the validity but also the high efficiency and accuracy of the proposed algorithm [5].

REFERENCES

[1] S. N. Atluri, *Alternate stress and conjugate strain measures and mixed variational formulations involving rigid rotations for computational analysis of finitely deformed solids with applic ation to plates and shells–I. Theory*, Comput. Struct., 1, (1983), 93-116.

[2] D. Peric, D. R. J. Owen and M. E. Honor, *A model for finite strain elasto-plasticity based on logarithmic strains*: Computational issues, Comp. Meth. Appl. Mech. Eng., 94, (1992), 35-61.

[3] G. Gabriel and K. J. Bathe, *Some computation issues in large strain elasto-plastic analysis*, Computers & Structures, (2/3), (1995), 249-267.

[4] J. C. Simo, R. L. Taylor, *A return mapping algorithm for plane stress elasto-plasticity*, Int. J. Num. Meths. Eng., 22, (1986), 649-670.

[5] M. Li, *Implementation of the finite deformation plastic theory – Tensor derivative and metal forging simulation*, Progress in Natural science, 13, (2003), 184-190.

COMPUTATIONAL MECHANICS (Abstracts)
WCCM VI in conjunction with APCOM'04, *Sept. 5-10, 2004, Beijing, China*

Fatigue Behavior of Metallic Forged Components: Numerical Implementation of Modified Criteria Taking into Account Grain Flow Orientation

N. Caillet *, P. -O. Bouchard, Y. Chastel

CEMEF, Ecole des Mines de Paris, Rue Claude Daunesse, BP 207, Sophia Antipolis, 06904, France
e-mail: nicolas.caillet@ensmp.fr, pierre -olivier.bouchard@ensmp.fr, yvan.chastel@ ensmp.fr

Abstract

Forging processes of metal components induce grain flow orientation, which gives rise to anisotropic mechanical properties. In practice, a designer tries to optimise a part component to use efficiently these specific properties and to ob tain superior fatigue strength for a given cyclic loading. The scope of this article is to insert this grain flow anisotropy into fatigue life calculation. We have focused on high cycle fatigue, and we concentrate on a description of the different mechanic al behaviors as a function of inclusion size in the metal. For the domain of inclusion controlled fatigue, a new multiaxial criterion has been developed on the basis of the Murakami equation which deals with uniaxial loading [1]. For complex loading paths the main difficulty comes from kinematic hardening of the metal during cyclic loading and calculating the displacement of the centre of yield surface is costly. A simplified methodology is presented to allow for efficient calculations of an equivalent residual stress state for a given loading path. For small inclusions, the Dang Van criterion [2] is used. Both models have been extended for anisotropic behaviours. Particularly, the first criterion uses apparent surface of inclusions, directly linked to strai n during forging and to the direction of normal stresses applied during cycling loading. Both criteria have been implemented in the finite element code Forge3®. This computational tool can use the information provided by forging simulations to account for subsequent anisotropic fatigue behavior. The definition of the appropriate mesh size with respect to the grain flow heterogeneity of a forged component and the fatigue strength calculation is discussed. A good orientation of stretched inclusions and their alignment may improve fatigue strength by reducing their apparent surface. For an initial size of inclusions, the forging process induces specific strain path which lead to inclusions distributions. Based on these final inclusions distributions and on the fatigue loading mode with respect to the inclusions orientations, one or the other of these two material behavior hypotheses might be satisfied. This is automatically dealt with the global model so that properties of a final part can be optimised since the first steps of the design.

REFERENCES

[1] Y. Murakami, M. Endo, *Effects of defects, inclusions and inhomogeneities on fatigue strength*, Fatigue, 16, (1994), 163-182.

[2] K. Dang van, B. Griveau, O. Message, *On a new multiaxial fatigue limit criterion: theorie and application*, Biaxial and Multiaxial Fatigue, EGF 3, Mechanical Engineering Publications, London, (1989), pp. 479-496.

COMPUTATIONAL MECHANICS (Abstracts)
WCCM VI in conjunction with APCOM'04, *Sept. 5-10, 2004, Beijing, China*

Stochastic Fracture Mechanics Investigation of Test Specimen Size Effect on the Endurance Limit of a Carburized 4320 Steel

Edward F. Punch [1*], Robert R. Binoniemi [2], Hong Lin [3]

[1] *Punch Software Solutions, 1211 Bishop Road, Grosse Pointe Park, MI 48230, USA*
[2] *Dana Corporation, Heavy Vehicle Systems and Technologies Group, P.O. Box 4097, Kalamazoo, MI 49003, USA*
[3] *Dana Corporation, Technology Development Group, 8000 Yankee Road, Ottawa Lake, MI 49267, USA*
e-mail: flynnpunch@provide.net, bob.binoniemi@dana.com, hong.lin@dana.com

Abstract

The S-N curve of steel usually displays an endurance limit, stresses below which will not cause fatigue failure. The endurance limit depends upon the size of a component, with larger ones exhibiting lower endurance limits [1, 2]. Since this size effect is a serious issue in fatigue design of large components, and since it is difficult to explain by the conventional stress -life or strain life theories, this paper analyzes the size effect using a stochastic fracture mechanics approach.

Fatigue cracks initiate at randomly distributed inclusions and surface defects under sufficiently high cyclic loads. Steel suppliers have established the size distribution of these inclusions by ultrasonic testing [3]. Based on fatigue growth of these randomly located and probabilistically sized defects , each failure can be thought of as one particular representation of a stochastic process. By considering a large population of failures, it is possible to establish the material's true lowest -bound endurance limit and evaluate the size effect.

Larger components have a greater volume of material, and a larger population of defects. Statistically, a larger population contains more extreme defect sizes, both large and small. The presence of larger defects leads to crack growth and failure at lower stress leve ls, hence a lower endurance limit.

In place of expensive and often time consuming experimental testing, a large population of failures can be generated by computer simulation, provided that accurate material properties and crack growth models are used. Treating defect size and location as random variables, the Monte Carlo technique economically generates statistical representations of the fatigue process. The lowest value can be taken as the endurance limit, or alternatively a "six sigma" confidence level c an be employed.

Using the efficient alternating finite element method and linear elastic fracture mechanics, the endurance limits of two cantilever beam specimens with high stressed material volume ratio of 6:1 are computed, and the results are compared wi th experimental test results. Moreover, sensitivity studies are performed on crack aspect ratio, initial defect size range, and crack growth threshold value ΔKth.

REFERENCES

[1] R. Kuguel, Am. Soc Test. Mater. Proc., 61, (1961), 732-748.

[2] B. C. Taylor, *Analysis of specimen size and residual stresses in modified Brugger specimens for bending fatigue*, Ph.D. thesis, Colorado School of Mines, USA, (2003).

[3] R. L. Leibensperger, J. D. Stover, *Inclusions, rolling contact fatigue and power dense transmission deisgn*, Int. Conf. On Low Residual Steels, organized by The Indian Inst. of Metals and Tata Steel, Jamshedpur, India, Feb. (1999).

COMPUTATIONAL MECHANICS (Abstracts)
WCCM VI in conjunction with APCOM'04, *Sept. 5-10, 2004, Beijing, China*

Fast Prediction of Crack Growth Life of MSD Structure by Numerical Method

Jianyu Zhang *, Rui Bao

School of Aeronautic Science and Technology, BeiHang University, Beijing, 100083, China
e-mail: jianyu_zhang124@sohu.com, qimugege@sohu.com

Abstract

The phenomenon of aging structures has focused attention on the problems of multiple -site damage (MSD). Multiple site damage (MSD) refers to multiple cracks of arbitrary length, commonly occurring along rows of fastener holes in the fuselages and wings of commercial and military aircraft. In isolation each flaw or crack may be safe. However, the cumulative effect of multiple, interacting cracks may significantly degrade the damage tolerance of the structure. Cracks caused by MSD are extremely difficult to detect and greatly reduce the residual strength, fatigue life and overall structural integrity of aircraft panels. It has become increasingly important to develop methodologies to predict the fatigue crack growth life in damaged fuselage structures in the presence of widespread fatigue damage. Developments of these methodologies for complex loading are essential to ensure the continued safe operation of the aging commercial transport fleet and aircraft for the next generation.

Over the years, stable crack growth in metallic materials has been studied extensively using elastic –plastic finite-element methods. These studies have extensively investigated various fracture criteria. And methods for predicting crack growth life of MSD structure have been developed. But most of the presented numerical methods must calculate the fatigue life cycle by cycle. These methods spend much time to calculate the stress distribution. It's unnecessary.

A simple numerical method, using the principles of fracture mechanics and the computation solutions, was developed for predicting the fatigue crack growth life of MSD structure is pres ent in this paper. Different geometrical configurations were adopted in the numerical example. An obvious crack interaction effect was found, but it occurred mainly in the last part of the fatigue crack growth lives. The relative size of MSD cracks depends significantly on scatter of the initial crack.

REFERENCES

[1] L. Ma, P. W. Lam, M. T. Kokaly, A. S. Kobayashi, *CTOA of a stable crack in a thin aluminum fracture specimen,* Eng. Frac. Mech., 70(3-4), (2003), 427-442.

[2] R. M. V. Pidaparti, M. J. Palakal, Z. A. Rahman, *Simulation of structural integrity predictions for panels with multiple site damage,* dvances in Eng. Software, 31 (2), (2000), 127-135.

[3] G. S. Wang. *Analyzing the onset of multiple site damage at mechanical joints,* International Journal of Fracture, 105(3), (2000), 209-241.

[4] R. M. V. Pidaparti, M. J. Palakal, *Fatigue crack growth predictions in aging aircraft panels using optimization neural network,* AIAA Journal, 36 (7), (1998), 1300-1304.

[5] L. Wang, F. W. Brust, S. N. Atluri, *The elastic-plastic finite element alternating method (EPFEAM) and the prediction of fracture under WFD conditions in aircraft structures .1. EPFEAM theory,* Computational Mechanics, 19 (5), (1997), 356-369.

[6] L. Wang, F. W. Brust, S. N. Atluri. Computational Mechanics, 19(5), (1997), 370-379.

[7] L. Wang, F. W. Brust, S. N. Atluri. Computational Mechanics, 20(3), (1997), 199-212.

COMPUTATIONAL MECHANICS (Abstracts)
WCCM VI in conjunction with APCOM'04, *Sept. 5-10, 2004, Beijing, China*

Axially Loaded Generalized Beam Element on a Two-Parameter Elastic Foundation with Semi-Rigid Connections and Rigid Offsets

I. E. Avramidis *, K. Morfidis

Department of Civil Engineering, Aristotle University of Thessaloniki, 54124 Thessaloniki, Greece
e-mail: avram@civil.auth.gr

Abstract

The problem of an axially loaded elastic member resting on an elastic foundation has been an important tool for modelling and analysis, especially in the design of building structures, where the superstructure-foundation-soil interaction has to be taken into account. In order to model soil behaviour, several approaches have been developed in the past. In the majority of the proposed solutions, the foundation-supporting soil is represented on the basis of the well-known Winkler-hypothesis. The discontinuous nature of Winkler model gave rise to the development of various forms of two-parameter elastic foundation models, in which the continuity is introduced by assuming the Winkler springs to be connected by a shear layer or a membrane or a beam. The two-parameter models describe the soil behaviour more accurately and yet remain simple enough for practical purposes.
Two other related problems encountered in every-day practice refer to the modelling 1) of rigid joints or, more generally, of structural elements which can be assumed to behave as rigid bodies, e.g., massive footings in reinforced concrete foundations on flexible ground, and 2) of flexible joint connections, e.g., in steel frame structures.
The objective of this paper is to address all these topics by means of a generalized finite beam-column element. The original concept of this element has been presented in [1] without taking axial loading effects into account. Here, stiffness matrices and load vectors are based on the exact analytical solution of the differential equation, which describes the problem of an axially loaded member resting on a two-parameter elastic foundation, while featuring rigid offsets at its ends. The connection of rigid offsets to the interior element is achieved by means of rotational springs. The proposed element, possess the ability to degenerate to various simpler elements: it is possible to separately ignore the rigid offsets, the semi-rigid connections and even the elastic support. This is accomplished by zeroing certain coefficients in the expressions of the stiffness matrix, or by forming their limit values. These abilities render the generalized element very useful in structural analysis computer programs. In addition to the stiffness matrices, equivalent element nodal load vectors for external load patterns are developed. The usefulness of the proposed element in modelling and analysis of reinforced concrete and steel structures is documented by numerical examples.

REFERENCES

[1] K. Morfidis, I. E. Avramidis, *Formulation of a generalized beam element on a two-parameter elastic foundation with semi-rigid connections and rigid offsets*, Computer & Structures, 80 (2002), 1919-1934.

COMPUTATIONAL MECHANICS (Abstracts)
WCCM VI in conjunction with APCOM'04, *Sept. 5-10, 2004, Beijing, China*

A Neuro-Fuzzy System for the Post-Earthquake Damage Assessment of Buildings

Martha L. Carreño[1], Omar D. Cardona[2], Alex H. Barbat[1*]

[1] *Department of Structural Mechanics, Technical University of Catalonia, Jordi Girona 1 -3, Modulo C1, Barcelona, 08034, Spain*

[2] *Department of Civil Engineering, University of Los Andes, 092858, Bogotá D.C., Colombia*

e-mail: liliana@cimne.upc.es, ocardona@uniandes.edu.co, alex.barbat@upc.es

Abstract

After a strong earthquake the state of the affected buildings should be assessed in order to define their habitability and reparability. Due to the extension of damage in the str uck area, it is not possible to make the evaluations of all the building only by expert engineers. Many inspectors without expertise and knowledge must be involved due to the need of massive survey and the necessity to have the diagnosis as soon as possibl e to aid the people affected. As a consequence, it is common the tendency of non-expert inspectors to aggravate or to underestimate the real level of damage. This paper develops the algorithmic basis of an integrated tool of computational intelligence tech nique to assist non-experts in the quick building damage diagnosis and occupancy decision -making. A hybrid neuro-fuzzy system is proposed, based on a special three -layer feed-forward artificial neural network and fuzzy rule bases [1-2]. The inputs of the system are fuzzy sets, taking into account that the damage levels of the structural components are linguistic defined by means of qualifications such as light, minor, moderate, average, severe, etc., which are very appropriate to handle subjective and incomplete information. A Kohonen unsupervised learning algorithm is used, as a first module, to assess the structural damage degree. The weights for each appraisal are obtained by the T -conorm of the triangular membership functions, which represent the input d amage level qualifications, and by fitting to a sigmoidal damage function of reference. For the training of the artificial neural network, damage evaluations made after the Quindío's earthquake in Colombia (1999) were used. A fuzzy rule bases module is then used for the habitability and reparability decision -making. It takes into account: 1) the building safety, obtained from the damage evaluation in structural and architectural components, 2) the site seismic effects and 3) the pre -existing conditions that can make the building more vulnerable such as the quality of construction materials, plant and height irregularities and bad structural configurations.

REFERENCES

[1] J. Jang, C. Sung, E. Mizutani, *Neuro-Fuzzy and Soft Computing, A Computational Approach to Learning and Machine Intelligence*, Prentice Hall, Inc., New Jersey, (1997).

[2] D. Rutkowska, *Neuro-Fuzzy Architectures and Hybrid Learning*, Physica Verlag, Heidelberg, (2002).

COMPUTATIONAL MECHANICS (Abstracts)
WCCM VI in conjunction with APCOM'04, *Sept. 5-10, 2004, Beijing, China*

Shapes of Submerged Funicular Arches: Allowance for Selfweight

Y. H. Chai [1], C. M. Wang [2*]

[1] *Department of Civil and Environmental Engineering, University of California, Davis, USA*
[2] *Department of Civil Engineering, National University of Singapore, Kent Ridge 119260, Singapore*
e-mail: yhchai@ucdavis.edu, cvewcm@nus.edu.sg

Abstract

Recently, Gavin and Reilly [1] conducted a study on submerged funicular arches. Upon solving the differential equation governing the shape of the submerged funicular arch numerically, they found that the shape of the arch is nearly parabolic when the hydrostatic pressure is small while the arch approximates that of a semicircular shape when the hydrostatic pressure is large. Wang and Wang [2] further examined the characteristics of submerged arches and showed that the arch shape may take on two solutions or even no solution, depending on the depth of water and arch dimensions. Fung [3] and Chai and Kenneth [4] derived closed-form solutions for the shape of weightless submerged funicular arches.

In this study, we derive approximate expressions for the shape of a funicular arch with span $2L$, height H, submerged at a water depth D and carrying its own weight. With the origin of the Cartesian coordinates at the arch apex and the y-axis directed in the gravity direction, the shape of the submerged funicular arch with allowance for its selfweight may be given by

$$\frac{y}{H}=\frac{\sqrt{\alpha_2{}^2+4\alpha_3+\cos\theta\left(2\alpha_1(\alpha_1+\alpha_2)-4\alpha_3-\alpha_1{}^2\cos\theta\right)}-\alpha_2-\alpha_1\cos\theta}{2\alpha_3+\alpha_1{}^2\cos\theta}$$

$$\frac{x}{H}=\frac{p_2\,F\left(\frac{\theta}{2},\frac{2p_1}{p_1+p_2}\right)-(p_1+p_2)E\left(\frac{\theta}{2},\frac{2p_1}{p_1+p_2}\right)+\sqrt{p_1+p_2}\,\alpha_1\sin\theta}{2\sqrt{p_1+p_2}\,\alpha_3}$$

where $F(\varphi,m)$ and $E(\varphi,m)$ are incomplete elliptic integrals of the first kind and second kind, respectively, θ the angle between the tangent on the arch and a horizontal x -axis, $\alpha_1=\gamma_a H/\sigma_o$, $\alpha_2=p_o H/N_a$, $\alpha_3=\gamma_w H^2/(2N_a)$, $p_1=2\alpha_1(\alpha_1+\alpha_2)-4\alpha_3$, $p_2=\alpha_2{}^2+4\alpha_3$, γ_a the unit weight of the arch, γ_w the unit weight of the water, N_a the axial compressive force at arch apex, p_o the hydrostatic pressure at apex, σ_o the compressive stress in the arch where a fully stressed condition is assumed.

REFERENCES

[1] H. P. Gavin, K. J. Reilly, *Submerged funicular arches*, Journal of Structural Engineering, ASCE, 126(5), (2000), 627-629.

[2] C. M. Wang, C. Y. Wang, *Funicular shapes for submerged arches*, Journal of Structural Engineering, ASCE, 128(2), (2002), 266-270.

[3] T. C. Fung, *Shapes of submerged funicular arches*, Journal of Engineering Mechanics, ASCE, 129(1), (2003), 120-125.

[4] Y. H. Chai, S. K. Kunnath, *Geometry of submerged funicular arches in Cartesian coordinates*, Journal of Structural Engineering, ASCE, 129(9), (2003), 1087-1092.

COMPUTATIONAL MECHANICS (Abstracts)
WCCM VI in conjunction with APCOM'04, *Sept. 5-10, 2004, Beijing, China*

Optimization of Steel Frames Considering Life Cycle Cost using Pushover Analysis

Michalis Fragiadakis*, Nikos D. Lagaros, Manolis Papadrakakis

Institute of Structural Analysis & Seismic Research, School of Civil Engineering, National Technical University of Athens, Zografou Campus, Athens, 15771, Greece
E-mail: {mpapadra, mfrag, nlagaros}@central.ntua .gr

Abstract

Severe damages caused by recent earthquakes in the United States and Japan made the engineering community to wonder about the effectiveness of modern design codes. Although the number of loss in terms of human lives was low, the economic loss was devastating. Given that the primary goal of contemporary seismic design is the protection of human life it is evident that additional performance targets should be considered. Recently the concept of performance -based design for structures subjected to seismic loading conditions was introduced. Combined with the progress made in the last two decades in the field of computational mechanics, the shift from design procedures based on conventional static analysis to more realistic nonlinear analysis procedu res is possible and also necessary. Structural optimization is the only rational procedure to assess the performance of those design procedures ensuring that the required performance targets are met.
The purpose of this work is twofold. The first objective is to present a new methodology for the optimal design of structures subjected to seismic loading considering inelastic behavior, while the second is to investigate the influence of life cycle cost together with initial material cost. The paper combines structural optimization procedures with inelastic static analysis also referred as pushover analysis. The constraints of the optimization problem are based on the provisions of European design codes [1][2], while additional constraints are imposed in order to control the load bearing and deformation capacity of the structure. For the solution of the optimization problem evolutionary algorithms and in particular the Evolution Strategies algorithm [3] that has been specially tailored to meet the specific chara cteristics of the problem at hand is employed. Life cycle cost is implemented in order to consider the impact of future earthquakes on the design adopted [4]. Since, the resulting problem is a pareto problem a number of optimum designs exist. The choice of the final design depends on whether a minimum cost or a maximum safety design is preferred.

REFERENCES

[1] Eurocode 3, *Design of steel structures, part1.1: general rules for buildings,* CEN, ENV 1993-1-1, (1992).
[2] Eurocode 8, *Design provisions for earthquake resistant structures,* CEN, ENV 1998-1-1/2/3, (1994).
[3] N. D. Lagaros, M. Fragiadakis and M. Papadrakakis, *Optimum design of shell structures with stiffening beams,* AIAA Journal, 42(1), (2004), 175-184.
[4] Y. K. Wen, Y. J. Kang, *Minimum building life-cycle cost design criteria. II: applications,* Journal of Structural Engineering, 127 (3), (2001), 338-346.

COMPUTATIONAL MECHANICS (Abstracts)
WCCM VI in conjunction with APCOM'04, *Sept. 5-10, 2004, Beijing, China*

Design of Tall Reinforced Concrete Buildings Using Genetic Algorithms

J. Kong *

Division of Building Science and Technology, City University of Hong Kong, Hong Kong, China
e-mail: bsjkong@cityu.edu.hk

Abstract

In the design of tall building frameworks, it is one of the most important tasks for the engineer to design an efficient structural layout for resisting the high lateral wind load and satisfying various architectural and building services requirements. Once such a layout is determined, member sizes should be minimized, particularly for vertical members of the framework, so that usable floor space can be maximized. Minimizing the weight of the superstructure also leads to a significant saving in foundation cost, particularly in areas with poor soil conditions. However, in the current design practice, such a minimization is usually carried out using the conventional trial-and-error procedure, which obviously is inefficient and not necessarily generating the optimal solution.

To facilitate the process of finding the optimal member sizes for a given structural framework, this paper presents the application of the genetic algorithm [1] to the design of tall reinforced concrete building framework under lateral loading. Cross-sectional dimensions of member groups are treated as discrete design variables and the method aims to minimize the weight of the structure subject to multiple inter-storey deflection constraints, based on the Hong Kong Design Code. To ensure that member sizes in the lower storeys would have larger or equal member sizes than the upper storeys, an additional size constraint is introduced in this study. These constraints are treated as penalty function during the evolutionary process. In addition to the usual selection, cross-over (2-point) and mutation operators of a genetic algorithm, a special growth operator and a pre-selection scheme are introduced and their effectiveness in accelerating the convergence of the algorithm and the minimal weight of the structure are investigated. To demonstrate the effectiveness of the method, present results are also compared with those obtained using optimality criteria method [2] and the usual trial-and-error procedure. The method will be further developed for the design of 3-D reinforced concrete building framework.

REFERENCES

[1] C. S. Krishnamoorthy, *Structural optimization in practice: Potential appli cations of genetic algorithms*, Journal of Structural Engineering and Mechanics, 11(2), (2001), 151-170.

[2] C. M. Chan, S. Sun, *Optimal drift design of tall reinforced concrete building framework*, in D. M. Frangopol, F. Y. Cheng eds. *Advances in Structural Optimization*, Proc. 1st US-Japan Joint seminar on structural optimization, Chicago, Illinois, USA, (1997), pp. 31-42.

COMPUTATIONAL MECHANICS (Abstracts)
WCCM VI in conjunction with APCOM'04, *Sept. 5-10, 2004, Beijing, China*

Form-finding Analysis of Membrane Structures Using the Force-Density Method

L. B. Li *, D. M. Wei

Department of Civil Engineering, South China University of Technology, Guangzhou 510640, China
e-mail: lilianbin201@sohu.com, weidemin@21cn.com

Abstract

The spatial membrane structure, constructed with soft materials with high tensile resistance and the affiliated bracket structures such as trusses, rigid columns, and cables etc, is a new type of structure developed from the mid-20th century. Several different construction methods can be used when tensioning the structure in order to make the soft membran e create spatial stiffness and eventually have a definite form. Membrane structures can be used as not only cover structures but also ordinary structures sustaining outer loads.

The design method for the membrane structures is very different from that of t he traditional structures. When designing traditional structures, the profile design must be made firstly, to determine the outline of the structures. Then the calculation can be made to access its safety. However, the first step of membrane structure des ign is form-finding, in which the initial shape of the structure with distributing prestress force is determined. In the following steps, the mechanical characteristics will be analyzed and the membrane-cut analysis will be made for the real-time constructions.

This paper primarily examined the force density method of the form-finding of the membrane structures. A series of theoretical deduction has been made and some useful conclusions have been obtained. Based on these deduction and conclusions, some FORT RAN programs have been composed. Using these FORTRAN programs, several typical samples are calculated and some useful results are acquired.

REFERENCES

[1] S. Z. Shen, *Membrane structure — an innovative spatial structure with bright prospect*, Journal of Harbin University of C. E. & Architecture, 32(2), (1986), 11-15, (in Chinese).

[2] S. M. Huang, L. Wei, S. Wang et al, *The shape-finding analysis of cable reinforced membrane structure*, Journal of Spatial structures, 5(4), (1999), 22-26, (in Chinese).

[3] Y. Xiang, J. Li, S. Z. Shen, *The initial form-finding design analysis of membrane structures*, Journal of Spatial Structures, 5(3), (1999), 19-27, (in Chinese).

[4] X. B. Ye, Z. L. Li, J. S. Wu, *Analysis theory of membrane structures by curved surface finite element method*, Journal of Building Structures, 9(5), (1998), 17-21, (in Chinese).

[5] Y. Xiang, S. Z. Shen, *A practical cutting design method of membrane structures*, Journal of Spatial Structures, 5(2), (1999), 46-50, (in Chinese).

[6] H. J. Schek, *The force density method for form-finding and computation of general network*, Comp. Mech. and Engng, 3, (1974), 115-134.

[7] L. B. Li, *The form-finding analysis of the cable membrane structures*, Taiyuan University of Technology, Taiyuan, China, (2002), (in Chinese).

COMPUTATIONAL MECHANICS (Abstracts)
WCCM VI in conjunction with APCOM'04, *Sept. 5-10, 2004, Beijing, China*

Seismic Spectrum Analysis of Chinese National Grand Theater

Zhu Li*, Xiuhua Guo, Wenfang Zhang

Taiyuan University of Technology, Taiyuan, Shanxi, 030024, China
e-mail: lizhu9999@vip.sina.com, gxhsgq.student@sina.com

Abstract

A first-rate building, Chinese National Grand Theater in Beijing is analyzed by using ANSYS 5.7.1. Its main structure is reinforced concrete shear wall, which is partially assisted by frames composed of concrete filled tubular columns and prestressed concrete beams. Considering the complexity, irregularity and importance, it is necessary to do a structural analysis to present its dynamic characteristic and provide some important data for structural design. Because of the dimension and complication of National Grand Theater, 3-Dimentional Finite Element method, instead of general simplifying method is adopted in structural analysis, thus the computation model is accord with the actual structure. The finite element model of this structure possesses near 100,000 degrees of freedom (DOF) and more than 500 types of members.

In this paper, modal and seismic spectrum analysis is performed to study its dynamic characteristic. According to modal analysis, most of modes are partial modes. There are many reasons to partial weak stiffness, such as sudden change of stiffness, big cantilever components, long -span thin slabs, long-span mega-beams and the floor slabs with big hole, and so on. Additionally some measures must be taken to decrease the whiplash tip effect on the top layer of Concert Hall, which can improve the stories' lateral displacement response.

When the modal characteristic is analy zed, this 3-D model is so complex that there are many secondary modes between two main modes and every mode is greatly correlative with its adjacent modes. Therefore, the more modes should be extracted and taken into account and then the main modes should be found and used in the spectrum analysis. With large data processing, this paper presents some main modes in each direction x, y and z. These modes can reflect the structural dynamic characteristic. Finally, in the process of spectrum computer, the combi nation of modes should be performed with "Complete Quadrics Combination" (CQC) method. Under these key steps above mentioned, the accuracy of the seismic response could be ensured. These experiences and methods are greatly useful for other vibration comput ation of this type of complex structure.

By the structural analysis, the dynamic characteristics of Chinese National Grand Theater are shown as following: high fundamental frequency; dense spectra of frequencies; obvious partial modes and many vertical modes. In spectrum analysis, the stories' shear force inclusive figures and stories' displacement inclusive figures have been arrived which provides some important parameters for structural design. These parameters showed following seismic abilities: The stories' stiffness versus the stories' height isn't relatively even; the whole structural stiffness is very great and the seismic abilities in all directions are equivalent.

COMPUTATIONAL MECHANICS (Abstracts)
WCCM VI in conjunction with APCOM'04, *Sept. 5-10, 2004, Beijing, China*

Least Weight Design of Submerged Spherical Domes

C. M. Wang, K. K. Vo *

Department of Civil Engineering, National University of Singapore, Kent Ridge, 119260, Singapore
e-mail: cvewcm@nus.edu.sg, g0306015@nus.edu.sg.

Abstract

As population and urban developments expand in coastal countries, city planners and engineers are pressured to shift some of the developments from the land to the sea. It is envisaged that floating and underwater cities will be constructed in this century due to land space limitation. For underwater cities, a dome may be used to create the living environment suitable for sustaining human activities underwater for a long time. This prospect prompted the authors to study the least weight design of submerged spherical domes.

Consider a submerged, spherical dome of radius R, subtended angle 2α and specific weight γ_s as shown in Fig.1. The dome is subjected to hydrostatic pressure and its own selfweight. Adopting a uniform strength design, the variation of the shell thickness h can be shown to be given by a power series of the form:

$$h(\phi) = h_0 \sum_{i=0}^{n} c_i \phi^{2i} = h_0 \left(c_0 + c_1\phi^2 + c_2\phi^4 + c_3\phi^6 + \ldots\right)$$

where the coefficients c can be calculated from the recursive formula:

$$\left(2i + 2 - \frac{R\gamma_s}{\sigma_0}\right) c_i = \frac{(-1)^i \gamma_w R(D + R\cos\alpha)}{(2i)!h_0\sigma_0} + \frac{(-1)^{i+1} 2^{2i-1} \gamma_w R^2}{(2i)!h_0\sigma_0} + \sum_{j=1}^{i} \frac{2(-1)^{j+1}(i+j+1)}{(2j+1)!} c_{i-1}$$

where

$c_0 = 1$,

σ_0 = allowable compression stress,

h_0 = dome thickness at the apex,

D = depth of water,

γ_w = specific weight of water.

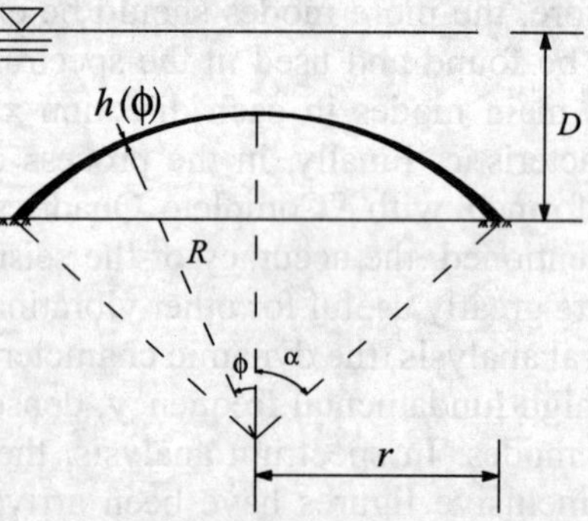

Fig. 1 Submerged spherical dome

For a family of uniform strength designs associated with a prescribed value of base radius r, we also investigate the optimal value of α which gives the minimum total weight of the submerged dome.

COMPUTATIONAL MECHANICS (Abstracts)
WCCM VI in conjunction with APCOM'04, *Sept. 5-10, 2004, Beijing, China*

Finite Deformation Stiffness of Sliding Elastomeric Bearings with Bolts

Jianchun Xiao [1*], Jianguo Nie [2], Changan Cai [1]

[1] *College of Civil Engineering & Architecture, Guizhou University of Technology, Guiyang, 550003, China*

[2] *Department of Civil Engineering, Tsinghua University, Beijing, 100084, China*

e-mail: xiaojianchun@tsinghua.org.cn, niejg@mail.tsinghua.edu.cn, caichangan@163.com

Abstract

Anchor bolts are set in some elastomeric bearings of large-span column-supported space steel roofs. Besides to help the bearing site-assembly, the bolts play the roles of sliding position limitation and vertical anchorage. To analyze the effect of bearings on nonlinear structural behavior, the bearing stiffness change is studied just before and after the elastomer pad is in contact with the bolts.

The discussion is base on the following assumptions: 1) the bolts being identical and uniformly distributed in the pad, 2) the deformation of pad and bolts being elastic, 3) the pad carrying the support vertical load, 4) the horizontal contact force between the pad and the bolts being dominant; neglecting the contact force on the other directions, 5) neglecting the effects of vertical displacement and bending deformation on shearing deformation of the pad.

For the bearing that the pad is glued with the top/bottom steel plates, three kinds of relation of pad and bolts (i.e. separation, point contact and line contact) are discussed separately and a unified horizontal stiffness formula is obtained with parametric analysis method, i.e.

$$K = \frac{G_r A_r}{t} + \frac{3E_b I_b}{2t^3(1+3\eta)}\left\{1+\tanh\left[\kappa\left(\frac{|P|t}{G_r A_r} - c\right)\Big/c\right]\right\}.$$

Based on the analysis results a stiffness expression for sliding elastomeric bearings is deduced, i.e.

$$K = \left(\frac{G_r A_r}{t} + \frac{3E_b I_b}{2t^3(1+3\eta)}\left\{1+\tanh\left[\kappa\left(\frac{|P|t}{G_r A_r} - c'\right)\Big/c'\right]\right\}\right)\left\{1+\tanh\left[\kappa\left(\mu N - \frac{G_r A_r c'}{t}\right)\Big/\mu N\right]\right\}.$$

To solve the computational problem caused by the bearing stiffness changes, an improved imbalance force vector formula in which use configuration C_2 as the reference frame of unbalanced force vector is proposed.

Finite elastic deformation analysis of a four-column-supported prestressed shallow lattice shell roof shows that the bolts have influence upon the computed results more significantly.

REFERENCES

[1] F. Naeim, J. M. Kelly, *Design of seismic isolated structures: from theory to practice*. John Wiley & Sons, New York, (1999).

[2] J. M. Kelly, *Earthquake-resistant design with rubber*, second edition, Springer-Verlag, London, (1997).

COMPUTATIONAL MECHANICS (Abstracts)
WCCM VI in conjunction with APCOM'04, *Sept. 5-10, 2004, Beijing, China*

Slippage Analysis of Continuous Cable

Zhihong Zhang [1*], Shilin Dong [2], Tamura Yukio [1]

[1] *Department of Architecture, Tokyo Polytechnic University, 243-0297, Japan*
[2] *Department of Civil Engineering, Zhejiang University, Hangzhou, Zhejiang, 310027, China*
e-mail: zhangzh@arch.t-kougei.ac.jp

Abstract

An algorithm for the analysis of continuous cable of tensile structure based on a two-node catenary cable element is proposed in this paper. For continuous cable in nonlinear finite element analysis, there are three conditions: the node force balance of the whole structure, the continuous cable condition, and the original cable length condition. A nonlinear finite element program in visual C++6.0 is developed, and two examples are presented to verify this method.

As the development of the science and technology, large-span structures enter into a new era due to the construction of Olympic stadiums and conference & exhibition centers. Large-span flexible structures gain more popularity such as cablenets, suspension bridges and tensegrity structures (cable dome) etc. Cable can be classified in different way, for example, classified into suspension cable and tensile cable for its sag, or into close cable and unclose cable, continuous cable and discrete cable. During construction the cable becomes tense step by step when it is prestressed. This paper presents an algorithm for the analysis of continuous cable.

Hypothesis: Cable is ideal soft and its material satisfies Hooke's law. There is no friction at the node when cable segments slip across.

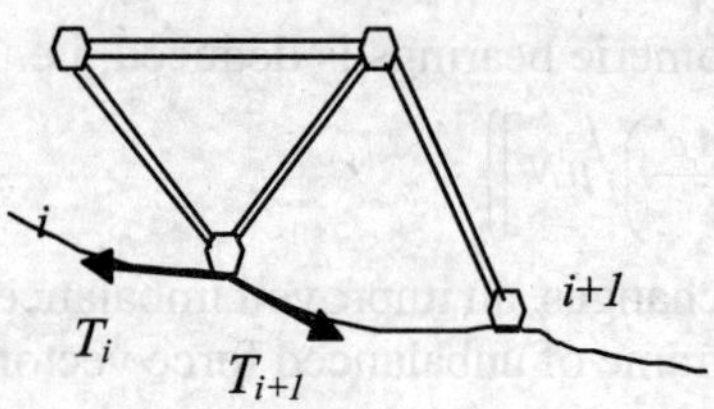

Fig. 1

Conditions:

1) The forces are in equilibrium at each node;
2) The original cable length is consistent, i.e. the total original length of all cable segments in a continuous cable should be equal to that we need in the construction;
3) It is assumed that $T_i = T_{i+1}$, as shown in Fig.1. But the direction of the two internal forces can be different.

For a cable segment of one continuous cable its original length changed due to the increasing prestress. In other words, this problem is equivalent to how to evaluate the amount of slippage of each cable segment. Discrete cable can be resolved by nonlinear finite element method, since there is not slippage on each node of cable. But for the continuous cable an additional condition as $T_j^k = T_i^{k+1}$ must be satisfied. The catenary equation is derived without taking account of the elastic elongation, so it is the exact equation to get the original length of cable segment, if the location of the node of cable segment and its internal force is known.

In this paper the error of the original length is distributed equally to each cable segment nearby. There are two numerical examples presented in this paper to validate this method.

COMPUTATIONAL MECHANICS (Abstracts)
WCCM VI in conjunction with APCOM'04, *Sept. 5-10, 2004, Beijing, China*

On Integrated CAD Approach to Design of Building Structures

T. Y. P. Chang *, J. H. Long

Department of Civil Engineering, Hong Kong University of Science & Technology, Hong Kong, China
e-mail: cepchang@ust.hk

Abstract

Structural design of a tall building involves several rather complex processes, or design activities, including data capture and modeling, conceptual design, structural analysis, structural detailing, and preparation of design documents. Although each of these process has been assisted by computer and CAD software tools, application of such tools are generally isolated from each other with no direct data connections. That is to say, data transfer among various design processes is usually done by manual means (paper drawings), which is both tedious and error prone. In this paper, an integrated CAD environment is emphasized under which direct data capture and structural modeling, design automation and structural optimization can be exercised to achieve a cost-effective design solution for a building structure. Discussions on several topics within the framework of CAD applications, i.e., system integration, direct data capture and structural modeling, building product modeling, structural optimization, and automated design checks are given. Also included in the paper are examples to demonstrate the utility of the proposed design approach under an integrated CAD framework.

Author Index